Soy, Globalization, and Environmental Politics in South America

Soy in South America constitutes one of the most spectacular booms of agro-industrial commodity production in the world. It is the pinnacle of modernist agro-industrial practices, serving as a key nexus in food–feed–fuel production that underpins the agribusiness–conservationist discourse of "land sparing" through intensification. Yet soy production is implicated in multiple problems beyond deforestation, ranging from pesticide drift and contamination to social exclusion and conflicts in frontier zones, to concentration of wealth and income among the largest landowners and corporations. This book explores in depth the complex dynamics of soy production from its diverse social settings to its transnational connections, examining the politics of commodity and knowledge production, the role of the state, and the reach of corporate power in everyday life across soy landscapes in South America. Ultimately, the collection encourages us to search and struggle for agroecological alternatives through which we may overcome the pitfalls of this massive transnational capitalist agro-industry.

This book was originally published as a special issue of *The Journal of Peasant Studies*.

Gustavo de L. T. Oliveira is a PhD candidate in the Department of Geography at University of California, Berkeley, USA. His dissertation analyses the political ecology of Chinese investments in Brazilian agribusiness and logistics infrastructure. He is a member of the BRICS Initiative for Critical Agrarian Studies.

Susanna B. Hecht is Professor in the Luskin School of Public Affairs and the Institute of the Environment at University of California, Los Angeles (UCLA), USA, and Professor of International History at the Graduate School of International Development Studies, Geneva, Switzerland. She is the author or editor of more than 16 books and numerous articles on the political ecology of tropical forests. Her book on Amazonian environmental history *The Scramble for the Amazon and the Lost Paradise of Euclides da Cunha* won the Melville Prize from the American Historical Association, and the Carl Sauer Award in Geography.

Critical Agrarian Studies

Edited by
Saturnino M. Borras Jr.
International Institute of Social Studies (ISS) in The Hague

Critical Agrarian Studies is the new accompanying book series to *The Journal of Peasant Studies*. It publishes selected special issues of the journal and, occasionally, books that offer major contributions in the field of critical agrarian studies. The book series builds on the long and rich history of the journal and its former accompanying book series, the *Library of Peasant Studies* (1973–2008), which had published several important monographs and special issues as books.

For a full list of books in this series please visit https://www.routledge.com/series/CAG.

Recent titles in this series include the following:

Soy, Globalization, and Environmental Politics in South America
Edited by Gustavo de L. T. Oliveira and Susanna B. Hecht

An Endogenous Theory of Property Rights
Edited by Peter Ho

Global Land Grabbing and Political Reactions 'from Below'
Edited by Marc Edelman, Ruth Hall, Saturnino M. Borras Jr., Ian Scoones and Wendy Wolford

Critical Perspectives in Rural Development Studies
Edited by Saturnino M. Borras Jr.

The Politics of Biofuels, Land and Agrarian Change
Edited by Saturnino M. Borras Jr., Philip McMichael and Ian Scoones

New Frontiers of Land Control
Edited by Nancy Lee Peluso and Christian Lund

Outcomes of Post-2000 Fast Track Land Reform in Zimbabwe
Edited by Lionel Cliffe, Jocelyn Alexander, Ben Cousins and Rudo Gaidzanwa

Green Grabbing: A New Appropriation of Nature
Edited by James Fairhead, Melissa Leach and Ian Scoones

The New Enclosures: Critical Perspectives on Corporate Land Deals
Edited by Ben White, Saturnino M. Borras Jr., Ruth Hall, Ian Scoones and Wendy Wolford

Soy, Globalization, and Environmental Politics in South America

Edited by
Gustavo de L. T. Oliveira and
Susanna B. Hecht

Routledge
Taylor & Francis Group
LONDON AND NEW YORK

First published 2018
by Routledge
2 Park Square, Milton Park, Abingdon, Oxon, OX14 4RN, UK

and by Routledge
711 Third Avenue, New York, NY 10017, USA

Routledge is an imprint of the Taylor & Francis Group, an informa business

British Library Cataloguing in Publication Data
A catalogue record for this book is available from the British Library

ISBN13: 978-1-138-29691-6

Typeset in TimesNewRomanPSMT
by diacriTech, Chennai

Publisher's Note
The publisher accepts responsibility for any inconsistencies that may have arisen during the conversion of this book from journal articles to book chapters, namely the possible inclusion of journal terminology.

Contents

Citation Information

The chapters in this book were originally published in *The Journal of Peasant Studies*, volume 43, issue 2 (March 2016). When citing this material, please use the original page numbering for each article, as follows:

Introduction

Sacred groves, sacrifice zones and soy production: globalization, intensification and neo-nature in South America
Gustavo de L. T. Oliveira and Susanna B. Hecht
The Journal of Peasant Studies, volume 43, issue 2 (March 2016) pp. 251–285

Chapter 1

Strategies and hybrid dynamics of soy transnational companies in the Southern Cone
Valdemar João Wesz Jr
The Journal of Peasant Studies, volume 43, issue 2 (March 2016) pp. 286–312

Chapter 2

Disappearing nature? Agribusiness, biotechnology and distance in Argentine soybean production
Amalia Leguizamón
The Journal of Peasant Studies, volume 43, issue 2 (March 2016) pp. 313–330

Chapter 3

Which territorial embeddedness? Territorial relationships of recently internationalized firms of the soybean chain
Clara Craviotti
The Journal of Peasant Studies, volume 43, issue 2 (March 2016) pp. 331–347

Chapter 4

The geopolitics of Brazilian soybeans
Gustavo de L. T. Oliveira
The Journal of Peasant Studies, volume 43, issue 2 (March 2016) pp. 348–372

Chapter 5

China's soybean crisis: the logic of modernization and its discontents
Yan Hairong, Chen Yiyuan and Ku Hok Bun
The Journal of Peasant Studies, volume 43, issue 2 (March 2016) pp. 373–395

Chapter 6
Different farming styles behind the homogenous soy production in southern Brazil
Bert Vander Vennet, Sergio Schneider and Joost Dessein
The Journal of Peasant Studies, volume 43, issue 2 (March 2016) pp. 396–418

Chapter 7
Soybean agri-food systems dynamics and the diversity of farming styles on the agricultural frontier in Mato Grosso, Brazil
Mateo Mier y Terán Giménez Cacho
The Journal of Peasant Studies, volume 43, issue 2 (March 2016) pp. 419–441

Chapter 8
Farming is easy, becoming Brazilian is hard: North American soy farmers' social values of production, work and land in Soylandia
Andrew Ofstehage
The Journal of Peasant Studies, volume 43, issue 2 (March 2016) pp. 442–460

Chapter 9
Green for gold: social and ecological tradeoffs influencing the sustainability of the Brazilian soy industry
Rachael D. Garrett and Lisa L. Rausch
The Journal of Peasant Studies, volume 43, issue 2 (March 2016) pp. 461–493

Chapter 10
On the margins of soy farms: traditional populations and selective environmental policies in the Brazilian Cerrado
Ludivine Eloy, Catherine Aubertin, Fabiano Toni, Silvia Laine Borges Lúcio and Marion Bosgiraud
The Journal of Peasant Studies, volume 43, issue 2 (March 2016) pp. 494–516

Chapter 11
Genetically modified soybeans, agrochemical exposure, and everyday forms of peasant collaboration in Argentina
Pablo Lapegna
The Journal of Peasant Studies, volume 43, issue 2 (March 2016) pp. 517–536

Chapter 12
'More soy on fewer farms' in Paraguay: challenging neoliberal agriculture's claims to sustainability
Laureen Elgert
The Journal of Peasant Studies, volume 43, issue 2 (March 2016) pp. 537–561

Chapter 13
The moving frontiers of genetically modified soy production: shifts in land control in the Argentinian Chaco
Lucía Goldfarb and Gemma van der Haar
The Journal of Peasant Studies, volume 43, issue 2 (March 2016) pp. 562–582

Chapter 14

Bolivia's soy complex: the development of 'productive exclusion'
Ben McKay and Gonzalo Colque
The Journal of Peasant Studies, volume 43, issue 2 (March 2016) pp. 583–610

Notes on Contributors

Catherine Aubertin is a researcher at the IRD (Institut de Recherche pour le Développement), France, and a visiting researcher at the IESA (Instituto de Estudos Socioambientais), Goiás Federal University, Brazil.

Marion Bosgiraud is an agronomic engineer specializing in international agricultural development, College of International Agro-Development, ISTOM, France.

Gonzalo Colque is a Bolivian researcher and Executive Director of Fundación TIERRA based in La Paz, Bolivia.

Clara Craviotti is a Research Fellow of the National Scientific and Technical Research Council (CONICET) and a Professor at the Latin American Social Sciences Institute (FLACSO), Argentina.

Joost Dessein is a Scientific Coordinator at the Social Sciences Unit of Flanders Research Institute for Agriculture, Fisheries and Food (ILVO), Belgium and Professor at the Department of Agricultural Economics of Ghent University, Belgium.

Laureen Elgert is an Assistant Professor of Environmental Policy and International Development at Worcester Polytechnic Institute (WPI), Massachusetts, USA.

Ludivine Eloy is a research scientist at the National Center for Scientific Research, Montpellier, France, and an Invited Researcher at the Center for Sustainable Development, University of Brasília, Brazil.

Rachael D. Garrett is a Giorgio Ruffolo Postdoctoral Research Fellow in Sustainability Science at Harvard University, Massachusetts, USA.

Lucía Goldfarb is a PhD researcher in International Development Studies at the University of Utrecht, The Netherlands.

Yan Hairong is Associate Professor at the Department of Applied Social Sciences at the Hong Kong Polytechnic University, Hong Kong.

Susanna B. Hecht is Professor in the Luskin School of Public Affairs and the Institute of the Environment at University of California, Los Angeles (UCLA), USA, and Professor of International History at the Graduate School of International Development Studies, Switzerland.

Ku Hok Bun is an Associate Professor in the Department of Applied Social Sciences at the Hong Kong Polytechnic University, Hong Kong.

Pablo Lapegna is an Assistant Professor of Sociology and Latin American and Caribbean Studies at the University of Georgia, USA.

Amalia Leguizamón is an Assistant Professor of Sociology at Tulane University in New Orleans, USA.

Silvia Laine Borges Lúcio is a PhD student at the Center for Sustainable Development, University of Brasília, Brazil.

Ben McKay is a PhD candidate at the International Institute of Social Studies (ISS) in The Hague, The Netherlands, and a research associate of the BRICS Initiative for Critical Agrarian Studies (BICAS), USA.

Mateo Mier y Terán Giménez Cacho currently researches the scaling-up of agroecology in México at El Colegio de la Frontera Sur, Mexico.

Andrew Ofstehage is a PhD candidate in Anthropology at the University of North Carolina at Chapel Hill, USA, and a Fulbright Scholar.

Gustavo de L. T. Oliveira is a PhD candidate in the Department of Geography at the University of California, Berkeley, USA, and a research associate of the BRICS Initiative for Critical Agrarian Studies (BICAS), USA.

Lisa L. Rausch is a postdoctoral researcher at the University of Wisconsin–Madison, USA.

Sergio Schneider is a Professor of Rural Sociology and Development Studies at the Federal University of Rio Grande do Sul (UFRGS), Porto Alegre, Brazil.

Fabiano Toni is a Professor and Graduate Coordinator at the Center for Sustainable Development, University of Brasília, Brazil, where he teaches environmental governance.

Gemma van der Haar is Assistant Professor at the chairgroup Sociology of Development and Change at the Wageningen University and Research Centre, The Netherlands.

Bert Vander Vennet is a PhD researcher in Rural Sociology at the Flanders Research Institute for Agriculture, Fisheries and Food (ILVO), Belgium, in collaboration with the Wageningen University and Research Centre, The Netherlands.

Valdemar João Wesz Jr is Assistant Professor at the Federal University of Latin American Integration (UNILA), Brazil, and a researcher at the Observatory on Public Policies for Agriculture (OPPA/CPDA/UFRRJ), Brazil.

Chen Yiyuan is Assistant Professor of Sociology at the College of Humanities and Development Studies, China Agricultural University, China.

Sacred groves, sacrifice zones and soy production: globalization, intensification and neo-nature in South America

Gustavo de L. T. Oliveira and Susanna B. Hecht

Soy has become one of the world's most important agroindustrial commodities – serving as the nexus for the production of food, animal feed, fuel and hundreds of industrial products – and South America has become its leading production region. The soy boom on this continent entangles transnational capital and commodity flows with social relations deeply embedded in contested ecologies. In this introduction to the collection, we first describe the 'neo-nature' of the soy complex and the political economy of the sector in South America, including the new corporate actors and financial mechanisms that produced some of the world's largest agricultural production companies. We then discuss key environmental debates surrounding soy agribusiness in South America, challenging especially the common arguments that agroindustrial intensification 'spares land' for conservation while increasing production to 'feed the world'. We demonstrate that these arguments hinge on limited data from a peculiar portion of the southern Amazon fringe, and obfuscate through neo-Malthusian concerns multiple other political and ecological problems associated with the sector. Thus, discussions of soy production become intertwined with broader debates about agrarian development, industrialization and modernization. Finally, we briefly outline the contributions in this volume, and identify limitations and fruitful directions for further research.

1. Introduction

Soy. Whether praised as the most efficient means to 'feed the world' or condemned for the manner in which it erases entire ecosystems beneath its seemingly endless green rows, soy has indubitably become one of the most important agroindustrial commodities in the world - serving as the nexus for the production of food, animal feed and hundreds of industrial products - and driving one of the most rapid landscape alterations (and enclosures) of the last 30 years. Soy production in South America now covers over 57 million ha, more than on any other continent (USDA 2015). Farmers and companies there are responsible for 54 percent of global production and 58 percent of total exports. This has taken place through the direct extension of highly capital and chemical intensive agroindustrial practices into areas that had been considered marginal for this kind of production system, such as the fringes of the Amazon, the Cerrado and Caatinga in Brazil, across Bolivia's and Argentinian Chaco forests, and parts of the Atlantic forests. The system also builds on the 'post-frontier' areas where tenure regimes have been stabilized and consolidated, intensifying production over pastures and replacing less profitable crops (as in the Brazilian Cerrado and the Pampas of Argentina and Uruguay). It has also involved the transformation of many complex woodland systems whose inhabitants lived from forest products, livestock and small-scale agriculture (Porro 2005; Sawyer 2008; Gordillo 2014). This process has converted huge tracts of some the world's most complex ecosystems, some with as many as 600 species per ha, into a monoculture with its symbionts with fragmented forest remnants along waterways, mountain-tops and some protected forests. The social life of agriculture and livelihoods in many places has been equally depleted. Yet less than six percent of all soy produced in the world is consumed directly as human food, and virtually all of it in South America is crushed to produce livestock feed and edible oil, as well as biodiesel and other industrial products (Highquest Partners and Soyatech 2011). Soy is no simple staple crop – it is an agro-industrial feedstock, a complex assemblage of technologies and techniques for the flexible implementation of its production across highly variable landscapes, a global network of machinery, warehouses, trucks and ships channeling commodity flows to multiple markets around the world, structured by an even broader diversity of institutions, social relations and practices. This is the 'material' soy that we can study and engage as land use, commodity, industrial input and oilseed.

But soy also embraces another set of meanings and symbols generated by the narratives of its development and the massive transformation of societies and landscapes that this has entailed in South America, presaging perhaps also transformations sought elsewhere in the tropics, particularly sub-Saharan Africa. As Norman Borlaug, a Nobel Laureate and one of the fathers of the first 'Green Revolution', claimed,

> Eventually the Cerrado technology or one similar to it will move to the llanos of Colombia and Venezuela, and hopefully into Central and Southern Africa where the soils are similar. This will bring tens of millions of hectares previously marginal for agriculture into high yield agriculture. (Borlaug 2006)

The symbolic work of soy as emblem of agrarian progress extends beyond its reality as an agricultural system into a signifier or icon of twenty-first-century rural development, from its genomics to its globalization. It carries the visual imprint of order and efficiency, the reality of a highly technified, 'fordist' style of production rather than the scruffy, atavistic and externally unintelligible features of locally complex agriculture that include substantial areas with woody species and differentiated landscape management. Even though the boom of soy production across South America has been implicated in multiple social and ecological problems that we discuss further on, its extension rests on several powerful narratives. One is simply the economic power of the crop that kept many of the new 'pink tide' governments in good financial standing and floated many of their redistributive policies, such as 'Bolsa Familia' and the expansion of the agricultural processing and service sector in new urban centers like Lukas do Rio Verde in Mato Grosso, Brazil (See Garrett and Rausch 2016, this volume; Weinhold et al. 2013; Richards, Pellegrina, and Spera 2015). With faltering industrial production and the global shifts associated with the rise of China's manufacturing industries, commodity development valorized a largely disarticulated export-led development model. Next, the simple Malthusian narrative is often advanced about expanding food production for a world of eight billion (Brown 2012; Smil 2001), and thus soy was positioned as an international food security question. Since most soy production is used as animal feed, the Malthusian framing increasingly has given way to a more truthful stance: soy attends to rising incomes in Asia and its demand for meat. The expansion is thus framed as an international market rather than a humanitarian concern. These powerful narratives about soy itself are complemented with symbolic content about what it means to be modern in rural areas.

The expansion of soy expresses most profoundly the image of the new developed and developing third-world state, even as, we show later, many of the previous structural vices of 'unmodern' agriculture were reproduced under the modernist mantle of soy. Its role in triggering urban and agrarian transitions was seen as salutary, again a mark of post-millennial modernism. These narratives and their allied environmental discussions provide tremendous strength to soy agribusiness in South America and beyond, which has increasingly drawn on depoliticized Malthusian, modernist and technocratic discourses to depict the sector as the solution for – rather than the driving force behind – the problems of equity, environment and the viability of future rural development in the tropics. Yet soy has also been central to the volatility of southern cone economies that structured their strategies around production and export of this commodity, and which are all at this moment in various degrees of recession.

We begin with a description of soy as material commodity, land-use practice, and nexus of structural relations between agroindustry, the state and new forms of finance. We then shift to the 'discursive' soy and its symbolism, contestations and implications. This then leads into our discussion of the contributions selected for this collection, our collective contribution to the literature, as well as its limitations and directions for further research.

2. Neo-nature and the soy complex in South America

2.1. *The shock of the new*

Soybeans were first brought to South America for agronomic experiments in the late nineteenth century, and small-scale production was established during the first half of the twentieth century by Japanese, Mennonite, Ukrainian and other immigrant farmers in southern Brazil, parts of Paraguay and the Argentinian pampas who rotated it with wheat as a nitrogen-fixing cover crop. In the second half of the twentieth century, soy became established as a major input for the vegetable oil and confined chicken industry in Brazil and Argentina, and later even in Colombia (Shurtleff and Aoyagi 2009; Goodman, Sorj, and Wilkinson 1987; Sorj et al. 2008). Since the 1970s, soy, and especially tropical soy, has evolved as the most dynamic legal agricultural segment in Latin America's rural economies and the fastest growing agroindustrial commodity in the world. Its continental magnitude is depicted on the map below (Figure 1) and the fast pace of its expansion is visible in Figure 2.

'Tropical soy' is a relatively new genomic agroindustry – a kind of 'neo-nature' – an assemblage of an exotic leguminous oilseed, selected to be amenable to mechanized planting and harvesting, adapted to longer photoperiods and higher temperatures, and able to grow in more acidic, low-phosphorus soil conditions than the temperate areas of China, the USA and Ukraine. A precondition for this 'tropicalization' of soy was the agronomic research and adaptation of soybean varieties to South American landscapes and (sub-)tropical climates undertaken by Brazilian and Argentinean state-owned agricultural research companies during the 1970s. They and domestic seed companies played leading roles until the late 1980s, when Brazil and Argentina served as 'incubators' for sub-tropical and tropical soybean production technologies that were then transferred to Paraguay, Uruguay and Bolivia. But the development of transgenic technologies by chemical companies in the United States and Europe has since displaced these public and domestic enterprises, even though they continue to play key roles in the development of soybean varieties

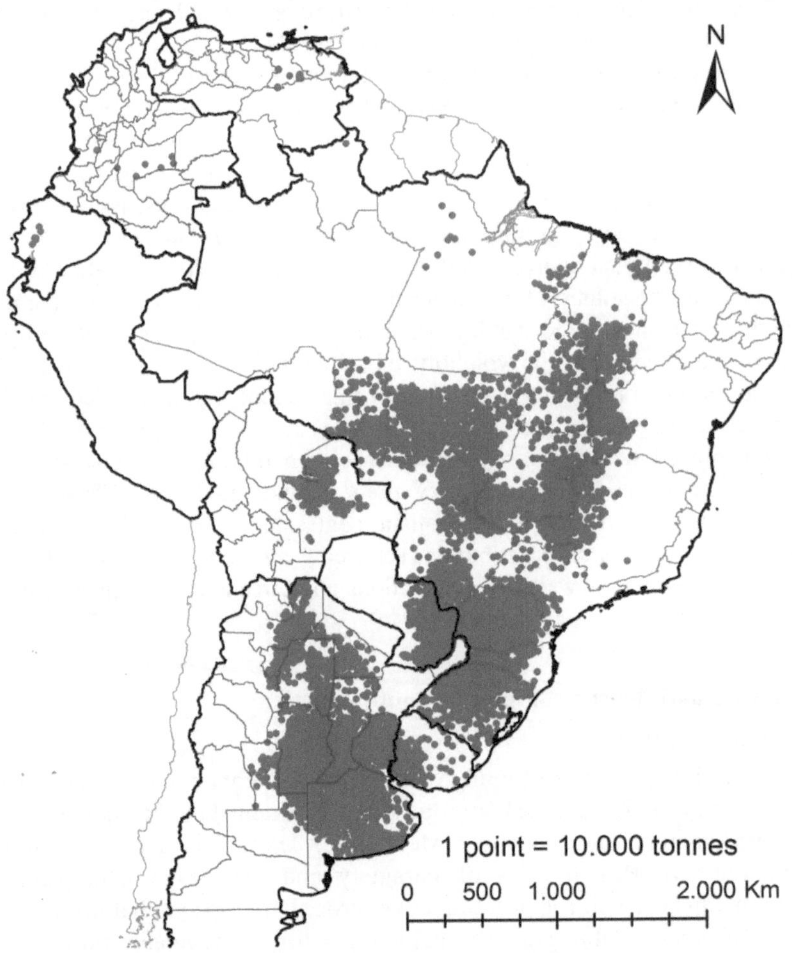

Figure 1. Map of soy production in South America by volume, 2013.
Source: Elaborated by Valdemar Wesz Jr., Ben McKay, Gonzalo Colque, Efrain Tinta, and the authors from multiple sources.

into which patented transgenic material from transnational companies is inserted, as well as multiplying and commercializing seeds with traits from transnational companies (Craviotti 2016, this volume). The soy boom that began in South America during the 1970s accelerated further in the 1990s into tropical areas, with policy shifts to export-oriented production, elimination of export tariffs and deregulation of banking sectors that facilitated foreign investments in processing and trade infrastructure (e.g. warehouses, crushing facilities and ports).

Transgenic or genetically modified (GM) varieties were approved in Argentina in 1995 and smuggled into Brazil, Paraguay and Bolivia until those governments also approved them between 2003 and 2005. During this period, a handful of transnational agrochemical companies began to dominate soybean seed and associated agrochemicals markets. Currently, the top three companies – Monsanto, Syngenta and DuPont/Pioneer – control over 55 percent of global soybean seed markets, and this concentration is even greater in

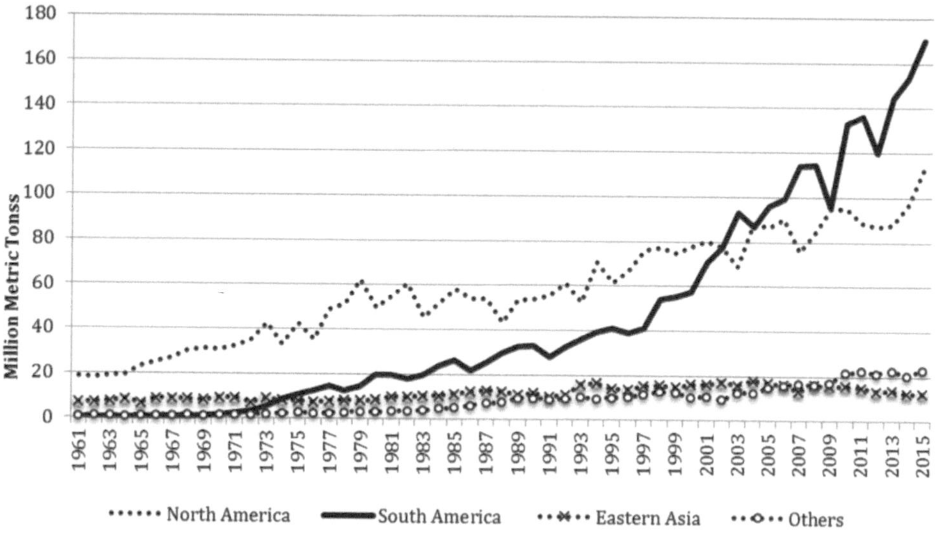

Figure 2. Soybean production volume by continent, 1961–2015.
Source: Elaborated by the authors from United States Department of Agriculture data.

South America where GM varieties predominate. The top five companies – Syngenta, Bayer, Basf, Dow and Monsanto – control 69.5 percent of global agrochemical markets, and the first three alone control over 49.1 percent of the agrochemical market in Brazil (EcoNexus and Berne Declaration 2013; Silva and Costa 2012). The ongoing merger of Dupont and Dow, worth USD 130 billion and ChemChina's acquisition of Syngenta would consolidate an already highly oligopolistic sector even more.

This key transformation in technologies and techniques began with GM soybean varieties modified to resist glyphosate-based herbicides (originally patented by Monsanto as RoundUp and RoundUp-Ready or RR soybean seeds). Agrochemical and soy agribusinesses asserted the RR technological packet would simplify production practices by reducing applications of agrochemicals and increasing productivity, thereby lowering production costs and increasing farmer profits. One of the central allurements of the tropical soy system is that it enabled no-till farming techniques that enable double cropping, placing it at a clear productivity advantage over temperate zone soy cultivation systems. Currently, GM seed use ranges from 88 percent in Brazil to 93 percent in Bolivia, 95 percent in Paraguay and over 99 percent in Uruguay and Argentina (WWF 2014). In its complete destruction of pre-existing natures, agrarian and otherwise, and its use of land essentially as a substrate, soy represents almost more completely than any other biotic production system the construction of a 'neo-nature'.

Yet the idea that GM use would be more productive and reduce applications of agrotoxics compared to conventional varieties has not been borne out everywhere and over the long term (Cotacora-Vargas et al. 2012; Altieri and Pengue 2006). Although soy systems are novel, 'neo' ecosystems in many ways, they are not exempt from evolutionary processes, and problems of resistance have emerged as pests have adapted to the selection pressure attending these monocultures. Indeed, many of these, such as Johnson grass in Argentina (Binimelis, Pengue, and Monterroso 2009), have become widespread and almost ineradicable pests, and a single field may have to be drenched with multiple treatments as much as 15 or 16 times per planting cycle.[1] This has exacerbated a technology treadmill that constantly requires new public and private research and development, and forces farmers to purchase newer varieties and inputs in order to maintain competitive levels of productivity. It has also been attended by panicky searches for short-term, highly toxic alternatives when existing solutions fail.

The leading agrochemical associated with GM soy is glyphosate. It is consumed more than any other agrochemical in Brazil, Argentina and other soy-producing countries in South America even though resistance to it is now widely reported (Nicholls and Altieri 1997; Binimelis, Pengue, and Monterroso 2009; Cerdeira, Gazziero, and Matallo 2011), and it has been identified by the World Health Organization (WHO) and multiple South American scholars as 'probably carcinogenic to humans' (Guyton et al. 2015; Silva et al. 2015), and associated with multiple other serious health problems (Marc et al. 2004; Richard et al. 2005; Benachour and Séralini 2009; Berger and Ortega 2010; Paganelli et al. 2010; Moreira et al. 2010; Oliveira et al. 2014). Even breast milk in soy production regions has been contaminated by agrochemicals (Palma et al. 2014). Aerial spraying of glyphosate and other agrochemicals is widely associated with collateral damage to rural residents and livestock, and destruction of non-target crops and other vegetation (Rulli 2007; Moreira et al. 2010; Cerdeira, Gazziero, and Matallo 2011; Arancibia 2013; Gordillo 2014). The emergence of glyphosate-resistant weeds has produced a frenzied search for more toxic herbicides such as atrazine and Dow's 2,4–D, a component of 'Agent Orange', the chemical weapon used by the US military in Vietnam, which has not yet been authorized in the US. Another dire example during the 2013–2014 season occurred when there was a major outbreak of *Helicoverpa hermigera* in Brazil, supposedly controlled by the pest-resistant *Intacta* GM varieties. Despite the widespread adoption of *Intacta* seeds, soy farmers across Brazil used a record amount of pesticides, even importing benzoate-based agrochemicals from Paraguay that are illegal in Brazil (cf. Cotacora-Vargas 2014). Although the Brazilian Soybean Producers Association (APRSOJA) and the national landowners association (Confederação Nacional da Agricultura – CNA) publicly defend the use of GM varieties and push the government to hasten approval of new and stronger agrochemicals, many soy farmers – including those in leadership positions in these associations – complain that 'we are being held *hostage* by the chemical companies', who 'do not have our interests at heart'.[2] Indeed, inputs (agrochemicals and seeds) already account for 37–47 percent of production costs for soybean in Brazil (Silva and Costa 2012), and they have certainly been increasing with the latest round of technological upgrades necessary to cope with resistant pests and weeds.

The highly standardized technological packet that characterizes GM soy cultivation assures product uniformity essential to intermediate processing and international markets (Goodman, Sorj, and Wilkinson 1987; Jepson et al. 2005; Sousa and Busch 2006), benefitting those companies downstream from farming that tightly integrate soy and other agroindustrial commodity processing, logistics and global markets. With the ability to shift sourcing and attend multiple markets, these companies made soy into one of the world's leading 'flex crops' sitting at the nexus between food, animal feed, fuel and myriad industrial inputs. This idea of 'flexing' is not necessarily new. Henry Ford, for example, who was

[1]Michael Coe, personal communication, 2015; Gustavo Oliveira, interviews with soy producers in Mato Grosso do Sul, Goiás, Tocantins, and Bahia states in Brazil, 2014 and 2015.
[2]Gustavo Oliveira, field notes from meetings of the Grain and Oilseed Chambers of APROSOJA/State Federation of Agriculture in Goiás, Bahia, and Rio Grande do Sul, Brazil, 2014 and 2015.

an eccentric booster of this crop as a diet supplement and industrial input, already imagined his car's non-metallic structures composed mostly of soy derivatives. But with the advent of soy-based biodiesel and the actual integration of soy as input for hundreds of processed foods and non-food products, soy 'flexing' is now becoming a real and powerful force shaping the sector (Borras et al. 2015; Hecht and Mann 2008; Oliveira and Schneider 2015; Grandin 2009). It is recasting continental production systems, agrarian regimes, agronomic infrastructures, finance systems, urban forms and networks, and migration and economic patterns according to what we can classify as a tropical high modernist 'neo-nature' (Scott 1998; Hecht 2005).

2.2. *The 'United Soybean Republic', or 'Soylandia'*

South America has surpassed North America as the world's leading soybean-producing region since 2003 (Figure 2), and Brazil now disputes the position of largest producer with the United States. Five countries in the 'southern cone' of South America (Brazil, Argentina, Paraguay, Uruguay and Bolivia) occupy half the spots on the list of the top 10 soybean producing countries, which concentrate the lion's share of total soy production around the world. Brazil and Argentina alone are currently responsible for 30 and 19 percent of world production. Although China and India are among the world's top 10 soybean producers, India never was a major exporter and China became a net importer decades ago. Now, virtually all their harvests are consumed domestically. Outside North and South America, the Ukraine is the only meaningful exporter, and even then it is only responsible for less than two percent of the total share of the international market. Presently, the five South American soybean-producing countries account for 58.3 percent of global exports (Table 1). But note that the following figures and table only indicate imports and exports of unprocessed soybeans, without considering soy oil and meal, and thus disguising somewhat the more significant participation of countries that supply international markets with greater ratio of soy oil and/or meal to whole beans, particularly Argentina.

The agroindustrial model of the 'green revolution' into which soy production became integrated in South America over the course of the twentieth century has its ideological and technical foundation in the US, but it wasn't until the neoliberal reforms of the 1990s in South America that domestic companies across the soybean complex, from seed production to soybean processing and international trade, were overtaken by the transnational trading corporations that had an oligopoly over US production technology and dominated market share and exports. Argentina, Brazil and the rest of South America set multiple policies to encourage export-oriented production and attract foreign investments to the sector, including tariff reduction, free trade policies and banking reform, practices that converged in the aggressive push by the leading US and European trading companies (ADM, Bunge, Cargill and Louis Dreyfus – collectively known as the ABCDs) to acquire local companies and make greenfield investments in soybean storage, processing, logistics and trade that have since made them dominant exporters from South America (Goldsmith et al. 2004; HighQuest Partners and Soyatech 2011). Until 1995, these companies only owned about 10 percent of soybean-crushing capacity in South America, but by 2002 they controlled about 50 percent of crushing capacity and 85 percent of whole-bean exports from South America, and this volume has since increased even further (Wesz Jr. 2016, this volume). These reforms also orchestrated private and foreign acquisitions of agricultural research and development, which shifted from state agencies and domestic seed

Table 1. Soybean area, volume and share of global production and exports in 2015.

	Country	Area (million ha)	Production (million metric tons)	Share of global production (%)	Share of global exports (%)*
1st	USA	33.4	106.88	33.54	35.96
2nd	**Brazil**	**32.1**	**96.20**	**30.19**	**44.53**
3rd	**Argentina**	**19.3**	**60.80**	**19.08**	**7.69**
4th	China	6.80	12.35	3.87	−63.75
5th	India	10.9	9.00	2.82	0.20
6th	**Paraguay**	**3.24**	**8.10**	**2.54**	**3.63**
7th	Canada	2.23	6.05	1.90	2.77
8th	Ukraine	1.80	3.90	1.22	1.91
9th	**Uruguay**	**1.33**	**3.11**	**0.96**	**2.24**
10th	**Bolivia**	**1.08**	**2.65**	**0.82**	**0.21**
11th	Russia	1.90	2.59	0.81	−1.35
12th	South Africa	0.69	1.06	0.33	−0.06
13th	Italy	0.32	0.90	0.28	n/a
14th	Nigeria	0.65	0.65	0.20	−0.08
15th	Indonesia	0.43	0.60	0.18	−1.67
16th	Serbia	0.18	0.44	0.14	−0.00
17th	Mexico	0.19	0.36	0.09	−3.25
18th	Japan	0.13	0.23	0.07	−2.38
20th	Iran	0.08	0.20	0.06	−0.76
20th	Burma	0.17	0.20	0.06	0.00
	Others	2.58	2.00	0.63	0.86

Source: USDA-FAS data, elaborated by the authors.
*Negative numbers are provided for net importers, indicating share of global imports.

companies in South America to a handful of leading seed-and-agrochemical companies from the United States and Europe.

The rapid rise and concentration of this technologically homogenous agro-ecosystem across state borders in the 'southern cone' of South America led transnational agribusinesses to imagine – and project the image of – a 'United Soybean Republic' over which they could preside (Figure 3; Grain 2013; Turzi 2011). This reflects a strategic response by North Atlantic-based agrochemical and commodity trading companies to capture the emerging shifts of production to South America from the United States, as well as consumption markets created by increased poultry and pig production in Europe and East Asia, particularly China, triggering a race for control over these new production sites and flows of this key agroindustrial sector (Figures 4 and 5; Oliveira and Schneider 2015).

While the technological homogeneity and corporate concentration of the sector has been characterized as an iconic mark of a global 'corporate food regime' distanced and disarticulated from local political, social and environmental realities (Turzi 2011; McMichael 2012; Clapp 2014), it is important to understand that there has been continued state engagement in this sector since its early implementation, including institutional development throughout the region in terms of government-subsidized credit, modernization of cadastral data and land titling, state-supported infrastructure development and machinery imports, and significant research and development funding (Hecht 2005; Fearnside 2007; Wolford 2008; Richards 2010; Urcola et al. 2015; Oliveira 2016, this volume). Although transnational seed and export companies dominate the sector, their business logic and operations

Figure 3. *Locus classicus* of the so-called 'United Soybean Republic'.
Source: Reproduced with permission from Grain 2013.

remain firmly rooted in concrete social and ecological relations, embedded rather than abstracted from their place.

There has already been considerable scholarship on how the ABCDs established a global monopsony (Goldsmith et al. 2004; HighQuest Partners and Soyatech 2011; Morgan 2000; Murphy, Burch, and Clapp 2012; Oliveira and Schneider 2015), but countering these tendencies was the earlier development of cooperatives in southern Brazil that pooled capital to purchase inputs, build storage and processing facilities, and increase their bargaining power vis-à-vis trading companies (Chase 2003; Fajardo 2005). Some of the largest scale farmers, such as Blairo Maggi in Brazil and Gustavo Grobocopatel in

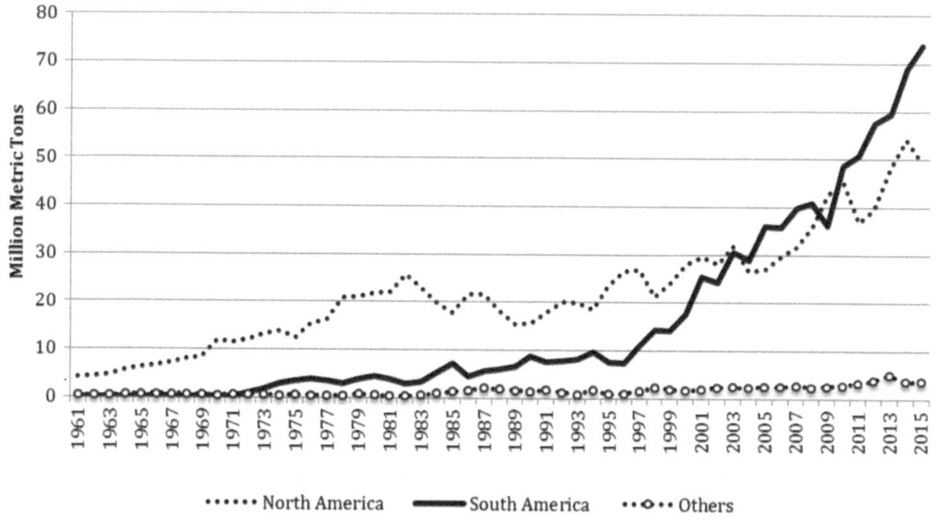

Figure 4. Soybean exports by continent, 1961–2015.
Source: Elaborated by the authors from United States Department of Agriculture data.

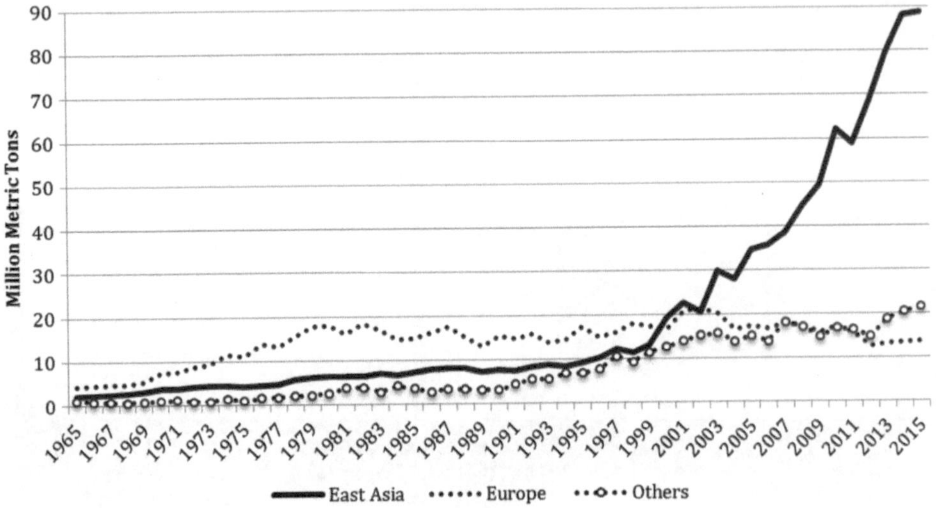

Figure 5. Soybean imports by continent, 1965–2015.
Source: Elaborated by the authors from United States Department of Agriculture data.

Argentina, have even been able to expand vertically into the construction and operation of their own trading operations (Table 2). But farmers are largely price takers in domestic and international soybean markets, because of the bottleneck associated with the concentration of a very few economic actors in the processing, logistics and marketing realms. A small minority of farmers have contracted for non-GM or even organic soy production in order to gain price premiums from niche markets (Vennet et al. 2016, this volume; Garrett, Lambin, and Naylor 2013; VanWey and Richards 2014), and policy incentives for biodiesel companies to source from small-scale farmers have also established a segmented market for some soy producers in southern Brazil (Oliveira and Schneider 2015). The predominant strategy for soy farmers to cope with their weak position relative to soybean crushers and trading companies has been to increase production and control the quality of their harvests. Since the intensive use of glyphosate and other herbicides drastically reduces the contamination of harvests with weeds and leaves, farmers now consider glyphosate-resistant GM varieties necessary in order not to incur price deductions at the point of delivery, deepening the technological treadmill and the farmers' need for finance.

2.3. *Financing 'Soylandia' and fragmenting farms*

Historically, soy and other cash crop production was financed largely by state-owned and subsidized national banks, but this role has largely shifted to agrochemical/seed providers and commodity trading companies. Partnerships between major seed and agrochemical input manufactures with major soybean trading companies (e.g. Monsanto with Cargill, Syngenta with Bunge) enable these agribusinesses up- and downstream from soy farms to effectively control the inputs and farming practices of most soy farmers across South America, and lock in prices and delivery of portions of their harvests through prearranged provision of fertilizer, pesticides, herbicides and seeds – a practice typical of contract farming more generally. Smaller and less capitalized farmers might commit as much as two thirds of their harvest to input/trading companies before the planting season, and even large and well-capitalized farmers frequently contract around a quarter of their

Table 2. Major *pools de siembra* and farm management companies in South America.

Name	Country/year established	Investors	Total ha controlled	Notes
Cresud/ Brasilagro	Argentina, 1936 (restructured in 1959 when listed on the Buenos Aires Exchange)	Originally established by Credit Foncier (Belgium), purchased by the Dolphin Fund in 1994 (operated by Eduardo Elsztain), who remain its major shareholders and primary managers. In 2006, it created Brasilagro with JP Morgan Whitefriars (USA) and Credit Suisse (Switzerland).	Owns 866,215 ha in Argentina, Brazil, Paraguay and Bolivia; 204,706 ha cultivated in 2015, including 77,179 ha on lease or concession.	Its main focus since 1994 has been the development of farmland for resale, and at its peak it owned over 1 million ha across four countries. Soybeans account for 56% of its crop production, while sugarcane and ranching cover 27% and 14%, respectively, of its land under production. In 2006, Cresud created Brasilagro as a public company listed on BM&F Bovespa to expand into Brazil.
Amaggi/Bom Futuro	Brazil, 1977, Bom Futuro becomes independent in 1993	Privately controlled by the Maggi family. The Amaggi Group is led by past governor and senator of Mato Grosso state, Blairo Maggi. Bom Futuro Group is owned by the Maggi Scheffer branch of the family, led by Blairo's cousin Eraí.	> 511,300 ha in Brazil; Amaggi: 252,300 ha (51,900 ha on lease); Bom Futuro: over 259,000 ha.	Largest private soybean producers in the world. The companies have also expanded to seed and fertilizer production, agricultural commodity storage and trading, fluvial navigation and hydroelectric energy production. Amaggi in particular participates in key joint ventures with Bunge and Louis Dreyfus for Amazon basin exports, and operates its own trading offices in Argentina, Paraguay, Norway, Switzerland and the Netherlands.

(*Continued*)

Table 2. Continued.

Name	Country/year established	Investors	Total ha controlled	Notes
Adecoagro	Argentina, 2002 (with the purchase of Pecom Agropecuária)	Soros Fund Management, Ospraie Management, Jennison Associates, Brandes Investment (USA), Stichting Pensioenfonds Zorg en Welzijn (Netherlands) and Qatar Holdings.	437,245 ha in Argentina, Brazil and Uruguay; (124,412 ha leased for sugarcane, and 55,797 ha for soy).	In 2004, it began expanding to Uruguay and Brazil and diversifying from soy to sugarcane, rice, and dairy production and processing. In 2011, it was listed on the New York Stock Exchange.
SLC Agrícola	Brazil, 1945 (as Schneider Logemann & Co., producing agricultural machinery)	Black Rock, Neuberger Berman, Fidelity SelectCo, DFA Emerging Markets, Global Thematic Partners, Dimensional Fund (USA), Verde Asset Mgt. (Brazil), Deutsche Bank (Germany/China), Credit Suisse (Switzerland).	384,070 ha in Brazil; (75,871 ha leased).	SLC began producing soy in 1977 when it still focused on agricultural machinery production and a joint venture with John Deere, then divested from machinery production in 1989, and became listed on BM&F Bovespa in 2007. In addition to agricultural production, SLC also aims to profit from bringing 'new' land into production for resale.
El Tejar/O Telhar	Argentina, 1987, moved headquarters to Brazil in 2013	Private company held by the Alvarado family, Altima Partners (UK) since 2006, the Capital Group (USA) since 2009, and other minor investors.	~300,000 ha primarily in Brazil; ~1,100,000 ha at its peak in 2010 across Argentina, Brazil, Paraguay and Uruguay (about 70% on leased land).	One of the creators of the *pool de siembra* model, the Alvarado family controlled the company until 2009, when majority shifted to its financial investors. It expanded into Brazil in 2003, buying its own farmland, but maintained leases for its expansion into Paraguay and Uruguay. However, the company abandoned this model, withdrawing from Paraguay in 2011, and from both Uruguay and Argentina in 2013, when it hired a Brazilian chief executive officer and shifted its headquarters to Brazil.

(Continued)

Table 2. Continued.

Name	Country/year established	Investors	Total ha controlled	Notes
TIAA-CREF Global Agriculture	Brazil, 2008, via Radar and Tellus: 2012, as TCGA	TIAA Global Ag (US), Cosan (Brazil), AP2 (Sweden), Caisse de Depot et Placement du Quebec, British Columbia Investment Management (Canada), Cummins UK Pension Plan Trustee, Greater Manchester Pension Fund (UK), New Mexico State Investment Council, Environment Protection Agency Pension Fund (USA).	> 230,000 ha in Brazil; Also owns land in the US and Australia, seeking investments in Chile, New Zealand, Central and Eastern Europe.	Led by one of the earliest and most aggressive pension funds to launch transnational farmland investments, the Teachers Insurance and Annuity Association – College Retirement Equities Fund (USA), it joined the Brazilian sugar conglomerate Cosan in 2008 to purchase and lease farmland for sugarcane and soy production. By 2012, it already owned 151,468 ha, when it began expanding towards the northern Cerrado states, where it has been accused of involvement in illegal land grabs and violent dispossession of peasants.
V-Agro	Brazil, 2003, as Brazil EcoDiesel (BED); changed its name in 2011 merging with Vanguarda and Maeda Agronegócios	BED was composed by BT Global, Zartman, Boardlock, and Carleton Towers (USA), Deutsche Bank (Germany), BMG (Saudi Arabia), Nelson Silveira, Bradesco, Fibra, Bonsucex, and Banco Fator (Brazil). Current main shareholders: Fim CP Veritas, Gávea Investmentos, Bonsucex, EWZ Holding, and Pollus Capital (Brazil).	224,317 ha in Brazil; (135,220 ha leased).	BED was the leading biodiesel producer in Brazil at the time of its merger with Vanguarda and Maeda (with the capacity for producing 700,000 m^3 of biodiesel per year); since then V-Agro has become listed on BM&F Bovespa, divested from biodiesel processing, and focused on soy production, and also began operations to develop farmland for resale.

(*Continued*)

Table 2. Continued.

Name	Country/year established	Investors	Total ha controlled	Notes
MSU	Argentina, 1985, 1997, as *pool de siembra*	Private company held and managed by the Uribelarrea family, with Stitchtings Pensioenfonds ABP (Netherlands) and other foreign minority investors since 2007.	210,000 ha in Argentina, Brazil, Paraguay and Uruguay.	Originally as Juamarita SA operating on 15,000 ha of private land, MSU began leasing farmland in 1997. In 2007, it created Santa Juana Ltd. as a land acquisition company to expand from Argentina to neighboring countries. In 2012 its main creditors were Banco Galicia and Hypotecario (Argentina), and Itaú (Brazil), accounting for about 42% of its loans. It has also established subsidiaries for agricultural aviation, poultry and egg production, farm management and bioenergy.
Los Grobo	Argentina, 1984	The Grobocopatel family are the major shareholders and primary managers, but multiple minority partners and joint ventures have been incorporated for specific operations, such as UPJ in Argentina, Tierra Roja in Paraguay, ADP in Uruguay, and Pactual Capital/Vinci Partners in Brazil.	Currently leases ~100,000 ha in Argentina; At its peak it leased ~320,000 ha in Argentina, Brazil, Paraguay and Uruguay in 2010/2011.	One of the creators of the *pool de siembra* model, it began leasing farmland in 1987. In 2004 it expanded to Uruguay, in 2005 to Paraguay, and in 2008 to Brazil. In 2013, it sold its Brazilian subsidiary to Mitsubishi (from Japan) and withdrew from Paraguay, and in 2015, it sold its participation in ADP to its Uruguayan partners. It has shifted its focus from agricultural production (only accounting for 15% of revenues by 2014) to focus on agroindustrial processing, financing and trading.

Source: Elaborated by the authors from company websites, stock exchange reports and other sources.

harvest in exchange for fertilizer and other inputs (Wesz Jr. 2016, this volume). The leading agrochemical, seed and trading companies now derive a very substantial portion of their income from such operations, and such financing also permits much closer monitoring of production, providing extremely valuable data for transnational trading companies. This is part of a broader process of global financial deregulation and financialization of farmland markets and agricultural production that has become the subject of an important new field of research (Murphy, Burch, and Clapp 2012; Fairbairn et al. 2014; Fairbairn 2014; Isakson 2014).

A still under-studied aspect of this process has been the creation of new vehicles for channeling multiple forms of capital into soy production. These emerged from the practices of land-leasing and outsourcing specific operational tasks (such as harvesting or spraying agrochemicals) that were beginning to take shape on the Argentinian Pampas during the 1980s (Gudynas 2008; Urcola et al. 2015; Leguizamon 2016, this volume). Where this process has taken root, the figure of the independent soy farmer has been splintered into multiple characters, as smaller scale landowners lease their farms to companies that hire agronomists, farm managers, planter/harvester operators and other specialized laborers to run the soy production system. These companies, known as *pools de siembra* or 'sowing pools', collect their capital from a large number and variety of investors, ranging from rural and urban individuals to institutional investors and financial corporations. Drawing on increasingly larger and more cross-regional and transnational financial flows, these companies merge agroindustrial and management expertise with rented farmland and hired labor to expand vertiginously across the continent. Now, literally hundreds of thousands of farm units are operated by a handful of companies that manage millions of hectares across South America. Table 2 provides illustrative samples.

The land-lease basis of Argentinian *pools de siembra* expanded to a certain extent into neighboring countries, but multiple institutional and economic challenges also led them to shift and transform strategies to accomplish similar goals of cross-regional expansion to reduce risks associated with climate variation, pest outbreaks and multiple institutional challenges/benefits in different places. Some *pools de siembra* shifted towards production on their own land (especially in Brazil), and others shifted into processing, logistics and trade, while still others acquire farmland only to develop it for re-sale, capturing a speculative boost in many cases. This is especially true when soy development 'leapfrogs' into more frontier areas (Gibbs et al. 2010; Meyfroidt et al. 2014; Richards, Walker, and Arima 2014; Graesser, Aide, and Ramankutty 2015). As management shifts to farm-service companies, even non-corporate contract farming relations are transforming, so the nature and 'autonomy of farming' are increasingly disciplined and structured by external forms of management based on the application of technological packages and farm-service logistics. Indeed, soy farm operations in South America have become so extensive and tightly interwoven with service provision that 'family farmers' from the Midwest of the United States who come to Brazil struggle more to deal with these new social relations of production than with new climates or agroecosystems (Ofstehage 2016, this volume). In Bolivia and Paraguay as well, where Argentinian *pools de siembra* did not take strong root, the technological treadmill and managerial logic of multi-unit operators has left many farmers who previously gained small (c. 50 ha) plots under colonization schemes with no option but to lease their land to neighboring soy farmers who gradually increased their land to a few hundred hectares. This trend towards farmland and wealth concentration has accelerated with the introduction of GM technologies, and aggravated the already unequal distribution of farmland, credit and capital in the eastern lowlands of Paraguay and Bolivia that results from their frontier form of colonization (McKay and Colque 2016, this volume; Elgert 2016, this volume).

In some areas of southern Brazil, relatively small soy farms (< 300 ha) continue to exist, yet they are only economically viable insofar as they are able to capture price premiums from niche markets or integrate soybeans as part of a more diversified farming system (Vennet et al. 2016, this volume). In the Cerrado region of central Brazil, most soy farms are medium (300–1000 ha), but large farms (1000–30,000 ha) account for the overwhelming majority of production. In some areas, for example, farms larger than 1000 ha account for over 90 percent of the planted area (Cacho 2016, this volume). Moreover, large-scale farm management firms – modeled upon Argentinian *pools de siembra* bur frequently owning more land than they lease – are quickly increasing their operations over the Cerrado region, with multiple farms that range from 10,000 to 30,000 ha. Consequently, they have become the preferred instrument for new financial actors – such as private equity and pension funds like TIAA-CREF – to make arm's-length investments in both farming and land speculation. These new financiers have no history in direct agricultural production, but found in *pools de siembra* and farm management companies a useful tool for diversifying their portfolios, capturing gains from these expanding commodity and land markets.

3. Monoculture, modernization and Malthus

Serious environmental questions and intense debates attend the expansion of soy production throughout South America. Several concerns have to do with the environmental impacts of the soy fields on their own sites, particularly as direct and indirect exposure to agrochemicals generates problematic levels of human and environmental poisoning (Altieri and Pengue 2006; Rulli 2007; GRR 2009; Berger and Ortega 2010; Moreira et al. 2010; Arancibia 2013; Palma et al. 2014; Silva et al. 2015). Further longitudinal and epidemiological studies remain to be carried out, but enough evidence has emerged for the WHO to list glyphosate as 'probably carcinogenic to humans' (Guyton et al. 2015).

A considerable literature has also explored more regional and indirect environmental effects of soy expansion, including the dynamics of run-off, deforestation and impacts on the critical Pantanal wetland and other ecosystems adjacent to major soy production areas (Silva and Sousa 2015b; Schulz, Ioris, and Glenk. 2015); the complex effect on regional hydrology, including stream flow dynamics and water warming (Coe et al. 2009; Deegan et al. 2011; Macedo et al. 2013; Neill et al. 2013); impacts on biodiversity through habitat loss, deforestation and ecosystem fragmentation (Steininger et al. 2001; Fearnside 2007; Carrero and Fearnside 2011; Redo, Millington, and Hindery 2011; Redo 2012; Soares-Filho et al. 2012; Oliveira, Costa, and Coe 2013; VanBeek, Brawn, and Ward 2014; Alencar et al. 2015); and the larger scale interactions between soy production zones, continental air moisture flows, deforestation and climate change (Sawyer 2008; Nobre et al. 2009; Malhi et al. 2009; Asner, Loarie, and Heyder 2010; Coe et al. 2013; Brando et al. 2014; Swann et al. 2015; Zhang and Castanho 2015). These regional impacts affect far more land than even the millions of hectares of direct conversions, and sit at the heart of the long-term problem of historical and current carbon emissions.

At this macro level, serious concerns are raised about what one might call 'disappearing biomes' underneath the expansion of soy. Most widely discussed is the southern Amazon transition zone infamous as the 'arc of deforestation' (Barlow and Peres 2008; Nepstad et al. 2014; Anadon, Sala, and Maestre 2014), but far more significant is the Cerrado itself, which has declined by 53–66.3 percent over the last few decades (Jepson 2005; Bianchi and Haig

2013; Beuchle et al. 2015), and its transition into the Caatinga – the dry forests of northeastern Brazil that declined by about 53 percent in a similar period (Santos and Leal 2011). The Paraguayan Atlantic forests have also been almost completely cleared with the expansion of industrial agriculture (Richards 2011), while more than 15.8 million and 9.67 million ha of dry forests have been cleared in the Paraguayan and Argentinian Chaco respectively (Macchi, Grau, and Marinaro. 2013; Gasparri, Grau, and Sacchi 2015; Vallejos et al. 2015; Volante and Paruelo 2015). Even highly remote forests like the Chiquitania in eastern Bolivia have declined by some 18 percent (Pinto-Ledezma and Mamani 2014; Baldi et al. 2015).

There are multiple processes of soy expansion, from direct clearing to intensification of existing land uses – 'writing over' with the enhanced 'ecological efficiency' of soy systems – and the indirect displacement of previous land uses, such as extensive ranching, into woodland landscapes (Marsik et al. 2011; Meyfroidt et al. 2014; Graesser, Aide, and Ramankutty 2015). These dynamics are driven by global and regional commodity markets (Fearnside et al. 2013; VanWey and Spera 2013), regulated by international commodity prices and currency exchange rates (Richards, Myers, and Walker 2012; Oliveira 2016, this volume), institutional configurations (Garrett, Lambin, and Naylor 2013), supply chain dynamics (Nepstad et al. 2013; Newton, Agrawal, and Wollenberg 2013), differentials in land prices (Mann and Kaufmann 2010; Richards 2015; Sauer and Leite 2012), tenure and access regimes (Barbier and Burgess 2001; Hecht 2005; Walker and DeFries 2009; Jepson et al. 2010; Larson 2011; Aldrich et al. 2012; Borner et al. 2014; Meyfroidt et al. 2014; Richards, Walker, and Arima 2014; Marinaro, Grau, and Zelaya 2015), and the restructuring of 'flex crop' processing companies (Oliveira and Schneider 2015). These also include enhanced infrastructure for export like roads, warehouses and ports, as well as continued agronomic extension, research and credit lines. Common to all analyses is the evidence that intensification of profitable land uses tends to enhance its spread rather than to confine it spatially, regardless of the mix of drivers (Hecht 2005; Morton et al. 2008; Rudel et al. 2009; DeFries, Rudel, and Hansen 2010). Yet in spite of its notable and extensive direct and indirect impacts, soy agribusiness is viewed very positively in policy circles and mainstream media, certainly in its contribution to national coffers, but also in symbolic terms. This dynamic has two central dimensions: first, the meaning of this expansion for the state, as we suggested earlier in this paper, through its economic and developmental symbolism pertaining to a modern state and its lands; and, second, an emergent environmental and moral discourse rather at odds with earlier vehement critiques of agribusiness.

Beginning with regional integration and state making over ideologically 'empty lands' that are in fact inhabited working landscapes,[3] soy has become the central means for contemporary state 'territorialization', expanding and integrating state presence into remote areas of national hinterlands. In a way, this is a continuation of the five-century-long process of colonization that rested primarily on extensive cattle ranching across South America. Yet this process began to change during the twentieth century through what Brazilians called the 'March to the West' and Bolivians dubbed the 'Conquest of the Orient', among other kindred slogans about territorial integration. These forays accelerated during the second half of the century with state-funded infrastructure, credit and land titling and colonization programs, and although 'green revolution' technologies began to be

[3]These landscapes often turned out to have been occupied by traditional peoples for many decades, and by some people for millennia.

implemented, most labor relations and production practices expressed rather atavistic arts and land use. While in many ways livestock achieved territorialization goals by transforming earlier frontiers (Hecht 1985; Gardner 2009; Rodrigues et al. 2009; Walker et al. 2009; Hoelle 2014), it lacked significant backward linkages (since both the grass and the animals were able to reproduce themselves) and limited forward linkages, remaining largely linked to regional administrative capitals and abattoir cities, while creating a great deal of displacement of peasants into urban slums (Gonçalves, Siqueira, and Hacon 2014). Given the low productivity of extensive ranching, land degradation and abandonment, low employment, and deep association with speculation and violence generated over decades, such land use came under sustained critique as a development strategy (Hecht and Cockburn 1989; Schmink and Wood 1992; Laurance et al. 2004), enabling the search for a substitute that could better accomplish state goals.

Soy, with its production complexity, global imprimatur, demand for specialized goods, and (few but some) skilled labor and services, produced a new kind of agroindustrial landscape as well as a new agroindustrial urbanism: unlike the earlier administrative centers or labor-depot urbanism of earlier South America frontiers (Browder and Godfrey 1997), soy cities are structured more around service provisioning, industrial processing and finance (VanWey and Spera 2013; Richards, Pellegrina, and Spera 2015). Places like Lucas do Rio Verde stand as iconic cases of this sort of urbanism (Hecht and Mann 2008; VanWey and Spera 2013; Richards and VanWey 2015), as they emerged 'from nothing' through complex migration flows and services in increasingly depopulated surrounding landscapes (Desconsi 2011; Rumstain 2012; Miranda 2012). From unruly forest and grasslands emerge ordered, fast-growing cities and fields as the incarnations of progress, a positive denouement for states that always saw their hinterlands as backward. In this sense, the agrarian modernism that soy agribusiness encapsulates is more than just a production system; it becomes a larger national symbol, and a feature of modern state making (cf. Scott 1998; Oliveira 2016, this volume).

This dynamic has also been attended by a new and powerful class of agribusiness managers and technocrats operating across regional, national and international politics (Gudynas 2008), who have become increasingly assertive over national political debates on agricultural and land-use policy. In Brazil, for example, the agroindustrial lobby encouraged politicians to roll back environmental restrictions in the Forest Code and provide amnesty and legitimation of previous illegal clearing (Soares-Filho et al. 2014). In lowland Bolivia, agribusinessmen – particularly in the soy sector – tried to secede from the rest of Bolivia in reaction to the electoral victories of the indigenous and leftist coalition for Evo Morales, and Paraguay's agricultural elite successfully organized to overthrow the country's first leftist president, Fernando Lugo, when he sought to redistribute farmland dedicated to soy agribusiness. These examples illustrate how soy agribusiness has inflected and influenced state policy not only through the territorialization of the state in agroindustrial frontiers, but through its capacity for putting political pressure on the state in favor of agribusiness interests.

One of the most important manners in which these agribusiness interests have become articulated has been in the discursive transformation of soy from a key driver of deforestation in the Amazon into the modern embodiment of the solution to the eternal problem in tropical development: combining environmental stewardship with development. The extraordinary attention to this dynamic had a dramatic effect, which we call the 'Amazon swerve'.

3.1. *The Amazon swerve: Mato Grosso exceptionalism and the 'land sparing' debate*

Soy production was heavily critiqued by conservationists due to the social and ecological impacts arising from monocultures that drench landscapes in glyphosate and pesticides, and the history of violent displacement and extensive deforestation that produced these landscapes in the first place (Hecht 2005; Pacheco 2006; Morton et al. 2008; Rudel et al. 2009; Zhouri 2010). The largest soy-producing state in Brazil, Mato Grosso, was reviled for its continuing clearings along the 'arc of deforestation' across the southern Amazon. But around 2004, deforestation began to decline in Mato Grosso and neighboring Pará states, falling about 80 percent below its 2004 high point (Nepstad, Stickler, and Almeida 2006; Hecht 2012; Nepstad et al. 2009; Nepstad et al. 2014;). We call this event 'the Amazon swerve', a process that radically changed the perception of soy within environmental circles (powerfully structured by international non-governmental organizations (NGOs) like World Wildlife Fund and Conservation International) from a leading landscape destroyer to a key tool for conservation, melding the modernist state and agroindustrial project with conservation, meaning 'conservation of the Amazon' and *not* the sacrifice zones to its south and east.

Resting theoretically on the idea of 'land sparing' through intensification, the idea originally advanced by Borlaug (even described as the 'Borlaug hypothesis') proposes that increased productivity will satisfy crop demands without extending production area, allowing other areas to be set aside for 'strong' conservation, largely protected from all human use, as opposed to the difficult and messy process of agro-ecological harmonization through diversified working landscapes, the so called 'land sharing' approach to biodiversity conservation (Peres and Zimmerman 2001; Laurance et al. 2012; Grau et al. 2013; Macchi, Grau, and Marinaro. 2013; Edwards et al. 2015). The discourse emerged with local and international mobilizations against deforestation, particularly of the Amazon, leading to consumer boycotts and NGO efforts at monitoring and enforcement of environmental regulations. Leading NGOs, state officials and major soy agribusinesses coalesced around a Roundtable on Sustainable Soy, later renamed the Roundtable on Responsible Soy, to set voluntary guidelines for production practices. Ultimately, concerns over deforestation and the advance of indigenous and other traditional community demands for protecting their territories (in the Amazon) produced a voluntary 'moratorium' on soy from newly deforested Amazonian landscapes. Direct forest conversion to soy production declined significantly in Mato Grosso state. As a major deforestation hot spot, the slowdown dramatically reduced Brazil's global forest destruction profile.

Against the backdrop of intense Amazonian deforestation that galvanized scholars and activists since the 1970s, this notable decline in clearing in that particular stretch of Amazon-Cerrado transition zone appeared to show that increases in production could truly become a conduit for conservation. This projected the 'land sparing' discourse as a general principle rather than a peculiarity of the Mato Grosso political dynamics, shaped by an atypical combination of public and private actors at many different scales (Arvor et al. 2012; Macedo et al. 2012; Brando et al. 2013; Stickler, Nepstad, and McGrath 2013; Nepstad et al. 2013; VanWey and Richards 2014; Alkimim, Sparovek, and Clarke 2015). In principle, soy intensification would solve food production and the environmental crises in one swoop, especially if strong set-asides (such as indigenous reserves and completely protected parks) could be established once 'good governance' was in place (Nepstad et al. 2006; Ceddia, Bardsley, and Sedlacek 2014). Thus, despite all its unusual attributes, Mato Grosso became the 'poster child' for the land-sparing model, suggesting its local dynamics are emblematic rather than exceptional.

There are several issues one can raise about the Mato Grosso case. First, as mentioned, it represents an extraordinarily dense assemblage of forces at multiple scales that limit the replicability of models based upon it. While the soy technology itself may be an immutable mobile, *sensu* Latour, institutions and governance are a mutable immobile (Latour 1987) and this Amazon soy lineup is exceedingly difficult to reproduce, particularly the conditions for regulation without repression that do not exist everywhere in South America. Next, the Amazonian focus has diverted attention away from the clearing footprint of soy across the continent, and even within Brazil, where conversion of other ecosystems (particularly the Cerrado) is far higher than in the Amazon. As mentioned earlier, estimates of Cerrado loss run as high as 66 percent, and Paraguay's Atlantic forest biome has been almost completely destroyed. Researchers of other woodland systems that have been ravaged by soy production point to the lack of conservation infrastructure and general inattention to these dry and open forest ecosystems where the intensification/expansion of production coalesces (Hecht 2005; Klink and Machado 2005; Santos and Leal 2011; Bianchi and Haig 2013; Volante and Paruelo 2015; Grau et al. 2015). These 'orphan' forest systems fall outside most conservation interest, marked by weaker national and international networks of concern, monitoring and institutional or practical mechanisms for regulation, so the central drivers of deforestation reflect the dynamics and volatilities of agricultural commodity prices, land rents and state concerns about regional integration. These regions – where soy production has in fact been expanding while the Amazon has become protected – are thus much more vulnerable to boom-and-bust dynamics than the hyper-institutionalized and monitored Amazonian landscapes where soy production was curtailed. As Pfaff and others have also pointed out, one person's forest transition is another's deforestation, and there is evidence that the tight environmental regulations, cadastral requirements, better monitoring and enforcement in the Amazonian fringe have triggered 'leakage' into other woodland systems elsewhere in Brazil, Bolivia, Paraguay and Argentina, operational dynamics that are obvious to cross-continent farm management companies and migration choices of small- and medium-scale soy farmers (Hecht 2005; Pfaff and Walker 2010; Richards 2011). Indirect land-use change (ILUC) dynamics that can only be surmised in remote-sensing models become explicit through field-site inspections and interviews with agribusiness managers and soy farmers, who say that their expansion plans are not curtailed by intensification of production, but rather conditioned by favorable institutional settings for the 'development' of 'new' lands in less regulated areas like the northeastern edge of the Cerrado in Brazil (e.g. southern Maranhão and Piauí states), where profits from intensified production are often reinvested.[4] After all, as a general rule, intensification of profitable land uses results in territorial expansion rather than its reduction in area (Hecht 2005; Rudel et al. 2009; DeFries et al. 2013). This is the Jevons Paradox of tropical soy.

The narrow focus on Amazonian clearing, and the unusual conjuncture that produced the decline in deforestation in a section of the southern Amazon, has had the effect of obscuring a more general recognition of the much more widespread clearing consequences of this land use. There is a rich debate on this topic which we cannot engage fully in this essay, but the unique emphasis on agricultural productivity has been taken to task for its lack of attention to evolutionary processes (resistance of weeds, pests, etc.) externalities associated with non-target victims of toxics, land development, carbon emissions, and

[4]Gustavo Oliveira, interviews with farmers and soybean producer associations in Brazil, 2014 and 2015.

other questions of equity and sustainability (Haslberger 2006; Altieri and Pengue 2006; Porto and Milanez 2009; Schiesari and Waichman 2013). Moreover, the emphasis on soy as a solution to rising food demand elides the differences between agroindustrial commodities and 'food', ignoring the fact that most 'flex crop' production has expanded in recent years to attend non-food demands for fuel and industry (Borras et al. 2015; HLPE 2011). This is very clear in the case of soy, which finds in concentrated animal feed operations (CAFOs) its most important demand, and in biodiesel, its most important new market. Detailed life-cycle analyses and calculations of total-factor productivity unequivocally demonstrate that grain- and soy-based CAFOs are extremely *inefficient* forms of producing protein for human food, and soy-based biodiesel is only carbon efficient when produced in small-scale diversified farming systems (Cavalett and Ortega 2010; Oliveira and Schneider 2015; Schneider 2014; Steinfeld et al. 2006; Weis 2010, 2013a, 2013b). These considerations, among a broader political ecology literature, challenge the 'land sparing' discourse that sustains favorable academic, policy and business attitudes towards soybean agribusiness. How can intensified production limit expansion given the difficulty of 'good governance' across the entire soy frontier of South America, and given that increased production does not aim to satisfy relatively inelastic human food needs, but rather boundless demands for fuels and extremely inefficient but profitable concentrated livestock production?

4. Contributions to the literature

The papers in this collection do not necessarily resolve these debates and controversies, but they certainly provide a more nuanced and detailed discussion of the social, political, economic and ecological settings of the contemporary boom of soybean production in South America. We sought cutting-edge research with a solid foundation in recent fieldwork and innovative analysis that may shed light on the production processes and outcomes of this controversial crop. The contributions in this volume expose continuities in the political and economic processes of the export-oriented insertion across South America but also highlight truly novel patterns of production and social relations that emerge from soy's neo-nature. We hope thereby to contribute to and perhaps even reframe some key debates on soy production in South America upon the material realities confronted by upcoming and leading scholars in the field.

This collection is divided into four thematic sections. The first explores the 'friction' (Tsing 2005) of globalization in new agroindustrial landscapes of soy production. The papers in this section tackle the common notion that the homogenization of production technologies/techniques, domination of input markets and commodity trading by transnational corporations, increasing involvement of international finance, privatization of research and infrastructure, and deregulation across South America has produced a 'Soylandia' or 'United Soybean Republic' that is abstracted away from local social and ecological particularities. In his contribution, Valdemar Wesz, Jr., draws on aggregate political economic data across the southern cone of South America and ethnographic fieldwork in Brazil to reveal how the leading trading companies adopt strategies to become 'hybrid' actors that incorporate transnational dynamics while simultaneously relying on local employees who are deeply embedded in the social fabric of soybean farmer communities to operate successfully. Similarly, in her contribution, Clara Craviotti claims that while operations of seed companies and *pools de siembra* have expanded from Argentina throughout the continent with little apparent regard for borders, they remain profoundly embedded in territorial dynamics and social relations, and, like the trading companies analyzed in the previous contribution, it is precisely this territorial embeddedness that generates their power over soy farmers. Finally, Amalia Leguizamon argues that attempts by financial actors and technological innovators in the soybean sector towards 'distancing' and 'abstracting' this production system from social and ecological specifics are never fully accomplished, and remain very much rooted in physical bodies, resources and practices. This on-the-ground reality of globalization finds very interesting correlations with Andrew Ofstehage's ethnography of US soy farmers operating in Brazil. Together, these contributions demonstrate that despite the technical homogenization and omnipresence of transnational actors at every link of soybean production chains, place-specific material relations – particularly social relations of trust and familiarity for trade and labor management – remain crucially important for their successful operations. Rather than treating the soybean complex as an abstract global force increasingly disarticulated from specific ecologies and social relations, therefore, we ought to investigate instead how both transnational and local actors actively maneuver to reconfigure their relations to each other and to the concrete landscapes where they operate.

Concluding this opening section, a pair of papers explores the political dynamics and stakes of the current trajectory of soybean agribusiness globalization, focusing on Brazil as its main export platform and China as its predominant destination. Gustavo Oliveira delivers a fine-grained historical account of the development of the sector in Brazil that reveals its role in the consolidation of agrarian capitalism during the so-called Cold War. Oliveira then argues the Brazilian soy sector has continued to be fundamentally shaped by state interests even while transnational companies have taken over production technologies and trade, particularly given its geopolitical significance in times of climate change, high and volatile commodity prices, and new intersections between agro-industrial markets and currencies that challenge the hegemony of the US state and dollar. Yan Hairong, Chen Yiyuan and Ku Hok Bun in turn provide a detailed account of how and why China's soybean production collapsed with the inflow of imports from the United States and South America, and how this soybean crisis in China has created ecological, economic and geopolitical concerns about dependence on transgenic seeds and imports. All contributions in this section reveal how domestic actors and histories in China and South America have profoundly conditioned the current restructuring of the soybean complex at a global scale, and how the highly uneven concentration of benefits to a few corporate actors and distribution of costs and harms among multiple peoples and landscapes have placed transnational soybean agribusiness central to debates over agrarian development, industrialization and modernization.

The second section of this collection focuses on ethnographic and interview-based case studies of soy producers across various landscapes in South America. The contribution by Bert van der Vennet, Sergio Schneider and Joost Dessein, and that of Mateo Mier y Teran, attempt to undermine the simplistic caricature that all soy production in Brazil takes place on large-scale corporate farms. The former research two areas in southern Brazil where relatively small-scale farms (< 300 ha) still exist in substantial numbers, while the latter researches an area of central Brazil where mid-scale properties (300–1000 ha) make up the majority of farms, but *not* the bulk of soybeans produced. Their contributions reveal that multiple and distinct social relations of production (what these authors call 'farming styles', following Jan Dowe van der Ploeg) enable these relatively small and medium soy farms to remain in production for the time being, in part because of the historical stability of land tenure regimes, and in part because of their ability to attend niche markets. However, in spite of the diversity of farm sizes and their associated relations of production, both contributions reaffirm that these soy production systems themselves are indistinguishable from the largest corporate farms, and they become entangled in the economic and technological treadmill that adds constant downward pressure on soy prices and upward pressure on production costs, consolidating the long-term trend towards farmland concentration and corporate control. Andrew Ofstehage's ethnography tackles these questions of scale obliquely through his discussion of the challenges faced by US soy farmers who left their relatively small-scale farms (< 300 ha) in the US to purchase and manage large-scale farms (> 1000 ha) in the Brazilian Cerrado. Even though the techniques and technologies of production are exactly the same as in the United States, these farmers must suddenly address questions of labor management and cultural adaptation unlike anything they had dealt with before. Revealing in intimate detail the manner in which US agribusinessmen enter Brazilian territory at the point of production, Ofstehage's contribution also speaks to broader questions of globalization and the formation of transnational actors.

The third section of our collection presents papers that engage the environmental politics and debates regarding soy production in South America that often stand for discussions of contemporary biotechnology and agroindustrialization in general. Here in our introduction, we have shown how agribusinesses, NGOs and state actors have projected a model of forest transition through 'land sparing' and 'low carbon' intensification of soy production. Yet the 'Amazon swerve' upon which this argument relies has restricted debates over soy production to models and patterns of deforestation alone, drawn from case studies that cannot account for continent-wide 'leakage', and glossed over an array of other social and ecological concerns within soy landscapes themselves. A more general balance of the social and environmental contributions and shortfalls of the soybean sector to 'sustainable development' is the goal of the contribution by Rachel Garrett and Lisa Rausch. They compare soy production with sugarcane and extensive ranching in Brazil, and argue that the agroindustrial intensification of the soybean sector has provided greater macroeconomic benefits to the Brazilian economy, particularly in the mid-size cities that serve as hubs of the sector in the Cerrado region. Yet they do not analyze several other significant social and ecological harms associated with soy production and agribusiness expansion, and their comparison presumes that agroindustrial soy, sugarcane or extensive ranching are the only viable production practices for these landscapes. Smaller scale, diversified production systems are rendered invisible (for contrast, see Perfecto, Vandermeer, and Wright 2009).

This problem of 'visibility' is central to the contribution by Ludivine Eloy et al. to this volume, as they bring an innovative political ecology of environmental policies in the Cerrados of Brazil to demonstrate how environmental managers identify and blame fire-dependent traditional production systems and undermine participative management of protected areas, while legitimating the expansion of soybeans around them and contributing to rural exodus. In other words, the frameworks of environmental 'virtue' and the production of knowledge about social and ecological harms and benefits must also be carefully researched, analyzed and critiqued. Similarly, Pablo Lapegna's ethnography of social movement and government relations on the regulation of soy production in Argentina reveals how environmental problems caused by the expansion of GM soybeans are experienced primarily through the exposure of rural populations to toxic agrochemicals. Lapegna argues that the institutional recognition of rural social movements and the performative actions of the authorities from national to local levels of government respond to cases of agrochemical exposure in manners that create barriers to collective action that could more effectively curtail the negative socio-ecological impacts of soy production. The ethnographic nuance of these contributions demonstrates how various actors who seek to transform soybean production practices navigate tremendously disparate and complex social and political realities. In concluding this section, Laureen Elgert's contribution draws from her research on soy production in Paraguay to reveal how discourses of 'sustainable development' exemplified by the Roundtable on Responsible Soy are better understood as means of protecting large-scale investments, rather than as means of achieving sustainability in agricultural commodity production (cf. Baletti 2014). Elgert proposes more inclusive alternatives than soybean production, marked as it is by the exclusion of small-scale producers through the concentration, and often foreignization, of farmland in Paraguay.

The fourth and final section of this collection gathers two contributions that investigate the expansion of soybean production into new agricultural frontiers of South America. Lucia Goldfarb and Gemma van der Haar's contribution draws on interview data and participant observation in Argentina's Chaco to review the multiple paths of migration and the relations of conflict and cooptation that enable well-capitalized soy farmers to gain territory from its previous occupants and displace competing farmers and land uses. How land becomes incorporated into the soy economy remains one of the most dynamic research arenas, especially as the debates over displacement become more acute. Similarly, in their contribution, Ben McKay and Gonzalo Colque trace the expansion of soy production in Bolivia's eastern lowlands through the lens of 'productive exclusion', whereby the development of the soy economy in Bolivia is itself responsible for the social and economic exclusion of the majority of the population living in these areas that were previously considered marginal to the economy and politics of the country.

5. Limitations and directions for further research

We recognize that many lacunae still remain. Given the complex migration patterns that produce soy frontiers in South America, the intersection between class and race/ethnicity has been investigated in the literature about access to land and labor relations on soy-producing regions (Desconsi 2011; Rumstain 2012). In this collection, several contributions touch on the racial/ethnic tensions and connections to the production of these agroindustrial landscapes (Ofstehage 2016, this volume; Cacho 2016, this volume; van der Vennet, Schneider, and Dessein 2016, this volume; Goldfarb and van der Haar 2016, this volume). Yet, at the household level, what does this massive change mean for those who migrate and those who stay, and for those integrated and those excluded from production, and what are its gender dynamics? Much of the literature remains relatively blind to the intersections with and particular problems of gender that imbue daily life in South America's hinterlands, government halls and corporate offices. These gender dynamics of labor, migration and urbanization remain the terrain of limited studies (Ferro 2013; Miranda 2012; VanWey and Spera 2013; Richards and VanWey 2015).

Another important limitation is reflected in the heavy focus on Brazil and Argentina among the contributions to this volume. Certainly these two countries account for the vast majority of soy produced in South America, and the extensive nature of Brazil suggests that growth of the sector in South America will take place primarily there. Yet the historical geography of soy production in Paraguay and Bolivia, and their idiosyncratic connections with soy farmers and agribusinesses across the region, certainly deserves more attention than we were able to give it in this collection. And, of course, as Uruguay has even surpassed Bolivia in 2011 to become the fourth largest producer on the continent, it is especially regrettable that it is not the focus of any paper in this collection. The trajectory of soy production in Uruguay raises important questions about the dynamics of soy agribusinesses across the continent. For example, why didn't soy production expand significantly in Uruguay until the 2000s, when the crop already boomed a few hundred kilometers north

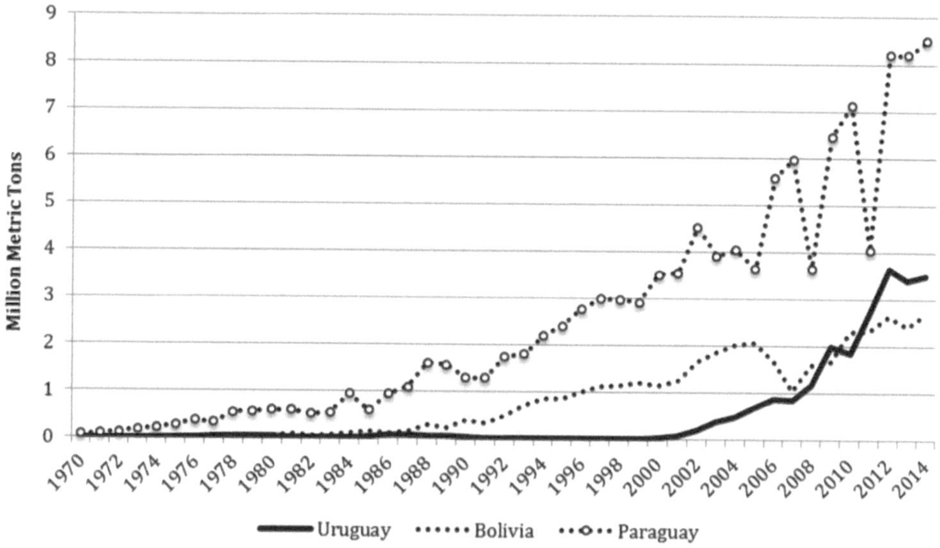

Figure 6. Soy production in Paraguay, Uruguay and Bolivia, 1970–2014.
Source: Elaborated by the authors from United States Department of Agriculture data.

and south of its borders in Brazil and Argentina? And why did this tendency change so sharply in the first decade of this millennium? (Figure 6).

There is of course excellent Uruguayan literature in Spanish discussing how Argentinian *pools de siembra* and other firms and farmers from the continent and beyond are participating in this dramatic expansion, and demonstrating how Uruguayans themselves transform these soybean operations and their own relations to the land. Oyhantçabal and Narbondo (2011) provide an extensive description and critique of the soybean agribusiness in Uruguay, identifying its main actors and showing how some collaborate with Argentinian and Brazilian investors to expand soy production not only in Uruguay, but also into in Venezuela and Colombia as well. We call for greater attention to and integration of the work of Paraguayan, Bolivian and Uruguayan scholars in Anglophone publications, as they not only complement but also transform our understanding of the operations of this major agroindustrial sector across the continent and around the world.

This need for greater attention to the distinct dynamics of the sector across the continent also applies to less traditional areas of soy cultivation in South America where expansion is nevertheless taking place (particularly Venezuela), where soy production has stagnated at low levels despite renewed interest in investments in the sector (namely Ecuador and Colombia), and where soy hasn't become incorporated into local agricultural production at all even though it is imported in increasingly larger amounts, such as Peru (Figure 7). Even a very limited account of the multiple trajectories of the soy sector in these countries suffices to demystify any illusion of uniform continent-wide expansion, and underscores the social and ecological 'friction' of each place. Take Ecuador, for example. Soy production began expanding there during the 1970s much like in the southern cone of the continent, but what caused this boom to turn into a sharp bust in the mid-1990s? And why has the recovery of the sector since stagnated when the country is the most advantageously located in South America to supply China, the world's largest (and growing) market for this commodity? The situation is similar in Colombia, but the sector never took root in Peru in the first

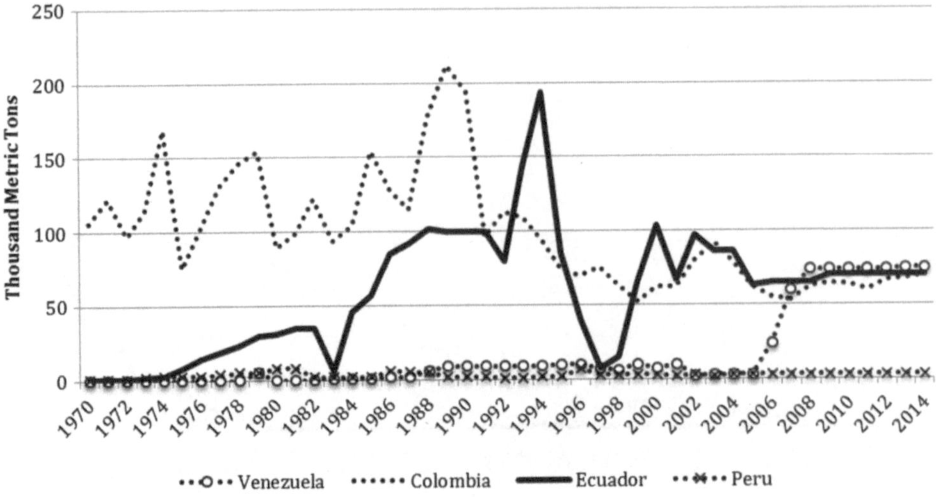

Figure 7. Soy production in Venezuela, Colombia, Ecuador and Peru, 1970–2014.
Source: Elaborated by the authors from United States Department of Agriculture data.

place. What are the alternatives or frictions that limited soy production from expanding further and faster in these Andean countries? A very different trajectory is witnessed in Venezuela, marked by its government import-substitution efforts to draw technologies and practices from Brazil in order to establish soy production over its eastern flatlands. The role of the state is particularly notable here, as government-led projects in the state of Anzoátegui account for the vast majority of all soy produced in the country, and where the Ministry of Agriculture and Land plans to bring an additional 32,000 ha under irrigation to produce over 250,000 tons of soy per year by 2019 (Ministry of Agriculture 2015). Still, private Venezuelan soybean agribusinesses, alongside those from Colombia and Peru, prioritize investments across borders in Bolivia rather than focusing on domestic production (cf. McKay and Colque 2016, this volume). What explains these different strategies and dynamics among Andean countries and agribusinesses? How are different actors negotiating different social networks and ecological landscapes as they assemble multiple investment practices and production technologies for expanding their soybean agribusinesses across the continent? This collection is only able to gesture towards these as important questions for further research and international collaboration.

A very prominent set of themes that could have been the subject of far more in-depth discussion are the transformations in the political economy and discursive practices of research and development in biotechnology. Ultimately, the continued expansion of the soybean agribusiness in South America depends on the narrative that the technological progress of the 'green revolution' rescued millions across the region from hunger and poverty, and that a new technological revolution based upon GM seeds and their associated agrochemicals is necessary to resolve our future need to increase food production. But if large-scale and long-term data on the adoption of GM technologies in the soy sector across the continent do not demonstrate rising yields and falling agrochemical use (Cotacora-Vargas et al. 2012; Cotacora-Vargas 2014), and we witness the aggravation of ecological harms and social exclusions already evident in previous 'green revolution' production systems, then agroecological alternatives to soy monocultures and GM technology must be sought (Foster 2000; Altieri and Pengue 2006; Mazoyer and Roudart 2006; Oliveira 2010;

Perfecto and Vandermeer 2010; Richards 2010; Patel 2013; Altieri, Nicholls, and Lana 2015).

This would require, of course, a fundamental reconstruction of the soy-livestock nexus that currently generates the greatest demand for this commodity as animal feed (Steinfield et al. 2006; Schneider and McMichael 2010; Weis 2010, 2013a, 2013b), and also a critical reconceptualization of the role of soybeans in the production of bioenergy and other industrial inputs, particularly biodiesel (Cavalett and Ortega 2010). The latter in particular may hold special importance, as the ability for agroindustrial processing firms to shift flexibly between various feedstocks and attend a growing number of multiple markets is becoming a key driver of global agroindustrial restructuring, a process that is being framed in agribusiness and development as a solution to the convergence of climate, energy and food crises, but which may in fact aggravate social-environmental harms already associated with the agroindustrial production of these flex crops (Borras et al. 2015). Ultimately, as Yan, Chen and Ku demonstrate in their contribution to this volume, these political-ecological questions and debates triggered by the contemporary dynamics of the soy sector lead us to re-examine the nature of globalization, the logic of modernization, the role of the state, the politics of science and knowledge production and the reach of corporate power in everyday life, and, most importantly, they encourage us to search and struggle for agroecological alternatives.

Disclosure statement

No potential conflict of interest was reported by the authors.

References

Aldrich, S., R. Walker, M. Caldas, and S. Perz. 2012. Contentious land change in the Amazon's Arc of deforestation. *Annals of the Association of American Geographers* 102, no. 1: 103–128.

Alencar, A., P. Brando, and F. E. Putz. 2015. Landscape fragmentation, severe drought, and the new Amazon forest fire regime. *Ecological Applications* 25, no. 6: 1493–1505.

Alkimim, A., G. Sparovek, and K. C. Clarke. 2015. Converting Brazil's pastures to cropland: An alternative way to meet sugarcane demand and to spare forestlands. *Applied Geography* 62: 75–84.

Altieri, M., C. Nicholls, and M. A. Lana. 2015. Agroecology and the design of climate change-resilient farming systems. *Agronomy for Sustainable Development* 35, no. 3: 869–890.

Altieri, M., and W. Pengue. 2006. GM soybean: Latin America's new colonizer. *Seedling*, 17, no.1: 13–17.

Anadon, J., O. Sala, and F. T. Maestre. 2014. Climate change will increase savannas at the expense of forests and treeless vegetation in tropical and subtropical Americas. *Journal of Ecology* 102, no. 6: 1363–1373.

Arancibia, F. 2013. Challenging the bioeconomy: The dynamics of collective action in Argentina. *Technology in Society* 35, no. 2: 79–92.

Arvor, D., M. Meirelles, Agnès Bégué, and Y. E. Shimabukuro 2012. Analyzing the agricultural transition in Mato Grosso, Brazil, using satellite-derived indices. *Applied Geography* 32, no. 2: 702–713.

Asner, G., S. Loarie, and U. Heyder. 2010. Combined effects of climate and land-use change on the future of humid tropical forests. *Conservation Letters* 3, no. 6: 395–403.

Baldi, G., J. Houspanossian, Adriel A. Rosales, Carla V. Rueda , and E. G. Jobbágy 2015. Cultivating the dry forests of South America: Diversity of land users and imprints on ecosystem functioning. *Journal of Arid Environments* 123: 47–59.

Baletti, Brenda. 2014. Saving the Amazon? Sustainable soy and the new extractivism. *Environment and Planning A,* 46, no. 1: 5–25.

Barbier, E., and J. Burgess. 2001. Tropical deforestation, tenure insecurity, and unsustainability. *Forest Science* 47, no. 4: 497–509.

Barlow, J., and C. Peres. 2008. Fire-mediated dieback and compositional cascade in an Amazonian forest. *Philosophical Transactions of the Royal Society B-Biological Sciences* 363, no. 1498: 1787–1794.

Benachour, N., and G.-E. Séralini. 2009. Glyphosate formulations induce apoptosis and necrosis in human umbilical, embryonic, and placental cells. *Chemical Research in Toxicology* 22: 97–105.

Beuchle, R., R. Grecchi, et al. 2015. Land cover changes in the Brazilian Cerrado and Caatinga biomes from 1990 to 2010 based on a systematic remote sensing sampling approach. *Applied Geography* 58: 116–127.

Berger, M., and F. Ortega. 2010. Poblaciones expuestas a agrotóxicos: autoorganización ciudadana en la defensa de la vida y la salud, Ciudad de Córdoba, Argentina. (Populations exposed to agrotoxins: citizen self-organization in defense of life and health, City of Cordoba, Argentina). *Physis: Revista de Saúde Coletiva* 20, no. 1: 119–143.

Bianchi, C., and S.M. Haig. 2013. Deforestation Trends of Tropical Dry Forests in Central Brazil. *Biotropica* 45, no. 3: 395–400.

Binimelis, R., W. Pengue, and I. Monterroso. 2009. 'Transgenic treadmill': Responses to the emergence and spread of glyphosate-resistant johnsongrass in Argentina. *Geoforum* 40, no. 4: 623–633.

Borlaug, N. 2006. In: *Cerrado*. Available at: www.worldfoodprize.org/en/laureates/20002009_laureates/2006_lobato_mcclung_paolinelli/cerrado/ [Accessed January 4, 2016].

Borner, J., S. Wunder, S. Wertz-Kanounnikoff, G. Hyman, and N. Nascimento 2014. Forest law enforcement in the Brazilian Amazon: Costs and income effects. *Global Environmental Change-Human and Policy Dimensions* 29: 294–305.

Borras, S., J. Franco, et al. 2015. The rise of flex crops and commodities: implications for research. *Journal of Peasant Studies* 43, no. 1: 9–115.

Brando, P., J. Balch, et al. 2014. Abrupt increases in Amazonian tree mortality due to drought-fire interactions. *Proceedings of the National Academy of Sciences of the United States of America* 111, no. 17: 6347–6352.

Brando, P., M. Coe, et al. 2013. Ecology, economy and management of an agroindustrial frontier landscape in the southeast Amazon. *Philosophical Transactions of the Royal Society B-Biological Sciences* 368, no. 1619: 20120152.

Browder, J., and B. Godfrey. 1997. *Rainforest cities: urbanization, development, and globalization of the Brazilian Amazon*. New York: Columbia University Press.

Brown, L. 2012. *Full planet, empty plates: The new geopolitics of food scarcity*. New York: WW Norton & Company.

Carrero, G., and P. Fearnside. 2011. Forest clearing dynamics and the expansion of landholdings in Apui, a deforestation hotspot on Brazil's Transamazon Highway. *Ecology and Society* 16, no. 2: 26–0.

Cavalett, O., and E. Ortega. 2010. Integrated environmental assessment of biodiesel production from soybean in Brazil. *Journal of Cleaner Production* 18, no. 1: 55–70.

Ceddia, M., N. Bardsley, and S. Sedlacek 2014. Governance, agricultural intensification, and land sparing in tropical South America. *Proceedings of the National Academy of Sciences of the United States of America* 111, no. 20: 7242–7247.

Cerdeira, A., D. Gazziero, and M. B. Matallo. 2011. Agricultural impacts of glyphosate-resistant soybean cultivation in South America. *Journal of Agricultural and Food Chemistry* 59, no. 11: 5799–5807.

Chase, J. 2003. Regional prestige: Cooperatives and agroindustrial identity in southwest Goiás, Brazil. *Agriculture and Human Values* 20, no. 1: 37–51.

Clapp, J. 2014. Financialization, distance and global food politics. *Journal of Peasant Studies* 41, no. 5: 797–814.

Coe, M., M. Costa, et al. 2009. The influence of historical and potential future deforestation on the stream flow of the Amazon River: Land surface processes and atmospheric feedbacks. *Journal of Hydrology* 369, no. 1–2: 165–174.

Coe, M., T. Marthews, et al. 2013. Deforestation and climate feedbacks threaten the ecological integrity of south-southeastern Amazonia. *Philosophical Transactions of the Royal Society B-Biological Sciences* 368, no. 1619: 20120155.

Cotacora-Vargas, G., P. Galeano, S. Agapito-Tenfen, D. Aranda, T. Palau, and R. Nodari. 2012. *Soybean production in the Southern Cone of the Americas: Update on land and pesticide use.* Tromso, Norway: Center for Biosafety–GenOk.

Cotacora-Vargas, G. 2014. *Sustainability assessment of genetically modified herbicide tolerant crops: the case of Intacta Roundup Ready 2 Pro soybean farming in Brazil in light of the Norwegian Gene Technology Act.* Biosafety Report 2014/02. Tromso, Norway: Center for Biosafety–GenOk.

Deegan, L., C. Neill, M. Victoria R. Ballester, Alex V. Krusche, Reynaldo L. Victoria, Suzanne M. Thomas , and E. de Moor . 2011. Amazon deforestation alters small stream structure, nitrogen biogeochemistry and connectivity to larger rivers. *Biogeochemistry* 105, no. 1–3: 53–74.

DeFries, R., M. Herold, M. N. Macedo, and Y. Shimabukuro. 2013. Export-oriented deforestation in Mato Grosso: harbinger or exception for other tropical forests? *Philosophical Transactions of the Royal Society B-Biological Sciences* 368, no. 1619: 20120173.

DeFries, R., T. Rudel, and M. Hansen. 2010. Deforestation driven by urban population growth and agricultural trade in the twenty-first century. *Nature Geoscience* 3, no. 3: 178–181.

Desconsi, C. 2011. *A Marcha dos 'Pequenos Proprietários Rurais': trajetórias de migrantes do Sul do Brasil para o Mato Grosso.* (The march of the 'small rural landowners': Trajectories of southern Brazilian migrants to Mato Grosso.) Rio de Janeiro: E-papers.

EcoNexus and Berne Declaration. 2013. *Agropoly: A handful of corporations control world food production.* Oxford: Berne Declaration & EcoNexus. Available at: http://econexus.info/publication/agropoly-handful-corporations-control-world-food-production [Accessed August 29, 2014].

Edwards, D., J. Gilroy, et al. 2015. Land-sparing agriculture best protects avial phylogenetic diversity. *Current Biology* 25, no. 18: 2384–2391.

Fairbairn, M. 2014. 'Like gold with yield': evolving intersections between farmland and finance. *Journal of Peasant Studies* 41, no. 5: 777–795.

Fairbairn, M., J. Fox, et al. 2014. Introductions: New directions in agrarian political economy. *Journal of Peasant Studies* 41, no. 5: 653–666.

Fajardo, S. 2005. As cooperativas paranaenses e o novo padrão de desenvolvimento agroindustrial. (The cooperatives from Paraná and the new standard of agroindustrial development). *Formação* 12, no. 1: 165–192.

Fearnside, P. 2007. Brazil's Cuiabá-Santarém (BR-163) Highway: The environmental cost of paving a soybean corridor through the Amazon. *Environmental Management* 39, no. 5: 601–614.

Fearnside, P., A. Figueiredo, and S. C. M. Bonjour. 2013. Amazonian forest loss and the long reach of China's influence. *Environment Development and Sustainability* 15, no. 2: 325–338.

Ferro, S. 2013. *Gènero y Propiedad Rural en la Repùblica Argentina.* (Gender and rural property in the Republic of Argentina.) 2. ed. Buenos Aires: UCAR-MAGyP.

Foster, J. B. 2000. *Marx's ecology: Materialism and nature.* New York: Monthly Review Press.

Gardner, B. 2009. Are livestock a troublesome commodity? *Geoforum* 40, no. 5: 781–783.

Garrett, R., E. Lambin, and R. L. Naylor. 2013. Land institutions and supply chain configurations as determinants of soybean planted area and yields in Brazil. *Land Use Policy* 31: 385–396.

Gasparri, N., H. Grau, and L. Valeria Sacchi. 2015. Determinants of the spatial distribution of cultivated land in the North Argentine Dry Chaco in a multi-decadal study. *Journal of Arid Environments* 123: 31–39.

Gibbs, H., A. Ruesch, M. K. Clayton, P. Holmgren, N. Ramankutty , and J. A. Foley 2010. Tropical forests were the primary sources of new agricultural land in the 1980s and 1990s. *Proceedings of the National Academy of Sciences of the United States of America* 107, no. 38: 16732–16737.

Goldsmith, P., B. Li, J. Fruin, and R. Hirsch. 2004. Global shifts in agro-industrial capital and the case of soybean crushing: implications for managers and policy makers. *International Food and Agribusiness Management Review* 7, no. 2: 87–115.

Gonçalves, K., A. Siqueira, and S. de Souza Hacon. 2014. Indicator of socio-environmental vulnerability in the Western Amazon. The case of the city of Porto Velho, State of Rondonia, Brazil. *Ciência & Saúde Coletiva* 19, no. 9: 3809–3818.

Goodman, D., B. Sorj, and J. Wilkinson. 1987. *From farming to biotechnology: A theory of agro-industrial development.* New York: Basil Blackwell.

Gordillo, G. 2014. *Rubble.* Durham, NC: Duke University Press.

Graesser, J., T. Aide, and N. Ramankutty. 2015. Cropland/pastureland dynamics and the slowdown of deforestation in Latin America. *Environmental Research Letters* 10, no. 3: 034017.

Grain. 2013. The United Republic of Soybeans: Take two. *Against the Grain*, June. Barcelona: Grain.

Grandin, G. 2009. *Fordlandia.* New York: Metropolitan.

Grau, H., R. Torres, Pedro G. Blendinger, Sofía Marinaro, and L. Macchi. 2015. Natural grasslands in the Chaco: A neglected ecosystem under threat by agriculture expansion and forest-oriented conservation policies. *Journal of Arid Environments* 123: 40–46.

Grau, R., T. Kuemmerle, and L. Macchi. 2013. Beyond 'land sparing versus land sharing': environmental heterogeneity, globalization and the balance between agricultural production and nature conservation. *Current Opinion in Environmental Sustainability* 5, no. 5: 477–483.

Grupo de Reflexión Rural (GRR). 2009. *Pueblos Fumigados: Informe sobre la problemática del uso de plaguicidas en las principales provincias sojeras de la Argentina.* Fumigated Peoples: Report on the problem of pesticide use in the main soy-producing provinces of Argentina. Available at: www.grr.org.ar/trabajos/Pueblos_Fumigados__GRR_.pdf

Gudynas, E. 2008. The new bonfire of vanities: soybean cultivation and globalization in South America. *Development* 51, no. 4: 512–518.

Guyton, K., D. Loomis, et al. 2015. Carcinogenicity of tetrachlorvinphos, parathion, malathion, diazinon, and glyphosate. *Lancet Oncology* 16, no. 5: 490–491.

Haslberger, A. 2006. Need for an 'integrated safety assessment' of GMOs, linking food safety and environmental considerations. *Journal of Agricultural and Food Chemistry* 54, no. 9: 3173–3180.

Hecht, S.B. 1985. Environment, Development and Politics - Capital Accumulation and the Livestock Sector in Eastern Amazonia. *World Development* 13, no. 6: 663–684.

Hecht, S.B. 2005. Soybeans, development and conservation on the Amazon frontier. *Development and Change* 36, no. 2: 375–404.

Hecht, S.B. 2012. From eco-catastrophe to zero deforestation? Interdisciplinarities, politics, environmentalisms and reduced clearing in Amazonia. *Environmental Conservation* 39, no. 1: 4–19.

Hecht, S.B., and A. Cockburn. 1989. *The fate of the forest: Developers, destroyers, and defenders of the Amazon.* New York: Verso.

Hecht, S.B., and C. Mann. 2008. How Brazil outfarmed the Americans. *Fortune* 157, no. 1: 92–106.

High Level Panel of Experts on Food Security and Nutrition (HLPE). 2011. *Price volatility and food security.* Report to the Committee on World Food Security of the Food and Agriculture Organization of the United Nations. Rome: CFS/FAO.

HighQuest Partners and Soyatech. 2011. *How the global oilseed and grain trade works.* St. Louis, MO: United Soybean Board/US Soybean Export Council.

Hoelle, J. 2014. Cattle culture in the Brazilian Amazon. *Human Organization* 73, no. 4: 363–374.

Isakson, S. R. 2014. Food and finance: The financial transformation of agro-food supply chains. *Journal of Peasant Studies*, 41(5): 749–775.

Jepson, W. 2005. A disappearing biome? Reconsidering land-cover change in the Brazilian savanna. *Geographical Journal* 171: 99–111.

Jepson, W., C. Brannstrom and R. Stancato. 2005. A Case of Contested Ecological Modernization: Governance of Genetically Modified Crops in Brazil. *Environment and Planning C: Government and Policy*, 23: 295–310.

Jepson, W., C. Brannstrom, et al. 2010. Access regimes and regional land change in the Brazilian Cerrado, 1972–2002. *Annals of the Association of American Geographers* 100, no. 1: 87–111.

Klink, C., and R. Machado. 2005. Conservation of the Brazilian Cerrado. *Conservation Biology* 19, no. 3: 707–713.

Larson, A. 2011. Forest tenure reform in the age of climate change: Lessons for REDD+. *Global Environmental Change-Human and Policy Dimensions* 21, no. 2: 540–549.

Latour, B. 1987. *Science in Action.* Cambridge, MA: Harvard University Press.

Laurance, W., A. Albernaz, et al. 2004. Deforestation in Amazonia. *Science* 304, no. 5674: 1109–1109.

Laurance, W., D. Useche, et al. 2012. Averting biodiversity collapse in tropical forest protected areas. *Nature* 489, no. 7415: 290–294.

Macchi, L., H. Grau, and S. Marinaro. 2013. Trade-offs between land use intensity and avian biodiversity in the dry Chaco of Argentina: A tale of two gradients. *Agriculture, Ecosystems & Environment* 174, no. 0: 11–20.

Macedo, M., M. Coe, et al. 2013. Land-use-driven stream warming in southeastern Amazonia. *Philosophical Transactions of the Royal Society B-Biological Sciences* 368, no. 1619: 20120153–20120153.

Macedo, M., R. DeFries, C. M. Stickler, G. L. Galford, and Y. E. Shimabukuro 2012. Decoupling of deforestation and soy production in the southern Amazon during the late 2000s. *Proceedings of the National Academy of Sciences of the United States of America* 109, no. 4: 1341–1346.

Malhi, Y., L. Aragão, et al. 2009. Exploring the likelihood and mechanism of a climate-change-induced dieback of the Amazon rainforest. *Proceedings of the National Academy of Sciences of the United States of America* 106, no. 49: 20610–20615.

Mann, M., R. Kaufmann, et al. 2010. The economics of cropland conversion in Amazonia: The importance of agricultural rent. *Ecological Economics* 69, no. 7: 1503–1509.

Marinaro, S., H. Grau, and P.V. Zelaya. 2015. Land tenure and biological communities in dry Chaco forests of northern Argentina. *Journal of Arid Environments* 123: 60–67.

Marc, J., R. Belle, et al. 2004. Formulated glyphosate activates the DNA-response checkpoint of the cell cycle leading to the prevention of G2/M transition. *Toxicological Sciences* 82, no. 2: 436–442.

Marsik, M., F. Stevens, et al. 2011. Amazon deforestation: Rates and patterns of land cover change and fragmentation in Pando, northern Bolivia, 1986 to 2005. *Progress in Physical Geography* 35, no. 3: 353–374.

Mazoyer, M., and L. Roudart. 2006. *A history of world agriculture: from the neolithic age to the current crisis*. New York: Monthly Review.

McMichael, P. 2012. The land grab and corporate food regime restructuring. *Journal of Peasant Studies* 39, no. 3–4: 681–701.

Meyfroidt, P., K. Carlson, et al. 2014. Multiple pathways of commodity crop expansion in tropical forest landscapes. *Environmental Research Letters* 9, no. 7: 1–13.

Ministry of Agriculture and Land of the Bolivarian Government of Venezuela. 2015. *Plan nacional de agricultura de riego y saneamiento de tierras*. (National Plan for Irrigated Agriculture and Land Reclamation.) Caracas: MPPAT/National Institute for Rural Development.

Miranda, H. 2012. The expansion of agriculture and its relationship with the urbanization process in the Northeast Region of Brazil (1990–2010). *Eure-Revista Latinoamericana De Estudios Urbano Regionales* 38, no. 114: 173–201.

Moreira, P., W. Pignati, et al. 2010. *Avaliação do risco à saúde humana decorrente do uso de agrotóxicos na agricultura e pecuária na região Centro-Oeste do Brasil.* (Evaluation of risk to human health due to agrotoxic use in agriculture and ranching in the Center-West region of Brazil.) Report 555193/2006–3, National Council for Scientific and Technological Development. Brasília: CNPq.

Morgan, D. 2000 [1979]. *Merchants of Grain: the power and profits of the five giant companies at the center of the world's food supply.* Lincoln, NE: iUniverse.

Morton, D., R. Defries, L. Giglio, W. Schroeder, and G. R. Van Der Werf. 2008. Agricultural intensification increases deforestation fire activity in Amazonia. *Global Change Biology* 14, no. 10: 2262–2275.

Murphy, S., D. Burch, and J. Clapp. 2012. *Cereal secrets: the world's largest grain traders and global agriculture*. Oxfam Research Report. Oxford: Oxfam.

Neill, C., M. Coe, et al. 2013. Watershed responses to Amazon soya bean cropland expansion and intensification. *Philosophical Transactions of the Royal Society B: Biological Sciences* 368, no. 1619: 20120425–20120425.

Nepstad, D., S. Irawan, et al. 2013. More food, more forests, fewer emissions, better livelihoods: linking REDD plus, sustainable supply chains and domestic policy in Brazil, Indonesia and Colombia. *Carbon Management* 4, no. 6: 639–658.

Nepstad, D., D. McGrath, et al. 2014. Slowing Amazon deforestation through public policy and interventions in beef and soy supply chains. *Science* 344, no. 6188: 1118–1123.

Nepstad, D., S. Schwartzman, et al. 2006a. Inhibition of Amazon deforestation and fire by parks and indigenous lands. *Conservation Biology* 20, no. 1: 65–73.

Nepstad, D., B. Soares, et al. 2009. The End of Deforestation in the Brazilian Amazon. *Science* 326, no. 5958: 1350–1351.

Nepstad, D., C. Stickler, and O. T. Almeida. 2006b. Globalization of the Amazon soy and beef industries: Opportunities for conservation. *Conservation Biology* 20, no. 6: 1595–1603.

Newton, P., A. Agrawal, and L. Wollenberg. 2013. Enhancing the sustainability of commodity supply chains in tropical forest and agricultural landscapes. *Global Environmental Change-Human and Policy Dimensions* 23, no. 6: 1761–1772.

Nicholls, C., and M. Altieri. 1997. Conventional agricultural development models and the persistence of the pesticide treadmill in Latin America. *International Journal of Sustainable Development and World Ecology* 4, no. 2: 93–111.

Nobre, P., M. Malagutti, D. Urbano, R. de Almeida, and E. Giarolla. 2009. Amazon deforestation and climate change in a coupled model simulation. *Journal of Climate*, 22, no. 21: 5686–5697.
Oliveira, G. de L. T. 2010. Prescrições agroecológicas para a crise atual (Agroecological prescriptions for the current crisis). *Revista NERA* 13, no. 16: 33–47.
Oliveira, G. de L. T., and M. Schneider. 2015. The politics of flexing soybeans: China, Brazil, and global agroindustrial restructuring. *The Journal of Peasant Studies* 43, no. 1: 167–194.
Oliveira, N., G. Moi, Ageo Mário Candido Silva, and Wanderlei Antônio Pignati. 2014. Malformações congênitas em municípios de grande utilização de agrotóxicos em Mato Grosso, Brasil. (Birth defects in municipalities with intensive use of agrotoxins in Mato Grosso, Brazil). *Ciência & Saúde Coletiva* 19, no. 10: 4123–4130.
Oliveira, L., M. Costa, and M. T. Coe. 2013. Large-scale expansion of agriculture in Amazonia may be a no-win scenario. *Environmental Research Letters* 8, no. 2: 024021.
Oyhantçabal, G., and I. Narbondo. 2011. Radiografía del agronegocio sojero: Descripción de los principals actors y los impactos socio-econónicos en Uruguay. (Radiography of the soybean agribusiness: Description of the main actors and socio-economic impacts in Uruguay.) Montevideo: REDES-AT.
Pacheco, P. 2006. Agricultural expansion and deforestation in lowland Bolivia: The import substitution versus the structural adjustment model. *Land Use Policy* 23, no. 3: 205–225.
Paganelli, A., V. Gnazzo, Silvia L. López, and A. E. Carrasco. 2010. Glyphosate-based herbicides produce teratogenic effects on vertebrates by impairing retinoic acid signaling. *Chemical Research in Toxicology* 23, no. 10: 1586–1595.
Palma, D., C. Lourencetti, et al. 2014. Simultaneous determination of different classes of pesticides in breast milk by solid-phase dispersion and GC/EDC. *Journal of the Brazilian Chemical Society* 25, no. 8: 1419–1430.
Patel, R. 2013. The long green revolution. *Journal of Peasant Studies* 40, no. 1: 1–63.
Peres, C., and B. Zimmerman. 2001. Perils in parks or parks in peril? Reconciling conservation in Amazonian reserves with and without use. *Conservation Biology* 15, no. 3: 793–797.
Perfecto, I., J. Vandermeer, and A. Wright. 2009. *Nature's matrix: Linking agriculture, conservation and food sovereignty*. London: Earthscan.
Perfecto, I., and J. Vandermeer. 2010. The agroecological matrix as alternative to the land-sparing/agriculture intensification model. *Proceedings of the National Academy of Sciences* 107, no. 13: 5786–5791.
Pfaff, A., and R. Walker. 2010. Regional interdependence and forest "transitions": Substitute deforestation limits the relevance of local reversals. *Land Use Policy* 27, no. 2: 119–129.
Pinto-Ledezma, J., and M. Mamani. 2014. Temporal patterns of deforestation and fragmentation in lowland Bolivia: implications for climate change. *Climatic Change* 127, no. 1: 43–54.
Porro, R. 2005. Palms, Pastures, and Swidden Fields: The Grounded Political Ecology of ?Agro-Extractive/Shifting-cultivator Peasants? in Maranhão, Brazil. *Human Ecology* 33, no. 1: 17–56.
Porto, M. and B. Milanez. 2009. Economic development axis and socio-environmental conflict generation in Brazil: Challenges to sustainability and environmental justice. *Ciência & Saúde Coletiva* 14, no. 6: 1983–1994.
Redo, D. 2012. Mapping land-use and land-cover change along Bolivia's Corredor Bioceanico with CBERS and the Landsat series: 1975–2008. *International Journal of Remote Sensing* 33, no. 6: 1881–1904.
Redo, D., A. Millington, and D. Hindery. 2011. Deforestation dynamics and policy changes in Bolivia's post-neoliberal era. *Land Use Policy* 28, no. 1: 227–241.
Richard, S., S. Moslemi, Nora Benachour, and Gilles-Eric Seralini. 2005. Differential effects of glyphosate and roundup on human placental cells and aromatase. *Environmental Health Perspectives* 113: 716–720.
Richards, D. 2010. Contradictions of the 'New Green Revolution': A view from South America's Southern Cone. *Globalizations* 7, no. 4: 563–576.
Richards, P. 2015. What drives indirect land use change? How Brazil's agriculture sector influences frontier deforestation. *Annals of the Association of American Geographers* 105, no. 5: 1026–1040.
Richards, P., H. Pellegrina, and S. Spera. 2015. Soybean development: The impact of a decade of agricultural change on urban and economic growth in Mato Grosso, Brazil. *Plos One* 10, no. 4: 0122510.
Richards, P., and L. VanWey. 2015. Where Deforestation Leads to Urbanization: How Resource Extraction Is Leading to Urban Growth in the Brazilian Amazon. *Annals of the Association of American Geographers* 105, no. 4: 806–823.

Richards, P. 2011. Soy, cotton, and the final Atlantic forest frontier. *Professional Geographer* 63, no. 3: 343–363.

Richards, P., R. Myers, R. T. Walker. 2012. Exchange rates, soybean supply response, and deforestation in South America. *Global Environmental Change-Human and Policy Dimensions* 22, no. 2: 454–462.

Richards, P., R. Walker, and E. Y. Arima. 2014. Spatially complex land change: The Indirect effect of Brazil's agricultural sector on land use in Amazonia. *Global Environmental Change-Human and Policy Dimensions* 29: 1–9.

Rodrigues, A., R. Ewers, C. Souza, A. Verissimo, and A. Balmford. 2009. Boom-and-bust development patterns across the Amazon deforestation frontier. *Science* 324, no. 5933: 1435–1437.

Rudel, T., L. Schneider, et al. 2009. Agricultural intensification and changes in cultivated areas, 1970–2005. *Proceedings of the National Academy of Sciences of the United States of America* 106, no. 49: 20675–20680.

Rulli, J. 2007. The refugees of the agroexport model. In *United Soya Republics: The truth about soya production in South America*, ed. J. Rulli. Buenos Aires: Grupo de Reflexión Rural, pp. 193–216.

Rumstain, A. 2012. *Peões no trecho: trajetórias e estratégias de mobilidade no Mato Grosso.* (Farmhands on the road: Trajectories and mobility strategies in Mato Grosso.) Rio de Janeiro: E-papers.

Santos, J., I. Leal, et al. 2011. Caatinga: the scientific negligence experienced by a dry tropical forest. *Tropical Conservation Science* 4, no. 3: 276–286.

Sauer, S., and S. Leite. 2012. Agrarian structure, foreign investment in land, and land prices in Brazil. *Journal of Peasant Studies* 39, no. 3–4: 873–898.

Sawyer, D. 2008. Climate change, biofuels and eco-social impacts in the Brazilian Amazon and Cerrado. *Philosophical Transactions of the Royal Society B: Biological Sciences* 363, no. 1498: 1747–1752.

Schiesari, L. A. Waichman, et al. 2013. Pesticide use and biodiversity conservation in the amazonian agricultural frontier. *Philosophical Transactions of the Royal Society B: Biological Sciences* 368, no. 20120378: 1–9.

Scott, J. 1998. *Seeing like a state: How certain schemes to improve the human condition have failed.* New Haven, CT: Yale University Press.

Schmink, M., and C. Wood. 1992. *Contested frontiers in Amazonia.* New York: Columbia University Press.

Schneider, M. 2014. Developing the meat grab. *Journal of Peasant Studies* 41, no. 4: 613–633.

Schneider, M., and P. McMichael. 2010. Deepening, and repairing, the metabolic rift. *Journal of Peasant Studies* 37, no. 3: 461–484.

Schulz, C., A. Ioris, and K. Glenk. 2015. Prospects for payments for ecosystem services in the Brazilian pantanal: A scenario analysis. *Journal of Environment & Development* 24, no. 1: 26–53.

Shurtleff, W., and A. Aoyagi. 2009. *History of soybeans and soyfoods in South America (1882–2009): Extensively annotated bibliography and sourcebook.* Lafayette, CA: Soyinfo Center.

Silva, M., and L. Costa. 2012. A indústria de defensivos agrícolas (The pesticide and herbicide industry). *BNDES Setorial* 35: 233–276.

Silva, J., A. Silva, Ricardo Dutra Aydos, and Inês Echenique Mattos 2015a. Correlação entre produção agrícola, variáveis clínicas-demográfcas e câncer de próstata: um estudo ecológico. (Correlation between agricultural production, clinical-demography variables, and prostate cancer: an ecological study). *Ciência & Saúde Coletiva* 20, no. 9: 2805–2812.

Silva, C., K. Sousa, et al. 2015b. Biodiversity and its drivers and pressures of change in the wetlands of the Upper Paraguay-Guapore Ecotone, Mato Grosso (Brazil). *Land Use Policy* 47: 163–178.

Smil, V. 2001. *Feeding the world: A challenge for the 21st century.* Cambridge, MA: MIT Press.

Soares-Filho, B., R. Silvestrini, Paulo Brando, Hermann Rodrigues, Ane Alencar, Michael Coe, Charton Locks, Letícia Lima, Letícia Hissa, and C. Stickler. 2012. Forest fragmentation, climate change and understory fire regimes on the Amazonian landscapes of the Xingu headwaters. *Landscape Ecology* 27, no. 4: 585–598.

Soares-Filho, B., R. Rajão, M. Macedo, A. Carneiro, W. Costa, M. Coe, H. Rodrigues, and A. Alencar. 2014. Cracking Brazil's Forest Code. *Science* 344, no. 6182: 363–364.

Sousa, I., and L. Busch. 2006. Standards and state-building: The construction of soybean standards in Brazil. In *Agricultural standards: The shape of the global food and fiber system*, ed. J. Bingen and L. Busch, Dordrecht, Netherlands: pp. 125–135.

Sorj, B., M. Pompermayer, et al. 2008. Camponeses e agroindústria: Transformação social e representação política na avicultura brasileira. (Peasants and agroindustry: Social transformation and political representation in Brazilian poultry production). Rio de Janeiro: Scielo/Centro Edelstein de Pesquisas Sociais.

Steinfeld, H., P. Gerber, et al. 2006. *Livestock's long shadow: Environmental issues and options.* Rome: FAO.

Steininger, M., C. Tucker, Timothy J. Killeen, Arthur Desch, Vivre Bell, and Peter Ersts. 2001. Tropical deforestation in the Bolivian Amazon. *Environmental Conservation* 28, no. 2: 127–134.

Stickler, C., D. Nepstad, and D. G. McGrath. 2013. Defending public interests in private lands: compliance, costs and potential environmental consequences of the Brazilian Forest Code in Mato Grosso. *Philosophical Transactions of the Royal Society B-Biological Sciences* 368, no. 1619: 20120160.

Swann, A., M. Longo, Lee, Eunjee, and Moorcroft, Paul R. 2015. Future deforestation in the Amazon and consequences for South American climate. *Agricultural and Forest Meteorology* 214-215: 12–24.

Tsing, A. 2005. *Friciton: An ethnography of global connection.* Princeton: Princeton University Press.

Turzi, M. 2011. The soybean republic. *Yale Journal of International Affaris,* 6, no. 2: 59–68.

United States Department of Agriculture (USDA). 2015. World Agricultural Production. Circular Series WAP 12-15. Washington, DC: USDA/FAS.

Urcola, H., X. de Sartre, Elverdin, Julio, and A. Christophe. 2015. Land tenancy, soybean, actors and transformations in the pampas: A district balance. *Journal of Rural Studies* 39: 32–40.

Vallejos, M., J. Volante, Laura M. Vale, M. Laura Bustamante, and J. Paruelo. 2015. Transformation dynamics of the natural cover in the Dry Chaco ecoregion: A plot level geo-database from 1976 to 2012. *Journal of Arid Environments* 123: 3–11.

VanBeek, K., J. Brawn, and M. P. Ward. 2014. Does no-till soybean farming provide any benefits for birds? *Agriculture Ecosystems & Environment* 185: 59–64.

VanWey, L., and P. Richards. 2014. Eco-certification and greening the Brazilian soy and corn supply chains. *Environmental Research Letters* 9, no. 3.

VanWey, L., S. Spera, et al. 2013. Socioeconomic development and agricultural intensification in Mato Grosso. *Philosophical Transactions of the Royal Society B-Biological Sciences* 368, no. 1619: 20120168.

Volante, J., and J. Paruelo. 2015. Is forest or Ecological Transition taking place? Evidence for the Semiarid Chaco in Argentina. *Journal of Arid Environments* 123: 21–30.

Walker, R., J. Browder, Cynthia Simmons, Ritaumaria Pereira, Marcellus Caldas, Ricardo Shirota, Sergio de Zen. 2009a. Ranching and the new global range: Amazonia in the 21st century. *Geoforum* 40, no. 5: 732–745.

Walker, R., R. DeFries, et al. 2009b. The expansion of intensive agriculture and ranching in Brazilian Amazonia. In *Amazonia and Global Change*, ed. M. Keller, M. Bustamante, et al., Washington, DC: American Geophysical Union, pp. 61–81.

Weinhold, D., E. Killick, and E. J. Reis. 2013. Soybeans, poverty and inequality in the Brazilian Amazon. *World Development* 52: 132–143.

Weis, T. 2010. The accelerating biophysical contradictions of industrial capitalist agriculture. *Journal of Agrarian Change* 10, no. 3: 315–341.

Weis, T. 2013a. *The ecological hoofprint: The global burden of industrial livestock.* London: Zed Books.

Weis, T. 2013b. The meat of the global food crisis. *Journal of Peasant Studies* 40, no. 1: 65–85.

Wolford, W. 2008. Environmental Justice and the construction of scale in Brazilian agriculture. *Society & Natural Resources* 21, no. 7: 641–655.

World Wide Fund for Nature (WWF). 2014. *The growth of soy: Impacts and solutions.* Gland, Switzerland: WWF International.

Zhang, K., A. Castanho, et al. 2015. The fate of Amazonian ecosystems over the coming century arising from changes in climate, atmospheric CO2, and land use. *Global Change Biology* 21, no. 7: 2569–2587.

Zhouri, A. 2010. Adverse forces in the Brazilian Amazon: Developmentalism versus environmentalism and indigenous rights. *Journal of Environment and Development* 19, no. 3: 252–273.

Strategies and hybrid dynamics of soy transnational companies in the Southern Cone

Valdemar João Wesz Jr

Economical liberalization, market globalization and soy expansion stimulated the advance of big transnational companies in the Southern Cone countries (Brazil, Argentina, Paraguay and Uruguay). Currently, the main corporations acting on the last links of the productive chain are ADM (Archer Daniels Midland), Bunge, Cargill and Dreyfus (the ABCD firms), global leaders in the soy trade. The objective of this contribution is to analyze the different strategies these companies articulate in the Southern Cone, and their dynamics in local space through market relations with local producers. The results show the rapid and intense process of denationalization of the firms in the soy productive chain as well as the high level of market internationalization and company concentration. In spite of this, this study shows that all transnational power of ABCD firms, which seems so abstract and intimidating when seen in the global scale, depends on its basis of the formation, maintenance and exploration of relations of proximity, trust and reciprocity with local actors (especially rural producers), including family friendship linkages.

Introduction[1]

From 1970 onward a new international order started in the political and institutional sphere, characterized by the emergence of a complex reconfiguration of the world's economy and the alteration of the nature and dynamics of production, consumption and markets (Friedmann and McMichael 1989; Santos 2002). In the agricultural and food-production sector, this process is expressed in the liberalization of international commerce and the performance of transnational companies looking to exploit the comparative advantages of this new order (Wilkinson 2009). In this context, heterogeneous and geographically dispersed actors and spaces are connected, and the dynamics of the relations between firms and ways of governance are modified (Gereffi, Humphrey, and Sturgeon 2005; Bonanno and Constance 2008; Clapp and Cohen 2009; Clapp and Fuchs 2009).

In Latin America's Southern Cone[2], the globalization process was intensified from the 1990s, during which a strong commercial and financial opening occurred, the results of which were the mutation of the companies' patrimonial structure, exemplified by the

[1]This contribution presents some of the results of the author's doctoral dissertation (Wesz Jr. 2014).
[2]In this paper, the Southern Cone includes the following countries: Brazil, Argentina, Paraguay and Uruguay.

expansion of foreign direct investment (FDI) and the increase in the number of mergers and acquisitions (M&A). Also, intensification of the firms' denationalization process, growth of markets' internationalization, business concentration and capital centralization occurred (Benetti 2004; Cohen 2007; De La Torre and Schmukler 2007).

In the soy market, the economic liberalization and market globalization process, and the legume production increase, stimulated the advance of the big transnational companies in the Southern Cone. This way, ADM (Archer Daniels Midland), Bunge, Cargill and Dreyfus, world leaders in agricultural trade, started to play an important role in the region's soy market, from the 1990s onward. It is important to notice that soy is today the main farming activity in economic (export value) and territorial (planted surface) terms in Brazil, Argentina, Paraguay and Uruguay.

The objective of this paper is to analyze the strategies and dynamics of the transnational companies ADM, Bunge, Cargill and Dreyfus (known as the ABCD firms) on the Southern Cone soy productive chain. In this way, it seeks to understand, through the different national and international transformations, the current configuration of the companies related to the region's soy market, which have led the world production of this bean. At the same time, it discusses the dynamics of these firms in the local space related to their commercial relations with rural producers. With this debate it aims at understanding how transnational companies' global practices are shifted, shuffled and adapted, and regain meaning through interactions with actors and the contexts that exist in the different scales (processes that produce hybrid organizations).

Besides using specialized academic literature on the subjects discussed, secondary data research was conducted for the different countries, mainly consulting the Brazilian Association of Vegetable Oils Industries (ABIOVE; Brazil), the Brazilian Institute of Geography and Statistics (IBGE; Brazil), the National Company of Supply (CONAB; Brazil), the Foreign Trade Secretariat (SECEX; Brazil), the Ministry of Agriculture, Livestock and Fishing (MAGyP; Argentina), the National Institute of Statistics and Census (INIDEC; Argentina), the Ministry of Economies and Public Finances (MECON; Argentina), the Ministry of Agriculture and Livestock (MAG; Paraguay), the Paraguayan Assembly of Cereals and Oilseeds Exporters and Traders (CAPECO; Paraguay) and the Ministry of Livestock, Agriculture and Fishing (MGAP; Uruguay). Also, information was collected from the media, especially newspapers and magazines, as well as reports, institutional bulletins and balance sheets of ADM, Bunge, Cargill and Dreyfus. Thirty-six qualitative interviews were held in 2012 and 2013 in Mato Grosso (the main Brazilian state for soy production) with representatives of ABCD companies and with rural producers (with regard to the latter, producers that plant more than 200,000 hectares of soy and farmers in rural settlements that produce the grain on 10 hectares were interviewed).

After this introduction, the soy expansion in the Southern Cone is presented, highlighting this market's features in the four countries. In the following, the presence of ABCD in the region is discussed with special attention in terms of the installed crush capacity and export volume. The main strategies of the ABCD firms are also analyzed. And, before the concluding section, the performance of these companies in the local space related to their commercial relations with rural producers is discussed.

The soy expansion in the Southern Cone

Until the mid part of the twentieth century, soy was an experimental crop in the Southern Cone countries, without economic and territorial importance. Between the 1950s and 1960s there was a first movement to stimulate production, mainly in Brazil, where soy started to be planted in rotation with wheat (Embrapa 2004). As shown in Figure 1, from 1970 onwards, there is a significant increase in the planted area, led by Brazil and, to a lesser extent, by Argentina. The decades of 1980 and 1990 were characterized by an increase in the planted surface, although there were significant recession periods, especially in Brazil, which entered an economic crisis (Delgado 2005). By the mid 1990s, what has been called the soy boom took place (in Uruguay this process came later, starting by the 2000s), the main characteristic of which was the high rate of planted area growth, which doubled in less than 10 years. In a general way, this expansion has continued until today, mainly in the Paraguayan and Argentinean cases, where the planted surface has had practically uninterrupted growth during the last 20 years (Figure 1). Today, the soy-planted area in the Southern Cone equals the sum of the territories of Germany, Portugal and Belgium.

The dynamics of soy in the Southern Cone is linked to the incorporation, on a global scale, of this legume in the production of vegetal oil and proteins for animal feed (swine, birds and cattle) that was boosted mainly by the USA after World War II (Du Bois, Tan, and Mintz 2008). The growth of the international demand for and the price of the grain, especially from 1973 onwards, stimulated countries in the Southern Cone to invest in the legume for export. It is important to highlight that the national states played a fundamental role in disseminating this harvest when they created several measures to encourage its production. In this sense, policies for the modernization of agriculture were decisive, for they relied on subsidized credit, technological innovation, minimal prices, modernization of the input and processing industry, the creation of new channels for distribution, etc. (Piñeiro and Moraes 2008; Heredia, Palmeira, and Leite 2010; Gras and Hernández 2013).

In parallel, the soy expansion received strong benefits from migration policies, especially in Brazil and Paraguay. In the former, the military government promoted the 'occupation of Cerrado' and the expansion of the agriculture frontier through several public policies (directed at land concession, infrastructure construction, agriculture modernization, territory occupation, fiscal incentives, etc.). One of its main goals was to extend farm production in the region to generate earnings and equalize the commercial balance via the increase of exports (Fernández 2007; Moreno 2007). In the Paraguayan case, the Stroessner government (1954–1989) tried to consolidate the agroexporting model through the incorporation of Brazilian people in its territory, so that they could extend

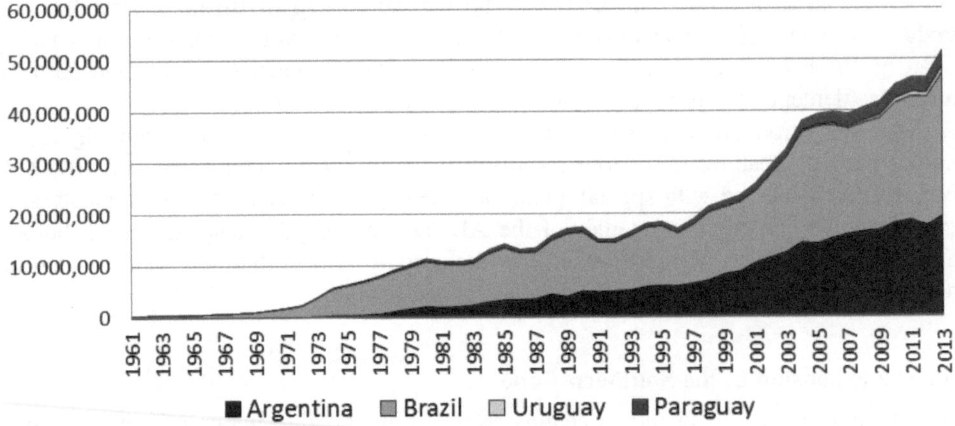

Figure 1. Soy-planted area in the Southern Cone (1961 to 2013).
Source: Faostat (2014).

areas of plantation destined for export (especially soy and wheat). For such, the law forbidding the acquisition of land by foreigners in the range of 150 kilometres from their borders was abolished, and it facilitated the land concession and financing of agricultural activities (Pappalardo 1995; Albuquerque 2005; Blanc, 2015). Today, it is estimated that 90 percent of Paraguayan soy is produced by Brazilians or their descendants (Revista Exame 2011).

In addition to sectorial actions of the different countries (such as rural credit, price, land and territorial order, agronomic research, technical assistance policies, etc.), a series of instruments were decisive for the construction of an environment favourable to soy expansion (such as labour, environmental, industrial, commercial, fiscal, tributary, energy, infrastructure and services policies, etc.). Still, domestic impulses cannot be analyzed as disconnected actions, but as a set of instruments mobilized to sustain an agroexport model of development that, during the years, has shifted and restructured when faced with the national and international political-economic context (Pires and Santos 2013).

It is necessary to recognize that public policies constructed by the different countries to maintain the agroexport model are linked to the context of economic, commercial and financial globalization, defended and stimulated by international organizations (especially the International Monetary Fund, the World Bank and the World Trade Organization) (Gilpin 2011). In parallel, the transnational companies deepen and intensify the process of market integration, since they are the main locus of accumulation and economic power controlling specific assets (capital, technology and management, organizational and market capacity) and defining private norms for market regulation (Clapp and Fuchs 2009). In brief, it can be said that the expansion of the soy frontier in the Southern Cone has been stimulated by the relations among national governments, transnational companies and institutions of global governance.

Gudynas (2008), Guibert et al. (2011) and Gras and Hernández (2013) also note how technological and socio-organizational transformations were also fundamental for the soy boom. Amongst the technical changes, the authors highlight the importance of direct seeding and transgenic variety introduction, which simplified the planting and management process, reducing implementation costs (despite the payment of royalties and the growing use of agrotoxics). On the socio-organizational front, there were information and communication structures, business management, new financial tools and the organization of the companies on business networks, that allowed the introduction of financial sources coming from other sectors, the expansion of actors' mobility, and the possibility of controlling bigger areas in different regions.

In the last few decades, the Southern Cone has become the main area of expansion of soy plantations (mainly Brazil and Argentina, which dominate more than 90 percent of the regional planted surface of the grain, but with a growing advance over the Paraguayan and the Uruguayan territories[3]). In 2013 these four countries accounted for more than half of the world's production (52 percent), while in 1970 they represented only 4 percent. From 1970 to 2013, the soy-planted area grew more than 34 times, while in the other producer

[3]In Bolivia there is a strong process of soy expansion, especially in the department of Santa Cruz, which is similar to the prevailing situation of the Southern Cone. However, Bolivia was not considered in this paper because the ABCD firms do not appear as strong there as they do in the other countries (even with ADM and Cargill acting there). In this country, Bolivian, Peruvian and Venezuelan agroindustries have a major presence (even because the main destination of soy is the Andean countries themselves) (Colque 2013). Besides, there is limited access to data and information on the companies.

countries, this growth was only two times (Faostat 2014). The data show the region's importance and leading role in soy cultivation.

Genok (2012) argues that, associated with this productivity growth through the years, since the 1990s there has been an expansion over native vegetation areas, and over other activities (such as summer cultivation and livestock). The strong hegemony soy has assumed on Southern Cone agriculture can be observed in the control that it exerts on the total arable lands, which has been greater than 40 percent since 2005. Two of every five arable hectares in the region are used for soy (Genok 2012).

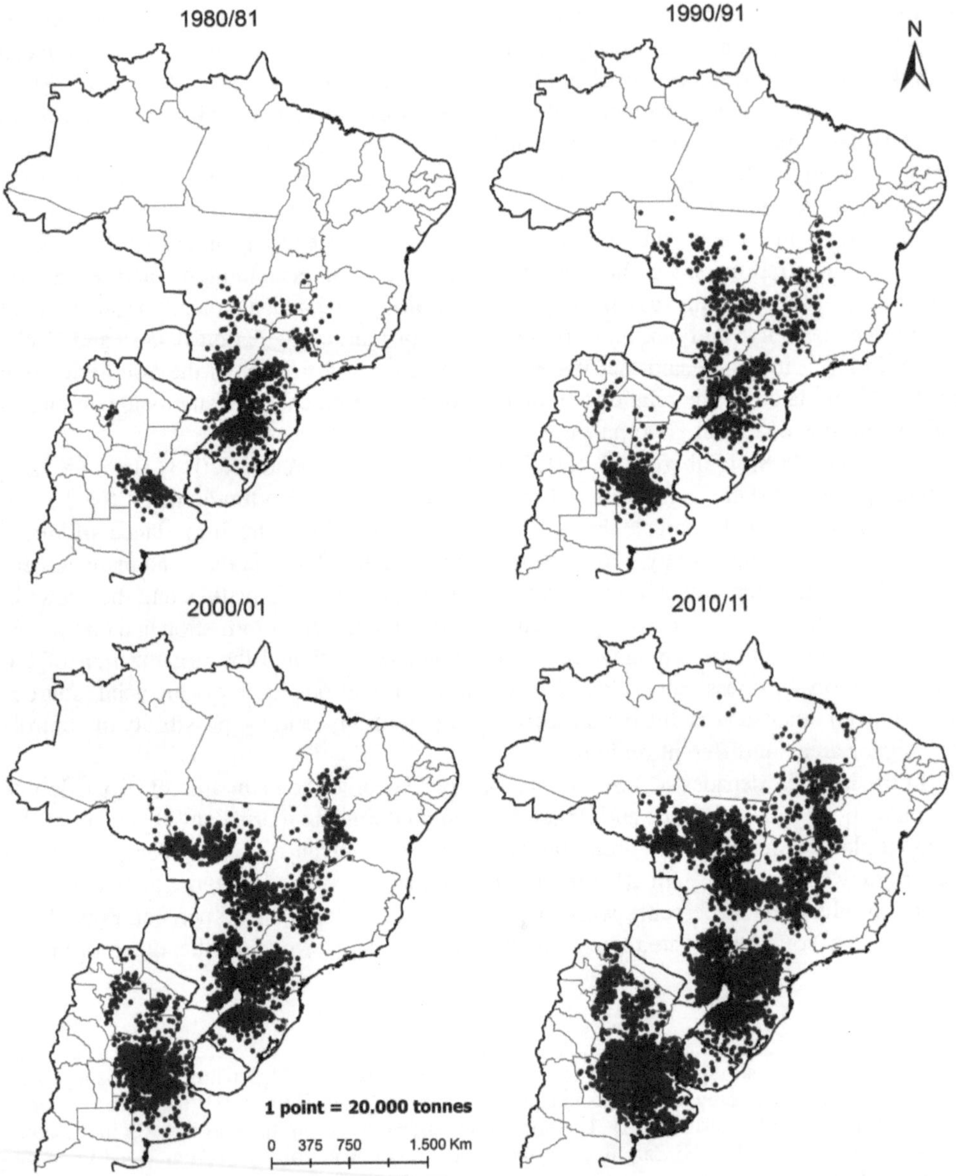

Figure 2. Soy production in the Southern Cone (Brazil, Argentina, Paraguay and Uruguay). Source: IBGE (2013); MAGyP (2013); MAG (2013); Capeco (2013); MGAP (2013), compiled by the author.

The territorial expansion of soy can be observed in Figure 2, which presents the grain production in Brazil, Argentina and Paraguay. This instrument allows us to comprehend, historically and spatially, the expansion of the legume in the different regions of the countries analyzed. While at the beginning of the 1980s, soy was present in a more significant way in the Brazilian South and the humid pampas of Argentina, through the years it is possible to observe the proliferation of new spots, such as the center-north of Brazil and Argentina, the Uruguayan west and the Paraguayan east (Figure 2[4]).

When analyzing the soy data in each country, the importance it has assumed over the last decade becomes unquestionable. In Brazil's case, the legume occupies a surface that is greater than 50 percent of temporary cultures and accounts for 9.4 percent of total exports. It is also the largest country exporting soy, in grain (CONAB 2013; SECEX 2013). In Argentina, soy has become the main farm activity, surpassing traditional cultures such as wheat, corn and sunflower. Over the last few years the grain has reached 60 percent of farming planted areas and is responsible for 25 percent of the total value of sales to foreign countries, and the country is the world leader in soybean meal and oil exports (INDEC 2013; Faostat 2014). In Paraguay, soy has been identified as 'the backbone of agribusiness', occupying 72 percent of arable lands in the country and accounting for 9.7 percent of national gross domestic product (GDP), and 42 percent of total exports (CIP 2013; Faostat 2014). In Uruguay, the legume took an 'explosive' leading role because it had such little expression in the beginning of the twenty-first century (0.1 percent of exports and 7.7 percent of planted surface), while a decade later it accounted for 16 percent of total exports and 86 percent of the cultivated land during the summer (UruguayXXI 2012; MGAP 2013).

Through the years, the installed capacity of soy agribusinesses grew significantly, especially in Argentina and Brazil, which concentrate an important part of the sector's enterprises. When analyzing the industrial situation of both countries, it is evident that between 1984 and 2010 there was an expressive growth in Argentina (350 percent), while in Brazil this was more modest (75 percent). With this, Argentinean industries practically equaled Brazilian ones in 2010, while by the mid-1980s they had reached only a little more than a third (Figure 3). In Paraguay and Uruguay, the soy-crushing infrastructure was much smaller than that of the neighbouring countries, because of the smaller production, but also because a big part of this commodity is exported without being processed, as will be shown later in this contribution.

[4]In relation to Figure 2, two comments must be made. First, concerning the spatial unity, where some equivalence between the four countries was sought. This is particularly important to keep proportionality where absolute values are used, instead of average values. For this, the unit 'micro-region' was used in Brazil, and the 'department' in Argentina, Paraguay and Uruguay. The second comment refers to the data, which were built based on different sources: Brazil – agricultural research by municipalities of the Brazilian Institute of Statistics and Economy (IBGE 2013); Argentina – Ministry of Agriculture, Livestock and Fishing (MAGyP 2013); Paraguay – Ministry of Agriculture and Livestock (MAG 2013) and Paraguayan Assembly of Cereal and Oilseeds Exporters and Traders (Capeco 2013); Uruguay – Ministry of Livestock, Agriculture and Fishing (MGAP 2013). In spite of the different statistical sources, the variable 'soy production' did not present comparability problems. As the analysis was made for a 30-year period, in some countries there were no data allowable by department (only national), as in the case of Paraguay (harvest 1980/1981) and Uruguay (harvest 1980/1981 and 1990/1991). In these two countries, the data were estimated from studies and news that portray the soy trajectory in these areas. This way, they are approximations that were made to allow a comparative spatial and historical analysis.

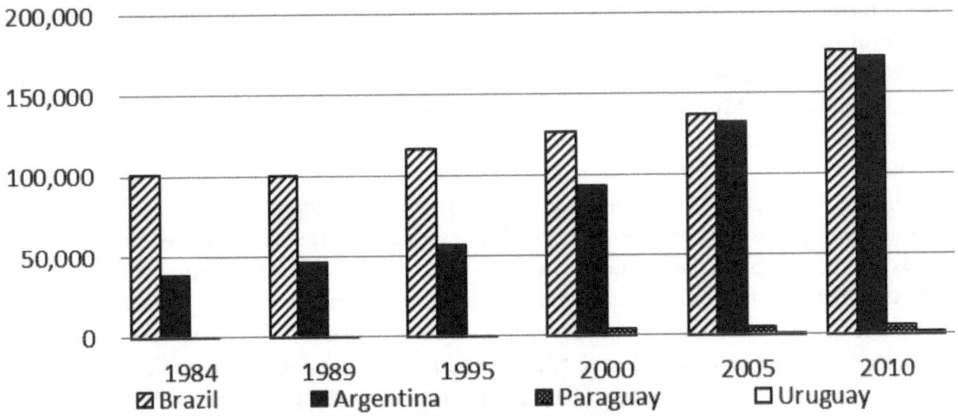

Figure 3. Installed capacity of the soy industry in the Southern Cone.
Note: Since there is no official information on the installed capacity in Paraguay and Uruguay, these data were estimated from studies and news from these countries.
Source: Abiove (2014); Wesz Jr. (2011); Hinrichsen (2013); Barbosa and Nogueira Jr. (2007) and media documents.

The slower pace of Brazilian installed capacity growth is the result, mainly, of the Kandir Law enactment in September 1996, which exonerated raw material exportation from the Tax on Distribution of Goods and Services (ICMS), maintaining the tributary burden on industrialized products. This, on the one side, allowed a greater competitiveness in Brazilian *in natura* farm exports and, on the other, sensibly reduced the viability of the agribusiness production for the external market. The Kandir Law results were immediate and, two years after its enactment, soy passed from 5 to 30 percent of the total exportations of this productive complex (Wesz Jr. 2011). Besides the tributary aspect, enterprises have preferred to invest in Argentina because there is a better infrastructure, especially for crop marketing; a better exchange rate for exports; and a regulatory framework which incentivizes grain processing in the country (Schavarzer and Tavosnanska 2007; INTA 2009).

Currently, the four countries present very different industrialization levels of soy production, the highest percentage being in Argentina where, in the last few years, an average of three quarters of the legume were crushed and converted into oil, meal and other by-products (MAGyP 2013; INDEC 2013). This is the case, when compared with the data from Paraguay (22 percent) and Uruguay (5 percent), because historically, there has been in Argentina a stimulus for the exportation of processed products, since meal and oil foreign sales are less taxed compared to the soybean exports. In Brazil, the industrialization level of soy production has dropped strongly, falling from 95 percent in 1995 to about half of that nowadays (Wesz Jr. 2011). The Kandir Law, described above, is basic to understanding this change in the soy allocation, with the increasing demand from China for the oilseed *in natura*.

Concerning the production allocation, soy keeps being mainly exported in the four countries of the Southern Cone. However, there are three important differences amongst these nations. The first one refers to the weight of domestic consumption over total production, which is larger in Brazil (around 25 percent of the harvest) and smaller in Paraguay and Uruguay (5 percent), while in Argentina there is an intermediate percentage (15 percent) (UruguayXXI 2012; SECEX 2013; INDEC 2013; Capeco 2013). In the Brazilian case, three factors define the high percentage of domestic consumption: (1) a strong influence of the government on the modification of diet and consumers' food habits, with an increased demand for vegetable oils which substitute animal fats (butter and fat) in the context of a growing population contingency; (2) the biodiesel demand increase since the enactment in 2004 of the National Program of Biodiesel Production and Use (PNPB), which determines a mandatory mix of 6 percent of biodiesel, made mainly from soy, in conventional diesel; (3) an increase in cattle, pig and poultry husbandry, with a growing consumption of soy meal (Wilkinson and Herrera 2010; Wesz Jr. 2011). In the case of Argentina, Uruguay and Paraguay, human consumption of refined soy oil is very low, and soy is used, domestically, for animal feeding (meal), although recently there has been an increase in the raw oil demand for biodiesel, since incentives for production and consumption of renewable energies have also increased (Rojas Villagra 2009; MECON 2011; MGAP 2013).

The second difference is linked to exported by-products, where there are very diverse situations. In Uruguay, 100 percent of the soy is sold *in natura*, which is justifiable because of the recent soy expansion in the country (over a little more than a decade) and because of the destination country, since China is purchasing two thirds of the Uruguayan soy and has a priority for the *in natura* grain to crush it in its own territory (UruguayXXI 2012). In Brazil and Paraguay, there is supremacy of the soybean, although there are oil and, especially, meal exports (SECEX 2013; Capeco 2013). In both countries there is a regulatory framework that favours exports with no aggregated value, because in sales abroad feedstock faces less taxation than manufactured products do (Rojas Villagra 2009; Wesz Jr. 2011). In Argentina, where exports of agroindustrialized products in the soy complex reach 70 percent of the total value (INDEC 2013), the regulatory framework encourages the processing of the grain inside the country, charging more taxes when the product *in natura* is exported (Figure 4; Fernandes Filho and Belik 2010).

The third difference refers to the allocation market of the soy complex, where China assumes a great importance in the total value of Uruguayan (67 percent), Brazilian (50 percent) and Argentinian (25 percent) exports, mainly because of the purchase of the soy *in natura*; in the Paraguayan case, marketing focuses mainly on Europe (which is also the main recipient of Argentinean meal) (UruguayXXI 2012; SECEX 2013; INDEC 2013; Capeco 2013). As it will be shown, this context, which includes different products and market allocations for soy, demands differentiated action strategies by the enterprises according to the country.

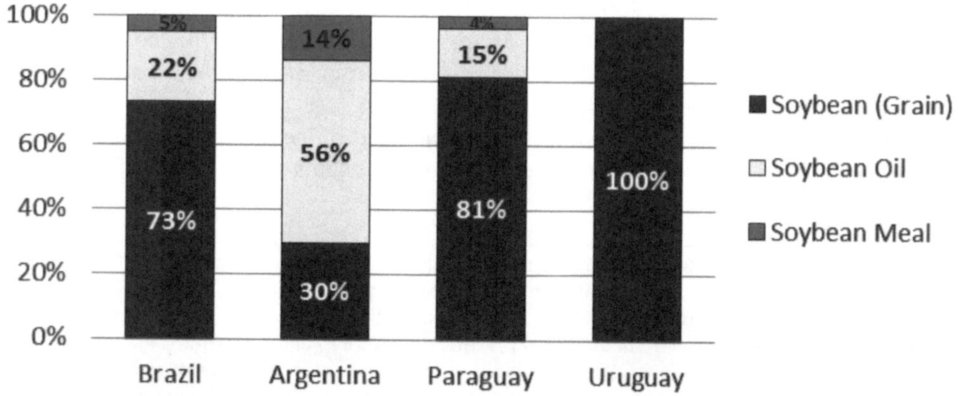

Figure 4. Export value (%) of the soy complex (grain, meal and oil) by country in 2011.
Source: UruguayXXI (2012); SECEX (2013); INDEC (2013); Capeco (2013).

ADM, Bunge, Cargill and Dreyfus (the ABCD firms) in the Southern Cone

Nowadays world cereal trade is controlled by four major transnational firms: ADM, Bunge, Cargill and Dreyfus (routinely called ABCD due to the coincidence of their initial letters). ADM (Archer Daniels Midland) began its activities in 1902, but appeared as a global actor in the 1970s. It is based in 75 countries, acting in the production of food ingredients, animal nutrition, and chemical and energy products. Bunge was founded in 1818 in the Netherlands, but in the early twentieth century it expanded to South America (where it is the major trader in cereal and the major producer of the region's fertilizers). It operates about 400 facilities in 40 countries specialized in the production of cereal, oilseed, sugar and ethanol. Cargill, which was founded in 1865 in the United States, controls a wide range of activities (farm services, risk management, food purchase and processing, etc.) in 66 countries. Louis Dreyfus was founded in 1851 in France. It differs from the previous three in that part of the company is still a family property, for they control about 20 percent of the shares. The firm is present in 90 countries, and its main focus is the feedstock and biofuel trade (Arroyo, Rama, and Rello 1985; Murphy, Burch, and Clapp 2012). These four companies are the leaders in the world trade of grain, controlling between 75 percent and 90 percent (The Guardian 2011).

ABCD took the leading role in the Southern Cone from the 1990s onwards, a period characterized by the economic, commercial and financial globalization in the region.[5] Until 1995, only Cargill and Dreyfus had industrial plants for soy crushing in Brazil and Argentina, controlling less than 10 percent of the installed capacity. From that year on, other companies, like Bunge and ADM, started to invest in industrial plants on the region. An initial strategy, common to both firms and applied in different countries, were mergers and acquisitions, through which they could advance over important companies in operation on the national market, changing the patrimonial structure of the endeavours.

For the ABCD firms, the mergers and acquisitions had as their main advantages: (1) eliminate competition, increasing the presence of the purchasing company on the market; (2) build production scales compatible with the new development standard, highly exigent regarding the permanent incorporation of technology; (3) secure the raw materials supply and the distribution channels of the final goods, already structured by the acquired company; and (4) enter quickly into the domestic markets, taking advantage of the economical context of that time (Benetti 2004). On their side, some national companies were changing the focus of their actions, with an interest in selling their soy businesses, while others were indebted (Castro 2002). In Brazil, the process of mergers and acquisitions was evident when Bunge acquired Ceval, Santista and Incobrasa, Dreyfus bought Gressy Lever, and ADM assumed the industrial soy plants from Sadia[6] (the same happened between Cargill and Matosul) (Wesz Jr. 2011).

After consolidating their control over a significant part of the market, these big transnational companies started to invest in the expansion of the units already existent and in the

---

[5]It is important to highlight that at the same time that the movements of globalization favoured the entrance of transnational firms in the Southern Cone, these corporations were one of the key elements in the transformation of the production of food to a financialized, globalized and complex business (McMichael 2009; Giménez and Shattuck 2011; Ahmed, Hamrick, and Gereffi 2014).

[6]It is worth highlighting that Ceval, Santista, Sadia, Incobrasa and Gessy Lever were among the six major food agroindustries in Brazil (Cargill was added to this group) that met 34 percent of the crush capacity in 1995. Therefore, mergers and acquisitions conducted by the ABCD firms happened over the market leaders that were, in most parts, national capital ones.

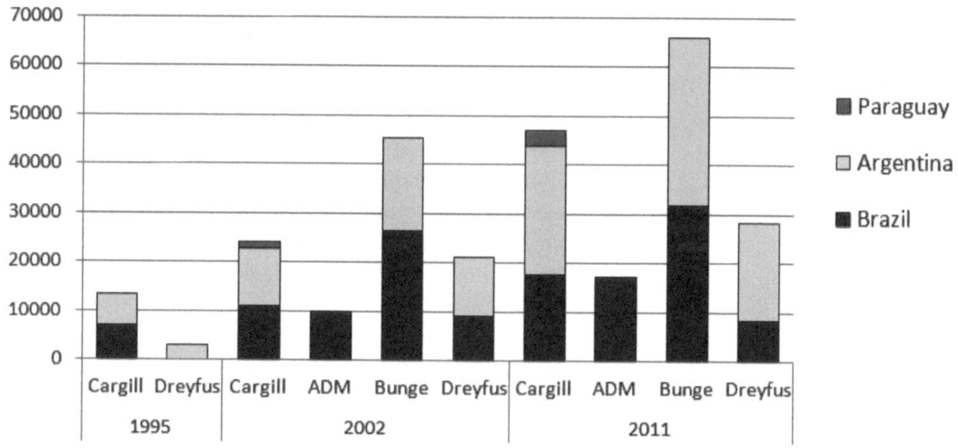

Figure 5. ADM (Archer Daniels Midland), Bunge, Cargill, Dreyfus installed soy-crushing capacity in 1995, 2002 and 2011 in Brazil, Argentina and Paraguay (tons/day).
Source: Wesz Jr. (2011), Hinrichsen (2013), companies' web pages and media documents.

building of endeavours in new areas (Pierri 2008; Rojas Villagra 2009; Wesz Jr. 2011). In 2002 it was already possible to observe the effect of mergers, acquisitions and investments, when Bunge, Cargill, ADM and Dreyfus started to dominate 50 percent of the crushing installed capacity in the Southern Cone, with a control superior to 100,000 tons/day. This process was quick and aggressive if we consider that in 1995 there were only two companies that controlled nine percent of the capacity (16,000 tons/day). In a very short time, in 2002, Bunge became the biggest company amongst the ABCD firms, dominating 20 percent of the processing capacity in the entire Southern Cone, even with a crushing capacity limited to Brazil and Argentina (Figure 5).

Although in the last few years the ABCD firms have expanded, in absolute terms, their installed capacity, reaching almost 180,000 tons/day in 2011, the percentage of industry control has remained at the same level as in 2002 (around 50 percent). The greatest company concentration is in the two countries that produce most part of the region's soy (Brazil and Argentina), while in Paraguay only Cargill has soy-crushing structure – although ADM, Bunge and Dreyfus are currently building industries in this country and, when they finish, the ABCD firms will start to control 80 percent of Paraguay's capacity (Chicago Tribune 2012). In Uruguay, although all of the companies act in the crop marketing, none has a crusher, since only 5 percent of the soy is industrialized in the country, as it was mentioned before.

Among the ABCD firms, Bunge keeps the leading role in crushing capacity, with more than 65,000 tons/day in 2011, followed by Cargill, which is present in Argentina, Brazil and Paraguay – although it is in the two first countries where 90 percent of its industrial structure is concentrated. Dreyfus, in the third position, has a capacity of 28,000 tons/day (with an 880 percent growth compared to 1995) and, like Bunge, only acts in Brazil and Argentina. For its part, ADM crushes soy only in Brazil, reaching 20,000 tons/day (Figure 5).

In recent times, there has been a new regionalization of the ABCD agroindustries, with a migration of production zones in Brazil and in Argentina to areas close to export channels. In the Brazilian case, the agribusiness dislocation from the South and Southeast to the country's Central-North is associated with the legume production increase in these new areas.

However, there are other reasons for this relocation, such as the fiscal incentive that is offered when feedstock is industrialized within the boundaries of the state that produced it (Kandir Law). Furthermore, companies' decisions are affected by the organization of the regional transportation system and the presence of a local demand for soy meal for livestock (Wesz Jr. 2011). In Argentina, the crushing capacity has concentrated close to seaports, mainly in Santa Fe Province, where the main cultivation area is located. Since most of the soy is exported and production and marketing areas are close, it is viable to concentrate the agribusiness in the port zones (Schavarzer and Tavosnanska 2007). In Paraguay, as well as in Argentina, the units are currently being built close to the port that is relatively close to the production region.

In terms of sales to foreign trade, there has been a significant increase during the last years. Between 2005 and 2011 exports increased from USD 12.2 to 33.6 billion FOB (Free On Board) (growing 174 percent), while the rest of the sales in the four countries increased at a slower pace (102 percent) (SECEX 2013; INDEC 2013; Capeco 2013; COMEX 2013). Figure 6 shows an important growth in these firms' exports from 2005 to 2007, but the downturn of international soy prices reduced the value in 2008, which grew again in the following years.

Although ADM and Dreyfus presented higher export growth, exceeding 200 percent between 2005 and 2011, Bunge and Cargill are the ones who dominate almost two thirds of the ABCD companies' total exports in the four countries (Figure 6). Brazil and Argentina represent a greater share of these companies' foreign trade, but in the last few years there has been a reduction in their share: in 2005 they accounted for 95 percent, and this fell to 91 percent in 2011. The greatest value mobilized by one company was that of Bunge in Brazil, which reached USD 6.5 billion FOB in 2011 (SECEX 2013), exceeding Paraguay's total exportations that same year, which reached USD 5.5 billion FOB (CIP 2013).

Although soy oil and meal exports are important for the ABCD companies, the *in natura* sales have gained much attention in recent years, mainly because of China's

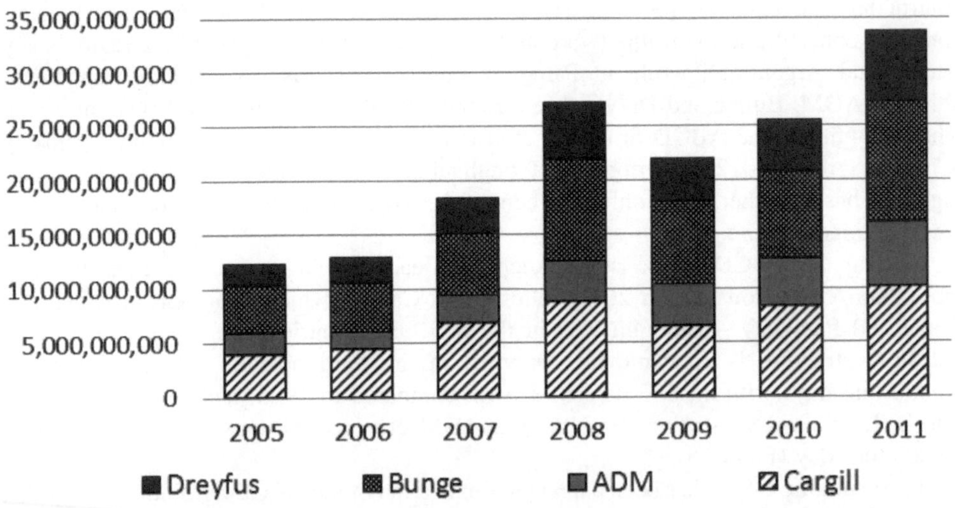

Figure 6. ADM (Archer Daniels Midland), Bunge, Cargill, Dreyfus total exports in the Southern Cone (in USD) – 2005 to 2011.
Source: SECEX (2013); INDEC (2013); Capeco (2013); COMEX (2013).

growing demand for the grain. Currently, it is estimated that Bunge, Cargill, ADM and Dreyfus control 85 percent of the total *in natura* soy exports in the Southern Cone (Wesz Jr. 2011; INDEC 2013; Capeco 2013; CIP 2013; COMEX 2013). In fact, this has been, during recent years, one of the main focuses of the companies, made possible by some countries' regulatory frameworks, as it happens in Brazil in Paraguay. This more attractive situation for *in natura* exports helps to explain the maintenance, in relative terms, of the ABCD companies' control on crushing capacity – which has remained around 50 percent for the last 10 years – while at the turn of the century, the necessity for industrial structure control was greater. In sum, *in natura* exports have gained importance for transnational companies because, besides reducing fixed capital investments, they have a growing demand and a favourable regulatory framework.

The participation that Bunge, Cargill, ADM and Dreyfus exercise on the Southern Cone total export value grew significantly through the years, passing from 6.8 percent in 2006 to 10 percent in 2011. This shows the strong marketing and economic power of the four transnational companies on the countries analyzed, which occurred rapidly, since until 1995 they presented a very modest action in the region. Being amongst the first exporters at the global level, it is evident that these companies not only have sectorial power (in the soy complex), but they are very important in foreign trade as a whole, including surplus generation.

Analyzing the Paraguayan case in isolation, the dimension that ABCD has assumed in this country is overwhelming. In 2006, these four transnational companies were responsible for 29 percent of the total exports, which was already a high value, but which grew more in recent years, reaching 47.2 percent in 2012 (CIP 2013). With this, almost half of the country's exports are controlled by a small number of multinational firms, who act principally in the soy market but also in other agricultural markets. The effect of the companies on economic stability and surplus generation is an important triumph of the companies when negotiating public investments in agribusinesses' strategic areas (plots for industrial plant building, infrastructure for production marketing, tax incentives, etc.). In this sense, these companies' economic and productive importance ends up turning the Paraguayan government into a trading hostage, using the conjuncture to negotiate public resources which contribute to their profits and their expansion in the grain market.

The main strategies of the ABCD companies in the Southern Cone

The increased control of ABCD companies over soy production in the Southern Cone as well as the increasing export value reflect the different strategies developed by these companies. In this direction, their consolidation and strengthening are linked to the distinct initiatives articulated by these actors in their contexts. It is important to notice that it is about not only a strategy, but several intrinsically articulated instruments that overlap and significantly reinforce each other.

One of the main strategies of ABCD companies is vertical integration, the main feature of which is the appropriation of different stages of the chain by only one company. This means that only one firm will be acting in the several phases of the productive process in a coordinated way – differing from previous times when that firm was only involved with one of the processes, such as the crushing or trade of the commodity. This strategy is linked to increasing investments that Bunge, Cargill, ADM and Dreyfus have made in different processes of the productive chain, such as the production and sales of inputs, financing and farm insurance services, technical advisory services, purchase of the beans, storage, transportation, industrialization and trade[7] (Pierri 2006; Souza 2007; INTA 2009; Rojas Villagra 2009; Oyhantçabal and Narbondo 2011; Wesz Jr. 2011).

To consolidate the verticalization strategy, the firms purchased large companies in the fertilization sector, through fusions and acquisitions, such as happened in the crushing segment. Simultaneously, ABCD firms continue their strong investments in the storage of soy through the purchase or leasing of warehouses that are usually located near the main areas of production or distribution. Besides, companies maintain major inversions in logistics and infrastructure with special attention to the construction and amplification of ports, waterways and railroads connecting regions that produce soy with the international market.

In some specific sectors, where ABCD companies do not have a major presence, they end up cooperating with other companies, as in the branch of seeds and pesticides. This was the case in partnerships established between Cargill and Monsanto, Bunge and DuPont, ADM and Syngenta, and Dreyfus and Genética Don Mario (partially purchased by Dow), among others. This way, with their own companies and partners, they can act on every segment of the productive chain.

Murphy, Burch, and Clapp (2012) argue that the presence of the same company in complementary sectors of agricultural production ends up performing a central role in decisions taken by the producers (what to grow, when, how, what amount and for which market). Verticalization allows these companies to monitor and participate in all farming processes, such as costing, providing inputs, capital and information, going through trade, industrialization and distribution of the production (Souza 2007). This kind of negotiation ends up making producers very dependent on the agroindustries, because a single actor becomes the main financial agent, input distributor, technical assistance agent and buyer of the product. Consequentially, the producer becomes 'immobilized' faced with the company's interests, creating a permanent relationship between them and preventing more independent actions by the soy producers (Rodrigues et al. 2009). And the more firm dependent producers become, the less farmer friendly contracts established with the companies will be, involving differences in interests rates, inputs' prices, deadlines for delivery of goods, and the general conditions of the contracts (Fernández 2009).

Analyzing the advantages of vertical integration for the ABCD firms, Souza (2007) points out that one of their motivations in maintaining and expanding this strategy is linked with financial issues. Gaspari (2004), in his turn, realizes a technological and marketing complementarity among the segments up and down the productive chain, which makes bigger returns through economies of scale and scope. Other operational advantages that stimulate verticalization are the achievement of synergies in transportation resulting from

[7]Stimulating these different sectors on a global scale demands large resources. In addition to directing an important share of profitability in these activities, companies mobilize resources from the financial market for such aims. This process, which has been called 'financialization', happens when private investment funds, companies of asset management, commercial banks and other financial institutions invest in futures markets of commodities, farmlands and agricultural production. ABCD firms buy and sell in the futures market and make use of financial instruments to mobilize resources, manage risks and increase their profits. For a deep analysis of this process, see Murphy, Burch, and Clapp (2012) and Salerno (2014).

guarantees of the return freight (when the vehicle that delivers the fertilizers returns with grains or industrialized products for exports). Besides, Farina and Zylbersztajn (1998) argue that companies take advantage of extra profits generated by governmental subsidies (exemption of taxation, for example).

Therefore, verticalization is one of ways the companies have found to increase their control over the chain participating in the different stages of the production process, increasing their profit margin, reducing production and transaction costs, minimizing risks and generating complementarities and synergies among the different sectors (Hendrickson, James, and Heffernan 2013). Even though this strategy is present in all the Southern Cone, results can be seen more distinctly in the Brazilian case, where ABCD firms controlled, in 2010, 65 percent of the national fertilizer production, 80 percent of the funding granted by the *tradings* for soy cultivation, 50 percent of the oilseed crushing capacity and 85 percent of the beans traded in the country (through purchase from the producers, cooperatives, input resales and minor companies) (Wesz Jr. 2011).

Another relevant strategy is about the production destination. As mentioned above, being able to unite domestic and international markets is very convenient, as well as the option of trading *in natura* and agroindustrialized goods (oil and soybean meals). These alternatives allow the increase of the companies' profitability because they expand their room for manoeuvre and make options according to the context assessment. In addition to these possibilities, ABCD firms have diversified their current strategic field, including the energy sector (Borras Jr., McMichael, and Scoones 2010; Fernandes, Welch, and Gonçalves 2010). This market's growth, good future perspectives and the use of soy as the main feedstock in the production of biodiesel in the Southern Cone are the main conditioning factors for investments in this activity. Besides, countries have already approved laws that establish a minimal percentage for the biodiesel and diesel mix (5 percent in Paraguay and Uruguay, 6 percent in Brazil and 9 percent in Argentina for 2014), guaranteeing a safe and structured market for the companies. In parallel, other public instruments for firms dedicated to agrofuel production were developed, such as tax-distinction/exemption, reduction in export rates, better financing conditions and the secured participation in auctions for sales. In the case of Brazil, these advantages are expanded when family farming is present in the productive chain (Wilkinson and Herrera 2010). In addition to the economic advantages the energy production aggregates, the involvement with this market allows the company to bring closer to its image themes of sustainability, green economy, social inclusion of family farming, etc. (McMichael 2010). In addition to biodiesel industries (concentrated in Brazil and Argentina), ABCD companies started investing in sugarcane ethanol, especially in Brazil (BiodieselBR 2013).

In spite of the vertical integration movements, ABCD companies are developing partnerships to expand even further their power and market share. Several examples of cooperation among ABCD can be mentioned, such as the articulation of Bunge, Cargill and Bunge (in addition to Brazilian Amaggi) in the constitution of a logistics company intending to participate in Brazil's railroads concession auctions, in order to build and operate new lines of distribution of the production (Estadão 2014). Another partnership was established between Bunge and Cargill to quadruple the embark capacity of soy and derivatives in Ilhéus, Bahia (Valor Econômico 2014). Another example is the joint venture between Bunge and Dreyfus in Paraguay; the companies united to build an agroindustry for processing soy in Villeta, with processing capacity of 3000 tons per day (Bunge 2010).

These partnerships are being established with large national companies, such as the joint venture between Dreyfus and Amaggi[8] to act with soy in some Brazilian states (Bahia, Maranhão, Piauí and Tocantins). In parallel, they are partners in the Terminal de Grãos do Maranhão (Tegram), in Itaqui Port (Brazil). Bunge and Amaggi, in addition to being partners in the project of the Terminal Graneleiro de Grãos (TGG) in Santos/São Paulo, signed a contract in 2003 that allows Amaggi to rent a share of its capacity at the Itacoatiara/Amazonas port to Bunge, enabling the transnational to distribute its products in Northern Brazil. By the end of 2013, both companies had released a new joint venture (Navegações Unidas Tapajós Ltda. – Unitapajós) that will be responsible for the distribution of grains from Mato Grosso by the waterway Tapajós-Amazonas up to Santarém/Pará (Valor Econômico 2014). In Argentina such partnerships are also happening, such as the union between Bunge and Aceitera General Deheza[9] in the construction of a corn-based ethanol plant (Bunge 2012). In Uruguay, Cargill and national Cereoil[10] formed a joint venture, becoming Crop Uruguay SA (Menéndez and Gulla 2013).

These partnerships among transnational and national leader companies show a shift in the relationship among firms: since the middle 2000s the former acquired or merged with the latter, and currently there are cooperation strategies (such as joint ventures). These examples show that in spite of 'disputing' the same market, some strategies of common interest end up being implemented collectively, expanding the large corporations' power in the face of the producers and minor companies. As the president of Amaggi Group stated in the journal *Valor Econômico* (2014), 'nowadays we are more ready to establish partnerships. In spite of being a family company, we have learned to deal with the competition'.

On the local level, these partnerships become even more evident. With this market strongly concentrated in a reduced number of firms, this situation enables cooperative relations among the different leader tradings. During field research in Mato Grosso (Brazil), it was possible to see that the exchange of information among the managers of different companies is very frequent and aims at mapping the amount produced in the domain of the companies, the individual demand of firms and the contracts each one of them has established with the producers. In this way, they build a framework with the available amount and its demand. Even though it was not possible to investigate this theme further, there are indications that these companies 'fraction' among themselves the local production based on the specific necessity of each one of them to achieve their goals, avoiding an increase in the product's regional price.

Besides, each firm has a register of the economic situation of the farmer (debts, payment conditions, creditors, regularity of loan payments, financial situation, real guarantees, etc.). This register is circulated among the companies. The negotiation with the producers only

[8]The Amaggi Group is a Brazilian company that started its activities in Paraná in 1977. Today it is the main firm with national capital in the soy market. The group is involved in different stages of the productive chain: agriculture (it planted 225,000 hectares for the 2012/2013 harvest), seed production, fertilizer distribution, crushing, trade, water transport and energy production. In addition to Brazil, it acts in Argentina, Paraguay, Switzerland, the Netherlands and Norway (Amaggi 2014). Blairo Maggi, the group's founder's son, was the governor of Mato Grosso state between 2013 and 2010, and was later elected senator.

[9]Aceitera General Deheza (AGD), created in Argentina in 1948, has industries for the production of oil, soybean meal and biodiesel, and also warehouses and port terminals aimed at exporting (AGD 2014).

[10]Cereoil was started in Uruguay by the 2000s, and its main business activities are the purchase of cereal and its exportation (Menéndez and Gulla 2013).

happens after they map the producer's background and other available information, as will be shown in the next section. There is also a blacklist that circulates among the firms' local managers, in which they track the farmers considered 'problematic' – that is, 'those that did not meet the contracts with tradings, diverting the production, questioning contracts, etc.' (Fernández 2009, 88).

As the regional manager of a large transnational company stated, 'the producer thinks we are enemies that also buy grain. But no. [...] We help ourselves so that we don't fall with a problematic producer'. Another manager stated: 'what doesn't fit me, doesn't fit my competitor either'. To broaden relations among the ABCD firms' employees in the municipalities, they organize 'soccer games only with the companies that buy soy and corn, without [the presence of] the producer and the [input] salesmen'. The statements make clear the different local initiatives of cooperation and partnerships that are developed among the tradings so that the price of the product they purchase does not increase, as well as protecting themselves from 'dishonest' strategies of the rural producers.

Neil Fligstein (1996, 2001), based on a political-instrumental perspective, states that in order to reduce the risk of competition for prices, actors use integration and diversification. Integration can be vertical (mergers with buyers or suppliers) or horizontal (fusion with competitor firms). It provides control of the market by a reduced number of companies and it eliminates a meaningful share of the threat of price decrease. Diversification, in its turn, seeks to minimize the dependency from a single product, expanding their means for survival. As highlighted earlier, these movements have occurred intensely during the last 20 years through vertical and horizontal integration of the soy market, and the firms have diversified their products (*in natura*, soybean meal, oil for human meals, biodiesel, etc.) and their destination (domestic market and exports).

Another important way to reduce the risk of competition by prices, argued by Fligstein (1996, 2001), is through cooperation among competitor companies, in such a way that they share markets. This presupposition breaks with the perspective of anonymous actors and indicates that stability can only be reached if they consider one another, understanding each other's goals, interests and values. The author denominates this strategy 'non-competitive forms of competition', considering that the major companies defend this position, most of the time, with collective and cooperated strategies – as could be seen in the partnerships among tradings. In this sense, ABCD firms try to create stable worlds through social solutions for competition.

To broaden even further their control over the cereal trade, ABCD firms search to influence and control the agrofood sector regulatory framework. According to Murphy, Burch, and Clapp (2012), companies do this via different channels, with direct pressure over governments (putting senior employees in high positions with an elevated degree of decision-taking, and/or employing those people to lobby for them). Besides, the companies are part of the Round Table on Responsible Soy Association (RTRS) that was started in 2004, and was intended to be a global platform composed of the main interested parties in the soy value chain. In it, ADM, Bunge, Cargill and Dreyfus try to lay down private rules and norms for soy (Wilkinson 2011). As in the case of biofuels, firms use their participation in RTRS to link their image with good farming practices, better work conditions, environmental responsibility, etc. Because it is an arena that also aggregates civil society and rural producers, they also try to show their predisposition to discuss strategic themes with other actors involved with and impacted by soy.

In sum, ADM, Bunge, Cargill and Dreyfus seek to control farming produce, prices, trade, logistics, financial inputs and regulatory framework as much as they can, defending themselves against future uncertainties. Beyond the already mentioned strategies, access to inside information results in an important asset. The fact that these firms control a large share of the world trade guarantees access to data they need (offers, demands, risks, etc.) while they promote a regulatory framework that benefits their business. It is important to highlight that when there is privileged knowledge about future situations, price volatility ends up being positive for companies that have access to that information (Murphy, Burch, and Clapp 2012).

Finally, it is important to highlight that most of the ABCD companies' strategies are exclusive to neither the Southern Cone nor the soy productive chain. On the contrary, they refer to initiatives and actions that are present in the different countries and sectors in which ADM, Bunge, Cargill and Dreyfus are involved.

In the next section, the insertion and performance of these firms in the local space is discussed, understanding that some processes can only be perceived, diagnosed and problematized in this field of analysis (Raynolds 2014). As such, it is approached by studying the way that the trade relations are constructed and maintained among ABCD companies and rural producers. This debates seeks to understand how global practices are adapted at different scales, forming hybrid organizations.

Trade relations among ABCD companies and rural producers

As could be seen above, a large part of the strategies developed by ABCD companies are similar. During field research in Mato Grosso, the companies themselves recognized that 'the credit policy is the same, the interest rate is the same and the soy price changes little, sometimes is higher, sometimes other companies' are higher. [. . .] Everything is very similar, not many things change'. This similarity between the companies' actions (added to the fact that soy is a commodity, has a reduced differentiation in its products and presents an internationally standardized price) raises a question: what makes a rural producer choose between one or another company when he sells his soy, for example?

The vertical integration, commented on in the previous section, ends up being decisive in this choice because it is an alternative that the company constructs and that aims, among other goals, at reducing the options for rural producers, making them depend on a single firm to keep a continuous relationship between them and not enabling more independent actions by the soy producers. In spite of this, vertical integration does not eliminate options for the producers, especially at the beginning of a new harvest.

For some authors (Paes Leme and Zylbersztajn 2008; Mazzoleni and Medeiros 2011; Monteiro et al. 2012, among others), closer to the transaction costs economics, the producer selects the firm basically by the price and by the conditions and terms of the contract (payment forms, deadlines, guarantees, etc.). However other studies have pointed that the relations built among different actors in the commodities market are much more complex and include, in addition to legal and economic elements, social, cultural, political, historic and spatial aspects (Smith 1985; Adams 2010; Hoelle 2012; Ofstehage 2013). In the interviews with soy producers the following question was asked: Why do you deal with company X? The answers were multiple. Most of the time, they didn't cite price and terms of the contract as the main motivation, or when these came up initially they were accompanied by other variables.

Among these variables, trust came up in all of the interviews, and the necessity for it was justified in the face of a previous context marked by several problems (not meeting previous agreements, default risks, etc.). According to a trading manager,

> At many times, when we trust the producer, some three days before the distribution I pay 100% of the production. When it's near the distribution and he needs some money, you can pay and send the money straight to his account. But what about the guarantee? '*É no bigode*' [a local expression meaning that he trusts his/her word], you only call to deposit, write a promissory note, and that's it. [...] It took the document [formal contract] four month to be ready, but the producer already had his money.

It is a consensus, however, that the appearance of legal instruments (contracts) has also brought more stability to the relationship between suppliers and clients. Still, there are transactions of an elevated financial value that are not made without the use of contracts, and there are actors who do not deal with some specific people or companies even with the whole contract apparatus is formalized. In this sense, it takes more than a contract to implement the deal, and trust and mutual creditability relations are always considered and appreciated by both parties. As Weber (1978 [1922]) stated, the importance of legal rules (contracts) when conducting economic transactions must not be exaggerated because sometimes it has less strength than social convention and moral rules. Therefore, the relationship between producers and companies is not limited by a written legality, occurring as a kind of interplay of mutual obligations based on trust between producers and certain company directors.

In spite of the trust in the company, the main locus of creditability is the figure of the manager, the technicians and the salesmen. Even though these employees have to know and obey the firm's conditions, they have some room for manoeuvre. Specific actions escape the normativity of the companies, creating informal rules that, rather than following the bureaucracy, adopt alternative paths through which oral compromises gain much importance because it is personalized trust relation in which the guarantee of someone 'giving their word' reinforces the credibility of the involved parties (Almeida 2013). In sum, it is about direct relationships based on personal/social guarantee (or, as an interviewed manager said, '*é no bigode*').

This trust relation, in some moments, ends up becoming a friendship because the company representatives build strategies to be closer to the farmers and to create a certain link, through fishing activities, cards and soccer games, field days, home visits, etc. According to Almeida (2013), salespeople are conscious that their personal behaviour results in business dividends, and they spare no effort in visiting their clients. Besides, the managers frequently visit technical commercial events as well as commemorative parties of the city's anniversary. They also participate in parties held in rural communities, donating resources for the promotion of events.

Boundaries between business and friendship are very sinuous in the relationships among producers and companies. But, in addition to friendships, family links gain importance because some negotiations occur due to some relative who works in the company. This makes part of the purchases or sales occur through the firm.

Besides, there is a kind of reciprocity logic among soy farmers and companies/managers/technicians. For a transnational company representative, 'there are many producers that keep the same company because it helped him or her in a difficult time, some moment of crisis'. The company, in its turn, values loyal producers, showing recognition and respect in this commercial relation. These elements, commonly called 'help', are very valued by both parties. But as soon as the producer stops selling his produce to the company, this creates resentment. In this sense, there is moral compromise and loyalty that must be cultivated during the years.

In this way, the history of the relationship is an important element in negotiations. In some cases, the sale of soy, for example, is immersed in a context of continuous interaction between producer and company, because frequently this relationship has lasted more than 30 years (it occurred via some national company, acquired by ADM, Bunge, Cargill or Dreyfus, but one that maintained the same manager who made relationships last for decades). It is interesting that this link is maintained even when some producers move from their region, and it also transcends generations: 'there are producers whose father we worked with in Paraná and when they came to Mato Grosso they kept selling to us'. These are links that are built and maintained during the years in spite of time and spatial shifts.

It also seemed fundamental for some producers that companies, managers and/or technicians value transparency (clarity in the business deals of the companies) and honesty (meeting pre-established agreements, independently of the institutionalization of the contract). The conditions of the service are also important, where the company needs to understand the features of each producer and his or her preferences and priorities to create a personalized relationship. Still related to the service conditions, producers cited ease and convenience as elements they consider relevant to define who to deal with, especially due to obtained benefits (less bureaucracy in access to funding), transaction simplification (verbally defined agreements with contract signature only occurring later on) and the possibility of getting more financial returns (better prices when a large volume of production is delivered).

The status that a certain relationship offers is also an important aspect in soy farmers choosing a firm. This can be seen when some companies give a prize for their 'best clients' (for their productivity, respect for the companies' norms, purchase of total amounts of inputs from the same firm, etc.) with a trip (frequently to the United States so that they can visit a region known for soy production) or with some hobby activities (in Mato Grosso, they are generally fishing activities). In this context, to receive a 'prize' aggregates a distinction in face of producers 'who were not selected for it'. In this sense, these are activities that give prestige to chosen farmers.

ABCD companies were also asked about the presence of a soy farmer profile that is more suitable to deal with and the criteria for selecting them (especially in the case of funding and early sale of products and services, the more risky transactions[11]). The mains aspects that came up in the interviews related to the economic situation of the producer, presence of machinery and land property (the size of the area must meet the volume of resources involved in the transaction). These dimensions are analyzed so that companies are secured when there is a problem in the relationship, having real guarantees. Soil and climate conditions are also checked, prioritizing flat, fertile areas with a good rate of rainfall because these features interfere in the development of the farm activity. Companies are also concerned with the background of the producer as regards the payment of debts, meeting contracts, etc. Besides, negotiations are built and assessed based on trust and reciprocity relations, the history of the relationship with the client, and their moral behaviour (honesty, loyalty, seriousness, etc.). In all interviews, the claim 'we don't deal with people we don't know' was unanimous.

In spite of soy being characterized as a standardized market (commodity), with an internationally established price, based on a global trade, dominated by transnational companies

[11]Obviously when the negotiation is specific as with prompt payment, the criteria do not gain relevance. However, most transactions do not occur this way currently.

and meaningful participation of large rural producers, this does not mean that this market is protected from local dynamics and immune to the influence and interference of social, historical, political and cultural issues. As several authors have pointed out (Gudynas 2008; Adams 2010; Peine 2010; Hoelle 2012), the soy market is not unrelated to social relations, cultural meanings, and ethical and moral values. Ofstehage (2013), analyzing the production of this grain by North American producers in Brazil, also highlights the importance of values, practices and social relations in the intensive, technified and financialized contemporary agriculture.[12]

The fact that the ABCD transnational corporations are linked with global strategies of their head offices, acting in different countries to maximize their final profitability, does not entail a complete disembeddedness of their actions. On the contrary, so that higher financial returns and incomes may be achieved, it is fundamental for these companies to be on a local basis through their affiliates. This presence occurs in different manners, from the physical presence (storage, silos, logistic structure, etc.) up to the employment of workers who are known in the regions and who have a large network of contacts, giving the company a good reputation. Not less important is the participation of technicians and managers in fairs, parties, balls, etc. In sum, the transnational firms themselves need to act locally to feed their high degree of insertion in global markets. The power of the large companies is, precisely, in articulating their strategies of accumulation in different scales, something that is difficult to achieve for small firms acting only at local and province levels.

This discussion brings us to the debate of 'hybrid organizations',[13] that has as one of its main focuses the understanding of how the transnational companies' global practices are mixed, modified and adapted, and gain new meanings when interactions between actors and the contexts present in different scales occur (Shimoni 2008). In this sense, this literature covers the governance models used by the companies and how specific patterns and norms, when implemented and disseminated, are influenced by the local reality composing hybrid forms (Wood Jr. 2010). Therefore, this framework searches to understand precisely this 'adjustment' of companies, especially the transnational ones, in activities, regions and cultures that are far from their local origin, implying, almost invariably, the adaptation of forms of governance (Brandsen, Karre, and Helderman 2009; Islam 2012).

The analysis of the 'adjustment' and 'immersion' of the transnational companies in the social local basis is interesting precisely because it shows contexts and situations that are very specific, creating norms and strategies in determined regions. An example of this is the criteria established by the ABCD companies for producers to access credit or to early-purchase inputs (when the deadline for payment is during the harvest). In a general way, companies have prerequisites that are similar in the different countries where they are located, and these are based, almost invariably, in tangible assets: the economic situation of the producer, presence of machinery, land property, etc.

However, a large number of local managers end up aggregating new criteria and they start selecting their clients based on local reputation, so that they avoid market relations with an eventual 'problematic client'. In the case of Mato Grosso, criteria that boost the

[12]This relation of interdependency between economic actions and other dimensions of society is close to discussions brought by classic authors in sociology (Weber 1978 [1922]; Durkheim 2014 [1893]), and by those from the new economic sociology (Granovetter 1985; Zelizer 1997; Fligstein 2001, among others).

[13]In the agrarian theme, the debate around hybrid livelihoods has gained attention focused especially on the relation of producers and rural families with the urban space, labour market, and global capital logics (Fairbairn et al. 2014; Hecht 2014; Toit and Neves 2014).

'good client' label were linked with the fact that he or she comes from a known family (as an interviewed person stated 'seriousness and honesty come from the cradle'); comes from the South of the country (which supposes some knowledge about market and production techniques); is loyal to deadlines, agreements and a single firm (not shifting to other companies frequently); is not involved in fights and problems in the community; participates in local events; and does not engage in morally reprehensible procedures, from financial investments (such as building a house before paying the creditors) to having addictions, etc.

For the ABCD firms' local representatives, the partial meeting of the requirements, as well as the trust relations among the involved parties, allows the establishment of verbal contracts (that are later 'legalized' to meet the demands imposed by the company's head office). In this sense, perceptions, values, qualifications and reputations based on multiple components are reflected in the way that business is held, carried out, and maintained on a local basis. And, more than rules, standards and concepts defined by the companies, these are the technicians', managers' and salespeople's perceptions and viewpoints in each region. The fact that the ABCD transnational companies are in dialogue with the values present in local social structures and with the daily practices of their clients points to the hybridity of these firms, whose negotiations are built in accordance with the social and cultural relations specific to each context.

Final comments

The results of this research show how quickly and intensely ADM, Bunge, Cargill and Dreyfus expanded throughout Brazil, Argentina, Paraguay and Uruguay. While at the beginning of the 1990s these four companies still had a small share in the south American market, they are currently the main firms, dominating around 50 percent of the four countries' installed crushing capacity and 85 percent of the soybean exportation (and 10 percent of the value of total exports in the Southern Cone). In the export sector, although processed products (oil and meal) are marketed, their major power is in grain sales, of which they control around 85 percent of the total. In addition, they account for 10 percent of the Southern Cone total sports. These numbers show the rapid and intense denationalization process of the firms in the soy productive chain, as well as the level of market internationalization and business concentration.

Through the last two decades, there has been a rapid and intense reconfiguration in soy agribusinesses, which reached a magnitude, proportion and dimension unknown in the last stages of the chain on the Southern Cone. In this scenario, Bunge, Cargill, ADM and Dreyfus became the four major grain crushers, besides acting in input production and sales, financial offers, technical advice, grain purchase, storage, industrialization, exports and domestic market sales. The proportion assumed by the companies of the different segments of the productive process suggests the constitution of a 'chain oligopoly' – that is, a small group of firms' control over a set of consecutive stages in the same sector (soy complex). To sustain and make moves in this 'oligopoly in chain' on a global scale, they have appealed to resources in the financial market. Another strategy that draws attention is the initiatives of partnerships among the ABCD firms, in addition to the cooperation with leader national companies. Another line of action is the agrofood sector regulatory framework, which they try to influence and control in different ways.

In sum, these are firms with a global presence, whose productive processes are internationalized, and whose organization is structured in corporations or conglomerates around the world. In spite of this, this work sought to show the centrality that local scale performs (and shall continue to perform) in ABCD companies' strategies, because it is where the relation between firms and producers (especially via technicians and managers) is built, maintained and feedbacked. In this space, in addition to the price and terms of the contract, the soy market is based on structures of regular interaction supported by trust, reciprocity, family links, friendship, loyalty and mutual compromise – even because, in the case of Mato Grosso, since there are small municipalities, this is an activity that has great visibility and 'everybody knows everybody'.

In spite of the movements of globalization, financialization and foreignization of the agrofood commodities markets, local space remains a strategic locus for the functioning of these economic activities. Besides, this study shows that all transnational power of the ABCD companies, which seems to be so abstract and intimidating when seen in global scale, depends, in its basis, on the formation, maintenance and exploration of a relation of proximity, trust and reciprocity with local actors (especially rural producers), going through family and friendship links.

With the current sectorial context – characterized by the increase in Southern Cone soy production, a growing international demand and a price higher than the historical average – it is most likely that the ABCD companies' power remains strong, and is even intensified with the processes of market internationalization, business concentration and firms' denationalization in the soy productive chain.

Despite the ABCD supremacy, some nationally owned firms have maintained and, in smaller numbers, have expanded their business, appearing amongst the main companies in the soy complex. This is the case for: (1) Aceitera General Deheza (AGD), Molinos Rio de la Plata and Vicentin in Argentina, which owns 22.7 percent of the crushing capacity and is responsible for more than 30 percent of oil and meal exports (Hinrichsen 2013; MAGyP 2013); (2) Barranca Erro and Cereoil in Uruguay, which in the last few years has controlled around 30 percent of the soy exports (COMEX 2013); (3) Amaggi and, to a lesser extent, Caramuru and Imcopa in Brazil, which own 10 percent of the soy-crushing capacity (Abiove 2014). The only exception happens in Paraguay, where the soy complex is dominated by the leader transnational companies and by Brazilian and Argentinean groups (Rojas Villagra 2009). However, as this paper highlighted, the presence of important national capital firms does not mean, necessarily, a free competition with ABCD companies for the same market, because the construction of cooperated initiatives in specific sectors and regions ends up limiting the entrance or growth of other companies.

Despite this, there is a growth of Asian investments in the last links of the soy chain – although these have not reached the current ABCD level in the Southern Cone (Oliveira and Schneider 2014). The Chinese state company Cofco, which since the beginning of 2014 has had a majority control of Noble and Nidera, is expanding in the four analyzed countries, especially in Argentina, where the sum of the values of these two companies corresponds to 10.7 percent of the installed crushing capacity, and an important part of the exports (10 percent of meal, 13 percent of oil and 15 percent of grain) (INDEC 2013; CIARA 2014). At the same time, there was an entry of Japanese trading to the sector, mainly in Brazil, such as Marubeni (through Gavilon), Sojitz (through CGG Trading and Cantagalo), Mitsui (through Multigrain) and Mitsubishi (through Ceagro and Sollus), which perform in the last links as well as in agricultural production, controlling more than 300,00 hectares of arable land (Valor Econômico 2014).

Another novelty is the capitalization of soy producers in the last few years, stimulated by the meaningful increase in the farm commodities price and by the financialization of agribusiness activities. This enabled some producers to invest and act more and more in activities out of the farm, such as the production and/or sale of inputs, storage, processing, trade, funding, etc. (Gudynas 2008; Deininger, Nizalov, and Singh 2013; Wilkinson and Pereira 2014). This applies to large groups of producers, as in the case of El Tejar, Los Grobos, Adecoagro, MSU and Cresud that controlled more than 2.5 million hectares in Brazil, Argentina, Paraguay, Bolivia and Uruguay, whose largest share of areas was destined for soy cultivation (Wesz Jr. 2014). These groups, because they mobilize a large volume of production, export directly through their own companies, not passing through ABCD (Guibert et al. 2011; Gras and Hernández 2013). In Mato Grosso, this strategy has also grown with the major rural producers' cooperatives that sell their produce to Asian buyers directly, even undertaking the logistic and transportation responsibility. In this sense, a process of verticalization of agriculture actors is evident, since they are searching to broaden their power and create more independence from the tradings, undertaking functions that were dominated almost exclusively by ABCD.

Finally, it is important to highlight that, even though this was not the focus in this work, different themes (food distribution, access to land, natural resources degradation, climate change, work conditions, environmental laws, agriculture price volatility, maintenance of family farming in the rural milieu, etc.) are directly affected by the distinct activities and strategies of the ABCD firms. In this sense, when thinking of rural development and food sovereignty, it is fundamental to bring to the debate the functioning of the current agro-food system, where ADM, Bunge, Cargill and Dreyfus play a central role.

Acknowledgements

This contribution presents some of the results of my doctoral dissertation (Wesz Jr. 2014). I would like to thank the National Counsel of Technological and Scientific Development (CNPq) and the Foundation for Research Support of the State of Rio de Janeiro (Faperj) for a scholarship and Sergio Pereira Leite by orientation for the supervision. I would also like to thank Susanna Hecht, Gustavo Oliveira and the reviewers who commented on this paper. Finally, I thank Ana Isabel Márquez Pérez and Bruno Azevedo Prado for translating the paper.

Disclosure statement

No potential conflict of interest was reported by the author.

References

Abiove – Associação Brasileira das Indústrias de Óleos Vegetais. 2014. *Capacidade Instalada da Indústria de Óleos Vegetais*. http://www.abiove.com.br

Adams, R. 2010. *Elite landowners in Santarém*: ranchers, gaúchos and the arrival of soybeans in the Amazon. PhD diss, Indiana University.

AGD – Aceitera General Deheza. 2014. *Nuestra identidad*: história. http://www.agd.com.ar/es/home.htm

Ahmed, G., D. Hamrick, and G. Gereffi. 2014. *Shifting governance structures in the wheat value chain: implications for food security in the Middle East and North Africa*. Center on Globalization, Governance & Competitiveness, Durham: Duke University.

Albuquerque, J.L.C. 2005. *Fronteiras em movimento e identidades nacionais*: a imigração brasileira no Paraguai. PhD diss., Universidade Federal do Ceará, Fortaleza.

Almeida, L. S. 2013. *Gaúchos, festas e negócios*: o agronegócio da soja no Meio-Norte matogrossense. PhD diss., Universidade Federal do Rio de Janeiro.

Amaggi. 2014. *Sobre o Grupo Amaggi*. http://www.grupoandremaggi.com.br

Arroyo, G., R. Rama, and F. Rello. 1985. *Agricultura y alimentos en América Latina: el poder de las transnacionales*. México D.C.: Universidad Nacional Autónoma del México.

Barbosa, M.Z. and S. Nogueira Jr. 2007. *As simetrias entre as agroindústrias de soja no Brasil e na Argentina. Revista de Economia Agrícola, São Paulo* 54: 87–107.

Benetti, M.D. 2004. Globalização e desnacionalização do agronegócio brasileiro no pós 1990. *Documentos FEE*, 61. Porto Alegre: FEE.

BiodieselBR. 2013. Notícias. http://www.biodieselbr.com/

Blanc, J. 2015. Enclaves of inequality: Brasiguaios and the transformation of the Brazil-Paraguay borderlands. *The Journal of Peasant Studies* 42, no. 1: 145–58.

Bonanno, A. and D.H. Constance. 2008. *Stories of globalization: transnational corporations, resistance and the corporate state*. University Park, PA: Pennsylvania State University Press.

Borras Jr., S., P. McMichael, and I. Scoones. 2010. The politics of biofuels, land, and agrarian change: editors introduction. *Journal of Peasant Studies* 37, no. 4: 575–92.

Brandsen, T., P. Karre, and J. K. Helderman. 2009. *The risks of hybrid organizations*: expectations and evidence. NISPACEE Conference. Budva.

Bunge. 2010. *Bunge y Louis Dreyfus Commodities anuncian una inversión conjunta*. http://www.bungeuruguay.com/ing/images/Prensa_Nov3_2010.pdf

Bunge. 2012. *Promaíz S.A.* http://www.bungeargentina.com/site/es/novedades/promaiz-sa

Capeco – Camara Paraguaia de Exportadores y Comercializadores de Cereales y Oleaginosas. 2013. *Estadísticas*. http://www.tera.com.py/capeco.

Castro, A.C. 2002. *Localização e identificação das empresas processadoras de soja, suas áreas de influência, preços e custos de transporte relacionados*. Rio de Janeiro: CPDA/UFRRJ.

Ciara – Camara de la Industria Aceitera de la Republica Argentina. 2014. *Estadísticas del Sector*. http://www.ciaracec.com.ar/ciara/bd/index.php

CIP – Centro de Importadores del Paraguay. 2013. *Ranking de exportadores e importadores*. http://www.cip.org.py/

Clapp, J. and M.J. Cohen, eds. 2009. *The global food crisis: Governance challenges and opportunities*. Waterloo, Ontario: Wilfrid Laurier Univ. Press.

Clapp, J. and D. Fuchs, eds. 2009. *Corporate power in global agrifood governance*. Boston: MIT Press.

Cohen, S.D. 2007. *Multinational corporations and foreign direct investment: avoiding simplicity, embracing complexity*. New York: OUP Catalogue.

Colque, G. 2013. Land appropriation on the frontier: changes in and struggles for access to land in Bolivia. PhD diss., International Institute of Social Studies.

Comex. 2013. *Relatórios de Comércio Exterior*. http://trade.nosis.com/pt/Comex.

Conab – Companhia Nacional de Abastecimento. 2013. *Série Histórica de Produção*. http://www.conab.gov.br

De La Torre, A. and S.L. Schmukler, eds. 2007. *Emerging capital markets and globalization: the Latin American experience*. Washington, DC: World Bank Publications.

Deininger, K., D. Nizalov, and S.K. Singh. 2013. Are mega-farms the future of global agriculture? Exploring the farm size-productivity relationship for large commercial farms in Ukraine. Policy Research Working Paper 6544, The World Bank.

Delgado, G.C. 2005. A questão agrária no Brasil, 1950–2003. In *Questão social e políticas sociais no Brasil contemporâneo*, ed. L. Jaccoud, 51–90. Brasília: IPEA.

Du Bois, C., C.B. Tan, and S. Mintz, eds. 2008. *The world of soy*. Urbana: University of Illinois Press.

Durkheim, E. 2014 [1983]. *The division of labor in society*. New York: Simon and Schuster.

Embrapa – Empresa Brasileira de Pesquisa Agropecuária. 2004. *Sistema de Produção 6*. http://www.cnpso.embrapa.br/download/publicacao.

Estadão. 2014. *Gigantes do agronegocio se unem para investir nas novas ferrovias*. http://economia.estadao.com.br/noticias/geral,gigantes-do-agronegocio-se-unem-para-investir-nas-novas-ferrovias-imp-,1144031

Fairbairn, M. et al. 2014. Introduction: New directions in agrarian political economy. *Journal of Peasant Studies* 41, no. 5: 653–66.

Faostat – División de Estadísticas de la Organización de las Naciones Unidas para la Alimentación y la Agricultura. 2014. *Estadísticas generales*. http://faostat.fao.org/.

Farina, E.M.M.Q. and D. Zylbersztajn. 1998. *Competitividade no Agribusiness Brasileiro*. São Paulo: IPEA.

Fernandes Filho, J.F. and Belik, W. 2010. *A política de tributação na exportação do complexo soja pelo Brasil*: transformação e resultados. Paper presented at the annual meeting for the Brazilian Society of Economics, Business and Rural Sociology, in Campo Grande, Brazil.

Fernandes, B.M., C.A. Welch, and E.C. Gonçalves. 2010. Agrofuel policies in Brazil: paradigmatic and territorial disputes. *Journal of Peasant Studies* 37, no. 4: 793–819.

Fernández, A.J.C. 2007. *Do Cerrado à Amazônia*: as estruturas sociais da economia da soja em Mato Grosso. PhD diss., Universidade Federal do Rio Grande do Sul.

Fernández, A.J.C. 2009. *Sociedade e economia do agronegócio: estudo exploratório do estado de Mato Grosso*. Rio de Janeiro: CPDA/UFRRJ.

Fligstein, N. 1996. *Markets as politics*: a political-cultural approach to market institutions. *American Sociological Review* 61: 656–73.

Fligstein, N. 2001. *The architecture of markets: an economic sociology of 21st century capitalist societies*. Princeton: Princeton University Press.

Friedmann, H. and P. McMichael. 1989. Agriculture and the state system: the rise and decline of national agricultures, 1870 to the present. *Sociologia Ruralis* 29, no. 2: 93–117.

Gaspari, C.A. 2004. *O paradigma biotecnológico e o processo de reestruturação das grandes empresas*: o movimento das bioindústrias na agroindústria brasileira. PhD diss., Universidade Estadual de Maringá.

Genok. 2012. *Producción de soya en las Américas: actualización sobre el uso de tierras y pesticidas*. Cochabamba: Virmegraf.

Gereffi, G., J. Humphrey, and T. Sturgeon. 2005. The Governance of Global Value Chains. *Review of International Political Economy* 12, no. 1: 78–104.

Gilpin, R. 2011. *Global political economy: Understanding the international economic order*. Princeton: Princeton University Press.

Giménez, E.H. and A. Shattuck. 2011. Food crises, food regimes and food movements: rumblings of reform or tides of transformation? *Journal of Peasant Studies* 38, no. 1: 109–44.

Granovetter, M. 1985. Economic action and social structure: the problem of embeddedness. *American Journal of Sociology* 91, no. 3: 481–510.

Gras, C. and V. Hernández. 2013. Los pilares del modelo agribusiness y sus estilos empresariales. In *El agro como negócio: produción, sociedad y territórios en la globalización*, ed. C. Gras and V. Hernández, 17–48. Buenos Aires: Biblios.

Gudynas, E. 2008. The new bonfire of vanities: soybean cultivation and globalization in South America. *Development* 51, no. 4: 512–18.

Guibert, M. et al. 2011. De Argentina a Uruguay: espacios y actores en una nueva lógica de producción agrícola. *Pampa* 7: 13–88.

Hecht, S.B. 2014. Forests lost and found in tropical Latin America: the woodland 'green revolution'. *Journal of Peasant Studies* 41, no. 5: 877–909.

Hendrickson, M., H. James, and W. D. Heffernan. 2013. Vertical integration and concentration in US agriculture. *Encyclopedia of Food and Agricultural Ethics* 1, 1–10.

Heredia, B.M.A., M. Palmeira, and S.P. Leite. 2010. Sociedade e Economia do "Agronegócio" no Brasil. *Revista Brasileira de Ciências Sociais* 25: 159–76.

Hinrichsen, J. J. 2013. *Annual Yearbook on Oilseeds Markets*. Buenos Aires, Argentina: Hinrichsen.

Hoelle, J. 2012. Black hats and smooth hands : elite status, environmentalism and work among the ranchers of Acre, Brazil. *Anthropology of Work Review* 33, no. 2: 60–72.

IBGE – Instituto Brasileiro de Economia e Estatística. 2013. Banco de dados agregados. http://www.sidra.ibge.gov.br.

Indec – Instituto Nacional de Estadística y Censos. 2013. Agricultora. http://www.indec.gov.ar/agropecuario/.

INTA – Instituto Nacional de Tecnologia Agropecuária. 2009. *Análisis de la cadena de la soja en la Argentina*. Buenos Aires: INTA.

Islam, G. 2012. Can the subaltern eat? Anthropophagic culture as a Brazilian lens on post-colonial theory. *Organization* 19, no. 2: 159–80.

MAG – Ministerio de Agricultura y Ganaderia. 2013. *Series Históricas de Cultivos Temporales*. http://www.mag.gov.py.

MAGyP – Ministerio de Agricultura, Ganadería y Pesca. 2013. Sistema integrado de información agropecuaria. http://www.siia.gov.ar/index.php.

Mazzoleni, E. M. and J. J. Medeiros. 2011. Governança e coordenação em cadeias de suprimento: proposta de um modelo tecnológico-relacional para análise no agronegócio. Paper presented at the annual meeting for the Brazilian Society of Economics, Business and Rural Sociology, Belo Horizonte, Brazil.

McMichael, P. 2009. A food regime analysis of the 'world food crisis'. *Agriculture and Human Values* 26, no. 4: 281–95.

McMichael, P. 2010. Agrofuels in the food regime, *Journal of Peasant Studies* 37, no. 4: 609–29.

Mecon – Ministerio de Economia y Finanzas Publicas. 2011. *Complexo oleaginoso. Serie Producción Regional por Complejos Productivos*. http://www.mecon.gob.ar/peconomica/docs/Complejo_Oleaginoso.pdf.

Menéndez, V. and J.C. Gulla. 2013. Cadenas Globales de Valor: los casos de la soja y el trigo. Documento de Trabajo, 89, Núcleo de Estudios Sociales Agrarios, Montevideo.

MGAP – Ministerio de Ganadería, Agricultura y Pesca. 2013. Estadísticas Agropecuarias (DIEA). http://www.mgap.gub.uy/.

Monteiro, G. F. A. et al. 2012. The role of empirical research in the study of complex forms of governance in agroindustrial systems. *Revista de Economia e Sociologia Rural* 50, no. 4: 667–82.

Moreno, G. 2007. *Terra e poder em Mato Grosso: política e mecanismo de burla (1892 a 1992)*. Cuiabá/MT: Entrelinhas e EDUFMT.

Murphy, S., D. Burch, and J. Clapp. 2012. *Cereal secrets*: the world's largest grain traders and global agriculture. Oxfam Research Report. Oxford: Oxfam.

Ofstehage, A. 2013. *Development and dispossession in Soylandia*: quotidian realities of the transnational corporate food regime. Society for Economic Anthropology Conference, St. Louis, USA.

Oliveira, G.L.T., and M. Schneider. 2014. The politics of flexing soybeans in China and Brazil. *Think Piece Series On Flex Crops & Commodities* 3: 1–20.

Oyhantçabal, G. and I. Narbondo. 2011. *Radiografia del agronegocio sojero: descripción de los principales atores y los impactos socioeconómicos en Uruguay*. Montevideu: Redes.

Paes Leme, M. F. and D. Zylbersztajn. 2008. Determinantes da escolha de arranjos institucionais: evidências na comercialização de fertilizantes para soja. *Revista de Economia e Sociologia Rural* 46, no. 2: 517–46.

Pappalardo, C. 1995. *Estrategias y políticas de desarrollo rural*. Asunción: El Lector.

Peine, E. 2010. Corporate mobilization on the soybean frontier of Mato Grosso, Brazil. In *Contesting development: critical struggles for social change*, ed. P. McMichael, 132–46. New York: Routledge.

Pierri, J. 2006. El boom de la soja. Un retorno al pasado? *Realidad Económica* 219: 53–63.

Pierri, J. 2008. El desempeño de las grandes empresas y las cooperativas en la "sojización". Los casos de Cargill y de la Asociación de Cooperativas Agrarias. *Documentos del CIEA*, Facultad de Ciencias Económicas de la UBA 3: 69–95.

Piñeiro, D. and M.I. Moraes. 2008. Los cambios en la sociedad rural durante el siglo XX. *El Uruguay del Siglo XX*: La sociedad. Ediciones de la Banda Oriental, Montevideo, 105–136.

Pires, M.J.S. and G.R. Santos. 2013. *Modelo agroexportador, política macroeconômica e a supremacia do mercado: uma visão do modelo brasileiro de exportação de commodities*. Brasília: IPEA.

Raynolds, L. T. 2014. Fairtrade, certification, and labor: global and local tensions in improving conditions for agricultural workers. *Agriculture and Human Values* 31, no. 3: 499–511.

Revista Exame. 2011. *O Paraguai é movido a soja*. http://exame.abril.com.br/revista-exame/edicoes/0986/noticias/movido-a-soja

Rodrigues, W. et al. 2009. Análise das estratégias de financiamento/comercialização dos produtores de soja da região de Santa Rosa do Tocantins. Paper presented at the annual meeting for the Brazilian Society of Economics, Business and Rural Sociology, in Porto Alegre, Brazil.

Rojas Villagra, L. 2009. *Actores del agronegocio en Paraguay*. Asunción: BASE Investigaciones Sociales y Diakonia – Acción Ecuménica Sueca.

Salerno, T. 2014. Capitalising on the financialisation of agriculture: Cargill's land investment techniques in the Philippines. *Third World Quarterly* 35, no. 9: 1709–27.

Santos, B.S. 2002. Os processos da globalização. In *A globalização e as ciências sociais*, ed. B. S. Santos, 2° ed., 25–104. São Paulo: Cortez.

Schavarzer, J. and A. Tavosnanska. 2007. El complexo sojero argentino. Evolución y perspectivas. *Documento de Trabajo*, 10, Centro de Estudios de la Situación y Perspectivas de la Argentina.

Secex - Secretaria de Comércio Exterior. 2013. *Indicadores e Estatísticas de Comércio Exterior*. http://www2.desenvolvimento.gov.br/sitio/secex/secex

Shimoni, B. 2008. Separation, emulation and competition: hybridization styles of management cultures in Thailand, Mexico and Israel. *Journal of Organizational Change*, v. 21, n. 1, 107–119.

Smith, G. 1985. Reflections on the social relations of simple commodity production. *The Journal of Peasant Studies* 13, no. 1: 99–108.

Souza, J.F.D. 2007. *Integração vertical e financeirização*: o caso da agroindústria processadora de grãos no Brasil. Msc. diss., Universidade de São Paulo.

The Guardian. 2011. *The global food crisis: ABCD of food* – how the multinationals dominate trade. http://www.theguardian.com/global-development/poverty-matters/2011/jun/02/abcd-food-giants-dominate-trade.

Toit, A. and D. Neves. 2014. The government of poverty and the arts of survival: mobile and recombinant strategies at the margins of the South African economy. *Journal of Peasant Studies* 41, no. 5: 833–53.

UruguayXXI. 2012. *Trigo y oleaginosas*. Montevideo: Departamento de Inteligencia Competitiva.

Valor Econômico. 2014. Notícias. http://www.valor.com.br

Weber, M. 1978 [1922]. *Economy and society: an outline of interpretive sociology*. Berkeley: University of California Press.

Wesz Jr., V.J. 2011. *Dinâmicas e estratégias das agroindústrias de soja no Brasil*. Rio de Janeiro: E-papers.

Wesz Jr., V.J. 2014. *O mercado da soja e as relações de troca entre produtores rurais e empresas no Sudeste de Mato Grosso (Brasil)*. PhD diss., Universidade Federal Rural do Rio de Janeiro.

Wilkinson, J. 2009. Globalization of agribusiness & developing world food systems. *Monthly Review* 61: 38–49.

Wilkinson, J. 2011. From fair trade to responsible soy: social movements and the qualification of agrofood markets. *Environment and Planning A* 43: 2012–26.

Wilkinson, J. and S. Herrera. 2010. Biofuels in Brazil: debates and impacts. *The Journal of Peasant Studies* 37, no. 4: 749–68.

Wilkinson, J. and P. Pereira. 2014. Brazilian soy: new patters of investment, finance and regulation. In: *Finance, Food and Farmland International Conference*, ISS and CESTRAD. The Hague, Holland.

Wood Jr., T. 2010. Organizações híbridas. *Revista de Administração de Empresas* 50, no. 2: 241–47.

Zelizer, V. A. R. 1997. *The social meaning of money*. Princeton: Princeton University Press.

Disappearing nature? Agribusiness, biotechnology and distance in Argentine soybean production

Amalia Leguizamón

A recent trend identified in the agro-food literature is that financialization in the global food system is further increasing the distance between farm and plate as well as abstracting physical commodities into market derivatives. How does food, a basic material need, become a commodity, a financial asset divorced from its physical form? This contribution explores the growing distance and abstraction of food from farm using Argentina's soy model as a case study. I explore the various drivers of distancing across the soy value chain in Argentina, including industrialization, globalization, corporatization and finance. I argue that the push for technological innovation by large-scale agribusinesses, in articulation with financial sector involvement, are both an example of and are instrumental in the process of distancing and abstraction identified in the agro-food literature. This paper also highlights how, despite agribusiness efforts to 'displace' and 'disappear' nature, these processes are never fully accomplished. I thus reflect on the socio-ecological contradictions that arise from the processes of distancing and abstraction which accompany the financialization of the corporate food system under neoliberal globalization.

1. Introduction

The biotechnology of genetically modified (GM) seeds is portrayed as the solution to a most pressing crisis: How to feed the world with increasingly scarce resources on an overpopulated, overheated planet (James 2014). In Argentina, the third largest global grower of GM crops (James 2014), this discourse has taken root (see Newell 2009). Agribusiness leaders express pride that, thanks to the adoption of the technological package of GM soy, Argentina has become a key global food producer (Grobocopatel 2010). On the other side, critics argue that the GM soy model is causing harmful effects on society and the environment (Giarracca and Teubal 2010). The humanitarian discourse of agribusiness masks the fact that technological innovation in soybean production, like the agro-industry in general, is motivated by profit maximization (Magdoff, Foster, and Buttel 2000; Magdoff and Tokar 2010). Soybeans have become particularly lucrative since the turn of the twenty-first century, as global demand grows and commodity prices rise (Teubal 2006, 2008). The trade – and growing speculation – in soy as a commodity in the global financial market puts into question the agribusiness claim to increase soy production to feed the world (Russi 2013).

How does food, a basic material need, become a commodity, a financial asset divorced from its physical form? This contribution explores the growing distance and abstraction of food from farm using Argentina's soy model as a case study. In the agro-food literature, 'distance' refers to the separation between farm and plate: the disconnection between decisions over natural resource use and over consumption (Kneen 1993; Princen 1997, 2002; Clapp 2014, 2015). This distance may be geographical (for example, when grapes are produced in Chile but consumed in the United States) or cultural, when consumers are ignorant of the social and environmental impacts of food production (e.g., when US shoppers are shielded from knowledge about the labor conditions of the grape growers).[1] In our globalized food system, farm-to-plate distance continually increases as multiple actors form new links in the commodity chain (Friedmann 1994; McMichael 2013). Increasing distance increases power asymmetries between producers and consumers, which allows producers to externalize and render invisible the social and ecological costs of their decisions (Princen 1997, 2002; Gould 2006).

As Clapp (2014, 2015) observes, the entry of financial capital into the global food system compounds 'distancing': financial actors and institutions – such as banks, hedge funds and mutual funds – enter commodity chains with a primary goal not of food production but of generating high returns. In financial markets commodity trading is devoid of real commodities; goods are not physically exchanged as they would be in, say, a local farmers' market. Derivatives, financial assets that derive their value from the performance of other underlying assets, are the extreme disembodiment of commodities (Russi 2013). This is how heightened financialization of agro-food systems 'abstracts' physical commodities and exacerbates distancing (Clapp 2014, 2015; Isakson 2014).

In the last two decades, Argentina has positioned itself in the global food system as a strategic provider of soybeans for the livestock complex (Teubal 2006, 2008; McMichael 2013). This is the result of a process of neoliberal agrarian restructuring that took place in the 1990s and accelerated drastically with the introduction of Roundup Ready™ (RR) soybeans in 1996 (Teubal 2006). Multinational seed and agrochemical giant Monsanto engineered RR soybeans to tolerate spraying with the glyphosate-based herbicide Roundup, which it also manufactures. The 'technological package' consists of planting RR soybeans using a no-tillage method and spraying with glyphosate as a primary weed killer. Argentine farmers were early and rapid adopters of the technological package of GM soy (Trigo and Cap 2003). Nowadays, 100 percent of soy planted in Argentina is RR (Trigo 2011).

The consequence has been a soy boom. Since 1997, an average of 726,000 hectares have been added to soy production yearly, reaching close to 20 million in 2015, or 52 percent of Argentine land under cultivation (MAGyP 2015). Record harvests have also been reported each year (Leguizamón 2014a), peaking at 61.4 million tons in 2015 (MAGyP 2015). Yet this production is not for domestic use. Ninety-six percent is exported as beans, meal, oil or biodiesel (ACSOJA 2015), making Argentina the third largest global producer and exporter of soybeans, after the United States and Brazil (USDA 2015).

At the forefront of this agrarian transformation is a relatively new actor in Argentine agriculture, 'sowing pool' agribusinesses (Murmis and Murmis 2012; Gras and Hernández 2009b, 2013, 2014). Sowing pools are collective investment funds for large-scale soy production that operate in Argentina and other South American countries, like Uruguay, Paraguay, Bolivia and Brazil. These agribusinesses have revolutionized Argentine agriculture

[1] As David Harvey (1990, 423) notes, 'The grapes that sit upon the supermarket shelves are mute; we cannot see the fingerprints of exploitation upon them or tell immediately what part of the world they are from'.

by promoting a new 'paradigm' of agricultural production based on novel arrangements for land and labor and constant technological innovation (Hernández 2009; Gras and Hernández 2009b, 2013, 2014).

This paper explores the various drivers of distancing across the soy value chain in Argentina, including industrialization, globalization, corporatization and finance. I argue that the push for technological innovation by large-scale agribusinesses engaged in soy production, in articulation with financial sector involvement, are both an example of and are instrumental in the process of 'distancing' identified in the agro-food literature (Kneen 1993; Friedmann 1994; Princen 1997, 2002; McMichael 2012, 2013; Russi 2013; Clapp 2014, 2015; Isakson 2014). I contribute to the agro-food literature by showing how the processes of distancing and abstraction happen from the ground up, starting with the farmer in the field. By exploring how labor, class relations and agricultural production are changing as a consequence of the cultivation of GM soy in Argentina, I trace the increasing distance between the farmer/producer and the food/product. This distance transforms agriculture from a food-producing activity into a commodity and, eventually, to a stock in the financial market. My case study provides empirical support for a trend recently identified in the agro-food literature whereby agriculture goes *beyond* the commoditization of food (and land) toward the utter dematerialization, decontextualization and financialization of nature (McMichael 2012; Russi 2013; Clapp 2014, 2015; Isakson 2014; Fairbairn 2014; Lawrence 2014).

This paper also highlights how, despite agribusiness efforts to 'displace' and 'disappear' nature, these processes are never fully accomplished. On the contrary, I show how these efforts are still very much rooted in physical bodies, resources and practices, and thus serve capitalist expansion by a process of accumulation through dispossession (Harvey 2003; Araghi 2009; McMichael 2012). Soybean production in Argentina relies on extensive agro-industrialization, a form of dispossession (Cáceres 2015) with growing social and ecological costs (Leguizamón 2014a, 2014b). In Argentina, the expansion of GM soy monocultures has been associated with violent peasant displacements (Lapegna 2013); increased cancer rates in rural and peri-urban populations due to agrochemical exposure (Berger and Ortega 2010; Carrizo and Berger 2012; Lapegna 2015); food insecurity (Teubal 2008); deforestation (Grau, Gasparri, and Aide 2005; Gasparri, Grau, and Gutiérrez 2013); soil nutrient depletion (Pengue 2005); and water, air and soil pollution (Binimelis, Pengue, and Monterroso 2009). Yet vast distance in a financialized global food system allows producers to disguise these costs (Princen 1997, 2002), making it hard to link the aforementioned consequences with the responsible parties (Gould 2006), and hence complicating the context of resistance (Clapp 2014).

For this work, I draw data from press sources, reports, websites and statistical databases, as well as 40 interviews conducted between 2009 and 2012 with Argentine rural actors, including peasants; rural workers; small, medium and large soy producers; rural contractors; and stockholders, chief executive officers (CEOs) and employees of sowing pool agribusinesses; in addition to rural inhabitants who do not profit directly from soy production. Following this introduction, I describe the context of soybean production in Argentina, focusing on the national and transnational links of the soybean chain. Sections 3 and 4 focus on socio-ecological change and inquire into how technological innovation transforms social relations and society-ecosystem dynamics (Schnaiberg 1980; Schnaiberg and Gould 1994; Gould, Pellow, and Schnaiberg 2008). In section 3, I show how the process of distancing starts on the farm. Here I look at how the adoption of new technologies removes physical labor from agriculture and contributes to distance and the commodification of food production. In section 4, I focus on how large sowing pool agribusinesses utilize technology to reduce their dependency on the natural conditions that make agriculture possible, most significantly land and labor. Thereby, agriculture is transformed and nature is, allegedly, 'displaced'. In the conclusion, I reflect on the socio-ecological contradictions that arise from the processes of distancing and abstraction which accompany the financialization of the corporate food system under neoliberal globalization (Harvey 2003; Araghi 2009; McMichael 2012, 2013; Clapp 2015).

2. The context of soybean production in Argentina

RR soybeans were approved for use in Argentina in 1996, making Argentina one of the first six countries to adopt GM crops for commercial use (James 1997). The introduction of GM biotechnology was the next obvious step in a history spanning two centuries of agrarian capitalism and technological innovation in the Argentine Pampas (Barsky and Gelman 2001; Hora 2012). The Pampas are a lowland central region extending across the provinces of Buenos Aires, La Pampa, Córdoba, Santa Fe and Entre Ríos. This region is characterized by vast plains of fertile land, and temperate weather, making it ideally suited for large-scale agriculture.

The Pampas have played the starring role in Argentina's rural export sector and served as the motor of its economy ever since Argentina gained independence from Spain in the early 1800s (Hora 2012). Starting in the 1880s, agricultural exports – primarily wheat, corn, flaxseed and livestock – contributed to building Argentina's *belle époque*, a time when Argentina matched the newly independent countries of Canada and Australia in growth rates and living standards. This first agro-export boom, which bust with the Great Depression of the 1930s, resulted from the expansion of the agrarian frontier over indigenous territory and from the introduction of agricultural machinery – mowers, harvesters and threshers – from the United States and Canada (Hora 2012). Pampas farmers, mostly newly arrived European immigrants, modeled their agrarian production after American capitalist agriculture, a model based on the constant modernization of agrarian technologies and methods and a focus on increasing productivity to yield short- and medium-term profits (Barsky and Gelman 2001; Gras 2009; Hora 2012).

Soybeans entered the Pampas in the late 1960s alongside the technologies of the Green Revolution, such as hybrid seeds, agrochemicals and tractors. Productivity increases resulting from the adoption of such technologies enabled the Pampas to recover its place as the motor of Argentina's economy via exports of oilseeds and grains (Barsky and Gelman 2001; Hora 2012). From the late 1990s onwards, soybeans became Argentina's 'star crop' (Hora 2012, 169), due to the introduction and rapid diffusion of the technological package of GM soy in a context of neoliberal economic restructuring (Teubal 2006, 2008; Newell 2009; Leguizamón 2014a).

Argentina's neoliberal restructuring, a process that started with the military dictatorship in the late 1970s and accelerated swiftly in the 1990s, entailed the return of an outward-looking development model based on nontraditional commodity exports. Facilitated by a global context of high prices and high demand, mostly from Asia, GM soy became farmers' preferred crop and the government's key accumulation strategy (Teubal 2006, 2008; Newell 2009; Leguizamón 2014a). The result was an unprecedented expansion of production – linked not necessarily to higher yields, but rather to larger areas under cultivation due to crop replacement and expansion of the agrarian frontier – followed by a growth in exports (Leguizamón 2014a). Between 2002 and 2012, soy and soy-derived exports (beans, meal and oil) represented, on average, 23 percent of Argentina's total international trade (Simoes 2015). In 2014, soy sector exports represented 28 percent of total Argentine exports and accounted for USD20 billion in foreign income (INDEC 2015). The year 2015 has seen a downturn in soybean prices due to falling oil prices: in September 2015 soybeans reached their lowest price since late 2008, at USD324/ton, compared to a peak of USD650/ton in September 2012 (*La Nación* 2015a, 2015b). The 2015 economic crisis in China, a primary buyer of Argentine soybeans, has further jeopardized the Argentine soy boom.

2.1. *The soybean chain*

The neoliberal restructuring of the 1990s led to a reorganization of Argentine agriculture to provide primary inputs for the global food system (McMichael 2013; Gras and Hernández 2014). As Otero (2012) notes, neoliberal restructuring in Latin America has led states, as instrumental actors, to promote the adoption of GM biotechnology as well as new regulations based on market principles that benefit large transnational corporations. In Argentina, the consequence has been that the soybean agro-export complex has become a privatized agribusiness regime dominated by national and transnational corporations (McMichael 2013). In what follows, I summarize the transnational links of the soybean chain in Argentina. As Friedmann (1994) argues, transnational agro-food chains contribute to distancing and blur the boundaries of production and consumption across sectors and across nations.

The global soybean chain is controlled by a handful of transnational corporations: the top six seed and agrochemical companies (Syngenta, Bayer, BASF, Dow AgroSciences, Monsanto and DuPont) control 76 percent of the global market share (ETC Group 2013). Four grain traders (ADM, Bunge, Cargill and Louis Dreyfus) control around 75 percent of the export market (EcoNexus 2013). Although these four firms, known as the ABCDs, still dominate world trade, Asian trading companies such as COFCO are emerging as important competitors in Latin American soybean production (Murphy, Burch, and Clapp 2012). The ABCDs together lost almost 20 percent of the Argentine soybean export market between 2005 and 2014, from 62.6 to 40.6 percent. Half of these lost exports were captured by the Chinese firm COFCO via its control of Nidera and Noble Group, which jointly account for 11.3 percent of 2014 exports (*Clarín 2015*).

ADM, Bunge, Cargill and Louis Dreyfus are multinational giants that also monopolize worldwide soybean processing capacity (USSEC 2011). The soybean milling and processing sector is highly concentrated as well, and shows upstream and downstream links across the production chain, from providing inputs and controlling purchases to farmers to dominating export sales (USSEC 2011). Through strategic alliances and joint ventures (e.g., Cargill with Monsanto, Bunge with DuPont), the four giant trade companies extend their grip on the value chain to the seed and agrochemical sector and thus control the entire global food system (EcoNexus 2013; ETC Group 2013).

The soybean chain in Argentina reflects the global trend of increased concentration and integration (Regunaga 2010; Turzi 2010). A small number of national and transnational companies monopolize the trading sector. Seven firms (Cargill, Noble, ADM, Bunge, Dreyfus, Toepfer and Nidera) control 85 percent of soybean exports. Eight firms (Cargill, Bunge, Dreyfus, AGD, Molinos Río de la Plata, Vicentín, and ADM) control 89 percent of soybean oil exports. Seven firms (Cargill, Bunge, AGD, Dreyfus, Molinos Río de la Plata, Vicentín and Nidera) control 92 percent of soybean protein meal exports (Regunaga 2010). The same traders and processors regularly participate in the commodity stock market (locally and, via their parent companies, internationally). This common practice enables both these companies and agricultural producers to hedge against the price fluctuations typical in commodity production. In the last few years, however, these agro-food giants, particularly the top four ABCDs, have also speculated in the financial market, inflating global food prices (Murphy, Burch, and Clapp 2012).

In Argentina, the largest exporters also own or operate local milling and processing plants, thereby controlling both the processing and export sectors. The six largest millers, Cargill, Bunge, Vicentín, Molinos Río de la Plata, Dreyfus and AGD, account for 87 percent of total soybean oil exports and almost 90 percent of total soybean meal exports (Regunaga 2010). Argentina has the most efficient vegetable oil processing complex in the world (Turzi 2010) and it is the largest global exporter of soybean oil and soybean meal – accounting for around 55 percent and 48 percent of global total exports, respectively (Regunaga 2010). Since 2007, the soybean-crushing sector has also specialized in biodiesel production for the domestic and export markets (Cohan 2012).

A key player in biodiesel processing and exporting in Argentina – and an exemplary case of distancing – is Renova, which in 2014 broke ground on the largest soybean-processing plant for biodiesel in the world, a USD480 million investment partly financed by a consortium of international banks (Télam 2014). Renova is a joint venture between three companies: Vicentín, Molinos Río de la Plata (one of Argentina's largest food producers) and Oleaginosa Moreno (a formerly family-owned Argentine agribusiness acquired in 1997 by the Swiss-based mining company Glencore) (Infocampo 2011). Glencore is 'one of the world's largest global diversified natural resource companies and a major producer and trader of more than 90 commodities' (Glencore 2015). In Argentina, Glencore controls La Alumbrera, an open-pit copper-gold mine in the northern province of Catamarca. In the agricultural sector it controls facilities for oilseed storage, processing and export logistics. In 2014, Glencore's global agribusiness branch grew more than 300 percent to account for 60–70 percent of its USD856 million in annual earnings (Terazono and Hume 2015).

By controlling processing facilities, storage and freight companies, and port facilities, trading companies like Cargill, the world's biggest grain trader, can control the whole soybean pipeline (Turzi 2010; Regunaga 2010). Operating in Argentina since 1947, Cargill is now the country's largest agro-food exporter. In addition to processing and trading in oilseeds (soybean, sunflower and rapeseed), malt and barley, the company engages in flour milling, beef production and financial services. Across the country, Cargill owns five export terminals, four oilseed-crushing plants, two malt plants, two beef-processing plants, seven flour mills, and a network of more than 50 elevators (Cargill 2015).

Transnational seed and agrochemical companies operate in Argentina as well. Frequent mergers, acquisitions, partnerships and joint ventures strengthen these multinational companies' grip on the soybean chain in Argentina and South America (Turzi 2010). Through various such agreements, firms that are ostensibly competitors collaborate to reduce competition and create cartels (ETC Group 2013). Examples of these practices abound. In 2008, Syngenta bought SPS, then the fifth-largest seed company in Argentina (Bertello 2008). Bayer CropScience acquired Argentine seed company FN Semillas in 2013 and, in 2014, bought Biagro Group, an Argentine company specializing in inoculants and biological seed treatments (SeedQuest 2013, 2014). In 2012, US-based GM biotech developer Arcadia Bioscience launched Verdeca, a 50–50 joint venture for soybean seed development with Argentina-based Bioceres, an agricultural investment group specializing in deregulation (ETC Group 2013). Capitalizing on the 2014 expiration of Monsanto's US patent for glyphosate-resistant soybeans, Verdeca is developing new GM seed varieties and obtaining regulatory approval for their use in South America, the United States, China and India (Dow Jones Newswires 2012). In 2013, Verdeca signed a collaboration agreement with GDM Seeds to develop and commercialize stress-tolerant soybeans (Verdeca 2013). Illustrating the close relationship among agribusinesses, GDM Seeds is part of the Argentine company Asociados Don Mario, the second largest GM soybean seed seller in Argentina and the largest in Brazil (Infocampo 2014a). Don Mario's seed market also extends to Uruguay, Paraguay, Bolivia, South Africa and the United States (Infocampo 2013a). In Argentina, Don Mario commercializes Monsanto's newest GM soybean variety Intacta RR2 Pro (Don Mario 2015). Don Mario also provides its seed genetics to Louis Dreyfus' Argentine seed branch, the third largest marketer of certified GM soybean seeds (Infocampo 2013b). Nidera, the largest seed seller in Argentina, has also adopted Monsanto's Intacta RR2 Pro germplasm to develop and sell its own Super Soy varieties (Nidera Semillas 2015).

The Argentine soybean seed market is not particularly profitable because certified seed sales are low and farmers do not make royalty payments (Delvenne, Vasen, and Vara 2013; Feeney and Berardi 2013). Unlike American farmers, Argentine farmers are not legally constrained to sign contracts and they are entitled under law UPOV 78 to save seeds for future replanting (Pierri and Abramovsky 2009). There is also a large black market in seeds, commercialized in unlabeled 'white bags'. Early on, a permissive intellectual property regime benefited GM companies by facilitating fast adoption of the technological package; now, however, companies are challenging the loss of revenue to the illegal seed market (see Delvenne, Vasen, and Vara 2013). The use of saved seeds combined with illegal seed trading accounts for 65 to 80 percent of the total soybean seed market. Feeney and Berardi (2013) estimate that for 2009–2010 the legal soybean seed market was only 35 percent of the total market. Other estimates are as low as 20 percent in 2014 (Infocampo 2014a). This is one reason firms expand their business across borders: the Brazilian seed market is particularly enticing as the use of certified seeds and royalty payments are close to 70 percent, accounting for sales of more than USD1.2 billion in 2014, compared to a meager USD240 million in Argentina (Infocampo 2014a).

This does not mean that seed companies are not profitable in Argentina. Seed companies benefit from widespread use of the technological package of RR soy, which requires increasing doses of agrochemicals. The major transnational seed companies, such as Monsanto, Nidera and DuPont also produce, import and distribute herbicides, insecticides and other weed-, pest- and disease-control agents in Argentina (Regunaga 2010). According to the Argentine association of agrochemical companies CASAFE (2012), in 2012 the agrochemical market rose to $2.4 billion, largely due to sales of the RR soy herbicide companion, glyphosate. Compared to the seed/agrochemical sector, the fertilizer sector is more concentrated and it is controlled by the large grain traders and oilseed millers Cargill, Bunge, Nidera, Dreyfus and AGD (Regunaga 2010).

To conclude, in this section I have shown how corporations operate across the different sectors of the soybean chain (seed, fertilizer and agrochemical inputs, processing, storing and trading) in Argentina, and how they have developed deep (but long) links with global and regional supply chains and markets (Friedmann 1994; McMichael 2013). Next, I explore how GM soy production has transformed agriculture and the social structure of rural Argentina. I show how technological innovation is a key to understanding these transformations. First, the adoption of new technologies transforms labor by replacing workers with machines (Schnaiberg 1980). This leads to rural displacement and depopulation in conjunction with the specialization and concentration of rural labor. New rural actors emerge, most notably sowing pools. My aim is to show how the distancing and abstraction processes identified in the agro-food literature (Russi 2013; Clapp 2014, 2015; Isakson 2014) unfold in Argentina.

3. From the ground up: distancing on the farm

The trend of rural depopulation and displacement in Argentina started at the turn of the twentieth century, alongside the history of mechanization of the Pampas region (Barsky and Gelman 2001; Hora 2012). However, this trend has accelerated with the expansion of GM soy monocropping. In 2010, rural residents accounted for 6.9 percent of the Argentine population (2.7 million people), down from 20.2 percent in 1970 (ECLAC 2014). Machines, fossil fuels and chemicals replace labor in large GM soy monocultures (Schnaiberg 1980; Leguizamón 2014a). Only 15 employees are needed to work a 1000-hectare soy farm, compared to 350 for a sugarcane plantation, or 1300 for a citrus orchard of that size (Selén 2014). Throughout the Global South, agro-industrialization has reduced the need for rural labor, displacing rural populations and creating what La Vía Campesina calls 'an agriculture without farmers' (McMichael 2012, 693).

The path of agro-industrial farming has been to minimize arduous physical labor in the field while increasing productivity and maximizing profits (Anderson 2009; for a critique, see Leguizamón 2012). The technological package of RR soybeans reduces labor to a minimum, a major factor in its enthusiastic adoption by farmers. Only a few *pasadas*, 'passes of a machine through the field', are needed: one for planting, another (or a few) for agrochemical spraying, and a final one for harvesting. As machines get bigger, they cover more ground in less time, reducing working time. Moreover, farmers describe working atop tractors and combines as a comfortable activity, almost leisurely compared to the backbreaking work of traditional farming – these machines come with hydraulic steering, stereo music systems, built-in refrigerators, air conditioning and computerized and global positioning system (GPS) controls.

Shielding the person who performs the actual task of cultivation from climate conditions in an air-conditioned tractor cabin increases the distance between the farmer and the activity of farming, which used to entail being close to the soil and the natural process of food-growing. This distance is actually even greater than it appears, because the person operating the machine is usually not the farm owner and likely is not even from the area. Agricultural production is performed by *rural contractors*, workers contracted to complete a specific task, such as sowing, fertilizer application, herbicide and pesticide spraying, or harvesting. Rural contractors typically own machinery but no land and thus travel from town to town, from province to province, planting, spraying and harvesting one farm, then moving on to the next (Lódola 2008). Recent studies estimate there are nearly 10,000 contractors, who perform 80 to 85 percent of total spraying and harvesting work in Argentina (Regunaga 2010).

Further separating those who control resources from those who perform the agricultural tasks and from nature are *production contractors*. Production contractors own neither land nor machinery but rent both, hiring other contractors to perform agricultural tasks (Bustamante and Maldonado 2009). Their added value resides in their knowledge of organizing production. Because of the highly technical specificities of using the technological package over extensive areas, agricultural production rests mostly in the hands of 'experts'. Farming knowledge, traditionally shared and passed down from generation to generation of farmers, is now taught at the university (Grosso and Albaladejo 2009). Those in charge of managing production are no longer 'farmers' but agronomists, engineers or masters in business administration. They organize and plan production from afar, using satellite reports, laptops and the Internet to control and monitor crops, soil humidity and inputs needed.

As a consequence, in the Pampas under the GM soy model, the role (and value) of agrarian labor has shifted from physical activity towards managerial tasks of organizing production, which call for a team of experts with combined knowledge and expertise beyond the productive aspects of agriculture. In interviews, large-scale soy producers agreed that in order to achieve 'efficiency' (i.e., to stay profitable), agrarian production requires a team that pools technical, managerial, commercial and financial expertise, and, more important, investment capacities. 'Sowing pools' emerged as investment funds to meet this need, particularly to bring in necessary resources after the Argentine crisis of 2001 (Bustamante and Maldonado 2009). An important characteristic of these sowing pools is that they bring urban inhabitants with no knowledge of agriculture into farming purely as financial investors. As stockholders, their goal is to maximize returns on their investment. Therefore, decisions about which crops to grow and which technologies to employ rely mostly on their potential to increase profits (Gould 2009; Isakson 2014). This exemplifies how financialization of the food system has distanced the controllers of agriculture from agriculture itself (Clapp 2014, 2015).

Next, I focus on the activities of large-scale sowing pool agribusinesses. These firms are both the cause and a manifestation of the ongoing agrarian transformation in Argentina, as they devise a new 'paradigm' of agricultural production (Hernández 2009; Gras and Hernández 2009b, 2013, 2014).

4. Disappearing nature: large agribusinesses and finance

Technological innovation has removed farmers from the physical labor of agriculture. New contracting and subcontracting arrangements further distance the farmer from the farm. This distance is compounded as new actors – mainly transnational agro-food corporations and financiers – enter the food system and take control of the farm (Clapp 2014, 2015). In this section I analyze how these actors operate in Argentina and how local agricultural practices have been transformed to fit the demands of the global food system.

Agricultural production in Argentina reflects global trends of increasing concentration, increasing integration and increased scale. The activities of large *agronegocios*, agribusinesses popularly known as *pools de siembra*, 'sowing pools', exemplify how Argentine agriculture has been reorganized to serve the global food system (Murmis and Murmis 2012; Gras and Hernández 2014). 'Sowing pool' refers to the pooling of land, capital and human resources for agricultural production. This novel mechanism for agricultural production is 'characterized by the fundamental role played by investment funds and by a business model whereby control of agricultural production is acquired by leasing large extensions of land … and by contracting teams for sowing, fumigating, harvesting, and transporting' (Gómez 2014, 11). Specifically, sowing pools are involved in 'leasing or providing property in trust, contracting machines and services, using technological packages based on modern machinery, heavily utilizing biotechnology or agrochemicals and incorporating digital systems and individuals who specialize in field selection, production, management, and marketing' (Murmis and Murmis 2012, 491).

While not all sowing pools engage in large-scale production, size does matter (Murmis and Murmis 2012; Gras and Hernández 2014; Cáceres 2015). Soy production in Argentina is highly concentrated by large agribusinesses that farm vast tracts of land (Gras and Hernández 2014). In 2010, only 2.6 percent of producers (approximately 1600 farmers) controlled more than 50 percent of soybean production, farming 9.34 million hectares in plots larger than 5000 hectares (Catacora-Vargas et al. 2012). According to the United Nations Food and Agriculture Organization (FAO), the five largest agribusinesses operating in Argentina – MSU, Los Grobo, El Tejar, Cresud and Adecoagro – control, via landownership or leasing, close to three million hectares of farmland in Argentina and neighboring

Uruguay, Bolivia, Paraguay and Brazil (Gómez 2014).[2] El Tejar controls 1.1 million hectares (Gómez 2014), and Cresud almost a million hectares in a mix of farming and farmland development in Argentina, Brazil, Paraguay and Bolivia (Cresud 2015), making these companies, respectively, the fourth and fifth biggest landowners in the world (Kazakhstan Newsline 2014). Adecoagro, controlled by the fund of US billionaire George Soros, owns 293,423 hectares (*New York Times 2015*), while Los Grobo harvested 320,000 hectares in 2011, its peak year (*Economist 2014*). The technological package of GM biotech along with developments in information technology[3] facilitate expansion and corporate concentration because they enable the farming of huge areas of land, and these giant firms can negotiate more favorable terms than smaller competitors can (Murmis and Murmis 2012; Gras and Hernández 2014).

The activities of these giant agribusinesses extend beyond farming through processes of vertical and horizontal integration to input supply, grain trading, food processing, transportation and investment in the stock market. Adecoagro, for example, engages in farmland development, crop production (oilseeds, cereals, sugar, coffee and cotton), cattle and dairy farming, and sugarcane processing for ethanol production (Adecoagro 2015). In addition to crop production and farmland development, Cresud participates in the livestock industry through a joint venture with Tyson (the world's largest meat processor) and US Cactus Feeders to create the 'first vertically integrated cattle project' in Argentina (SEC 2007). Via subsidiary IRSA, Cresud also engages in urban real estate development of shopping malls and luxury hotels (Cresud 2015). Los Grobo developed an innovative verticalization strategy called *modelo asociativo en red*, 'associative network company' (Los Grobo 2015). The model is based around building a network of input and service providers, including landowners, agronomists, contractors and branch managers, so that instead of directly hiring employees or owning land or machinery, the company operates through land leases and third-party contracting (Bell and Cintra 2010).

Los Grobo's associative network model represents a shift away from the materiality of farming (i.e., direct ownership and control of resources) and toward leasing, contracting and specialization in the managerial/organizational aspects of production. Particularly since 2012, years which witnessed slimmer profit margins for soy and rising tensions with the government, Los Grobo has further transformed its business by slashing direct agricultural production (hectares under lease for production are at an all-time low) and focusing instead on services, input provision, seed development (via companies Bioceres and Verdeca) and agrochemicals (via its 2013 acquisition of Agrofina) (Bertello 2013). Its stated goal is to expand vertical integration to provide a 'one-stop shop' for farmers (*Economist 2014*).

Acquisitions grow businesses but they are not cheap. Recent strategies for financing agribusiness expansion have involved partnerships and participation in the stock market. To raise capital to purchase Agrofina, Los Grobo sold shares on the Buenos Aires Stock Exchange (Infocampo 2014b). It also sold its Brazilian unit, Los Grobo Ceagro, to Japan's biggest trading company, Mitsubishi (Rosendo González 2014). Increasingly,

[2]Sowing pools are not counted as a category in the rural census; therefore, there is no official data about them. Data is estimated based on press and Internet sources (in some cases, the companies' own websites). These figures are compatible with those reported in similar peer-reviewed articles, such as Cáceres (2015); Gras and Hernández (2014); and Murmis and Murmis (2012).

[3]Information technology (IT) developments that have greatly facilitated geographical expansion include communication technologies (phone, fax and the Internet), GPS, and software applications for risk analysis and resource and production planning. For examples of how Los Grobo applies these technologies, see Bell and Cintra (2010).

Argentine agribusinesses sell shares locally on the Buenos Aires and Rosario stock exchanges, and internationally on the São Paulo and New York exchanges. In a matter of years, historically family-owned businesses like Los Grobo, MSU and El Tejar have rapidly transformed into very complicated corporate structures that include international commercial banks, mutual funds, hedge funds and private equity investors. For example, El Tejar, founded in 1987 by an association of Argentine farming families, is now 40 percent controlled by the New York- and London-based hedge fund Altima Partners, while the US private equity firm Capital Group owns another 15 percent stake. Capital injections from these financial sponsors have helped El Tejar expand its business by purchasing farmland in Brazil (Terazono and Webber 2013). At USD800 million, Altima is one of the largest investment funds in global agriculture, with holdings in companies operating in Australia, Eastern and Central Europe, South America and sub-Saharan Africa (Luyt 2013).

Cresud and Adecoagro offer other revealing examples of how financial capital gets injected into agricultural production. Cresud was a 63-year-old public company when Eduardo Elsztain acquired it in 1994, partly with funds from George Soros (Goldberg et al. 2013). Cresud was already traded on the Buenos Aires stock market and, after Elsztain's acquisition, a new rights offering raised USD64 million for land purchases. In 1997, Cresud undertook a global public offering for 43 percent of its shares on the NASDAQ Stock Exchange (Goldberg et al. 2013). Since then, share sales on NASDAQ have been frequent sources of financing for Cresud's expansion. The company also relies on partnerships and joint ventures to raise revenue. For example, in June 2011, Cresud signed a joint venture agreement to purchase land and cultivate soybeans for Heilongjiang Beidahuang Nongken Group, China's largest farming company (Grain 2012). Similarly, in 2006, Cresud founded BrasilAgro in partnership with a Brazilian real estate company and the investment fund Tarpon (Goldberg et al. 2013). Currently, Cresud holds 39.6 percent of BrasilAgro's shares; JP Morgan Whitefriars (a subsidiary of JP Morgan Chase) holds 10.5 percent; Brazilian billionaire Elie Horn, 5.6 percent; and Brazil's Banco Fator, 5.3 percent (*Valor Económico 2013*). In Paraguay, Cresud owned half of Cresca, a 50–50 joint venture with Argentine agribusiness Carlos Casado, but sold its shares in December 2013 to its partnership BrasilAgro, allegedly to bypass foreign currency restrictions in Argentina (El Cronista 2013).

Adecoagro was founded in Argentina in 2002 by a group of Argentine and Brazilian entrepreneurs sponsored by global institutional investors (Adecoagro 2015). Since 2010, the company has operated as a *societé anonyme*, the equivalent of a US corporation under Luxembourg law. In 2011, Adecoagro went public on the New York Stock Exchange, in an offering of more than USD400 million (DealBook 2011). Among Adecoagro's largest shareholders are Soros Fund Management; Al Gharrafa Investment Company, a unit of Qatar Holding; Ospraie Management, a US-based hedge fund also deeply invested in mining projects; mutual funds managed by US investment advisory firms such as Jennison Associates and Brandes Investment Partners; and PGGM Vermogensbeheer BV, a Dutch pension fund manager (CNN Money 2015).

The preceding examples show the pervasiveness of financial capital in agriculture, as well as the growing interest of the financial sector in farmland (McMichael 2012; Fairbairn 2014). Another emerging trend is the involvement of sovereign wealth funds (Luyt 2013), as investments by Qatar Holdings in Adecoagro, and by China's Beidahuang in Cresud, illustrate.

The Argentine 'agribusiness model' represents a paradigm shift in agricultural production (Hernández 2009; Gras and Hernández 2009a, 2013, 2014). Los Grobo has been instrumental in devising and implementing this 'new agricultural paradigm', summarized below by the Wave project, funded by BNP Paribas:

> Los Grobo, one of the largest agro-food producers in Argentina, operates according to an elegantly simple principle: no land, no machines, not even a tractor. Everything is rented.... According to Los Grobo CEO Gustavo Grobocopatel, 'This model is based on knowledge rather than ownership, and it will change the agricultural paradigm'. By adopting open innovation, creating a pool of more than 120 experts with a very wide range of skills, and operating as a network, these 'landless' South Americans are the new wave of innovators. (Wave 2015)

Los Grobo's novel approach to agriculture is disseminated across the country via groups like the Argentine No-Till Farmers Association (AAPRESID), and agricultural fairs like Expoagro. It is a 'paradigm' in the sense that it is a comprehensive worldview consisting of social, structural, technological, cultural and symbolic dimensions (Hernández 2009; Gras and Hernández 2009a, 2013). From the perspective of agribusiness, the value of agriculture lies not in the natural conditions that make it possible but in the expert know-how to manage and organize production, and in the scientific knowledge to develop new agrarian technologies (e.g., no-tillage machinery and GM biotechnology). Agribusiness elevates knowledge as a means of production more valuable than land itself. Whereas the demands of neoliberal globalization push for efficiency and flexibility, land, being fixed, means immobility. Therefore, in the new agribusiness paradigm, land becomes a drag, a hindrance to the mobility of capital (Hernández 2009).

Large producers in Argentina thus aim to accomplish a fabulous feat: making land disappear from the farming equation. How is it possible to 'overcome' land as a fixed factor of agricultural production? How do what the Wave project calls the 'landless' become the largest of agrarian producers?[4] Large sowing pools accomplish this, first, by leasing land (and machinery and labor) instead of owning it. Land leases, generally on one- to three-year contracts, afford flexibility: once the contract ends, producers are free to move elsewhere. Small landowners benefit as well, because it is more profitable to rent their land than to farm it themselves – and risk losing it altogether when they are unable to compete with large pools. Leasing land also allows pools to become bigger, expanding production over greater areas (Gras and Hernández 2014). This comes to show that land ownership has become less relevant, though control over land is still crucial. Large-scale farming then becomes in some cases akin to land grabbing (Murmis and Murmis 2012; Cáceres 2015).

Second, in another strategy to achieve flexibility, large agribusinesses extend production into other countries. Farming in different climates, environmental and political, accomplishes another main objective of soy producers, to reduce variability and uncertainty. Whether the challenge is variable weather and rainfall cycles or government-imposed export quotas and taxation, expanding production across several countries diversifies risk and stabilizes production output. Planting soybeans also facilitates these objectives as soybeans are 'flex crops' with multiple and flexible uses, as food, fuel, animal feed and building material (Borras et al. 2012). A regionally integrated South American soybean chain, where soybeans can be interchangeably produced in Paraguay or Brazil then processed in and shipped from Argentina (Turzi 2010), closes the cycle. Flexibility,

[4]Gustavo Grobocopatel, CEO of Los Grobo, was the first to call himself a 'landless'. Grobocopatel's statement was a deliberate verbal provocation of peasant activists by misappropriating the *sem terra* name from the Brazilian Landless Rural Workers' Movement (MST).

efficiency and predictability, required characteristics of just-in-time global capitalism, are thereby achieved. Farming, once constrained within the limits of the natural environment, now expands over a 'virtual territory' (Hernández 2009, 53–54).

The agribusiness approach towards agriculture and nature described in this section respond to, and are part of, the growing financialization of land and food crops around the world (Fairbairn 2014; Isakson 2014; Clapp 2014). The financialization of farmland is the latest frontier in the financialization wave, a process that accelerated after the rise of food prices in 2007 (Lawrence 2014). As Fairbairn (2014) notes, the farmland investment boom is not merely speculative (i.e., speculating on financial returns from land appreciation). Investors' interest also relies upon farmland's productive capacity; that is, engaging in agricultural production for potential return in commodities. Those consortiums solely interested in land's productive capacity, like Los Grobo, engage in a strategy Fairbairn (2014) identifies as 'lease-operate', which is preferred over landownership. Compared to 'own-lease out' and 'own-operate' investment strategies, 'lease-operate' gives the highest risk-return. Under a 'lease-operate' strategy, agribusinesses speculate not on the value of land but on the value of crops, which explains the preference for flex crops (Isakson 2014).

However profitable 'lease-operate' agricultural production may be, there are important implications to short-term contracts. As Fairbairn (2014, 16) notes, 'aside from the obvious impact this has on the structure of agriculture, it also reduces the farmer's incentive to use sustainable practices by removing his or her stake in future productivity'. Here is a key consequence of 'distancing': the ability of decision-makers to remove themselves from the social and ecological consequences of their decisions (Princen 1997, 2002; Gould 2006). Los Grobo's self-proclaimed emphasis on 'knowledge-based' agriculture, on innovation and on 'elegantly simple' principles (Wave 2015) is a 'shading' strategy (Princen 1997, 2002) that obscures the violence of dispossession perpetrated against the rural and indigenous poor (Lapegna 2013, forthcoming; Cáceres 2015). Shading allows decision-makers to present themselves as providers of solutions, not creators of environmental and human ills; thus, it becomes harder for activists at the end of the soy chain to pinpoint who is responsible for the negative consequences of agro-industrial production (Gould 2006; Clapp 2014).

5. Conclusion

Multinational agribusinesses present GM crops as a solution to alleviate world hunger and poverty in the context of the challenges of climate change. Versatile 'flex crops' fit this discourse well, because they could be alternatively directed to address any aspect of the food, fuel, fiber and climate crisis (Borras et al. 2012; McMichael 2012; Isakson 2014). Flexible GM crops like soy are thus touted as representing the promise to disconnect agricultural production from its natural conditions: allowing production to increase despite drought, land scarcity, excessive heat and other extreme climatic conditions. Argentine soy producers are major promoters of this belief; they claim knowledge – not land, climate or labor – is the most essential component of agricultural production. They are optimistic that scientific and technological innovation will offer solutions to the current socio-ecological crisis – and that Argentina, via GM biotech adoption, will be a key player in feeding the world.

In this contribution, I have explored how, by promoting a 'knowledge-based' agricultural paradigm for the production of profitable flex crops, agribusiness aims to disconnect and abstract agriculture from its natural origins as a food-producing, body-sustaining activity (Araghi 2009). By highlighting the many drivers of distancing across the soy value chain in Argentina, I have shown how large-scale agribusinesses engaged in soybean production, in articulation with global financial investors, are instrumental to the process of distancing and abstraction identified in the agro-food literature.

The distancing and abstraction of food and farm resulting from a 'knowledge-based' agribusiness model in particular, and from the financialization of the corporate food system under neoliberal globalization in general, has significant socio-ecological costs, however. Despite its 'virtual' appeal, the GM soy model has entailed – contrary to the rhetoric of agribusiness – an intensification of fossil fuel- and chemical powered-agro-industrial practices. Labor-replacing technologies within the logic of increasing accumulation in actuality decrease social benefits and increase ecological risk (Schnaiberg 1980; Schnaiberg and Gould 1994; Gould, Pellow, and Schnaiberg 2008). In Argentina, critics point out the 'dark side' of the expanding GM soy model (Lapegna 2013), claiming it is an unsustainable practice (Teubal 2008; Giarracca and Teubal 2010; Leguizamón 2014a; Cáceres 2015).

The agribusiness emphasis on knowledge is attractive: potentially, knowledge does not decrease from being shared, unlike scarce resources necessary for agricultural production, such as land and water. The agro-industry would have us believe that future developments in GM biotechnology promise to increase yield per hectare and enable cultivation on drought-prone and marginally fertile land. Yet technology cannot escape the fact there is no agriculture without land, and there is only so much land on this planet. The current wave of financial capital infusion in farmland is a response to the fact that land is a fixed asset with limited availability (Fairbairn 2014). The rush for land and water, and the enclosure of the commons in general, is nothing more than capital's survival strategy of accumulation by dispossession and displacement disguised by the abstractions of knowledge (Harvey 2003; Araghi 2009; McMichael 2012). Distancing and agribusiness promotion of an agricultural model based on knowledge and innovation render invisible the truth that the GM soy model feeds upon the violence of dispossession, by robbing indigenous and peasant families of their territories, endangering the health of populations exposed to agrochemical drift, and undermining the sustainability of the environment (Pengue 2005; Cáceres 2015; Lapegna forthcoming). Effectively linking decision-makers to the consequences of their decisions – that is, making elites responsible for the consequences of their actions, becomes more challenging as distance increases, but is crucial to socio-ecological sustainability (Gould 2006; Clapp 2014, 2015).

Acknowledgements

This contribution is based on work supported by two Doctoral Research Grants and a Dissertation Fellowship granted by The Graduate Center of the City University of New York (CUNY). I'm deeply grateful to Ken Gould, Justin Myers, Monica Guerra, Sam Cohn, Gustavo Oliveira, Susanna Hecht, JPS editor Jun Borras and two anonymous reviewers for their tremendous insight, comments, suggestions and support. Many thanks also to Kirsteen Anderson for her copy-editing help.

Disclosure statement

No potential conflict of interest was reported by the author.

Funding

The Graduate Center, City University of New York (CUNY).

References

ACSOJA (Asociación de la Cadena de la Soja Argentina). 2015. Argentina exporta a más de 100 países su producción de soja. http://www.acsoja.org.ar/nota.asp?cid = 1504.

Adecoagro. 2015. Company's website. http://www.adecoagro.com.

Anderson, J. 2009. *Industrializing the Corn Belt: Agriculture, technology and the environment, 1945–1972*. DeKalb: Northern Illinois University Press.

Araghi, F. 2009. Accumulation by displacement: Global enclosures, food crisis, and the ecological contradictions of capitalism. *Review (Fernand Braudel Center)* 32, no. 1: 113–46.

Barsky, O., and J. Gelman. 2001. *Historia del agro argentino: Desde la conquista hasta fines del siglo XX*. Buenos Aires: Grijalbo Mondadori.

Bell, D., and S. Cintra. 2010. *Los Grobo: Farming's future?* HBS Case No. 9-511-088. Boston: Harvard Business School Publishing.

Berger, M., and F. Ortega. 2010. Poblaciones expuestas a agrotóxicos: Autoorganización ciudadana en la defensa de la vida y la salud, ciudad de Córdoba, Argentina. *Physis: Revista de Saúde Coletiva* 20, no. 1: 119–43.

Bertello, F. 2008. La multinacional Syngenta se quedó con un semillero de soja. *La Nación*, November 11.

Bertello, F. 2013. Grobocopatel invierte en agroquímicos. *La Nación*, June 4.

Binimelis, R., W. Pengue, and I. Monterroso. 2009. 'Transgenic treadmill': Responses to the emergence and spread of glyphosate-resistant johnsongrass in Argentina. *Geoforum* 40, no. 4: 623–33.

Borras, S.M. Jr., J.C. Franco, S. Gómez, C. Kay, and M. Spoor. 2012. Land grabbing in Latin America and the Caribbean. *Journal of Peasant Studies* 39, no. 3–4: 845–72.

Bustamante, M., and G. Maldonado. 2009. Actores sociales en el agro pampeano argentino hoy: Algunos aportes para su tipificación. *Cuadernos Geográficos* 44: 171–91.

Cáceres, D. 2015. Accumulation by dispossession and socio-environmental conflicts caused by the expansion of agribusiness in Argentina. *Journal of Agrarian Change* 15, no. 1: 116–47.

Cargill. 2015. Company's website. http://www.cargill.com/worldwide/argentina/.

Carrizo, C., and M. Berger. 2012. *Citizens' rights and environmental genocide. Environmental Justice* 5, no. 2: 105–10.

CASAFE (Cámara de Sanidad Agropecuaria y Fertilizantes). 2012. Mercado argentino de productos fitosanitarios, 2012. http://www.casafe.org/biblioteca/estadisticas/.

Catacora-Vargas, G., P. Galeano, S. Agapito-Tenfen, D. Aranda, T. Palau, and R. Nodari. 2012. *Soybean production in the Southern Cone of the Americas: Update on land and pesticide use*. Cochabamba, Bolivia: Virmegraf.

Clapp, J. 2014. Financialization, distance, and global food politics. *Journal of Peasant Studies* 41, no. 5: 797–814.

Clapp, J. 2015. Distant agricultural landscapes. *Sustainability Science* 10, no. 2: 305–16.

Clarín. 2015. La exportación agrícola sigue dominada por los extranjeros. March 22.

CNN Money. 2015. Adecoagro S.A. http://money.cnn.com.

Cohan, L. 2012. *El aporte de la cadena de la soja a la economía argentina*. Tigre, Argentina: Wingord.

Cresud. 2015. Company's Website. http://www.cresud.com.ar/.

DealBook. 2011. Soros bets the farm. *New York Times*, January 13.

Delvenne, P., F. Vasen, and A.M. Vara. 2013. The 'soy-ization' of Argentina: The dynamics of the 'globalized' privatization regime in a peripheral context. *Technology in Society* 35, no. 2: 153–62.

Don Mario. 2015. Company's Website. http://www.donmario.com.

Dow Jones Newswires. 2012. Argentina JV aims to break big biotech's grip on transgenic soy. February 28.

ECLAC (UN Economic Commission for Latin America). 2014. Databases and statistical indicators. http://estadisticas.cepal.org.

EcoNexus. 2013. *Agropoly: A handful of corporations control world food production*. Zurich, Switzerland: Berne Declaration (DB) & EcoNexus.

El Cronista. 2013. Cresud planea expansión agropecuaria en la región vía Brasil por el cepo cambiario. December 20.

ETC Group. 2013. Putting the cartel before the horse … and farm, seeds, soil, peasants, etc. Who will control agricultural inputs? Communiqué no. 111.

Fairbairn, M. 2014. 'Like gold with yield': Evolving intersections between farmland and finance. *Journal of Peasant Studies* 41, no. 5: 777–95.

Feeney, R., and V. Berardi. 2013. Seed market segmentation: How do Argentine farmers buy their expendable inputs? *International Food and Agribusiness Management Review* 16, no. 1: 17–40.

Friedmann, H. 1994. Distance and durability: Shaky foundations of the world food economy. In *The global restructuring of agro-food systems*, ed. Philip McMichael, 258–76. Ithaca, NY: Cornell University Press.

Gasparri, N.I., H.R. Grau, and J. Gutiérrez Angonese. 2013. Linkages between soybean and neotropical deforestation: Coupling and transient decoupling dynamics in a multi-decadal analysis. *Global Environmental Change* 23, no. 6: 1605–14.

Giarracca, N., and M. Teubal. 2010. Ayer la Rural, hoy Grobo. *Página/12*. March 28.

Glencore. 2015. Company's website. http://www.glencore.com/.

Goldberg, R., A. Segel, G. Herrero, and A. Terris. 2013. *Cresud S.A., farmer or real estate developer?* HBS Case No. 211-011. Boston: Harvard Business School Publishing.

Gómez, S., ed. 2014. *The land market in Latin America and the Caribbean: Concentration and foreignization*. Santiago, Chile: Food and Agriculture Organization of the United Nations.

Gould, K.A. 2006. Promoting sustainability. In *Public sociologies reader*, ed. Judith Blau and Keri E. Iyall Smith, 213–30. Lanham, MD: Rowman & Littlefield.

Gould, K.A. 2009. Technological change and the environment. In *Twenty lessons in environmental sociology*, ed. Kenneth A. Gould and Tammy L. Lewis, 95–106. New York: Oxford University Press.

Gould, K.A., D.N. Pellow, and A. Schnaiberg. 2008. *The treadmill of production: Injustice and unsustainability in the global economy*. Boulder, CO: Paradigm.

Grain. 2012. Land grabs dataset. http://www.grain.org/article/entries/4479-grain-releases-data-set-with-over-400-global-land-grabs.

Gras, C. 2009. Changing patterns in family farming: The case of the Pampa Region, Argentina. *Journal of Agrarian Change* 9, no. 3: 345–64.

Gras, C., and V. Hernández. 2009a. El fenómeno sojero en perspectiva: Dimensiones productivas, sociales y simbólicas de la globalización agrorrural en la Argentina. *In Gras and Hernández* 2009b: 15–37.

Gras, C., and V. Hernández, eds. 2009b. *La Argentina rural: De la agricultura familiar a los agronegocios*. Buenos Aires: Biblos.

Gras, C., and V. Hernández. 2013. *El agro como negocio: Producción, sociedad y territorios en la globalización*. Buenos Aires: Biblos.

Gras, C., and V. Hernández. 2014. Agribusiness and large-scale farming: Capitalist globalisation in Argentine agriculture. *Canadian Journal of Development Studies* 35, no. 3: 339–57.

Grau, H.R., N.I. Gasparri, and T.M. Aide. 2005. Agriculture expansion and deforestation in seasonally dry forests of north-west Argentina. *Environmental Conservation* 32, no. 2: 140–48.

Grobocopatel, G. 2010. El modelo del poroto. *Página/12*. April 10.

Grosso, S., and C. Albaladejo. 2009. Los ingenieros agrónomos y la 'nueva agricultura': Des/reterritorialización de la profesión. *In Gras and Hernández* 2009b: 117–34.

Harvey, D. 1990. Between space and time: Reflections on the geographical imagination. *Annals, Association of American Geographers* 80, no. 3: 418–34.

Harvey, D. 2003. *The new imperialism*. New York: Oxford University Press.

Hernández, V. 2009. La ruralidad globalizada y el paradigma de los agronegocios en las pampas gringas. *In Gras and Hernández* 2009b: 39–64.

Hora, R. 2012. La evolución del sector agroexportador argentino en el largo plazo, 1880–2010. *Historia Agraria* 58: 145–81.

INDEC (Instituto Nacional de Estadísticas y Censos). 2015. Complejos exportadores: evolución año 2014. *Informe de prensa 08/05/15*. http://www.indec.gov.ar/.

Infocampo. 2011. Renova: el joint venture que invierte para ser líder en biodiesel. October 21. http://infocampo.com.ar/.

Infocampo. 2013a. Don Mario: La semillera argentina que continúa expandiéndose en la región. March 16.

Infocampo. 2013b. LDC Semillas, la empresa del grupo Louis Dreyfus para apostar a la genética de soja. December 7.

Infocampo. 2014a. Brasil: Afirman que el 43% del área sembrada de soja es con genética argentina. September 4.

Infocampo. 2014b. Los Grobo vuelve a la bolsa para financiarse. March 28.
Isakson, S.R. 2014. Food and finance: The financial transformation of agro-food supply chains. *Journal of Peasant Studies* 41, no. 5: 749–75.
James, C. 1997. *Global status of transgenic crops in 1997*. Ithaca, NY: ISAAA.
James, C. 2014. *Global status of commercialized biotech/GM crops: 2014*. Ithaca, NY: ISAAA.
Kazakhstan Newsline. 2014. Three Kazakhstani companies are among 10 world largest landowners. Factiva. August 21.
Kneen, B. 1993. *From land to mouth: Understanding the food system*. Toronto: University of Toronto Press.
La Nación. 2015a. Los precios del petróleo y la soja se derrumban mientras colapsan los mercados. August 24.
La Nación. 2015b. La soja y el maíz se negociaron con saldo negativo en la Bolsa de Chicago. September 2.
Lapegna, P. 2013. The expansion of transgenic soybeans and the killing of indigenous peasants in Argentina. *Societies without Borders* 8, no. 2: 291–308.
Lapegna, P. 2015. Genetically modified soybeans, agrochemical exposure, and everyday forms of peasant collaboration in Argentina. *Journal of Peasant Studies*. doi:10.1080/03066150.2015.1041519.
Lapegna, P. Forthcoming. Agricultural boom, subnational mobilization, and variations of violence in Argentina. In *Clientelism and violence in subnational Latin American politics*, ed. Tina Hilgers and Laura Macdonald.
Lawrence, G. 2014. Financialization. *Journal of Peasant Studies* 41, no. 3: 421–26.
Leguizamón, A. 2012. Review of Industrializing the Corn Belt: Agriculture, technology and the environment, 1945–1972 by J. L. Anderson. *Food, Culture and Society* 15, no. 1: 148–51.
Leguizamón, A. 2014a. Modifying Argentina: GM soy and socio-environmental change. *Geoforum* 53: 149–60.
Leguizamón, A. 2014b. Roundup Ready Nation: The political ecology of genetically modified soy in Argentina. PhD diss., The Graduate Center, City University of New York.
Lódola, A. 2008. *Contratistas, cambios tecnológicos y organizacionales en el agro argentino*. Santiago de Chile: CEPAL.
Los Grobo. 2015. Company's Website. http://www.losgrobo.com.ar/.
Luyt, I. 2013. *Emerging investment trends in primary agriculture: A review of equity funds and other foreign-led investments in the CEE and CIS region*. Rome: FAO.
Magdoff, F., J.B. Foster, and F. Buttel, ed. 2000. *Hungry for profit: The agribusiness threat to farmers, food, and the environment*. New York: Monthly Review Press.
Magdoff, F., and B. Tokar, ed. 2010. *Agriculture and food in crisis: Conflict, resistance, and renewal*. New York: Monthly Review Press.
MAGyP (Ministerio de Agricultura, Ganadería y Pesca). 2015. Sistema integrado de información agropecuaria. http://www.siia.gov.ar/series/.
McMichael, P. 2012. The land grab and corporate food regime restructuring. *Journal of Peasant Studies* 39, no. 3–4: 681–701.
McMichael, P. 2013. *Food regimes and agrarian questions*. Canada: Fernwood Publishing.
Murmis, M., and M.R. Murmis. 2012. Land concentration and foreign land ownership in Argentina in the context of global land grabbing. *Canadian Journal of Development Studies* 33, no. 4: 490–508.
Murphy, S., D. Burch, and J. Clapp. 2012. *Cereal secrets: The world's largest grain traders and global agriculture*. Oxford: Oxfam International.
New York Times. 2015. Company info: Adecoagro S.A. http://topics.nytimes.com/top/news/business/companies/adecoagro-sa/index.html.
Newell, P. 2009. Bio-hegemony: The political economy of agricultural biotechnology in Argentina. *Journal of Latin American Studies* 41: 27–57.
Nidera Semillas. 2015. Company's Website. http://www.niderasemillas.com.ar/
Otero, G. 2012. The neoliberal food regime in Latin America: State, agribusiness transnational corporations and biotechnology. *Canadian Journal of Development Studies* 33, no. 3: 282–94.
Pengue, W.A. 2005. Transgenic crops in Argentina: The ecological and social debt. *Bulletin of Science, Technology & Society* 25, no. 4: 314–22.
Pierri, J., and M. Abramovsky. 2009. Legislaciones de patentes de semilla y uso de insumos en la producción de soja en la Argentina y en Estados Unidos 1990/2006. *Realidad Económica* 244: 88–117.

Princen, T. 1997. The shading and distancing of commerce: When internalization is not enough. *Ecological Economics* 20, no. 3: 235–53.
Princen, T. 2002. Distancing: Consumption and the severing of feedback. In *Confronting consumption*, ed. Thomas Princen, Michael Maniates and Ken Conca, 103–32. Cambridge, MA: MIT Press.
Regunaga, M. 2010. *Implications of the organization of the commodity production and processing industry: The soybean chain in Argentina*. Washington, DC: World Bank, Latin America and the Caribbean Region.
Rosendo González, P. 2014. Argentina's Grobo seeks M&A as agribusiness consolidates. *Bloomberg Business*, September 3.
Russi, L. 2013. *Hungry capital: The financialization of food*. Winchester, UK; Washington, DC: Zero Books.
Schnaiberg, A. 1980. *The environment, from surplus to scarcity*. New York: Oxford University Press.
Schnaiberg, A., and K.A. Gould. 1994. *Environment and society: The enduring conflict*. New York: St. Martin's Press.
SEC (Securities and Exchange Commision). 2007. Form 6–K for Cresud, SA. http://www.sec.gov/Archives/edgar/data/1034957/000119312507168315/d6k.htm.
SeedQuest. 2013. Strengthened presence in Latin American soybean market – Bayer CropScience to acquire Argentinian seed company FN Semillas S.A. December 5. http://www.seedquest.com/
SeedQuest. 2014. Bayer CropScience to strengthen global seed growth business with acquisition of Biagro group. March 6.
Selén, L. 2014. Soja y poder económico. *Télam*. http://www.telam.com.ar/informes-especiales/1-soja-y-poder-economico.
Simoes, A. 2015. *The observatory of economic complexity*. http://atlas.media.mit.edu/.
Télam. 2014. Cristina inauguró la planta de procesamiento de soja más grande del mundo. October 21.
Terazono, E., and N. Hume. 2015. Time in the sun for Glencore's agribusiness. *Financial Times*, March 4.
Terazono, E., and J. Webber. 2013. Argentine agribusiness moves to Brazil. *Financial Times*, April 25.
Teubal, M. 2006. Expansión del modelo sojero en la Argentina: De la producción de alimentos a los commodities. *Realidad Económica* 220: 71–96.
Teubal, M. 2008. Genetically engineered soybeans and the crisis of Argentina's agriculture model. In *Food for the few: Neoliberal globalism and biotechnology in Latin America*, ed. Gerardo Otero, 189–216. Austin: University of Texas Press.
The Economist. 2014. Farming without fields. January 4.
Trigo, E.J. 2011. *Fifteen years of genetically modified crops in Argentine agriculture*. ArgenBio (Argentine Council for Information and Development of Biotechnology). http://www.argenbio.org/.
Trigo, E.J., and E.J. Cap. 2003. The impact of the introduction of transgenic crops in Argentinean agriculture. *AgBioForum* 6, no. 3: 87–94.
Turzi, M. 2010. The political economy of soybean production in Brazil, Argentina, and Paraguay. PhD diss., John Hopkins University, Baltimore, MD.
USDA (United States Department of Agriculture). 2015. Production, supply, and distribution (PS&D) database. http://www.fas.usda.gov/.
USSEC (US Soybean Export Council). 2011. *How the global oilseed and grain trade works*. Chesterfield, MO: US Soybean Export Council.
Valor Económico. 2013. Brazil: Buy-back for BrasilAgro. Factiva. September 5.
Verdeca. 2013. Verdeca and GDM Seeds collaborate to develop and commercialize stress-tolerant soybeans. December 19. http://www.verdeca.com/news/.
Wave. 2015. Los Grobo. http://www.wave-innovation.com/en/los-grobo.html/.

Which territorial embeddedness? Territorial relationships of recently internationalized firms of the soybean chain

Clara Craviotti

This contribution focuses on the territorial relationships established by companies of Argentinian origin that have extended their scope of action to other MERCOSUR countries (mainly Brazil) in the first stages of the soybean chain. Although some features of grain production and particularly of soybean enable a temporary commitment of these firms with local spaces, they develop some kind of foundations as a necessary condition to operate. A relevant side of their forms of territorial embeddedness is the strong selectiveness of local spaces and actors promoted by their behavior. Some local actors are in fact included; in the long term, however, their capacity to capture value is constrained since they are not able to control the terms and conditions of their incorporation.

Introduction

In the last two decades, several agri-food activities have shown strong dynamism in MERCOSUR countries. Their growth is related to the switch from state-led efforts to build domestic industries and accumulation around national markets that predominated in the middle part of the twentieth century, to 'outward-oriented development' involving export-promotion strategies (Robinson 2012). It occurred in a period of falling trade barriers, the emergence of the World Trade Organization (WTO) WTO and the policy prescriptions associated with the 'Washington Consensus' (Gereffi 2014). In this context, a new division of labor emerged on a global scale, where the agricultural sector of these countries is increasingly engaged in the production of *flex crops* that can be used as food, feed and biofuels (Borras et al. 2011).

Soybean is a crop with such characteristics and will be a central point of consideration in this contribution. In MERCOSUR countries, the soybean chain is the most integrated to the world trade: in 2010 nearly 80 percent of Argentinian soy was destined to the international markets, whilst in Brazil the figure was 65 percent (CIARA n.d.; ABIOVE n.d.). With nearly 47 million hectares devoted to this crop that same year (Catacora Vargas et al. 2012), MERCOSUR is the larger production area in the world, with China and the European Union as main destinations.

Internally, the expansion of soybean has implied changes in the regional division of production; it has caused the displacement of other crops and the expansion of the agricultural frontier, as well as migration processes and urban changes. In Brazil, the cultivation of soybean started in the south (particularly in the states of Rio Grande do Sul and Paraná) and expanded since the 1990s to the centre-west and centre-north of the country, to the

Cerrados and part of the Amazon region. In Argentina, soybean was initially cultivated in the Pampean region, but in the last 20 years it has spread to the northwest and northeast. However, the first area still represents 85 percent of the cultivated area (SIIA n.d.).

This expansion was accompanied by land grabbing processes that have been considered elsewhere (Murmis and Murmis 2012; Wilkinson, Rydon, and Di Sabbato 2012) and will not be the focus of this contribution. Instead, I aim to explore the type of relationships established with local spaces and actors by firms that hold a central position as drivers of the soybean expansion in the production stage, and the production of the seeds required for cultivation. MERCOSUR represents an interesting scale for such analysis, since some leading Argentinian firms adopted a multinational structure in the last decade and started activities in neighboring countries, with a special focus on Brazil.

Drawing upon previous studies, a collection of articles published in Argentinian newspapers and magazines during 2005–2014, relevant documents, in-depth personal interviews with key informants and statistical data, the aforementioned issue will be explored at an aggregate level through a comparative lens, considering similarities and differences in the strategies deployed by prominent firms of both stages. The comparative dimension will allow us to show that their behavior does not entail full-fledged delocalized relations. These agents need some kind of territorial foundations, so a selective inclusion of local spaces and actors takes place. The latter, in turn, implies uneven access among different social groups and individuals who are placed in distinct ways in relation to these interconnections.

The paper is organized as follows. Some key theoretical concepts will be briefly presented next as a basis for analyzing the *modus operandi* of a handful of leading Argentinian firms in two stages of soybean production, a task that will be done in the third section. Finally, some concluding remarks will be presented.

Territorial embeddedness, organizational schemes and selective incorporation

Earlier perspectives on globalization usually placed emphasis on its relationships with deterritorialization processes – that is, with the emergence of global systems that are able to free themselves from the specific determinations of a particular territory (Gallicchio 2003). The ability of global supplier networks to partition tasks and distantiate from the natural and organic geographies for which they are responsible (Morgan, Marsden, and Murdoch 2006) and from the social processes that make possible the production and distribution of a given commodity led to the proposal of homogenization as the inevitable outcome, and in the end brought about a certain disregard of the specific conditions of production. The latter are particularly important in the agri-food sector considering the fact that, except for some cases, activities need to be carried out on land and have a substantial relationship with soil. A great number of tasks still require large numbers of workers, their involvement and their cooperation. Consequently, materiality and territoriality are important (Bridge 2008).

On the other hand, some geographers such as Haesbaert (2011) have put into question the existence of a deterritorialization process, stating that there is no destruction of territories without their configuration on new foundations, and proposing multiterritoriality as an alternative concept. The latter implies the ability of a given actor to access different territories simultaneously. Haesbaert alerts about the complexity of multiterritoriality, pointing out that territories in this century are always network-territories, because they are related in varying degrees to hierarchical or complementary flows. Network-territories are spatially discontinuous, dynamic (with different degrees of mobility) and more susceptible to overlap. While zone-territories are dominated by continuity and co-presence, network-territories are characterized by complex relationships of absence–presence. This duality has both tangible and intangible effects that must be carefully acknowledged.

Alongside this strand of thought, a growing field of studies has developed that deals with the analysis of the relationships between the production, trade and consumption of goods on a global scale (Dannenberg and Kulke 2014). The global commodity chains (GCC) approach and its later development, global value chains (GVC; e.g. Gereffi, Humphrey, and Sturgeon 2005) are probably some of the most well known. Both are concerned with the governance of global value chains and the ways in which corporate power can shape the distribution of profits and risk. Yet the global production network (GPN) perspective has aimed to overcome some of the shortcomings attributed to this approach, namely its lack of consideration of the institutional and social contexts out of which firms arise and in which they are embedded (Henderson et al. 2002). It also points to the social processes involved in producing goods and services (and to reproduce capital, knowledge and work) as well as to the existence of horizontally (not only vertically) organized flows. Hence, it is less deterministic than the GVC approach in terms of the conceptualization of the relations between actors. It takes into accounts that inter-firm networks link societies with significant social and institutional variations, and different capabilities of the state to manage the economy. Territoriality – the geography of GCC – is explicitly considered. A comprehensive framework that analyzes territorial, social and network embeddedness is thus proposed (Hess 2004).

Taking this perspective as a point of departure, it is important to study how GPNs constitute and are reconstituted by the economic, social and political schemes of the places where they inhabit. GPNs are multiscalar and are constructed by agents with power asymmetries (Henderson et al. 2002). The differences between agents, their respective capacities (in terms of access to variable combinations of resources), their practices (how they exert power) and the nature of the relations between them are important issues of research. Territorial embeddedness – understood as the degree of commitment of an actor to a particular place – is an important question since the creation and capture of value depends on it. The answer will depend on the analysis of specific, although not necessarily place-bounded, conditions.

This perspective considers that GPN are not necessarily de-territorialized. However, corporate actors have the ability to transcend boundaries between territories, while others are more limited – and therefore constrained – by their spatial contexts (Henderson et al. 2002). In the same vein, other authors argue that large firms are *aterritorial* because they have the capacity to transcend regions (Rallet and Torre 2004). If the territorial adscription of the production processes can be managed by global strategies, this brings about the possibility of using different territories in the most efficient way, taking advantage of specific local conditions (Delgado Cabeza and Gavira Alvarez 2006).

Network production is particularly suitable for global value chains since it allows deverticalization through subcontracting as well as flexibility in operation. In this case the ownership of resources is not as important as their control. Networks can also be seen as relational processes and structures in which, and through which, power is exercised (Dicken et al. 2001, 90). Again, territorial embeddedness remains an open issue since some networks are relatively more localized and dependent on the advantages of territorial agglomeration, while others are controlled at a distance, when the key actors are spatially distanciated from the sites where events happen (Dicken et al. 2001, 96). The fact that production networks cross nation-state boundaries means that territories are inserted into networks, so there is a mutually constitutive process: while networks are embedded within territories, territories are, at the same time, embedded into networks (Dicken et al. 2001, 97).

Although a network type of production is frequently seen in the industrial sector, it can also be found in agri-food. In MERCOSUR countries, this type of organizational scheme emerged for grain production in the 1990s–2000s, bringing about a separation between land ownership and management. This phenomenon implies that the opportunity cost of land is incorporated in the decision-making process, and thus the capitalist logic of production enters fully into the primary sector (Bisang and Anllo 2014).

An important issue is the consideration of who is excluded from such networks and why. Firms in the agri-food sector may unfold different logics of territorialization; they can construct a fluctuating territory whereby producers are to some extent replaceable, or they can develop closer and frequent ties with them (Margétic 2006). Although a large number of studies has focused on the potential role of contracting in linking small holders to agribusiness firms (Dannenberg and Kulke 2014), additional evidence is needed on their conditions of exclusion and inclusion, particularly on the positive and negative implications of the complex ways in which they are incorporated into these chains (Du Toit 2009).

Besides, although belonging to a network implies complementarity and some kind of stability in relationships so as to reduce uncertainty (Hemsing 2002), the latter does not imply equality among network constituents, as well as immovability in its conformation. Some of these aspects will be illustrated in the following sections.

Regional expansion and territorial embeddedness of soybean production in the MERCOSUR countries

Preliminary considerations

A general overview of the soybean global production chain involves the following stages: input providers for primary production; soybean producers; services providers; crushers, international traders; producers of related food products, bio-fuels and other industrial inputs; main international importers (Regunaga 2009). In the MERCOSUR countries, downstream stages are deeply concentrated and coordinated; large crushing firms are also traders (some even are input providers and carry out production). International firms such as Bunge, Cargill, ADM and Dreyfus are the lead firms in these stages.[1]

Upstream, two important stages are the production of seeds as inputs for cultivation and the production of soybean itself. Before considering the behavior of some of the leading actors in both stages, it is worthwhile noting some features of the crop that impact on its embeddedness: first, soybean can be sown and harvested in less than six months and activities are mechanized, so they are relatively ephemeral. Second, the location of the crop may vary to some extent from one year to another thanks to its adaptability to different agro-ecological conditions (from subtropical to colder areas, and from more humid to drier areas) through the so-called maturity groups of the seeds.[2] While certain environmental parameters still must be respected, technological development has extended the possibilities

[1]However, national companies have a greater weight in Argentina than in Brazil, since they control over 30 percent of exports of flour and soybean oil (Wesz Jr. 2014a).

[2]The latter defines the type and speed of growth that a soybean cultivar will have. Maturity groups are defined numerically from 00 to VIII.

of carrying out soybean production in more extended latitudes. So production is rather dispersed and encompasses an extended area.

Regarding the location of processing facilities, in Argentina they are more geographically concentrated than in Brazil.[3] However, their very existence depends on the policies adopted by the national governments. In Brazil, the 1996 Kandir Law eliminated a tax that affected exports of raw materials and kept the tax burden on industrialized products (Wesz 2014a). Consequently, in 2010, 65 percent of the soybeans exported were non-processed (ABIOVE 2010). On the contrary, in Argentina, taxes encourage local processing. That same year 69 percent of exported soybeans were processed (57 percent as flour and 12 percent as oil).

Focusing on production-related activities, it can be stated that the regulatory environment affected the expansion of the crop in both countries. This can be seen in the case of genetically modified (GM) seeds, the main component of the technological package employed today.[4] In Argentina their use was approved by a public body (the National Commission for Knowledge and Use of Biodiversity – CONABIA) formed by members of the state, the industry and the scientific community that failed to represent the incipient public debate on the issue, so the first GM soybeans were released in 1996. The area sown with herbicide-tolerant soybeans increased from less than one percent of the total soybean planted area in 1997 to more than 90 percent in 2002 (Trigo and Cap 2003). This widespread adoption was facilitated by the existence of an intellectual property system that enabled the non-patentability of the gene and by the autogamic nature of the seed, whose reproduction does not alter its initial characteristics (Bisang and Sztulwark 2007).

While this was happening in Argentina, Brazil did not allow GM seeds until 2003. Nevertheless, producers of the southern states of the country obtained seeds illegally from Argentina. Before the final approval in 2005 of the soybean resistant to glyphosate by Law 11,105 (the Biosafety Law), the Brazilian government had already authorized grain marketing of GM soybean in 2003/2004 and 2004/2005 through provisional measures (Fuck and Bonacelli 2009). Ten years later, it is estimated that GM soy occupies 92 percent of the area planted with this crop in Brazil.

Second, in both countries there have been scarce policies regarding the negative aspects (environmental and social) associated with the territorial expansion of the crop. For instance, in 2007 Argentina approved legislation that protects native forests, defines three categories of conservation areas, establishes the types of activities that can be carried out in each of them and decentralizes the responsibility to establish zoning to provinces. This legislation was enacted after much of the deforestation was already completed, and, in some instances, procedures were resisted and the zoning requirements were not followed (REDAF 2012). In Brazil, the Forest Code approved in 2012 granted an amnesty to landowners who deforested illegally before 2008, reduced the area to be reforested, and introduced conservation measures that could pave the way for commoditizing standing forests (Soares Filho et al. 2014).

[3] In Argentina, most of the soybean is produced near the ports (within a distance of around 200 to 300 kilometres). Although storage facilities are located in small- and medium-sized towns, crushing facilities are geographically concentrated (Regunaga 2009). The metropolitan area of the city of Rosario in Argentina holds 83 percent of the installed capacity of the oil industry. Over 75 percent of agribusiness exports are shipped through its port terminals, and 90 percent of soybean exports (Bisang and Anllo 2014). The area also handles soybean production coming from Paraguay through the Parana River.

[4] The other components are no-till farming (direct sowing) and glyphosate.

Changes in soybean production and the emergence of transnational production firms

Soybean expansion is also associated with changes in the agrarian structure of MERCOSUR countries, which in turn affect the embeddedness of agrarian production. Although soy is carried by about 73,000 farmers in Argentina (an estimate of 25 percent of total producers according to the latest data available) nearly half of the production is in the charge of a smaller number (6 percent) who in 2008 planted individually more than 500 hectares (ONCAA 2008 in Regunaga 2009).

Some soybean producers live in places distant from where they develop farming activities. Their *modus operandi* helps to explain their detachment from the sites of production. More precisely, in the Argentinian grain sector a network-based system of production has developed, which some analysts have likened to the one prevailing in the most dynamic manufacturing industries (Bisang, Anlló, and Campi 2008). This form of production is exemplified by sowing pools (*pools de siembra*) that were constituted by the mid-1990s and expanded in the 2000s. These involve a flexible use of factors of production (land, labor and machinery) in each production cycle. The firms that organize sowing pools do not necessarily own the land that they operate but, rather, lease it on a short-term basis and outsource farm tasks through machinery contractors. Hence, sowing pools are able to 'translate' the flexibility of a product capable of various uses (such as soybean) to their business model; they are asset-light companies with low investments in fixed capital. Their core assets consist of their ability to reap the benefits of a large scale, their access to funding sources to obtain working capital – some of them have even created trust funds to attract investors – and their ability to coordinate and supervise productive, commercial and financial tasks.[5]

Flexibility may be seen not only in pools' organizational structure and interaction with other actors, but also in their primary mechanism of access to land: leasing.[6] The flip side of this phenomenon is the emergence of a stratum of small-scale rentiers, particularly in the Pampas region. The neoliberal policies adopted in the 1990s in Argentina led several small farmers to quit the activity and rent out their land. Some of them turned to offering their work as machinery contractors (Craviotti and Gras 2006); their retreat from production amplified the land market for firms willing to rent land. The demand for machinery contractors and other services was viewed as positive by some local actors who minimized the concentration of production brought about by pools. According to a farm organization's representative, 'pools took and gave work, and that helped many people to move forward'.[7]

[5]These large production firms also develop upstream agreements with trading and crushing companies to better manage logistics and to implement forward sales for risk management (Regunaga 2009).

[6]It should be pointed out, however, that the increase in leasing is driven not only by sowing pools but also by other farmers of the country who expanded their operations through this mechanism, mainly in grain production.

[7]In other areas of Argentina, another type of phenomenon has arisen associated with the dynamism of export agriculture. The fact that some peasants do not hold legal titles on the land they inhabit opened the door to different ways of gaining control of land that range from voluntary purchase to violent evictions. A recent study by the Ministry of Agriculture documented existing conflicts regarding land tenure (Bidaseca et al. 2013) and indicated that almost 80 percent of them involved holders (*poseedores*), namely people who, while they have been occupying their plots of land for decades, do not have legal titles to them. The Argentinian Civil Code recognizes the possibility of acquiring property rights by those who have lived on the same land for more than 20 years, and made improvements on it. But the acquisition of these rights requires a Prescription Trial, a procedure not always available to small farm holders and peasants. There have been, however, some recent institutional

The firms that organize the most important sowing pools are not traditional landowners of the country, although their Chief Executive Officers (CEOs) often come from farm families (Murmis and Murmis 2012). They have constructed a strong narrative as innovative entrepreneurs, distancing themselves from the traditional landowner class of the country (Hernández 2007).[8] Qualitative changes are clearly associated with this way of farming. First, because of the large operating capital required and the mobilization of this capital through trust funds and other types of financial instruments, pools can be seen as part of the process of financialization of agriculture. Second, they are instances of the decoupling of farm ownership and farming. Third, they operate on a large scale and adopt an industrial type of production based on the use of modern technology (such as precision agriculture) and agrochemical input (Craviotti 2015).

Over time, some of the firms that organize large sowing pools have developed a complex structure which involves not only grain production but also other activities.[9] Another aspect of their development has been their ability to evaluate and take advantage of different local conditions to *manage* their territorial adscription. Some of them started operations in other MERCOSUR countries; in the 2010/2011 campaign, big firms of Argentinian origin operated together about 700,000 hectares only in Brazil (Wesz 2014b). Regional expansion allowed them to diversify climatic and political risks associated with institutional contexts (Gras and Sosa Varrotti 2013). It was also related to their need for scale to diminish operating costs and the difficulties in accessing land for rent in Argentina (Manciana, Trucco, and Piñeiro 2009; Bisang and Anllo 2014).

The fact that the agents who control these transnational firms usually live far away from production sites and develop their activities in multiple MERCOSUR locations makes the issue of their degree of embeddedness particularly relevant. Technological developments such as geo-referenced information and online communication imply the possibility of simultaneously controlling plots located in different places, and facilitate the complex mix of presence–absence sustained by Haesbaert (2011).

On the one hand, a large number of aspects support the *aterritoriality* of this corporate agriculture. According to Guibert et al. (2011), the actors of this production model are flexibly associated with particular places: they move more and stay for less time in the same place. They generate non-contiguous territories and interstices of excluded spaces, whose features leave them outside the dynamic of the system (Reboratti 2003).

Regarding local impacts, this strand of thought maintains that there is a weakening of the local dimension of agricultural activity due to the lesser involvement of these actors in economic, social and political life. Former 'social spaces' have been transformed into mere 'productive spaces' because there is no need for developing social activities in them (Albaladejo 2013). In the same vein, it is argued that these firms sustain relationships with local actors and spaces as long as they contribute to their business. Relationships would also vary according to the stage of the chain: for commercial and financing activities, they resort to non-local agents; for transport and production to local people (Gras and Hernández 2013).

innovations – such as the approval of the 27,118 Law on 'Historical Reparation of Family Farming' in 2014, which includes the suspension of eviction trials for three years.

[8]This is a deeply controversial issue in the Argentinian literature, with well-known perspectives stressing the long-lasting importance of the landowner class based on the property of land (see Basualdo 2010).

[9]These firms used most of their profits for horizontal growth through leasing greater amounts of land. However, some of them have also invested in vertical integration activities and, to some degree, in the acquisition of land (Murmis and Murmis 2010).

The expansion of soybean would also remove diversity from the territories involved: homogenization in both actors and spaces takes place because the first are fewer and apply a similar productive organization in different locations (Guibert et al. 2011; Bisang and Anllo 2014).

However, these firms not only constitute the conditions of the places where they operate, they are also reconstituted by them (Henderson et al. 2002). They were not able to transfer mechanically to other countries their business model based on outsourcing productive tasks to machinery contractors. When compared with Argentina, in Brazil there is no such offer of service providers, so they established contracts with local farmers whereby they offered them a fee for their land and a fee for their machinery, and even a fee for their work (Bell and Scott 2010; Wesz Jr 2014a). This kind of arrangement enabled firms of non-local origin to take advantage of farmers' practical knowledge about local productive conditions. As time has passed, some of these firms have also bought land in Brazil to carry out grain production directly (Wesz Jr 2014b).[10]

Large production firms also need some sort of material foundations at the local level. In Argentina, one of these firms organizes its operation through production modules or 'clusters'. Each of them has a technician in charge of coordinating the tasks in plots which may belong to different landowners. The firm also has commercial offices where input items and grains are sold, which in turn are supported by regional service centers with grain storage facilities and agrochemical warehouses (Ordoñez and Nichols 2003; Bell and Scott 2010; Ederer 2013).

A relevant side of their territoriality is the role played by land. Although some approaches stress that in these firms, soil is a mere 'item' of the production function, not even the most important one (Bisang and Anllo 2014), they have departments that assess the quality of soils before decisions to buy or rent are taken (Gras and Sosa Varrotti 2013). Usually, part of the land operated belongs to their shareholders (Azcuy Ameghino 2007; Murmis and Murmis 2014). A survey of the most outstanding firms in Argentina (Manciana, Trucco, and Piñeiro 2009) indicated that although they diversified risks by planting in different places, the Pampean region was prioritized. Besides the better quality of soils, the latter could be related to the location of processing and exporting facilities, which impacts on transport costs.

Regarding the immaterial foundations of this corporate agriculture, the success of large production firms would also be related to their capacity to establish long-term relations with service providers, whom they support acting as collateral regarding credit institutions (Guibert and Sili 2011). Long-term relations are also sought with landowners who lease their land to them (Bell and Scott 2010; Ederer 2013).

Another immaterial foundation of this type of agriculture is its frequent reliance on local people who act as social brokers. Concerning the *modus operandi* of one of these Argentinian firms in one district of the Pampas region, evidence gathered by Vertiz indicates that the company's policy was for its technicians to live in the places where they worked and to favor the employment of local workers so as to strengthen its relations with communities. Regarding the operation of the same firm in the Mato Grosso state of the Cerrados, Wesz

[10]One of these lead firms, which operated in 240,000 hectares in Brazil in 2010/2011, profited from the situation of indebtedness of some producers to buy large tracts of land (Wesz Jr. 2014). The latter implied a change in its business model. Gras and Sosa Varrotti (2013) argue that this strategy was facilitated by the introduction of foreign capital in its shareholding structure, and was associated with the interest of institutional investors in land due to the rise in the prices of commodities.

(2014a) indicates that, recognizing the fact that the company was new in Brazil, it chose to hire local technicians. To rent their land, producers valued not formal contracts but their acquaintance with people they perceived as honest; on the other hand, the firm also needed the technicians' knowledge of reliable local producers. No participation in local events or communication of the firm's activities in the local media was, however, found.

Taking all these elements together, it could be stated that these firms need some kind of material and immaterial foundations in the local spaces where they carry out their activities. It could also be hypothesized that their territorial embeddedness varies according to the type of strategy pursued, e.g. horizontal or vertical growth, the latter implying a greater investment in local facilities and employment.

In any case, a selective network is formed in which, and through which, power is exercised (Dicken et al. 2001). Regarding the conditions of inclusion of local actors, it is clear that a reduced number participate in these networks and hold a subordinate position in them. Large production firms have a very centralized structure in the planning process, the provision of inputs and the decisions regarding renting land (Manciana, Trucco, and Piñeiro 2009). Some of them set operation protocols for each zone and crop that imply a reduced *room for manoeuvre* for local technicians, an issue that may bring about inefficiencies. In some cases the contractual status of these technicians has changed in recent years, from being firms' employees to receiving fees according to the number of hectares supervised (Vértiz 2014). Machinery contractors are always paid by hectare or by the volume of grains collected, but large firms are more able than other farmers to negotiate prices because of the amount of land they operate, among other reasons.

Besides managing conditions of inclusion, another important issue is the power of these firms to eventually *disconnect* local actors. Changing prices for soybean in the global markets and rising costs in Argentina have led them to decrease drastically the amount of land leased in the last few years and to change the business model adopted, triggering a process whose consequences are barely known.[11] If the number of land contracts has diminished, the terms of those persisting have probably changed. The retreat of these firms from some spaces has undoubtedly impacted on their network – basically the technicians that supervise the tasks, machinery contractors and other workers.[12]

However, effects may be ambiguous and contingent on the features of local places. The presence of a large firm in the Pampean region has the power of regulating the prices of land, impacting negatively on local farmers who want to lease.[13] Reduced competition in the land market could be beneficial for them. In the north of Argentina, on the contrary, the retreat of large firms could help to stop deforestation and other negative impacts on local peasants.

Changes on the input stage: transnational seed firms

The seed industry has played a key role since many decades ago in Argentina because crop production is less intensive than in developed countries, and improved seeds have been

[11]For instance, *Los Grobo* operated 86,000 hectares in the 2011/2012 campaign in Argentina and reduced them to 45,000 in 2012/2013 (*Infocampo*, 20 to 27 March 2015). *El Tejar* reached nearly 300,000 but only operated 30,000 in 2013. It also moved its headquarters to Brazil (*La Nación*, 18 April 2013).

[12]One of these lead firms reported in previous years 400 leasing contracts, 3800 service providers, 49 storage plants and 1004 workers (Los Grobo 2011).

[13]As stated for other countries with a strong tradition in grain production (Sommerville 2013), farmers may benefit from the rapid appreciation of land parcels associated with a dynamic agricultural sector, but this simultaneously constrains their ability to expand their operations.

traditionally the most important tool to increase productivity (Regunaga 2009). The sector also plays an important role in technology transfer and in financing this key input to farmers. In the case of GM soybeans, there has been a division of tasks between local firms and the main international seed companies, most of which are also producers and distributors of chemical products. Transgenic traits have been for the most part developed by the latter, while some Argentinian firms hold a leading role in the development of the varieties in which those genes are inserted. These firms do crossings for developing new varieties and, having selected the most suitable ones, they produce the *basic seeds* that will be later multiplied and delivered to farmers, either directly or by other agents.[14] Today, two firms control 90 percent of the Argentinian soybean market and cover the entire range of the varieties grown in the country; one of them is a local company.

The organization of the lead Argentinian firm shows the importance of its agreements with international biotech companies for incorporating transgenic traits to its varieties, and also with a myriad of agents that are involved in producing seeds for the company. On the one hand, the firm produces basic seeds in own and rented farms. These units, which in 2014 covered about 18,000 hectares, were located in the Pampean region (for short-group varieties) and in the north of Argentina (in the provinces of Tucumán and Salta), for large-group varieties. However, about 70 percent of the land for producing these seeds was located in the north of Buenos Aires and nearly 60 percent in only one district, where it accounted for nearly 10 percent of the area devoted to grain production. The amount of land involved and the prices paid – which are higher for seed production because of its requirements – implies that the firm defines the values of rented land at the territorial level. In this sense, the consequences on local producers who want to expand their scale of operation are similar to the existence of a large sowing pool.

Productive tasks are delegated to machinery contractors under the firm's supervision. They are a relatively stable group working for several years for the company, and most of them belong to the same area where its headquarters are located.

Besides the production of basic seeds under its direct supervision, outsourcing allows the firm to expand with less investment, and to diminish risks. So it has established agreements with *co-operators* who buy basic seeds and multiply, classify and sell them to farmers paying royalties to the company, which in turn controls the quality of the seeds produced. Cooperators have land and processing plants of their own, although they also rent land and facilities to third parties, a situation which ends up setting a 'network of networks'. According to different sources, between 80 and 150 co-operators would be linked to the firm in Argentina, although a small group multiplies the majority of the seeds; most of them belong to the Pampean region.

Second, the firm has also developed agreements with about 50–60 *multipliers* (farmers). In these cases, the firm sells them the basic seeds and then buys the multiplied seeds at a 'full' price (without transport and marketing costs), eventually with a bonus depending on the distance of the farmer to the shipping facilities. In practice, only a portion of the seeds produced is acquired by the firm because it overstates the planted area to be able to choose afterwards depending on the quality of the seeds and market situation. Although

[14]From a purely technical standpoint, the production of seeds for cultivation involves various stages that require different skills and scales (Anlló, Bisang, and Stubrin 2011): the development of new genetic material, the multiplication and processing of the seeds (drying, cleaning, classification and bagging), storage and commercialization. These activities can be carried out by the same agent (through vertical integration) or outsourced.

it manages a different over-sizing ratio in each region, it usually takes about 30 percent of the seeds produced. The company signs annual contracts with these farmers whereby all costs and risks are borne by them.

There are commercial aspects involved (price and payment terms) in these agreements, but also other issues. The firm carefully chooses the machine contractors and the multipliers (whether co-operators or farmers). In the case of the latter, it started from an initial base of producers who purchased seeds from the company, and later selected a group based on a set of criteria: the characteristics of their holdings (plots bigger than 30 hectares and with the best type of soils), their machinery (availability of 'axial' harvesters that provide less mechanical damage to grains) and their personal features, which can be synthesized in showing the 'profile of a seed producer' (that is, the farmer is involved in operation and decision-making, his/her team is reliable in respecting the best-suited moments to carry out the different tasks, and he/she is careful in cleaning the soil and the machinery employed). With these requirements, the firm formed a nucleus where trust and close relationships are central, although based on contracts. According to a firm's technician,

> If you revise the suppliers base there are loyal producers, who highly appreciate our brand, it ends up creating a very stable bond [...]. Thanks to the confidence that we have because there was always a transparency between the parties, when a situation arises in the field, these producers grab the phone and call you [....] and the same from us.

In the network formed for seed production, agents have different access to resources and position, and, consequently, different possibilities of negotiating their conditions of inclusion and permanence. Co-operators are in a better situation than multipliers since they fulfill multiple roles (production, classifying and selling of seeds), have a greater economic size and are better organized.

The differences in agents' possibilities can be seen in the outcome of the restructuring process carried out by the firm in the last year. The firm's strategy has been to increase outsourcing through co-operators and to transfer to them the multiplication agreements previously established with farmers, such that in 2014 outsourcing represented 65 percent of its seed production. This decision was influenced by the increase in costs and lower soybean prices, compared to the boom of previous years. To facilitate the transition to a more outsourced scheme, the company supported its co-operators in the management of their relationships with the producers–multipliers. The narrative of the firm's technician illustrates the firm's capacity (and will) to control those who participate in the network, and at the same time, to keep in hand a strategic asset - the multipliers - to stop them being captured by other competing firms in the seed industry:

> I cannot put these multipliers, which are the pick of the selection, I cannot lose them and leave them in the street for anyone to catch them. What we did, these multipliers that were strategic for us and still are, that they change to make seeds to the co-operators we have chosen [instead of making seeds directly for the company].

The rest of the seed production activities that are related to the industrial phase itself are carried out in facilities located near the production plots. The company had three classifying plants in the Argentinian Pampean region and a stake in a fourth plant located in the northeast of the country, but also hired out processing services to third parties (these may be the same co-operators, as their plants have idle capacity).

Regarding its workforce, the firm mobilizes different categories of workers for different tasks. For sorting and spraying the seeds and for the crossings involved in the development of new varieties, it employs temporary, 'unqualified' staff. Crossings are carried out in the summer and are a very demanding job due to the environmental conditions. The firm resorts to crews coming from the northern province of Santiago del Estero who are hired every year for a period of less than two months. The need to bring people from the north of the country is justified by the fact that they are more 'accustomed' to high temperatures, and refers to the presence of a flexible workforce involving the mobilization of vulnerable social categories, and to the constitution of a segmented labor market (Pedreño et al. 2015). On the contrary, the administrative and commercial staff has a permanent employment relationship with the company and live in the city where its headquarters are located, where it is considered one of the main employers.

With the marked devaluation of the Argentine currency in 2002, local seed companies started to display an internationalization strategy with soybean varieties as flagship products. Other countries of MERCOSUR, particularly Brazil, were the main destinations of their operations. It could be hypothesized that the relocalization of some of their major customers – i.e. sowing pools – influenced their expansion, since the two stages – seed production and production of soybeans – are closely related.

Social embeddedness is important for these firms, so they set agreements or partnerships with local entrepreneurs. In the case of the leading seed firm, it created a society in Brazil with local partners the same year (2003) that the Brazilian government authorized the commercialization of GM soybeans in the state of Rio Grande do Sul. This partnership allowed the firm to register varieties in two or three years, to afterwards license them to a network of multipliers. So it has established its operations in Brazil with the same business model it pursues today in Argentina, based on the outsourcing of production.

The expansion of the firm in Brazil was followed by Bolivia, Paraguay and South Africa. The United States was one of the last destinations: the country was portrayed by the firm as the finishing touch to strengthen within the 'big players' of the activity. However, Brazil is viewed as the market with the highest potential, since it is the second largest country in production of soybean after the United States, and has vast areas in the Cerrados where production can be expanded significantly.[15] The latter also seems to be an important pulling factor in the case of transnational production firms.

Along with its internationalization, the company developed the concept of 'Yields with no Borders'. The vision is that political divisions are of secondary importance; only latitude and climate matter and, accordingly, the most suitable varieties for each territory. However, other features continue to stress the importance of territoriality. The headquarters of the company in Brazil are located in Londrina, a place that the firm considered appropriate to visit the multipliers situated in the north and south of the country. For the purposes of research, the area is a transition, which enables producing specific varieties for the two regions. It should be added that the Empresa Brasileira de Pesquisa Agropecuaria (EMBRAPA) unit that evaluates the adjustment of soybean cultivars to the different climates and soils of the country is also located there.

In Brazil the firm expansion followed the expansion of the agricultural frontier led by soybean, where there was a process of relocalization of production and processing facilities

[15]According to Trigo et al. (2009), the Cerrado region – an area covering approximately 204 million hectares of land (or 24 percent of Brazil's entire land area) – has an estimated 40 to 50 percent under productive use.

from the southern to the centre-west states of the country.[16] So its first warehouses were located in the south, and the most recent ones in the Cerrados.

Research is conducted in all of the countries where the company has commercial presence. Yet the firm decided to install its largest laboratory of molecular markers in Brazil, with the idea of providing services to all its research stations. The main advantage of this strategy is the large number of samples that can be analyzed daily at a cost significantly lower than through a large number of field plots. Thus, biotechnology enables companies to reduce the incidence of the territorial issue, to increase their scale of operation and to reduce costs by shortening the time required to develop varieties.

From a broader point of view, the differences in results and growth prospects of the Brazilian market could generate a process of relocalization of the firm in the future. The view of key informants is that Brazil is pushing the firm's growth, while Argentina is losing relative importance. The so-called business climate matters in this process of delocalization/relocalization, because local interest rates are lower in Brazil, and also the country risk, if the firm seeks access to foreign funding. Government agribusiness policies also seem to be an important element at play for Brazil. Another factor that attracts the interest of Argentinian seed companies is the higher recognition of plant breeding rights.[17] Nevertheless, and when compared with Argentina, there is a greater weight of other private actors with whom the seed company competes and negotiates. The importance of big transnational companies in developing soybean varieties is higher in Brazil, and due to the volume traded, co-operators have much more power.

Weighing all these factors, the firm does not rule out a possible move of its headquarters to Brazil. In fact, it is channeling most of its investments there (e.g. the laboratory of molecular markers) and just over half of its permanent staff. The number of its research plots has increased in Brazil when compared with the rest of the countries, and now represents 50 percent of the total.

Differences in institutional contexts seem particularly relevant for defining the territorial scope of these recently transnationalized seed firms, as was the case with large sowing pools. The inclusion of 'new' territories is connected to the disconnection of others, and may bring about important consequences for firms' productive networks.

Final remarks

The comparative analysis of key actors in the initial stages of soybean production – the production of seeds for cultivation and the production stage itself – showed the important role of national groups in strengthening an agricultural production model based on flex crops and in reinforcing the leading role of the regional bloc in the global division of labor.

Although some features of grain production, and particularly of soybean, impact on its territorial embeddedness, in practice these agents develop some kind of tangible and intangible foundations as a necessary condition to operate in local spaces. It was also suggested that different ways of territorialization may be deployed according to the different strategies pursued; e.g. horizontal or vertical expansion. Invisible, immaterial foundations at the local

[16]According to Wesz Jr. (2014), the states of Mato Grosso do Sul, Goias, Mato Grosso, Bahia, Amazonas and Piauí, which together held 25 percent of the processing capacity of the country at the beginning of the 2000s, grew to 45 percent in 2009.

[17]The company estimated that 38 percent of the seeds sown in Argentina involve Plant Breeding Rights. In Brazil, 60 percent; in Uruguay, 100 percent; in Bolivia, 65 percent; in Paraguay, 40 percent.

level are important for firms of non-local origin, so they frequently draw on local technicians and farmers, visualized as social brokers for accessing key resources and as a source of legitimation. Corporate social responsibility activities are also carried out in selected spaces. *Aterritoriality* is perhaps a misleading word for their behavior, since it drives us to the absence of territoriality rather than to the different modalities that are employed for inserting different territories into their productive networks, exerting presence and profiting from local realities.

A relevant side of territorial embeddedness, however, is the selectiveness of local spaces and actors that these firms promote, and the kind of relations they establish with them. A key feature for the success of some recent transnationalized firms in MERCOSUR has been their ability to establish networks to obtain complementary assets, achieve flexibility and diversify risks. Networks could also be seen as legitimation mechanisms (Martins and Bevilaqua 2013). However, and following Massey (1994), different social groups and different individuals are placed in very distinct ways in relation to these interconnections, and this has consequences at the territorial level. Agents have different access to resources, hold a different position and, consequently, have different possibilities of negotiating their conditions of inclusion and permanence. The dimension of control exercised by lead firms seems to prevail above other components, either in the production or in the seed production stage.

Firms' strategies lead to new forms of uneven development between those who are participating in their networks and those who are not, but there are also internal hierarchies inside them. Some local actors are in fact included and may obtain some benefits of their incorporation; in the long term, however, their capacity to capture value is constrained since they are not able to control the terms and conditions of their incorporation. Other impacts at the local level are the displacement of other actors who sustain other logics of reproduction.

Territorial embeddedness also has a dynamic component: local actors and spaces that are strategic for the firms will be safeguarded, while others that are at the margin of the system will be subjected to unstable relations and even may be left behind when contexts change. So far, little is known about the particular places that continue being fundamental in the re-deterritorialization process practiced in the last few years by lead firms at the MERCOSUR level, and of the variable consequences of their retreat from other spaces.

Certainly, further research is needed on the modalities of territorial embeddedness of recently transnationalized firms and their evolution over time in specific areas. Two different lines of inquiry seem promising in this respect: the capacity to profit from the particularities of localities, and the degree to which they (still) are constrained by them.

Acknowledgements

The author would like to thank key informants for their support during field work, and the BRICS Initiative for Critical Agrarian Studies which provided a small grant for carrying out research. She would also like to give thanks for the suggestions and comments of two anonymous reviewers of *JPS*. The usual disclaimers apply.

Disclosure statement

No potential conflict of interest was reported by the author.

References

ABIOVE (Brazilian Association of Vegetable Oil Industries). n.d. Statistics. www.abiove.org.br

Albaladejo, C. 2013. Dinámica de la inserción territorial de la agricultura pampeana y emergencia del agribusiness. In *El agro como negocio: producción, sociedad y territorios en la globalización*, ed. C. Gras and V. Hernández. Buenos Aires: Biblos. pp. 67–95.

Anlló, G., R. Bisang, and L. Stubrin. 2011. *Las empresas de biotecnología en Argentina*. Colección Documentos de proyecto. Santiago de Chile: Naciones Unidas- CEPAL.

Azcuy Ameghino, E. 2007. 'Pruebe a nombrar de memoria cinco empresas que estén explotando campos … ' Propiedad y renta de la tierra en Argentina a comienzos del siglo XXI. *Revista Interdisciplinaria de Estudios Agrarios* 26/27: 123–40.

Basualdo, E. 2010. Los propietarios de la tierra y las economías de escala, sustentos del paradigma sojero en la Argentina. *Desarrollo Económico* 50, no. 197: 3–32.

Bell, D., and C. Scott. 2010. *Los Grobo: farming's future?* Harvard Business School Prod. 511088.

Bidaseca, K., A. Gigena, F. Gómez, A. Weinstock, E. Oyharzábal, and D. Otal. 2013. *Relevamiento y sistematización de problemas de tierra de los agricultores familiares en la Argentina*. Buenos Aires: Ministerio de Agricultura, Ganadería y Pesca de la Nación, PROINDER.

Bisang, R., and G. Anllo. 2014. *Impactos territoriales del nuevo paradigma tecno-productivo en la producción agrícola argentina*. Documento de trabajo 5. Buenos Aires: Instituto de Economía Política de Buenos Aires.

Bisang, R., G. Anlló, and M. Campi. 2008. Una revolución (no tan) silenciosa. Claves para repensar el agro en Argentina. *Desarrollo Económico* 48, no. 190/191: 165–207.

Bisang, R., and S. Sztulwark. 2007. Tramas productivas de alta tecnología. El caso de la soja transgénica en Argentina. In *Estructura productiva y empleo. Un enfoque transversal*. Buenos Aires: Ministerio de Trabajo Empleo y Seguridad Social. pp. 181–224.

Borras, S., J. Franco, S. Kay, and M. Spoor. 2011. *El acaparamiento de tierras en América Latina visto desde una perspectiva internacional más amplia*. Santiago: FAO.

Bridge, G. 2008. Global production networks and the extractive sector: governing resource-based development. *Journal of Economic Geography* 8: 389–419.

Catacora Vargas, G., P. Galeano, S. Zanona Agapito, D. Aranda, T. Palau, and R. Onofre Nodari. 2012. *Soybean production in the Southern Cone of the Americas. Update on land and pesticide use*. Cochabamba: Vimegraf.

CIARA (Argentinian Oil Industry Chamber). Statistics. n.d. www.ciara.com.ar.

Craviotti, C. 2015. Argentina's agri-food transformations in the context of globalization: Changing ways of farming. In *Handbook of the international political economy of agriculture and food*, ed. A. Bonanno and L. Busch. Chetelham-Northampton: Edward Elgar. pp. 78–96.

Craviotti, C., and C. Gras. 2006. De desafiliaciones y desligamientos: Trayectorias de productores familiares expulsados de la producción en la región pampeana argentina. *Desarrollo Económico* 46, no. 46: 117–34.

Dannenberg, P., and E. Kulke. 2014. Editorial: Dynamics in agricultural value chains. *Die Erde* 145, no. 3: 121–6.

Delgado Cabeza, M., and L. Gavira Alvarez. 2006. Agricultura y trabajo rural en la globalización. *Revista Española de Estudios Agrosociales y Pesqueros* 211: 21–61.

Dicken, P., P. Kelly, K. Olds, and H. Yeung. 2001. Chains and networks, territories and scales: towards a relational framework for analyzing the global economy. *Global Networks* 1, no. 2: 89–112.

Du Toit, A. 2009. *Adverse incorporation and agrarian policy in South Africa*. http://repository.uwc.ac.za/xmlui/bitstream/handle/10566/65/duToit_Adverse2009.pdf?.

Ederer, M. 2013. *Los Grobo. Agribusiness Value Systems for the future*. European Food and Business Seminar.

Fuck, M. & M. Bonacelli. 2009. Sementes geneticamente modificadas: (in)segurança e racionalidade na adoçao de transgenicos no Brasil e na Argentina. *Revista CTS* 12, no. 4: 9–30.

Gallicchio, E. 2003. *El desarrollo económico local. Estrategia económica y de construcción de capital social*. Montevideo: CLAEH.

Gereffi, G. 2014. A global value chain perspective on industrial policy and development in emerging markets. *Duke Journal of Comparative and International Law* 24, no. 3: 433–58.

Gereffi, G., J. Humphrey, and T. Sturgeon. 2005. The governance of global value chains. *Review of International Political Economy* 12, no. 1: 78–104.

Gras, C., and V. Hernández. 2013. Los pilares del modelo agribusiness y sus estilos empresariales. In *El agro como negocio: producción, sociedad y territorios en la globalización*, ed. C. Gras and V. Hernández. Buenos Aires: Biblos.

Gras, C., and A. Sosa Varrotti. 2013. El modelo de negocios de las principales megaempresas agropecuarias. In *El agro como negocio: producción, sociedad y territorios en la globalización*, ed. C. Gras and V. Hernández. Buenos Aires: Biblos. pp. 215–236.

Guibert, M., S. Grosso, P. Arbeletche, and M. Bellini. 2011. De Argentina a Uruguay: Espacios y actores en una nueva lógica de producción agrícola. *Pampa* 7: 13–38.

Guibert, M., and M. Sili. 2011. Argentine: expansion agricole et dévitalisation rurale. In *Dynamiques des espaces rurales dans le monde*, ed. Y. Jean and M. Guibert. Paris: Ed. Armand Colin. pp. 338–351.

Haesbaert, R. 2011. *El mito de la desterritorialización. Del 'fin de los territorios' a la multiterritorialidad*. México: Siglo XXI.

Hemsing, B. 2002. Perspectivas sobre desarrollo economic localizado. *EURE, Revista de Estudios Urbanos y Regionales* 28, no. 82: 33–61.

Henderson, J., P. Dicken, M. Hess, N. Coe, and H. W. Yeung. 2002. Global production networks and the analysis of economic development. *Review of International Political Economy* 9, no. 3: 436–64.

Hernández, V. 2007. El fenómeno económico y cultural del boom de la soja y el empresariado innovador. *Desarrollo Económico* 47, no. 187: 331–65.

Hess, M. 2004. Spatial relationships? Towards a reconceptualization of embededdness. *Progress in Human Geography* 28, no. 2: 165–86.

Infocampo, 20–26 March 2015. Agrofina les salva los números a Los Grobo.

La Nación, 18 April 2013. Desinvierte uno de los grupos líderes en granos.

Los Grobo. 2011. *Reporte de Sustentabilidad 2009/2010* (mimeo).

Manciana, E., M. Trucco, and M. Piñeiro. 2009. *Large-scale acquisition of land rights for agricultural or natural resource-based use: Argentina*. Buenos Aires: CEO.

Margétic, C. 2006. Agro-industries et territories: Recherches en Nord-Pas-de-Calais. *ESO* 25: 67–72.

Martins de Moura, L., and J. Bevilaqua Marin. 2013. Rede empresarial: a estratégia da produção de sementes de soja transgênica em Goiás. *Interações* 1, no. 14: 21–36.

Massey, D. 1994. A global sense of place. In *Space, place and gender*, ed. D. Massey. Minneapolis: University of Minnesotta Press, pp. 146–156.

Morgan, K., T. Marsden, and J. Murdoch. 2006. *Worlds of food: Place, power, and provenance in the food chain*. Oxford: Oxford University Press.

Murmis, M., and M. R. Murmis. 2012. El caso de Argentina. In *Dinámicas del mercado de tierras en América Latina y el Caribe. Concentración y extranjerización*, ed. F. Soto Barquero and S. Gómez. Santiago de Chile: FAO.

Ordoñez, H., and J. Nichols. 2003. *Agronegocios. Escenarios turbulentos, economías emergentes. Argentina. Caso Los Grobo*. Universidad de Buenos Aires- Texas A&M University.

Pedreño, A., C. de Castro, E. Gadea, and N. Moraes. 2015. Sustainability, resilience and agency in intensive agricultural enclaves. *Revista de Despoblación y Desarrollo Agrario* 18: 139–60.

Rallet, A., and A. Torre. 2004. Proximité et localisation. *Economie rurale* 280: 25–41.

Reboratti, C. 2003. Desarrollo regional y territorios privatizados. In *Territorio, sociedad y región: perspectivas desde el desarrollo regional y local*, ed. C. Acuña, A. Riella. Montevideo: Departamento de Sociología, Facultad de Ciencias Sociales, Universidad de la República, pp. 81–93.

REDAF (Red Agroforestal Chaco Argentina). 2012. *Bosque Nativo en Salta: Ley de Bosques, análisis de deforestación y situación del Bosque chaqueño en la provincia*. Santa Fe, Argentina: REDAF.

Regunaga, M. 2009. The soybean chain. Implications of the organization of the commodity production and processing industry. Case Studies in Latin America and the Caribbean Region. Buenos Aires.

Robinson, W. 2012. Global capitalism theory and the emergence of transnational elites. *Critical Sociology* 38: 349–63.

SIIA (Integrated System of Agricultural Information, Argentina). n.d. www.siia.gov.ar

Soares-Filho, B., R. Rajao, M. Macedo, A. Carneiro, W. Costa, M. Coe, H. Rodrigues, A. Alencar. 2014. Cracking Brazil's Forest Code. *Science* 344, no. 6182: 363–4.

Sommerville, M. 2013. *Financializing Prairie Farmland. Farmland Investments Funds and the Restructuring of Family Farming Systems in Central Canada*. LDPI Working Paper 38. Land Deal Politics Initiative.

Trigo, E., and E. Cap. 2003. The Impact of the Introduction of Transgenic Crops in Argentinean agriculture. *Agbioforum* 6, no. 3: 87–94.

Trigo, E., E. Cap, V. Malach, and F. Villareal. 2009. *The case of zero–tillage technology in Argentina*. IFPRI Discussion Paper. Washington: IFPRI.

Vértiz, P. 2014. *Estrategias de persistencia de la producción familiar tambera ante al avance del modelo de los agronegocios: el caso de Chascomús y Lezama*. Master Thesis., Maestría de Procesos Locales de Innovación y Desarrollo Rural, Universidad Nacional de La Plata.

Wesz Jr., V. 2014a. O Mercado da soja no Brasil e na Argentina. Semelhanças, diferenças e interconexões. *Seculo XXI* 4, no. 1: 114–61.

Wesz Jr., V. 2014b. *O mercado da soja e as relaçoes de troca entre produtores rurais e empresas no sudeste de Mato Grosso (Brasil)*. Ph. D. Thesis., Universidade Federal de Rio de Janeiro.

Wilkinson, J., B. Rydon, and A. Di Sabbato. 2012. El caso de Brasil. In *Dinámicas del mercado de tierras en América Latina y el Caribe. Concentración y extranjerización*, ed. F. Soto Barquero and S. Gómez. Santiago de Chile: FAO.

The geopolitics of Brazilian soybeans

Gustavo de L. T. Oliveira

Soybean plays a major role in the development of Brazilian agribusiness, and in turn in Brazilian geopolitical power as well. It is a pillar of Brazil's insertion into a 'new multipolar world order' as basis for much Brazilian land ownership in neighboring countries, for the extension of political influence in Africa, and it is especially important for balancing trade with Brazil's new primary commercial partner, China. Yet the US dollar and North Atlantic transnational companies still control global soybean markets and production technologies. In a context marked by booming but volatile commodity prices, food crises, riots and revolutions in food-importing countries, a global rush for farmland, and severe droughts and climate change, the soybean agribusiness in Brazil takes on new and crucial geopolitical significance. I trace the geopolitical role it has served in consolidating the 'green revolution' in Brazil, and raise questions about the intersection between agroindustrial markets and currencies: could agricultural commodities serve geopolitical functions (and thus contestation) similar to those 'petro-dollars' have served since the 1970s? These considerations show how the political ecology of soybean shapes and is shaped by inter-regional and global-scale processes, and reveals new directions for research on the emerging geopolitical landscape of our century.

Introduction

Soybean production, processing and trade have played a major role in the recent development of Brazilian agribusiness. Agribusiness as a whole is responsible for about 20 percent of Brazil's gross domestic product (GDP; CNA 2014a), and although Brazil is also the world's top exporter of sugar, coffee, orange juice and poultry meat, and the world's second biggest exporter of beef and maize, soybeans are the most economically important and strategically significant agroindustrial commodity produced and exported. Supplying 40 percent of the international market, Brazil has matched the United States to share global leadership in soybean exports. Soybean and its products (meal, oil and their derivatives) accounted for 12.9 percent of all Brazilian exports in 2013, second only to iron ore (13.4 percent), and significantly more than the next main export products, such as petroleum (7.2 percent), meat products (6.1 percent), land vehicles (5.8 percent), machinery (5.3 percent) and sugar and its derivatives (5.0 percent) (UN–Comtrade 2014). Soybean agribusiness is an illustrative case of broader changes in how Brazil is situated in the global political economy, but as a 'flex crop' that serves as the nexus of a new grain–livestock–fuel complex, there is also something distinctive about soybean that draws special attention to its geopolitical significance (Oliveira and Schneider forthcoming). In this contribution, I demonstrate how the development of the soybean agroindustrial complex has become a pillar of Brazil's insertion into a 'new multi-polar world order' (BRICS 2009).

Soybean cultivation is the basis for much Brazilian land ownership in neighboring countries, particularly Paraguay and Bolivia (Borras et al 2012; Galeano 2012; Urioste 2012). It is instrumental for the extension of political and economic influence in Africa (Cabral and Shankland 2013; Chichava et al 2013; Clements and Fernandes 2013). And it is especially important for balancing the growth of imports from Brazil's new primary commercial partner, China (Acioly, Pinto, and Cintra 2011; Jenkins and Barbosa 2012; Oliveira 2010). Yet the US dollar and North Atlantic transnational companies still control global soybean markets and production technologies: the three largest transnational seed companies (Monsanto, DuPont and Syngenta) control 55 percent of the global soy seed market, and, along with the three other leading agro-chemical companies (Bayer, Basf and Dow Chemical), they control 76 percent of global pesticide and herbicide markets (EcoNexus and Berne Declaration 2013; Silva and Costa 2012). Moreover, four transnational trading companies (ADM, Bunge, Cargill and Luis Dreyfus) control about 80 percent of international soybean trade, and in South America they control 50 percent of installed crushing capacity and 85 percent of soybean exports (Wesz Jr. 2014). The Chicago Board of Trade (CBOT) remains the main forum in which international soybean prices are set, hosting as well the largest share of soybean future market trading and hedging. These are familiar markers of the current 'corporate food regime', structured by the well-documented and widely discussed economic power and political influence of transnational agribusiness corporations (McMichael 2012; Murphy, Burch, and Clapp 2012; Patel 2008; Wilkinson 2009). Although this 'corporate food regime' is correctly associated with neoliberalization, many are apt to discount the persistent and intimate relations between the state and agribusiness development:[1] 'National borders are losing ground to a corporate-driven model of territorial organization', suggests Mariano Turzi (2011, 59). 'The new model is dictating production conditions and infrastructural developments; rearranging the geoeconomic space throughout Argentina, Bolivia, Brazil, Uruguay, and Paraguay into a single, unified "Soybean Republic"' (Turzi 2011, 59).

Even though soybean production technology and international trade remain largely controlled by North Atlantic transnational corporations, the diplomatic–military implications of extensive Brazilian-owned soybean farms in neighboring countries in South America cannot be ignored. Moreover, tax collection, international balance of payments and currency exchange rates are of utmost importance to the governance of the soybean sector in Brazil (as elsewhere), and for its *geopolitics* as well. Ultimately, this paper links the well-known history of the soybean sector in Brazil with new and understudied questions arising out of the political ecology of agribusiness and the contested terms of globalization in this new century. In this context marked by booming but volatile commodity prices, food crisis, riots and revolutions in food-importing countries, a global rush for farmland, severe droughts and climate change, the soybean complex in Brazil takes on new and crucial geopolitical significance – a significance that we are only beginning to understand.

[1]There are notable exceptions, of course, especially within the recent 'land grab' literature (e.g. Wolford et al. 2013). Yet as I argue in my contribution to that special issue (Oliveira 2013), the 'land grab' framework is insufficient (and often inadequate) to capture the agroindustrial dynamics and geopolitical significance of these developments, especially in places such as Brazil and China, where domestic agribusinesses are becoming increasingly more powerful (cf. Oliveira and Schneider forthcoming; Sommerville, Essex, and Le Billon 2014; Wilkinson 2009).

But first, a brief note on the theorization of geopolitics. Although I call attention to the role and interests of the Brazilian state in the soybean complex, my purpose is nevertheless to contribute to a *critical* geopolitics that situates power not in the hands of sovereign states alone, but in more relational ways that traverse a spectrum of scales and struggles. Thus, by focusing on a globally interconnected agroindustrial sector and its relations within the interstate system, this analysis avoids the 'territorial trap', i.e. the geographical assumptions of traditional geopolitics and international relations that naturalized 'the exercise of state power through a set of central political institutions' and homogenized 'the clear spatial demarcation of the territory within which the state exercises its power' (Agnew 1994, 53). Investigating the soy complex in Brazil accordingly reveals how the geopolitical imperatives of the Brazilian state operate through private actors at home and abroad, and underscores their alliances and frictions with foreign states as well as transnational companies. This study therefore considers these multiple relations simultaneously in terms of conflicts between states, competition between companies and struggles between classes within Brazil and across the globe.

Still, the global soy complex (characterized by extensive transnational corporate control in inputs and trade) appears to serve as a textbook representation of the 'growth of modes of exercising power that do not depend on imperative coordination by a territorialized state apparatus and that are independent of its borders on whatever scale they exist' (Jessop 2002, 352; cf. Turzi 2011). Nevertheless, I suggest that the distinct and sometimes contradictory 'logic of state' and 'logic of capital' driving the Brazilian soybean complex are increasingly assembled into a relatively coherent project through which Brazilian agribusinesses and state defy policies and practices that favor EU- and US-based agroindustries and trading corporations, the predominance of the US dollar in international markets, and US state hegemony. Soybeans, in other words, are a key instrument with which Brazilian state and corporate interests cultivate political and economic power, seeking to establish themselves as leaders in a 'new multipolar world order'.

Immigrant seeds and roots in the Green Revolution (1900–1970)

Soybeans were first brought to Brazil in 1882 for agricultural research purposes, but it was only after 1908 that Japanese migrants began cultivating them consistently for their own consumption in southeastern Brazil (Shurtleff and Aoyagi 2009). This fact appears to carry little geopolitical importance, but it literally planted the seeds for a major transformation of the global agroindustrial regime and the role of Brazil in the contemporary interstate system. Even as an immigrant subsistence crop, soybean was already implicated in processes that recur in its geopolitical history. First of all, the incipient transcontinental cultivation of this nutritious legume helped the emigration of poor peasants from Japan (much akin to contemporaneous emigration from Europe), which reduced political pressures for land and wealth redistribution in their places of origin, and established human connections that would later prove instrumental to the global explosion of soybean production. This dynamic would then be replicated internally within Brazil by southern soybean farmers – with Japanese funding – who 'occupied' the central highlands of Brazil during the second half of the twentieth century and now generate the bulk of Brazilian soybean production and exports (more on which below).

This early role of soybeans in (immigrant peasant) agriculture in Brazil also permitted the 'pacification' and commercial integration of the hinterlands, i.e. the establishment of 'legible' state subjects and production processes over spaces hitherto contested by indigenous peoples and other 'unruly' *sertanejos* or country folk (Estevam 2004; Scott 1998). This process began over the São Paulo and Paraná plateaus in southeastern Brazil from the 1900s to 1940s, where Japanese and European immigrants settled and established a commercially oriented production system distinct from the regimes of ranching *latifúndia* and sugar or coffee plantations that had dominated Brazilian agriculture since colonial times. Prior to their colonization, these regions were actively contested frontier zones, the stage of major revolts and land struggles, whether these were domestic struggles between different classes, like the 'war of *contestado*', or international struggles against neighboring countries, such as the Triple Alliance war against Paraguay (Fausto 1994; Foweraker 1981). These joint state-making and capitalist imperatives – the reduction of popular pressures for land and wealth redistribution in densely populated places, and the 'pacification' and integration of Brazil's hinterlands through the settlement of a commercially oriented 'farmer' class – undergirded the colonization practice that became known as the 'March Towards the West' and enabled the domestic harnessing of technologies and techniques of the global phenomenon that was dubbed the 'green revolution'.

The 'March Towards the West' was a state policy established under president Getúlio Vargas (1930–1945, 1951–1954) whereby the federal government and its state-government interveners promoted commercial agriculture as a means towards internal stability by providing an outlet for landless and land-poor peasant migrants, increasing the supply of cheap food for the growing industrial proletariat in southeastern Brazil, and increasing foreign exchange to finance import-substitution industrialization. Moreover, these imperatives linked the occupation of the 'empty' *sertão* (hinterlands) with fortifying external security, as nineteenth-century border conflicts with Bolivia, Argentina and the 'stop-gap' countries of Paraguay and Uruguay still weighed heavily in the politico-military concerns of the Brazilian elite. These geopolitical imperatives for the expansion of commercial agriculture onto the central highlands were explicitly projected by the leading strategists of the Brazilian military who served under Vargas's administrations and who then assumed direct control of the Brazilian state through a military coup in 1964 (Freitas 2004). Most prominent among this military intelligentsia was certainly General Golbery do Couto e Silva, who promoted the expansion of commercial agriculture for the territorial occupation of central and northern Brazil against threats of neighboring state incursions and rural communist uprisings (Couto e Silva 1957).

During the early and middle decades of the twentieth century, however, soybean was not yet a prominent commercial crop in Brazil. With the exception of Japanese migrants, soybeans simply did not feature in Brazilian diets, and there was little commercial demand for them in the emergent oilseed industry of the country (Sousa and Vieira 2008). But soybean is a nitrogen-fixing legume, so commercial farmers began planting it in Rio Grande do Sul during the 1940s as green fertilizer to restore soils degraded by wheat and maize production, as these were reaching limits of productivity and pushing the agricultural frontier into uncultivated lands of the Cerrado ecosystem (Foweraker 1981; Hasse and Bueno 1996; Shurtleff and Aoyagi 2009). This crisis of productivity had become widespread across all areas of the world where extensive grain monocultures had been established, and it became disastrous when conjoined with severe droughts and soil erosion, as was the case in the US with the infamous 'dust bowl' of the 1930s (Mazoyer and Roudart 2006; Worster 2004). This crisis triggered processes of state reconfiguration, capitalist accumulation, transnational agricultural investment and technological innovation, dubbed the 'green revolution'. Most narratives of this process, particularly those that regard it positively, limit themselves to technical aspects (development and extension of fertilizer, irrigation, improved hybrid seeds, state support and credit), but we must underscore its geopolitical dimensions as well in the context of the so-called 'Cold War'.

The term 'green revolution' itself was coined in order to contrast with the 'red revolution' that communists were leading all around the world, and the geopolitical intent of orchestrating massive technical transformations to grain production in places like Mexico, India and the Philippines were explicit (Patel 2013). After all, communists were winning because landless peasants and poor workers were rising up against absentee or oppressive landlords and poverty wages, so unless agricultural productivity could be rapidly boosted and capitalist/state rule could be assured over the countryside, the international capitalist system – and the role of the US in leading it – were under grave threat of annihilation. This was particularly true in places where peasant-led revolutions had recently taken place (e.g. Mexico), and where peasant uprisings were growing in the proximity of the recently victorious communists in China (e.g. India and the Philippines), so it was not due to technical or philanthropic considerations that these were the places where the joint efforts of the US government, the Rockefeller Foundation and a series of agribusiness corporations came together with full strength to make a 'green revolution' (Patel 2013).

Between the 1940s and 1954, soy production grew 10-fold in Brazil, from around 10,000 to 100,000 metric tons per year (Shurtleff and Aoyagi 2009), and the country entered a period of rapid economic growth and massive social upheaval. The 'March Towards the West' was culminating with the construction of a new capital from the ground up in the center of the country, while new tractors, fertilizers and hybrid-seed technologies were allowing grain monocultures to expand vertiginously, displacing peasants and spewing workers from the countryside even faster than import-substitution industrialization created employment opportunities in the cities, which began to mushroom with slums. Agricultural workers, who had not benefited from the extension of labor rights to industrial workers during the Vargas era, began organizing themselves into semi-legal Peasant Leagues, often with support from the Communist Party of Brazil. They organized massive and widespread strikes for better wages and working conditions in sugarcane plantation zones, they also resisted displacement – sometimes violently – against land grabs by agrarian elites and sometimes they even launched occupations demanding land redistribution (Bastos 1984; Motta and Zarth 2008). Merely 18 days after the democratically elected president João Goulart declared the government would undertake agrarian reform (13 March 1964), the military overthrew him in a coup and installed a dictatorship that repressed the Peasant Leagues, outlawed the Communist Party and secured the continuation of the technical transformations of the 'green revolution' in the Brazilian countryside. As was explicitly advocated by the military intelligentsia and most sectors of the domestic capitalist class, Brazil had to remain the bulwark against communism in South America (Ianni 1979).

The geo-politician General Golbery do Couto e Silva was instrumental not only to the conceptual vision and justification for this coup and the consolidation of the 'green revolution' in Brazil, but also in its practical development through the Brazilian state *and agroindustrial corporations*. While still a colonel in 1954, Couto e Silva was the lead author of an influential 'colonel's manifesto' that critiqued the Vargas administration – particularly João Goulart, who was then Minister of Labor – for raising the minimum wage and extending labor rights. This manifesto sparked a 'national security' debate that contributed to the destabilization of the Vargas administration and forced the resignation of Goulart (Freitas 2004). Five years later, upon the resignation of president Jânio Quadros, Couto e Silva again led the military in an attempt to prevent João Goulart (then vice president) from assuming the presidency. Interestingly, Goulart was in China at that moment negotiating the reestablishment of diplomatic relations with Brazil, something which Couto e Silva claimed was an indication of Goulart's 'communist sympathies', but which ironically became key for the contemporary boom of capitalist agroindustry in Brazil (enabled in great part by the growing Chinese demand for soybeans). Once the top military brass succeeded in overthrowing Goulart in 1964, Couto e Silva created and directed the National Information Services, which coordinated efforts to suppress communist propaganda and popular uprisings. Consequently, his influential 'doctrine of national security' began to shift from international defense of the borderlands against (implausible) neighboring incursions to focus upon the 'pacification' of landless peasants and poor rural workers who were being recruited into an endogenous communist movement (Couto e Silva and Franco 1967). Finally and very significantly for our current discussion, Couto e Silva took leave from the federal government to assume the presidency of Dow Chemical's subsidiary in Latin America from 1972 to 1974, one of the main companies that provided fertilizers and pesticides for commercial agriculture in Latin America, thus illustrating the intimate connections between the geopolitical and economic interests of the 'green revolution' in Brazil, and the alliance of (mainly US) foreign capital with the Brazilian state in the context of the 'Cold War' (Evans 1979; Freitas 2004). The geopolitical and economic interests of Dow Chemical extended beyond Brazil as well, as Couto e Silva explained that Brazil is 'a territorially satisfied nation', but with diplomatic–military responsibilities to lead the 'Latin, Catholic, and underdeveloped world' into alliance with the United States and the rest of 'Western civilization' (Couto e Silva 1981).

Between 1954 and 1969, soy production grew 10-fold again in Brazil, crossing the threshold of 1 million metric tons (mmt) (Shurtleff and Aoyagi 2009). This growth came from curtailed losses of productivity in the south, and the early expansion to Paraná and the Cerrado on the central highlands. Soybeans were not yet the crown jewel of Brazilian agribusiness, but the roots of this sector had already been planted as part of a fundamentally geopolitical project of state and agrarian capitalist alliance, under the hegemony of the US and its agribusiness companies (Evans 1979; Morgan 2000), serving moreover as a military and agroindustrial platform for sub-imperialist control over the rest of South America (Marini 1972; Flynn 2007). Starting in the 1970s, however, a series of agroindustrial and geopolitical transformations converted soybean in Brazil from a useful green fertilizer and oilseed of secondary importance into one of the most important globally traded commodities, creating new diplomatic and economic opportunities for corporations and state agents from emergent economies, especially Brazil.

Agroindustrial and geopolitical restructuring (1970–2000)

Soybeans have rapidly become the nexus of a new grain–livestock–fuel agroindustrial complex that has its origins in the United States, yet the growth dynamics of this agroindustrial complex also reveal the diminishing political and economic clout of the United States in the face of competition from net-agricultural-importing countries (such as Japan and China) and agroindustrial exporters among the major emergent economies, particularly Brazil (Oliveira and Schneider forthcoming). This transformation expresses the uneven development of capital, flagging the 1970s as a key turning point for familiar reasons: competition between US industries and the reconstructed Japanese and western European economies led to stagflation in the global North, which combined with an energy crisis (arranged by the Organization of the Petroleum Exporting Countries – OPEC) to transform international patterns of capital investment and circulation; consequently, global North financiers began tightening capital flows to 'developing countries', which triggered massive foreign debt crises (especially in Brazil) that forced a shift from import-substitution national development to 'neoliberalism', i.e. privatization of state industries and services, and deregulation of foreign investments and trade for export-oriented commodity production (Harvey 2005). But to understand why a focus on soybeans is especially useful to analyze this restructuring, we must first comprehend the history of its use as livestock feed.[2]

During the 1920s, US researchers discovered that adding vitamins A and D to livestock feed would allow year-round confinement, and by the 1940s, antibiotics began to be added as well to limit high mortality in these nascent concentrated animal feeding operations (CAFOs). Although soybean had been used primarily as a green fertilizer and occasionally as input for the production of edible oil in the US, wartime shortages during the 1940s expanded its use for edible oil and encouraged its use for the production of livestock feed as well. This oilseed, high in digestible protein and low in fiber, was then recognized as 'a virtual protein pill' for concentrated livestock, perfect in combination with maize on the farm level (where soybeans helped replenish the nitrogen taken up by maize) and also in CAFOs as cheap high-energy *and* high-protein livestock feed (Boyd 2001; Mason and Singer 1980). As agricultural and processing capacity in Europe and Japan were destroyed due to the war, the US became a major exporter of soybean oil *and* CAFO technologies during the 1950s. Then, as edible oil production recovered in Europe and Japan (along with the rest of their economies and per capita incomes), the strong demand for soybean oil was replaced by a strong demand for soybean meal for the new Japanese and European CAFOs. Consequently, soybean steadily displaced wheat and matched maize as one of the main agricultural products and exports of the US by the 1960s.[3] During this time, US soybean production and subsidized exports served the combined political and economic interests of Midwestern farmers and the US government in relieving a 'peacetime' surplus that would have drastically deflated prices, satisfying a growing middle class in the global North with more and 'cheaper'[4] meat in their diets, and provided a valuable 'carrot' for the US government to coax non-aligned countries (eager for 'cheap' food imports to facilitate their industrialization/urbanization) into the US sphere of influence. This geopolitical role of US agroindustrial exports, or 'food power', has been extensively studied (e.g. Friedmann 1982; McMichael 2012; Morgan 2000; Patel 2013), so now I turn to the specific (and prominent) role of soybean as a key instrument for Brazil's emergent agribusiness and state interests to challenge US hegemony.

The beginning of the crisis of US hegemony in the agribusiness sector may be marked by an apparent geopolitical victory over the Soviet Union that shifted the 'Cold War' from tension to détente: namely, the spectacular and unprecedented US grain exports to the USSR that began in 1973. Proximate causal factors were major droughts in the Soviet Union that reduced grain stocks, but the major long-term factor has been the great expansion of livestock feed demand for CAFOs in the Soviet Union, Europe and Japan. With the

[2]'The expansion and industrialization of grain and oilseed production and the expansion and industrialization of livestock production have been mutually reinforcing: ... increasing livestock populations were not only about increasing value-added opportunities in animal flesh, milk, and eggs, but were also about transforming structural grain and oilseed surpluses from a deflationary millstone into a steadily growing source of low margin earnings for large-scale farmers, processors and traders' (Weis 2013, 73).

[3]Between 1941 and 1966, soybean acreage in the US increased by more than 500 percent, while production increased almost 800 percent, providing soybean meal for livestock feed much more cheaply and at a larger scale than ever before (Boyd 2001).

[4]The 'cheapness' of this meat production is in fact very deceiving, resting on an illusion of efficiency that ignores enormous state subsidies in the US and western Europe, and obfuscates even more dramatic social and environmental costs that are 'externalized' (Cavalett and Ortega 2010; Steinfeld et al. 2006; Weis 2013).

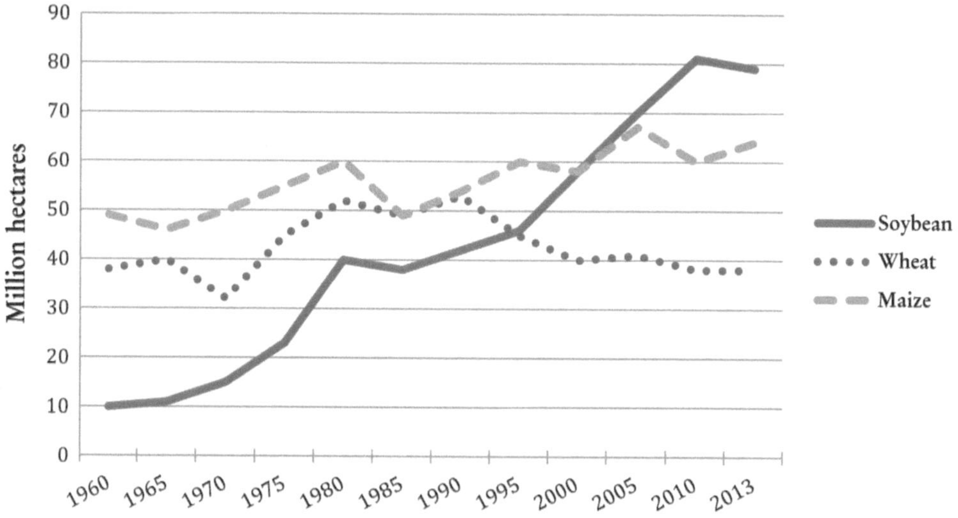

Figure 1. Area harvested in the western hemisphere, 1960–2012.
Source: Brown (2012), from the United States Department of Agriculture (USDA).

post-war US surpluses drained and global-scale restructuring of agroindustrial markets due to the 'Cold War' détente, the US government imposed embargos on soybean exports. This further destabilized international grain markets and encouraged major soybean and grain importers – particularly Japan and western Europe – to join with emergent soybean and grain exporters – especially Brazil and Argentina – to expand production for international markets (Morgan 2000; Patel 2013). Globally, between 1961 and 2009, soybean production area increased by 400 percent, and, combined with increases in yield per hectare, soybean output grew more than 800 percent. Only 6 percent of all soybean produced globally is consumed directly as human food, while 98 percent of the remainder is crushed for livestock feed production (HighQuest Partners and Soyatech 2011; WWF 2014). Figures 1 and 2 illustrate the fast increase of soybean production in the western hemisphere (mainly the US, Brazil, Argentina, Canada and Paraguay), and the growing share of the Brazilian contribution to international markets.

This expansion of soybean agribusiness in Brazil reflects and reinforces a state-making logic of capital accumulation and geopolitical maneuvering. In order to expand soybean production in Brazil, for example, branches of the federal government's National Company of Agricultural Research (EMBRAPA) were created for the development of commercial agriculture in the Cerrado region (Embrapa-Cerrados, est. 1971) and for soybean research and development in particular (Embrapa-Soja, est. 1975), devising significant interventions in the pH and nutrient availability of the soils that rendered it profitable for the intensive production of soybean varieties adapted to the Brazilian climate (CPAC 1975).[5] Given the Japanese state's intention to increase the availability of grain and oilseeds on international markets, their interests came together with the Brazilian state's to expand

[5]Funding for Embrapa-Cerrados (CPAC) came from the federal government's Development Program for the Cerrado (POLOCENTRO), which also financed road improvement, rural electrification, grain storage units, land clearing, production, commercialization, capital goods acquisition and harvest expenses in the Cerrado (Jepson, Brannstrom, and Filippi 2010).

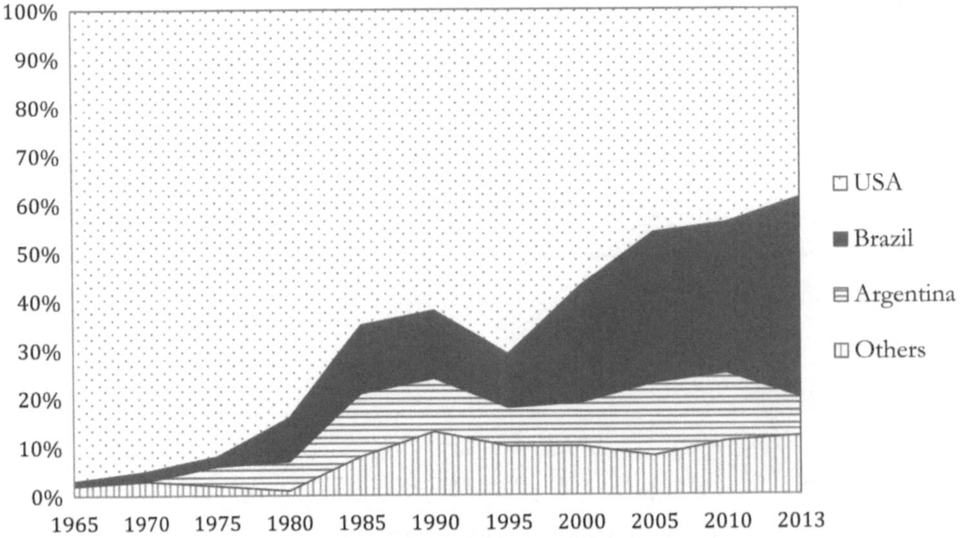

Figure 2. Share of global soybean exports (1965–2013).
Source: USDA (1965–1970; 1975–1995; 2000–2013).

agribusiness in the central highlands. The Program of Directed Settlement of the Alto Paranaíba (PADAP), for example, granted public land for the establishment of four agricultural settlements in the Cerrado region of Minas Gerais state in 1973. The Japanese International Cooperation Agency (JICA) provided credit and technical support for soybean and other grain production under PADAP, which also included the construction of transportation, energy and communication infrastructure (Hollerman 1988; Yoshioka 1992). The beneficiaries were small-scale market-oriented farmers from the southeast of the country, among whom 80 percent were Japanese immigrants or their descendants (Pires 2000). This was not coincidental, as JICA was building upon personal connections that reached back to the early days of Japanese migration to Brazil.

The success of PADAP encouraged further negotiations between JICA and the Brazilian state in 1975, which resulted in the expansion of the PADAP model into a region-wide Japanese–Brazilian Cooperation Program for the Development of the Cerrado (PRODECER) (Martins and Pellegrini 1984).[6] Moreover, this program had impacts far beyond the directly benefited farmers, since its infrastructure improvements dovetailed with other public policies (especially production credit by the Bank of Brazil) and, in effect, subsidized private investments by southern Brazilian soybean farmers, whose migration into the Cerrado established a tendency towards farmland concentration and agroindustrial integration across the soybean sector in Brazil (Jepson, Brannstrom, and Filippi 2010; Hecht

[6]Three stages have since been implemented in Cerrado regions, which have benefited 758 farmers, settled in 21 projects across seven states, producing mostly grains over a total area of 353,748 hectares (MAPA 2002). Currently, negotiations are stalled over a projected fourth stage that would expand the program into the Cerrado-Amazon transition zones of Maranhão, northern Mato Grosso, Rondônia and Pará, establishing an additional 60 settlement projects over an additional 41,000 hectares (Jepson, Brannstrom, and Filippi 2010), most likely because the focus of Japanese–Brazilian agroindustrial cooperation has shifted to Mozambique, with the ProSAVANA program discussed below.

2005). In the first 5 years of the 1970s, Brazilian soybean production grew 10-fold yet again, reaching 10 mmt (Shurtleff and Aoyagi 2009).

Thus, Brazilian and Japanese geopolitical interests coincided in expanding production, processing (especially by crushers in Japan) and consumption of soybeans (by CAFOs in Japan and Brazil) through new trade and investment circuits that explicitly sought to sidestep US corporate and state control. Furthermore, these efforts to diversify commercial partners in the soybean complex are not limited to a bilateral relation between Brazil and Japan, but characterize a global trend that increased during the 1990s with the devaluation of South American currencies and reduction of commodity export taxes in Brazil and Argentina, and then accelerated further with China's reduction of import tariffs and entry into the World Trade Organization (WTO) in 2001. Between 1990 and 2014, the US share of world soybean production declined from 50 percent to 31 percent as Brazil's share increased from 18 percent to 31 percent, and the US share of world soybean crush declined from 37 percent to 19 percent, while Brazil's share increased from 15 percent to 16 percent and China's share jumped from 4 percent to 29 percent (Oliveira and Schneider forthcoming). China is now the largest and fastest growing market for soy and its products, particularly due to rising demand for pork and poultry CAFOs (Peine 2013; Schneider 2011). Brazilian agribusinesses – and their partners in the Brazilian state – are increasingly drawing upon strategic partnerships with companies and government agencies beyond the US and EU to consolidate their joint economic *and political* interests in weakening the economic power of the global North, and increasing their influence over the construction of the emerging global food and agroindustrial order (Hecht and Mann 2008; Wilkinson 2009). These are increasingly reflected in the sub-imperialist role of Brazilian soybean agribusiness in South America, its influence through 'South–South cooperation' for the expansion of commercial agriculture in Africa, and its relations with China and other countries of the global South through bilateral agreements and multilateral forums of international governance such as the WTO.

Sub-imperialism and South–South cooperation (1970–present)

The notion that Brazil plays a 'sub-imperialist' role in South America hearkens back at least to the conflicts with Argentina that resulted in the creation of Uruguay as an independent 'stop-gap' state (1811–1828), the Triple Alliance war against Paraguay (1864–1870) and the conflict with Bolivia over Acre state (1899–1903), but it gained new traction during the military dictatorship period and context of the 'Cold War' as discussed above (Marini 1972).[7] Paraguay not only suffered the most violent Brazilian military incursion ever, but it also remains the main site of Brazilian sub-imperialist expansion, as soybean production constitutes the predominant economic activity of *Brasiguaios*, the Brazilian term for its migrants who dominate eastern Paraguay. This process began with the introduction of commercial soybean plantations during the 1970s, which was facilitated by the close geopolitical ties between the military dictatorships in both countries. These ties were institutionalized in the 'Friendship and Cooperation Treaty' of 1975, which protected the presence of Brazilian soybean farmers in eastern Paraguay (Galeano 2012; Zibechi 2009).

[7]Like Marini (1972), I use the term 'sub-imperialism' to designate a state practice of extending political and economic power beyond its borders, while remaining relatively subordinate to a greater imperialist power in the broader political and economic context.

The recent democratization process in both Brazil and Paraguay did little to alter the geopolitical terms of this cross-border soybean complex. In 2008, for example, the center-left democratically elected president Lula enacted Decree 6.592, which regulates the National Mobilization System dedicated to confronting 'foreign aggression'. The decree defines foreign aggression as 'threats or injurious acts that harm national sovereignty, territorial integrity, the Brazilian people, or national institutions, *even when they do not constitute an invasion of national territory*' (Brazil 2008, emphasis added). The broader context of this decree was the Bolivian nationalization of the Brazilian-owned oil and gas industry in 2006, but the more proximate incentives were increasingly militant agitation by landless peasants and indigenous peoples in Paraguay for redistribution of Brazilian-owned soybean farms. The aftermath of this decree aptly characterizes the ambiguous and sometimes contradictory geopolitical imperatives of Brazilian agribusiness and state agents. On the one hand, the Brazilian state sent a strong signal that expropriation of Brazilian-owned soybean farms in Paraguay would produce strong diplomatic and economic (if not military) reactions. But, on the other hand, when Paraguayan and *Brasiguayan* agribusiness leaders joined conservative politicians in Paraguay to depose Fernando Lugo in 2012 (a center-left president in Paraguay who proposed agrarian reform), the Brazilian government condemned the coup and temporarily suspended Paraguay from Mercosul (The Economist 2012). This allowed the entrance of Venezuela into the trade bloc, which had been obstructed by the Paraguayan legislature.

Despite this ambiguity and some contradictions, Brazilian soybean agribusiness and state interests coincide in a broader process of 'South–South cooperation' that has South American integration as its main pillar (Ribeiro and Ramos 2009). Brazil leads 'South–South cooperation' to assemble political, infrastructural and economic measures for integration that unsurprisingly benefits most directly the state-owned Petrobras oil company and a few major Brazilian construction conglomerates, the majority of their projects financed by the Brazilian National Bank for Economic and Social Development (BNDES; Flynn 2007; Robinson 2008). This also facilitated the expansion and consolidation of Brazilian agribusiness interests – especially soybean production – in Paraguay, Bolivia and Uruguay (Borras et al 2012; Galeano 2012; Urioste 2012). Brazilian soybean agribusiness is even taking advantage of reforms in Cuba to make inroads there, where Lula recently brought Senator – and biggest soybean producer of the country – Blairo Maggi to promote Brazilian agribusiness technology and production relations. Lula's own words are very informative about 'South–South cooperation' as a geopolitical alliance between Brazilian state and private agribusiness interests with (sub-imperial) ambitions across Latin America and Africa:

> I find that now we begin a new form of cooperation, beyond the desire of the Cuban comrades for the [Brazilian] EMBRAPA technicians to stay longer, which I think the Brazilian government will allow. Senator Maggi told the ministers of Agriculture *and Defense* that we are disposed to receive as many Cuban technicians as you desire to accompany the production cycle [in Mato Grosso], to increase the productivity of maize and soy .... It was very important to bring this businessman here, he [Blairo Maggi] is the biggest producer of soy in Brazil, he has all the technologies needed to help you. In addition to EMBRAPA, which is a very high quality scientific institution, the Brazilian *private* sector also has important scientific institutions, and we want to do everything possible to help not only Cuba, but also other Latin American and African countries to develop. And the best way to help is to take knowledge, make it so that people have access to modern technology, that's it! (Instituto Lula 2014, emphases added)

Thus, Brazilian presence and influence in these countries must be considered beyond the territorial control of Brazilian soybean farmers alone, as prominent in Paraguay and

Bolivia. In fact, even in these countries, the extension of Brazilian agroindustrial technologies and expertise through EMBRAPA and private institutions plays a fundamental role (Mackey 2011; The Economist 2010), as well as the extension and pavement of highways by Brazilian construction companies (financed by the BNDES) to facilitate exports of soybeans and other agricultural and mineral commodities through Brazil (Arellano-Lopez 2012).

Shifting focus to Africa, the literature reveals growing attention to Brazilian agribusiness *and* state interests under the rubric of 'South–South cooperation' (Amanor 2013; Cabral and Shankland 2013; Cesarino 2013), as was also evident in Lula's quote above. During the last decade, the Brazilian state has made considerable efforts to increase diplomatic and economic collaboration with various African states: it increased official visits by the head of state from zero-to-two per administration to 12 during Lula's two terms (2003–2011), increased its embassies in the continent from 18 to 37 (of 54 African states), and expanded trade from USD 5 billion in 2002 to over USD 56 billion in 2010. Brazilian investments by private and state-owned enterprises have also grown significantly, crossing the USD 10 billion threshold in 2009, but these remain largely concentrated in petroleum, mining and infrastructure sectors (Barbosa, Narciso, and Biancalana 2009).

An important characteristic of growing Brazilian economic and political influence in the global South, especially Africa, has been its articulation through a new discourse that replaces the global North's 'development aid' with 'South–South cooperation' or 'horizontal collaboration'. Celso Amorin, Brazilian Minster of Foreign Affairs from 2003 to 2011, said: 'South–south cooperation is a diplomatic strategy that originates from an authentic desire to exercise solidarity toward poorer countries. At the same time, it helps expand Brazil's participation in world affairs' (quoted in Dauvergne and Farias 2012, 909).[8] Indeed, national interests are evident in Brazilian agroindustrial cooperation and investments in Africa, most notably in the sugarcane sector in Angola, in cotton production and processing in Benin, Burkina Faso, Chad and Mali, and in the grain–livestock–fuel complex in Mozambique (Cabral and Shankland 2013; Cesarino 2013; Inoue and Vaz 2012). The project in Mozambique in particular has justifiably received the most attention, as it appears to be larger than all other Brazilian agroindustrial investments or 'collaboration' in Africa.

Formally called 'Triangular Cooperation for Agricultural Development of the Tropical Savannah in Mozambique', the program is more commonly known as ProSAVANA, and involves a three-way cooperation agreement between Mozambique, Brazil and Japan. The trilateral collaboration involving Japan follows directly upon the Japanese–Brazilian PRODECER program discussed above, with the Brazilian Cooperation Agency (ABC) channeling the Ministry of Agriculture to recreate the commercial success of PRODECER in the 'Nacala Corridor' of Mozambique (Chichava et al 2013; Clements and Fernandes 2013). Assembling Japanese capital, Brazilian agroindustrial technology and Mozambican land and labor, the project serves primarily the combined interests of the net-grain importing Japanese state and its general trading companies (*sogo shosha*), which will be able to

[8]'Brazilian governments have consistently emphasized that Brazil's foreign assistance is an alternative to traditional donor relations, with their hierarchical treatment of the recipient. Brazil has instead sought to go beyond hierarchy and national interests by emphasizing the language of "partnership" and "collaboration" … . At the same time, Brazil's provision for technical cooperation has been related, though often indirectly, to its broader foreign policy objectives [such as] opening and consolidating markets for Brazilian goods and services' (Inoue and Vaz 2012, 530).

control and profit much more from Mozambican soybean and grain exports than they can now from Brazilian or other exports. ProSAVANA attends as well the combined interests of the Brazilian state and agribusinesses, which expand their influence and profits while literally externalizing the negative social and ecological costs generated by PRODECER. Mozambicans themselves are not passive in this process, as domestic elites play a crucial mediating role in this trilateral collaboration, while peasants and landless workers actively resist the implementation of the project (Fairbairn 2013). Popular criticism in Mozambique increased especially after a leaked program master plan showed no measures to avoid the replication of the negative social and ecological consequences of PRODECER (Justiça Ambiental et al. 2013).

Despite the attention to Brazilian state engagements with the development of agribusiness in Africa, the role of *private* Brazilian agribusinesses and actors in Africa remains largely understudied, unlike the situation in South America discussed above. A major challenge to undertaking this research is that private Brazilian agribusiness actors in Africa are not usually landed proprietors engaged directly in production, as is the case in Paraguay or Bolivia, but rather technical consultants 'piggybacking' on the institutional connections established by Brazilian state agencies, especially the state-owned agricultural research company EMBRAPA (cf. Mackey 2011; Cesarino 2013). In my interviews with soybean agribusiness professionals in Brazil, I often encountered reports from my interviewees that they themselves or their colleagues are increasingly taking on private consultancies in Africa (especially in Angola, Mozambique, Ghana and Sudan); yet most evidence of this private 'South–South cooperation' remains anecdotal. Whether and how these private agribusiness collaborations in Africa articulate with Brazilian state interests remains to be investigated more carefully.

It becomes increasingly clear that Brazilian agribusiness leaders – and those in the soybean sector prominent among them – are new protagonists of global agroindustrial production networks and multilateral governance institutions. This follows upon a transformation of the very character of soybean farmers in Brazil (and Argentina). Unlike the entrepreneurial soybean family farmers from the Midwest of the US, the largest and most politically powerful Brazilian soybean farmers have expanded to the scale and organizational structure of veritable agribusiness enterprises: 'The classic image of poor farmers and rich ranchers is replaced by one of rural managers, most of them with university-level education, living in cities and specialized in business management. The MBAs [Master of Business Administration] are replacing farmers' (Gudynas 2008, 515). A highly integrated agroindustrial sector like soybean in Brazil, therefore, has produced a new class of Brazilian agribusiness professionals that increasingly operate in and through the Brazilian state to engage multilateral governance institutions such as the WTO. Sugarcane-ethanol producers and cotton producers (who always rotate their cotton fields with large amounts of soybean and maize) successfully won WTO arbitrations during the 2000s against US-government subsidized agribusinesses. Thereby, they developed such diplomatic savvy and technical expertise at the WTO that they were able to define the G20[9] position during the Doha Round, portraying their particular commercial interests (as competitors against the US and EU in agroindustrial export markets) as a generalized interest of the global South (Hopewell 2013; Margulis 2014).

[9]The Group of 20 developing countries established at the WTO in 2003 under the leadership of Brazil, South Africa, and India; not to be confused with the Group of 20 major economies.

This growing expression of Brazilian geopolitical power, largely through the soybean sector and other agribusinesses, might not be in the interest of several small and less industrialized countries in Latin America, Africa and Asia, but it does find decisive resonance with China's growing political and economic interests, as evidenced by the progressively closer ties between the Brazilian and Chinese governments and the mushrooming trade between these countries. About 48 percent of all Brazilian exports to China are constituted of soybeans and soy oil, and Brazil supplies half of all Chinese soybean imports (UN–Comtrade 2014). In order to administer this growing trade, the Brazilian and Chinese central governments have been holding increasingly more frequent government talks at all levels, including yearly presidential BRICS[10] summits since 2009, and establishing multiple bilateral agreements under the umbrella of a Joint Action Plan to coordinate economic, cultural and political exchange in the future decade. Yet the Brazilian manufacturing sector has been stridently opposed to the continuation of the sharply asymmetric trade between Brazil and China, whereby Brazil exports almost exclusively raw commodities but imports an extensive array of manufactured products. They and many economists warn about overvaluing the Brazilian currency and the 'threat of de-industrialization' in Brazil, leading to the loss of manufacturing competitiveness and market access domestically, as well as in Latin America and Africa (Jenkins and Barbosa 2012; Powell 2011). Others, however, point out that the Brazilian state could harness foreign reserves and infrastructure investments from China to upgrade its industrial capacity (shifting from crude oil, iron ore and soybean exports to the expansion of its petrochemical, steel, food processing and meat processing industries) and thereby also finance the extension and improvement of social services (Acioly, Pinto, and Cintra 2011; Oliveira 2010). Although the number of actual Chinese partnerships accomplishing this 'upgrading' remains relatively small, especially in agribusiness sectors (CEBC 2012, 2013), Brazil's massive exports to China have indeed strengthened the Real and built up foreign currency reserves, which in turn have allowed the Brazilian state to pay down its foreign debt and significantly improve its international credit rating (Inoue and Vaz 2012). The geopolitical implications of these financial and monetary effects of Brazilian soybean exports to China are discussed in the following section.

Soy-dollars in an age of 'scarcity' (1995–present)

Given the increasingly significant intersections of climate change, agroindustrialization and volatile commodity prices (Borras et al. 2014; Moreira 2013), the geopolitical implications of Brazilian soybean agribusiness extrapolate the direct territorial and politico-economic considerations raised above, and find their greatest expression as an instrument for the Brazilian state and agribusinesses to expand their power in the intersecting volatility of climate change and currency markets.

Consider the growing trope that climate change is inaugurating an 'age of scarcity' in which control over unstable food production and dwindling water resources becomes the cornerstone of any state's national security interests. This vision, which had some traction during the 1970s due to the (OPEC-organized) energy crisis and ensuing high food prices, regained popularity after the 1990s due to greater effects and awareness of climate change, and became widespread during the late 2000s when a conjunction of financial, energy and environmental crises hurled agricultural commodity and food prices to record highs:

[10]Brazil, Russia, India, China and South Africa

> With anxious markets sustaining one shock after another, food has quickly become the hidden driver of world politics. And crises like these are going to become increasingly common. The new geopolitics of food looks a whole lot more volatile – and a whole lot more contentious – than it used to. Scarcity is the new norm. (Brown 2012, 23)

Consequently, control over (and profit from) soybeans in Brazil and neighboring countries in South America, which have the greatest potential for agroindustrial expansion, takes on new and crucial geopolitical significance:

> The strategic importance of the Soybean Republic [sic] is that it contains a 'vital space' that secures access to food and water resources. In the coming decades, demographic and climate trends will determine the rise of these resources to a higher degree of importance than oil, as importing countries intensify their quest for food resources. (Turzi 2011, 66)

We must be careful, however, with this discourse that naturalizes scarcity as a given condition of our times. Scarcity cannot be merely viewed as a 'natural' phenomenon; it is embedded in social relations with particular historical geographies. Above all, the current 'scarcity' results from particularly inefficient forms of resource use, as has been repeatedly shown to be the case of soybean production for biodiesel and livestock feed, that largely benefits powerful state and corporate actors (Cavalett and Ortega 2010; Steinfield et al. 2006; Weis 2013). This is not to deny that climate change and environmental degradation are negatively affecting agricultural production around the world, and that domestic and transnational capital flows into the Brazilian soybean complex (and other farmland and agroindustrial investments around the world) amount in large part to water-seeking land grabs (Franco, Mehta, and Veldwisch 2013).[11] But rather than seeing this as a neo-Malthusian 'fact' due to climate change, we must analyze this 'scarcity' in terms of anthropogenic, political and economic struggles taking place simultaneously as conflicts between states, competition between companies and struggles between classes all around the world (Sommerville, Essex, and Le Billon 2014).

Therefore, it is more accurate to regard the problems associated with climate change in terms of price volatility and uneven access to food and resources across *class* divisions rather than naturalized scarcity ramping up simple Westphalian state competition. Doing so reveals more than coincidental similarities between climate-induced uncertainties and financial speculation on volatile commodity and currency markets. After all, the commodity price spike of 2007–2008 occurred primarily because of the massive inflow of speculative capital that was divesting from the real estate bubble in the US and finding no safe haven in the insolvent financial institutions of Wall Street, London or in US dollar reserves. This financial restructuring triggered in turn a sudden rush for farmland and agroindustrial investments that seeks to diversify production factors of agroindustrial commodities, thereby transforming both financial *and* ecological volatility from structural vulnerabilities to opportunities for windfall profits (Grain 2008; HLPE 2011). In other words, expanding and broadening one's portfolio with transnational operations enables an agribusiness company (and financial institutions lending to or owning stocks in agribusiness companies) to make extra profits from its operations in Brazil, for example, when there are major droughts or financial crises in the US, and vice versa. In addition, large-scale commodity

[11] 'Water grabs' and their articulation with agroindustrial and geopolitical interests must also consider the dredging of rivers and waterway infrastructure construction for agribusiness exports (e.g. Mamone 2007).

trading companies and agroindustrial input providers (who profit largely as well from the provision of credit and future market contracts with agricultural producers) are able to speculate in the CBOT and regional boards elsewhere through hedging and future markets to generate scandalous profits precisely due to this record volatility in commodity markets (Murphy, Burch, and Clapp 2012).

So far, these transformations seem to reinforce the notion that transnational agroindustrial corporation interests are the ultimate force shaping the current corporate food regime over and above states' geopolitical concerns (McMichael 2012; Turzi 2011), yet these studies have largely neglected the significance of currency markets and their intimate relations with agroindustrial commodities. Perhaps economists have largely insulated themselves from these debates as well, but not without warning about the growing importance of international monetary politics:

> International relations, political as well as economic, are being dramatically reshaped by the increasing interpenetration of national monetary spaces. Market-driven currency competition alters the distribution of resources and power around the globe. It generates mounting tensions and insecurities – potential threats to global stability as well as promising opportunities for cooperation. (Cohen 1998, 3)

An incipient effort at combining these insights is taking place in the 'land-use change' literature (Arcand, Guillaumont, and Jeanneney 2008; Richards et al. 2012). Considering Brazilian soybeans, for example, it is abundantly clear to anyone working in or observing the sector that the currency exchange rate between the US dollar and the Brazilian Real is of utmost importance: after all, a major part of soybean production costs in Brazil are in dollar-denominated inputs such as imported seeds, pesticides and fertilizers, while prices in the international markets to which half of all Brazilian soybean production is devoted are set in US dollars at the CBOT.[12] Consequently, the soybean complex in Brazil not only moves billions between US-dollar and Brazilian-Real currency markets each year, but it also significantly influences the Brazilian state's financial and monetary policy, helping to generate a positive balance of trade and foreign reserves as long as dollar-denominated soybean exports outpace dollar-denominated input imports. By the same token, Brazilian monetary policy also affects the ability of US and other soybean exporters to obtain favorable terms in international markets, as evidenced by the linkages between the devaluation of the Brazilian Real and a contraction of US soybean exports (Andino, Mulik, and Koo 2005; Richards et al. 2012). Thus, even though agroindustrial producers, processors and traders have strong incentives to operate transnationally, doing so ultimately requires favorable monetary policies (in addition to favorable fiscal policies, labor and environmental regulations, etc.), which enables states to exercise great control over agroindustrial development for their own geopolitical purposes.

[12] 'A major obstacle to agricultural expansion in recent years has been the rise in the local currency, the Real. Although a stronger Real results in cheaper inputs such as imported fertilizer, chemicals and machinery, it makes Brazilian farmers less competitive on the international market and cuts into already-tight profit margins. Soybeans provide a clear example of the impact of currency variations on competitiveness. When the Real was devalued at the end of the 1990s, making Brazilian products more competitive on the global market, the area planted to soybeans expanded rapidly, peaking in 2004. When the Real started to strengthen in 2003, the expansion of soybean acreage halted. (There is a lag between the decision to plant, which is influenced, among other things, by currency expectations, and the actual planting. Since 2007, acreage has increased again despite the strong Real as a result of factors such as vibrant demand from China and low corn prices)' (EIU 2010, 12).

There is an extensive literature on the international political economy of money and 'commodity currencies' that is very useful for deeper and more nuanced considerations of the geopolitics of soybeans in Brazil (and agroindustries more generally). 'Commodity currencies' are currencies of countries where agricultural and/or mineral commodities constitute an important share of exports, and thus their exchange rates are largely dependent on commodity prices (Cashin, Céspedes, and Sahay 2004; Chen and Rogoff 2003; Clements and Fry 2008). Predictably, the greater share that exports represent for a country's overall economy, the more pronounced will be this effect. Brazil's economy is large and quite heterogeneous, yet agroindustrial and mineral commodities are still its main exports, and the dynamics of 'commodity currencies' certainly apply to the Brazilian Real (Frischtak 2012). But there is far more than meets the (economist's) eye regarding the geopolitical implications of this relationship between commodities and currency markets. After all, disparity in currency exchange rates ultimately signifies differential terms of trade, which amounts to a transfer of wealth from commodity-importing to commodity-exporting countries in times of high commodity prices, and a reversal of this transfer of wealth in times of low commodity prices. These transfers of wealth operate far beyond the particular sectors engaged in international trade, affecting the operations of all companies and the basic cost of living for people in all but the most protectionist countries. Yet these effects are contradictory across classes, as local currency devaluation and interest rate hikes sustain competitive exports but also slash real wages and (especially when combined with tax increases and cuts in welfare spending) in effect transfer wealth from workers to domestic and transnational elites, especially financial and corporate elites in the United States operating through US-dollar-dominated commodity markets (Cohen 1998; Kirshner 2008; Spiro 1999).

This effect becomes even more pronounced when considering the role of international reserve currencies. These are currencies commonly used in international transactions and consequently held in significant quantities by state governments and institutions operating in international markets as part of their foreign exchange reserves. Corporations and individuals living in a country that issues a reserve currency gain major benefits from this arrangement, since they are able to gain greater profits from their exports, purchase imports more cheaply than those from other states and elude balance-of-payment crises since they pay for imports with their own currency. These benefits ultimately allow a country like the US to extract over USD 100 billion per year through this 'exorbitant privilege' of issuing the current international reserve currency (Eichengreen 2010; Gourinchas and Rey 2007). Even beyond the direct profits made by US-based transnational trading companies from controlling Brazilian soybean exports to China, therefore, the US government and US-based corporate and consumer interests benefit enormously from the maintenance of the US dollar as the default currency for the international soybean trade – especially now that Brazil has surpassed the US as the largest soybean exporter in the world. This bears striking resemblance to the political power and economic rents the US extracts from 'petro-dollars' that flow from oil exporting states through Wall Street, a process that gained much attention during the 1970s when the US ceased being the world's leading petroleum exporter but maintained economic profits and geopolitical control over the oil flow by issuing the international reserve currency *and assuring that Middle Eastern oil would be sold in US dollars* (Magdoff and Sweezy 1981; Kirshner 2008; Spiro 1999). We might ask, therefore, how agroindustrial commodity production and trade might serve geopolitical functions (and thus contestation) similar to those that 'petro-dollars' are known to have served since the 1970s. We should research, therefore, the politics of 'soy-dollars' between the US and its major competitors, i.e. Brazil and Argentina, and the major consumers of these exports, i.e. Europe and China.[13]

Brazilian soybean production increased from 10 mmt in 1975 to 25 mmt in 1994, and then more than tripled to 82 mmt in 2013 (Shurtleff and Aoyagi 2009). During this same period, CBOT prices for soybeans spiked from around USD 250 to 1000/bushel during the crises of the 1970s, then gradually plateaued around USD 500/bushel during the 1990s. After 2003, however, volatility increased again, and since 2012 prices have averaged USD 1500/bushel (Trading Economics 2014). It might be expected that, despite US 'exorbitant privilege', the recent commodity 'super cycle' has been very positive for Brazil. Yet the geopolitical implications of holding a 'commodity currency' are not so straightforward for a country that also hosts a broad industrial and manufacturing sector and an enormous domestic market, such as Brazil. High commodity prices and the strengthening of the Brazilian Real in relation to the US dollar might benefit the soybean sector and other commodity exporters, yet they also render Brazilian industrial and manufactured exports far less competitive in international markets and increase the tendency towards domestic inflation (Frischtak 2012; Jenkins and Barbosa 2012). Therefore, there is a constant battle in Brazilian corporate and policy circles over whether to loosen or tighten monetary policy at any given moment, with labor unions and representatives of the manufacturing sector pitted against the agribusiness, mining and petrochemical sectors. As long as Brazil remains dependent upon a foreign currency for its international exchanges, there is no way to escape this tension and, consequently, the expansion of regional trade blocs like Mercosul and currency swaps with major trade partners (such as Argentina and China) has become an increasingly attractive strategy for the Brazilian state to reduce this pressure (Carvalho 2008; Leahy 2013).

The most interesting and understudied geopolitical implication of Brazilian soybean arises precisely through the manner in which it serves as an instrument for Brazil to honor currency swaps with its new primary commercial partner and largest soybean importer, i.e. China, as this is seen as the harbinger of a shift of international reserve currencies. After all, these are relatively stable over many decades, but they do shift along with broader political and economic transformations (Arrighi and Silver 1999). It took several decades from the late nineteenth century until the end of World War II for the US dollar to displace the British pound as the main global reserve currency, and there is growing evidence that the US dollar is steadily losing its hegemony in international currency markets (Carbaugh and Hedrick 2009; Eichengreen 2010). Despite its current crisis, the most likely contender that could replace the US dollar as the main international reserve currency in the near future is the euro (Chinn and Frankel 2007), yet the Chinese yuan is also displaying signs that it could 'internationalize' along with the Chinese economy and become one of the main reserve currencies by 2030 (Chen and Cheung 2011; Jaeger 2010). At such a point, the US dollar, the euro and the yuan might share the status of reserve currencies, globally or in regional blocs, for an indeterminate and perhaps transitional period (Eichengreen 2010; Fischer 2011). The primary mechanisms for this transformation are bilateral currency swap agreements, which China has been pursuing aggressively with its neighbors and commercial partners in the global South (Zhang Jianhua, Chief Executive Officer of Bank of

[13]It is more accurate to address the politics of 'grain-dollars' rather than 'soy-dollars' alone, since all soybean producers also rotate their production with other crops such as maize, cotton and wheat, and the major transnational input and trading companies that control soybean technologies and trade also deal in maize, wheat and other grains.

China Brazil, quoted in Frischtak and Soares 2012; cf. Liao and McDowell Forthcoming). In March 2013, for example, Brazil and China established a currency swap fund between their central banks equivalent to USD 30 billion:

> Though the agreement is expected to be of little use in day-to-day trade, the deal will guarantee the flow of Brazil's growing soy, iron ore and other exports to China and China's imports of manufactured goods to Brazil regardless of global financial conditions. 'The purpose of this swap is that, independent of the conditions prevailing in the international financial market, we will have $30bn available which would represent eight months of exports from Brazil to China and 10 months of imports to Brazil from China', said Alexandre Tombini, president of Brazil's central bank in a press conference on the sidelines of the BRICS summit in South Africa. (Leahy 2013)

As long as the Chinese yuan is not a floating currency and the volume of soybeans flowing through the boards of trade in Brazil or China remains significantly lower than the CBOT, these swap agreements are merely preventive measures in case another financial crisis in the global North freezes US dollar markets, as in late 2008. But over the long term, as more trade and investments take place between Brazil, China and other countries in the global South, such agreements may indeed enable these states to pursue more independent politico-economic policies that challenge more directly the hegemony of the US dollar and the US government in the inter-state system (Kirshner 2008; Spiro 1999). For example, Russian prohibition on agricultural imports from the US, EU, Canada, Australia and Norway were enthusiastically welcomed by Brazilian (and Latin American) state and agribusiness interests. Brazilian agricultural export growth to Russia will be constituted primarily of meat products, the main 'aggregated value' industry that drives soy production upstream (CNA 2014b). Thus, by expanding soybean production and controlling its processing and trade within Brazil, its neighboring South American countries and even in distant African states, the Brazilian state and agribusinesses, along with other emergent powers, are cultivating political and economic power to establish themselves as leaders in a 'new multipolar world order'. In this way, the CBOT may lose its hegemonic control over soybean international trade to the boards of Dalian in China, Rosario in Argentina and São Paulo in Brazil (Treviño 2005).

Conclusion

In this contribution, I demonstrated the manner in which soybean cultivation has played important geopolitical roles, from its origins in Brazil at the hands of Japanese migrants and through its early use as green fertilizer, to facilitate the expansion of commercial agriculture and consolidate the 'green revolution' in Brazil. I then described the interrelated processes of agroindustrial and geopolitical restructuring that transformed soybean into one of the most important commodities traded in international markets, and began to undermine US hegemony in global agroindustrial assemblages and the inter-state system. I then reviewed the sub-imperialist role that Brazilian soybean production (as well as technology and expertise) plays in Latin America and Africa, alongside new dynamics of 'South–South cooperation' in international trade and governance. Finally, I raised several hitherto underexplored questions about the intersections between global agroindustrial production and currency exchange rates and reserves. I suggest that studying the politics of 'soy-dollars' might raise several new questions and important insights that remain unnoticed in current geopolitical studies of Brazil, as well as most critical agrarian studies of global agroindustrial assemblages or food regimes. Of course, in order to properly develop the concepts of 'soy-dollars' or 'grain-dollars', it would be necessary to undertake a much more extensive analysis in a separate paper of its own.

Although I have argued that the soybean complex in Brazil is particularly strategic and significant for contemporary geopolitics, the phenomenon analyzed – agroindustrial development as part of state-making and power projection – is not unique to soybeans in Brazil. Instead, this paper contributes to a broader study of the interrelations between states and 'global assemblages of agro-power' (Sommerville, Essex, and Le Billon 2014), which may be further developed for other key agroindustrial sectors of Brazil and for a larger set of countries as well. Ultimately, these geopolitical considerations must be overlaid on traditional agrarian questions in Brazil and beyond: how are we to analyze and evaluate the contradictory process whereby the power and influence of emergent agribusiness and state actors begin to successfully challenge US hegemony and obtain political and economic gains for national development, but do so through the domestic consolidation and sub-imperialist expansion of an agroindustrial production system that produces such serious social and ecological injustices? Will Brazil and other emerging economies succeed in establishing a 'new multipolar world order', and how will this new order be any more just and democratic if it also rests upon the concentration of power and wealth in the hands of a 'multipolar' elite? These and other questions must be further explored in order to deepen the study of the geopolitics of Brazilian soybean as not only an instrument of contestation between states or competition between companies, but also as a terrain of struggle between different classes and class fractions in Brazil and around the world.

Acknowledgements

I would like to thank Julie Klinger, Greg Thaler, two anonymous reviewers, the participants of the American Association of Geographers Annual Conference 2013 and the participants of a symposium at the Federal University of Goiás who provided valuable feedback on earlier versions of this paper.

References

Acioly, L., E. Pinto, and M. Cintra. 2011. China e Brasil: oportunidades e desafios [China and Brazil: opportunities and challenges]. In *A China na nova configuração global: impactos políticos e econômicos*, ed. R. Leão, E. Pinto and L. Acioly, [China in the new global configuration: political and economic impacts], 307–50. Brasília: IPEA.

Agnew, J. 1994. The territorial trap: The geographical assumptions of international relations theory. *Review of international political economy* 1, no. 1: 53–80.

Amanor, K. 2013. South–South cooperation in context: Perspectives from Africa. *FAC Working Paper No. 54*. Brighton: Future Agricultures Consortium.

Andino, J., K. Mulik, and W. Koo. 2005. The impact of Brazil and Argentina's currency devaluation US soybean trade. *Agribusiness & Applied Economics Report No. 574*. Fargo, ND: Center for Agricultural Policy and Trade Studies, North Dakota State University.

Arcand, J.-L., P. Guillaumont, and S. Jeanneney. 2008. Deforestation and the real exchange rate. *Journal of Development Economics* 75: 242–62.

Arellano-Lopez, F. 2012. Conflicting land use agendas: Environment, indigenous land rights and development in central Bolivia. The case of the Isiboro-Sécure Indigenous Territory and National Park. Paper presented at the LDPI International Conference on Land Grabs II, Cornell University, 17–19 October. http://www.cornell-landproject.org/download/landgrab2012 papers/arellano.pdf

Arrighi, G., and B. Silver. 1999. *Chaos and governance in the modern world system*. Minneapolis: University of Minnesota Press.

Barbosa, A., T. Narciso, and M. Biancalana. 2009. Brazil in Africa: Another emerging power in the continent? *Politikon* 36, no. 1: 59–86.

Bastos, E. 1984. *As ligas camponesas* [The peasant leagues]. Petropolis, RJ: Editora Vozes.

Borras, S.M., Jr, J.C. Franco, S. Gómez, C. Kay, and M. Spoor. 2012. Land grabbing in Latin America and the Caribbean. *Journal of Peasant Studies* 39, no. 3–4: 845–72.

Borras, S., J. Franco, R. Isakson, L. Levidow, and P. Vervest. 2014. Towards understanding the politics of flex crops and commodities: Implications for research and policy advocacy. *Think Piece Series on Flex Crops and Commodities, No.1.* Amsterdam: Transnational Institute (TNI).

Boyd, W. 2001. Making meat: Science, technology, and American poultry production. *Technology and Culture* 42, no. 4: 631–64.

Brazil. 2008. Decree number 6.592. Brasilia: Presidency of the Republic, 2 October.

Brazil-China Business Council (CEBC). 2012. The commercial relation between Brazil and China in agribusiness. *Carta Brasil-China No. 6.* Rio de Janeiro: CEBC.

Brazil-China Business Council (CEBC). 2013. *Chinese Investments in Brazil from 2007–2012.* Rio de Janeiro: CEBC.

BRICS. 2009. *Joint Statement of the BRIC Countries' Leaders.* Yekaterinburg, Russia, June 16. http://archive.kremlin.ru/eng/text/docs/2009/06/217963.shtml (accessed July 30, 2014).

Brown, L. 2012. *Full planet, empty plates: The new geopolitics of food scarcity.* New York: WW Norton & Company.

Cabral, L., and A. Shankland. 2013. Narratives of Brazil-Africa cooperation for agricultural development: new paradigms? *FAC Working Paper No. 51*, Brighton: Future Agricultures Consortium.

Carbaugh, R., and D. Hedrick. 2009. Will the US dollar be dethroned as the main reserve currency? *Global Economy Journal* 9, no. 3: 1–14.

Carvalho, D. 2008. Sistema de pagamentos em moeda local: aspectos jurídicos da nova alternativa para remessas de valores entre o Brasil e a Argentina [System of payments in local currency: juridical aspects of the new alternative for value exchange between Brazil and Argentina]. *Revista da Procuradoria-Geral do Banco Central do Brasil* 2, no. 2: 199–224.

Cashin, P., L.F. Céspedes, and R. Sahay. 2004. Commodity currencies and the real exchange rate. *Journal of Development Economics* 75, no. 1: 239–68.

Cavalett, O., and E. Ortega. 2010. Integrated environmental assessment of biodiesel production from soybean in Brazil. *Journal of Cleaner Production* 18, no. 1: 55–70.

Center for Research on the Cerrado (CPAC). 1975. Cerrado: filosofia e programa do centro de pesquisa [Cerrado: philosophy and program of the research center]. *Cerrado* 7, no. 27: 22–6.

Cesarino, L. 2013. Brazil as an emerging donor in Africa: from discourse to the frontline. African Studies Association Annual Meeting Paper. http://ssrn.com/abstract=2236885 (accessed July 30, 2014).

Chen, X., and Y-W. Cheung. 2011. Renminbi going global. *China & World Economy* 19, no. 2: 1–18.

Chen, Y., and K. Rogoff. 2003. Commodity currencies. *Journal of International Economics* 60, no. 1: 133–60.

Chichava, S., J. Duran, L. Cabral, A. Shankland, L. Buckley, T. Lixia, and Z. Yue. 2013. Brazil and China in Mozambican agriculture: Emerging insights from the field. *IDS Bulletin* 44: 101–15.

Chinn, M., and J. Frankel. 2007. Will the euro eventually surpass the dollar as leading international reserve currency? In *G7 current account imbalances: Sustainability and adjustment*, ed. R. Clarida, 283–335. Chicago: University of Chicago Press.

Clements, E.A., and B.M. Fernandes. 2013. Land grabbing, agribusiness and the peasantry in Brazil and Mozambique. *Agrarian South: Journal of Political Economy* 2, no. 1: 41–69.

Clements, K.W., and R. Fry. 2008. Commodity currencies and currency commodities. *Resources Policy* 33, no. 2: 55–73.

Cohen, B. 1998. *The geography of money.* Ithaca, NY: Cornell University Press.

Couto e Silva, G. 1957. *Aspectos geopolíticos do Brasil* (Geopolitical aspects of Brazil). Rio de Janeiro: Biblioteca do Exército Editora.

Couto e Silva, G. 1981. *Planejamento estratégico* (Strategic planning). Brasília: Editora da Universidade de Brasília.

Couto e Silva, G., and A. Franco. 1967. *Geopolítica do Brasil* (Geopolitics of Brazil). Rio de Janeiro: Editora José Olympio.

Dauvergne, P., and D. Farias. 2012. The rise of Brazil as a global development power. *Third World Quarterly* 33, no. 5: 903–17.

EcoNexus and Berne Declaration. 2013. *Agropoly: A handful of corporations control world food production.* http://econexus.info/publication/agropoly-handful-corporations-control-world-food-production (accessed August 29, 2014).

Economist Intelligence Unit (EIU). 2010. *The global power of Brazilian agribusiness*. London: The Economist-Accenture.

Eichengreen, B. 2010. *Exorbitant privilege: The rise and fall of the dollar and the future of the international monetary system*. Oxford: Oxford University Press.

Estevam, L. 2004. *O tempo da transformação: estrutura e dinâmica da formação econômica de Goiás* (The time of transformation: structure and dynamics of the economic formation of Goiás). Goiânia: Editora da Universidade Católica de Goiás.

Evans, P. 1979. *Dependent development: The alliance of multinational, state, and local capital in Brazil*. Princeton: Princeton University Press.

Fairbairn, M. 2013. Indirect dispossession: Domestic power imbalances and foreign access to land in Mozambique. *Development and Change* 44, no. 2: 335–56.

Fausto, B. 1994. *História do Brasil* [History of Brazil]. São Paulo: Editora da Universidade de São Paulo.

Fischer, C. 2011. Currency blocs in the 21st century. *Discussion Paper Series 1: Economic Studies No. 12*. Frankfurt am Main: Deutsche Bundesbank.

Flynn, M. 2007. Between sub-imperialism and globalization: A case study in the internationalization of Brazilian capital. *Latin American Perspectives* 34, no. 6: 9–27.

Foweraker, J. 1981. *The struggle for land: A political economy of the pioneer frontier in Brazil from 1930 to the present day*. Cambridge: Cambridge University Press.

Franco, J., L. Mehta, and G. Veldwisch. 2013. The global politics of water grabbing. *Third World Quarterly* 34, no. 9: 1651–75.

Freitas, J. 2004. *A escola geopolítica Brasileira* [The Brazilian school of geopolitics]. Rio de Janeiro: Biblioteca do Exército Editora.

Friedmann, H. 1982. The political economy of food: The rise and fall of the postwar international food order. *American Journal of Sociology* 88: S248–86.

Frischtak, C. 2012. A monetary tsunami? Brazil in the cross-fire of new style currency wars. In *New challenges for the global economy, new uncertainties for the G-20*, ed. I. Atiyas, 13–7. Washington, DC: Brookings Institution.

Frischtak, C., and A. Soares. 2012. *Yuan Internationalization. Carta Brasil-China Special Edition No.5*: Rio de Janeiro: CEBC.

Galeano, L. 2012. Paraguay and the expansion of Brazilian and Argentinian agribusiness frontiers. *Canadian journal of development studies/Revue canadienne d'études du développement* 33, no. 4: 458–70.

Goldsmith, P., B. Li, J. Fruin, and R. Hirsch. 2004. Global shifts in agro-industrial capital and the case of soybean crushing: Implications for managers and policy makers. *International Food and Agribusiness Management Review* 7, no. 2: 87–115.

Gourinchas, P.-O., and H. Rey. 2007. From world banker to world venture capitalist: The U.S. external adjustment and the exorbitant privilege. In *G7 current account imbalances: Sustainability and adjustment*, ed. R. Clarida, 11–55. Chicago: University of Chicago Press.

Grain. 2008. *Seized! The 2008 land grab for food and financial security*. Barcelona: Grain.

Gudynas, E. 2008. The new bonfire of vanities: Soybean cultivation and globalization in South America. *Development* 51, no. 4: 512–8.

Harvey, D. 2005. *A brief history of neoliberalism*. Oxford: Oxford University Press.

Hasse, G., and F. Bueno. 1996. *O Brasil da soja: abrindo fronteiras, semeando cidades* [The Brazil of soy: opening frontiers, seeding cities]. Porto Alegre: Ceval Alimentos/L & PM Editores.

Hecht, S. 2005. Soybeans, development and conservation on the Amazon frontier. *Development and Change* 36, no. 2: 375–404.

Hecht, S., and C. Mann. 2008. How Brazil outfarmed the American farmer. *Fortune*, 19 January, 92–105.

High Level Panel of Experts on Food Security and Nutrition (HLPE). 2011. *Price volatility and food security*. Report to the Committee on World Food Security of the Food and Agriculture Organization of the United Nations. Rome: CFS/FAO.

HighQuest Partners and Soyatech. 2011. *How the global oilseed and grain trade works*. http://www.unitedsoybean.org/wp-content/uploads/2013/07/RevisedJan12_GlobalOilSeedGrainTrade_2011.pdf (accessed July 30, 2014).

Hollerman, L. 1988. *Japan's economic strategy in Brazil: Challenge for the United States*. Lexington, MA: Lexington Books.

Hopewell, K. 2013. New protagonists in global economic governance: Brazilian agribusiness at the WTO. *New Political Economy* 18, no. 4: 603–23.

Ianni, O. 1979. *Ditadura e agricultura: o desenvolvimento do capitalismo na Amazônia* [Dictatorship and agriculture: the development of capitalism in the Amazon]. Rio de Janeiro: Editora Civilização Brasileira.

Inoue, C., and A. Vaz. 2012. Brazil as 'Southern donor': Beyond hierarchy and national interests in development cooperation? *Cambridge Review of International Affairs* 25, no. 4: 507–34.

Instituto Lula. 2014. Blairo Maggi e Lula em Cuba para ampliar produção de alimentos (Blairo Maggi and Lula in Cuba to expand food production). São Paulo: Instituto Lula, 26 February. http://www.institutolula.org/blairo-maggi-e-lula-em-cuba-para-ampliar-producao-de-alimentos/#.Ux3tDeddWMP (in Portuguese) (accessed July 30, 2014).

Jaeger, M. 2010. Yuan as a reserve currency: likely prospects & possible implications. *Deutsche Bank Research Briefing*, 16 July. Frankfurt am Main: Deutsche Bank.

Jenkins, R., and A. Barbosa. 2012. Fear for manufacturing? China and the future of industry in Brazil and Latin America. *The China Quarterly* 209, no. 2: 59–81.

Jepson, W., C. Brannstrom, and A. Filippi. 2010. Access regimes and regional land change in the Brazilian Cerrado, 1972–2002. *Annals of the Association of American Geographers* 100, no. 1: 87–111.

Jessop, B. 2002. The crisis of the national spatio-temporal fix and the tendential ecological dominance of globalizing capitalism. *International Journal of Urban and Regional Research* 24, no. 2: 343–60.

Justiça Ambiental et al. 2013. Leaked ProSAVANA Master Plan confirms worst fears. http://www.grain.org/article/entries/4703-leaked-prosavana-master-plan-confirms-worst-fears (accessed July 30, 2014).

Kirshner, J. 2008. Dollar primacy and American power: what's at stake? *Review of International Political Economy*, 15, no. 3: 418–38.

Leahy, J. 2013. Brazil and China agree currency swap. *Financial Times*, 26 March.

Liao, S., and D. McDowell. Forthcoming. Redback rising: China's bilateral swap agreements and RMB internationalization. *International Studies Quarterly*.

Mackey, L. 2011. Legitimating foreignization in Bolivia: Brazilian agriculture and the relations of conflict and consent in Santa Cruz, Bolivia. Paper presented at the LDPI International Conference on Land Grabs, IDS University of Sussex, 8 April. http://www.iss.nl/fileadmin/ASSETS/iss/Documents/Conference_papers/LDPI/23_Lee_Mackey.pdf (accessed July 30, 2014).

Magdoff, H., and P. Sweezy. 1981. The U.S. dollar, petrodollars, and US imperialism. In *The deepening crisis of U.S. capitalism*, ed. H. Magdoff and P. Sweezy, 94–106. New York: Monthly Review.

Mamone, I. 2007. Water and geopolitics: a closer look at the Paraguay-Paraná Waterway program. Paper presented at the 37th St. Gallen Symposium 'The power of natural resources', St. Gallen, Switzerland, 31 May.

Margulis, M. 2014. Trading out of the global food crisis? The world trade organization and the geopolitics of food security. *Geopolitics* 19, no. 2: 322–50.

Marini, R. 1972. Brazilian subimperialism. *Monthly Review* 23, no. 9: 14–24.

Martins, P., and B. Pelegrini. 1984. *Cerrados: uma ocupação Japonesa no campo* (Cerrados: a Japanese occupation in the countryside). Rio de Janeiro: Editora Codecri.

Mason, J., and P. Singer. 1980. *Animal factories*. New York: Harmony Books.

Mazoyer, M., and L. Roudart. 2006. *A history of world agriculture: From the neolithic age to the current crisis*. New York: Monthly Review.

McMichael, P. 2012. The land grab and corporate food regime restructuring. *Journal of Peasant Studies* 39, no. 3–4: 681–701.

Ministry of Agriculture, Livestock and Supply (MAPA). 2002. *Programas de cooperação em desenvolvimento agrícola entre Japão-Brasil na região do Cerrado no Brasil* (Agricultural development cooperation programs between Japan and Brazil in the Cerrado region of Brazil). Brasília: MAPA.

Moreira, H. 2013. A formação da nova geopolítica das mudanças climáticas [The formation of the new geopolitics of climate change]. *Sustentabilidade em Debate* 4, no. 1: 275–92.

Morgan, D. 2000 [1979]. *Merchants of Grain: The power and profits of the five giant companies at the center of the world's food supply*. Lincoln, NE: iUniverse.

Motta, M., and P. Zarth. 2008. *Formas de Resistência Camponesa, vol. II* [Forms of Peasant Resistance, vol. II]. São Paulo: NEAD.

Murphy, S., D. Burch, and J. Clapp. 2012. *Cereal secrets: The world's largest grain traders and global agriculture*. Oxfam Research Report. Oxford: Oxfam.

National Confederation of Agriculture and Livestock of Brazil (CNA). 2014a. Agronegócio no Brasil (Agribusiness in Brazil). http://agroinvestbrasil.com.br/agronegocio-brasil (in Portuguese) (accessed July 30, 2014).

National Confederation of Agriculture and Livestock of Brazil (CNA). 2014b. Rússia é mercado promissor para o agronegócio brasileiro (Russia is a promising market for Brazilian agribusiness). http://www.canaldoprodutor.com.br/comunicacao/noticias/russia-e-mercado-promissor-para-o-agronegocio-brasileiro (in Portuguese) (accessed September 3, 2014).

Oliveira, G. de L. T. 2013. Land regularization in Brazil and the global land grab. *Development and Change* 44, no. 2: 261–83.

Oliveira, G. de L. T., and M. Schneider. Forthcoming. The politics of flexing soybeans: China, Brazil, and global agroindustrial restructuring. *Journal of Peasant Studies*.

Oliveira, H. 2010. Brasil e China: uma nova aliança não escrita? [Brazil and China: a new unwritten alliance?]. *Revista Brasileira de Polítca Internacional* 53, no. 2: 88–105.

Patel, R. 2008. *Stuffed and starved: The hidden battle for the world food system*. New York and London: Melville House.

Patel, R. 2013. The long green revolution. *Journal of Peasant Studies* 40, no. 1: 1–63.

Peine, E. 2013. Trading on pork and beans: Agribusiness and the construction of the Brazil-China-soy-pork commodity complex. In *The ethics and economics of agrifood competition*, ed. H. James Jr, 193–210. New York: Springer.

Pires, M. 2000. Programas agrícolas na ocupação do Cerrado [Agricultural programs in the occupation of the Cerrado]. *Sociedade e Cultura* 3, no. 1: 111–31.

Powell, D. 2011. The dragon's appetite for soy stokes Brazilian protectionism. *Financial Times*, 10 October.

Ribeiro, P., and F. Ramos. 2009. From Mercosur to Unasur? Geopolitics of South American integration and the role of Brazilian diplomacy. Paper presented at the Joint International Meeting meeting of the ISA - ABRI, Rio de Janeiro, 22 July. http://www.allacademic.com/meta/p381229_index.html (accessed July 30, 2014).

Richards, P.D., R.J. Myers, S.M. Swinton, and R.T. Walker. 2012. Exchange rates, soybean supply response, and deforestation in South America. *Global Environmental Change* 22, no. 2: 454–62.

Robinson, W. 2008. *Latin America and global capitalism: A critical globalization perspective*. Baltimore: John Hopkins University Press.

Schneider, M. 2011. *Feeding China's pigs: Implications for the environment, China's smallholder farmers, and food security*. Minneapolis and Geneva: Institute for Agriculture and Trade Policy.

Scott, J. 1998. *Seeing like a state: How certain schemes to improve the human condition have failed*. New Haven, CT: Yale University Press.

Shurtleff, W., and A. Aoyagi. 2009. *History of soybeans and soyfoods in South America (1882–2009) Extensively annotated bibliography and sourcebook*. Lafayette, CA: Soyinfo Center.

Silva, M., and L. Costa. 2012. A indústria de defensivos agrícolas [The pesticide and herbicide industry]. *BNDES Setorial* 35: 233–76.

Sommerville, M., J. Essex, and P. Le Billon. 2014. The 'global food crisis' and the geopolitics of food security. *Geopolitics* 19, no. 2: 239–65.

Sousa, I., and R. Vieira. 2008. Soybeans and soy foods in Brazil, with notes on Argentina: Sketch of an expanding world commodity. In *The world of soy*, ed. Du Bois, C., C. Tan, and S. Mintz, 234–56. Chicago: University of Illinois Press.

Spiro, D. 1999. *The hidden hand of American hegemony: Petrodollar recycling and international markets*. Ithaca, NY: Cornell University Press.

Steinfeld, H., P. Gerber, T. Wassenaar, V. Castel, M. Rosales, and C. de Haan. 2006. *Livestock's long shadow: Environmental issues and options*. Rome: FAO.

The Economist. 2010. The miracle of the cerrado. Brazil has revolutionised its own farms. Can it do the same for others?. *The Economist*, 26 August.

The Economist. 2012. Out Lugo. *The Economist*, 25 June.

Trading Economics. 2014. Soybeans: Historical option data. http://www.tradingeconomics.com/commodity/soybeans (accessed August 29, 2014).

Treviño, L. 2005. Development and volume growth of organized derivatives trade in emerging markets. *Ensayos Revista de Economia*, 24, no. 2: 31–82.

Turzi, M. 2011. The soybean republic. *Yale Journal of International Affaris* 6, no. 2: 59–68.

United Nations International Trade Statistical Database (UN-Comtrade). 2014. United Nations commodity trade statistics database. http://comtrade.un.org (accessed July 30, 2014).

United States Department of Agriculture (USDA). 1965–1970. Fats and oils situation. Multiple issues. http://usda.mannlib.cornell.edu/MannUsda/viewTaxonomy.do?taxonomyID=17 (accessed July 30, 2014).

United States Department of Agriculture (USDA). 1975–1995. World Agricultural Supply and Demand Estimates. Multiple issues. http://usda.mannlib.cornell.edu/MannUsda/viewTaxonomy.do?taxonomyID=17 (accessed July 30, 2014).

United States Department of Agriculture (USDA). 2000–2013. Oilseeds: World Markets and Trade. Multiple issues. http://usda.mannlib.cornell.edu/MannUsda/viewTaxonomy.do?taxonomyID=17 (accessed July 30, 2014).

Urioste, M. 2012. Concentration and 'foreignisation' of land in Bolivia. *Canadian Journal of Development Studies/Revue canadienne d'études du développement* 33, no. 4: 439–57.

Weis, T. 2013. The meat of the global food crisis. *Journal of Peasant Studies* 40, no. 1: 65–85.

Wesz, V. Jr. 2014. *O mercado de soja e as relações de troca entre produtores rurais e empresas no Sudeste do Mato Grosso (Brasil)* (The soy market and exchange relations between farmers and firms in the southeast of Mato Grosso (Brazil)). Doctoral dissertation. Graduate Program of Social Sciences, Federal Rural University of Rio de Janeiro (UFRRJ).

Wilkinson, J. 2009. The Globalization of agribusiness and developing world food systems. *Monthly Review* 61, no. 04: 38–49.

Wolford, W., S.M. Borras, R. Hall, I. Scoones, and B. White. 2013. Governing global land deals: The role of the state in the rush for land. *Development and Change* 44, no. 2: 189–210.

World Wide Fund for Nature (WWF). 2014. *The growth of soy: Impacts and solutions*. Gland, Switzerland: WWF International.

Worster, D. 2004. *Dust bowl: The southern plains in the 1930s*. New York: Oxford University Press.

Yoshioka, Y. 1992. Development of agricultural policy in post-war Japan. In *Agriculture and trade in the Pacific*, ed. Coyle, W., D. Hayes, and H. Yamauchi, 91–100. Boulder, CO: Westview Press.

Zibechi, R. 2009. *Is Brazil creating its own backyard? Americas Program Report*. Washington, DC: Center for International Policy.

China's soybean crisis: the logic of modernization and its discontents

Yan Hairong, Chen Yiyuan and Ku Hok Bun

Lester Brown sounded an alarm in 1995: who will feed China? Against this backdrop, this contribution examines China's dramatic turn from having been a soybean net exporter, up until the mid-1990s, to being the biggest importer of genetically modified (GM) soybeans, since 2000. With China's growing soybean imports, domestic soybean production has experienced a drastic fall, creating an outcry about a 'soybean crisis' in China. This paper examines competing interpretations about China's soybean imports and how a wide arrange of heated debates and critical reflections have emerged about China's position in globalization, the role of the state in food security, the safety of GM foods, consumer rights, what constitutes scientific authority, and the power of transnational corporations. In these debates, Chinese critics have very different views about the US and South America, where significant GM soybeans are produced for export to China.

Introduction

China is the place of the domestication and emanation of soybean (Wang and Li 2000). Until the mid-1990s, China was not only self-sufficient in soybean, but was also a net exporter (Wang 2013). However, in the year 2000, China surpassed Europe to become the biggest soy importer on the global market. Its imports, over 70 million tons in 2014, accounted for 57.7 percent of global soybean trade and about 80 percent of China's soybean consumption (BBC 2015). In 2012, China imported 44, 41, 10 and 5 percent respectively from the USA, Brazil, Argentina and other producing countries (China Soybean Industry Association [CSIA] 2014, 49).

Soybean has become a key crop implicated in the changing political economic relations between China, the US and South America (Oliveira and Schneider 2014). Soybeans, not aircraft, are now the US's top export to China (Global Post 2014). As the US was already a major producer and exporter of soybeans before 2000, China's growing imports are in tandem with South America's fast expansion in soy production since the mid-1990s. In Brazil, soybean export serves as an instrument to balance the country's booming imports from China. China thus surpassed the US to become Brazil's largest trade partner in 2009 (Oliveira 2015, 17). Some observers in China, aware of the dominant role played by US-based transnational corporations in the global soy complex, perceive the situation to be that 'South America produces soybeans, China buys soybeans, and the US sells soybeans' (Zhou 2014; Guo 2012).

China's importing of genetically modified (GM) soybeans from various countries – but patented by Monsanto and a few other transnational companies – has given rise to different narratives and debates. It has become a key component of Lester Brown's widely popular China-threat narratives 'Who will feed China?' and 'Can the world feed China?' (Brown 1994; Earth Policy 2014).[1] While this type of narrative represents China as an avaricious monolith scrambling for food, it evades the fact that the colossal soybean trade has made it difficult for soybean farmers in China to sustain their production, creating a situation known in China as 'foreign soybeans chasing off domestic soybeans'. The planting of soybeans in all of China has been falling since 2008. By 2013, the total planting area had dropped 24 percent from that of 2008. China's main traditional soybean-producing province Heilongjiang has experienced the most dramatic decline in soybean sown acreage, prompting local people's congress representatives to repeatedly appeal to the central government for measures of protection (Zhang 2014). In 2013, the province saw the planting area drop 42 percent compared with the peak of 2008 (CSIA 2014, 4–5) (Figure 1).

Amid growing imports around 2004–2005, the soybean industry in China encountered an unprecedented crisis, with 70 percent of its domestic enterprises collapsing at one time. This crisis was turned into opportunity by global corporations such as ADM, Bunge, Cargill and Louis Dreyfus, to re-finance and control 70 percent of soybean-processing capacity in China (Ministry of Agriculture 2009, 159; Zang 2013, 24; Wang 2013, 95–96).[2] In 2009, nine of the top 10 soy processing enterprises in China were wholly or partly owned by foreign capital (Zhao 2009, 118). The dire situation of declining soybean farming and a depressed domestic soybean industry in the most recent decade has given rise to a critical narrative about 'the death of Chinese soybeans', or the soybean crisis. It challenges the more dominant narrative in China that justifies GM soybean imports as inevitable and beneficial for a populous country that is short on land resources.

The Chinese state has long pursued a policy of food security, a legacy from the Cold War era of a US-led embargo against China. In 1996, a Chinese government white paper, entitled *China's grain issues* (中国的粮食问题), set the rate of grain self-sufficiency at 95 percent (Zhang 2012). In 2013 it has been adjusted to 100 percent in staple food grain (rice and wheat), 90 percent in grains and 80 percent in overall food (China Radio International 2014). Among the crops that China imports, soybean is the most liberalized item.

The massive import of GM soybeans, in the context of China's entry into the WTO (World Trade Organization) and the growing presence of global agribusiness in China, has challenged the long-held principle of self-sufficiency. It has opened up heated debates about food security, consumer rights and inequality, scientific authority, the relationship between corporate interest and science, food as commodity or public good, the paradigm of development, socialism and capitalism. The issue has spurred debates and activism in China that have gone well beyond the trade issue. This paper will provide a discussion of various narratives of China's soybean imports and the points of debate. Critiques and activism regarding China's import of GM soybeans also serve as a backdrop for the notion of food sovereignty to be introduced to the public. This paper

[1]Earth Policy (2014) provides a context for how the essay by Lester Brown became a US media sensation and how it was received in China. When Brown's essay 'Who will feed China' was published in 1995, China was still an exporter of soybeans, but China's soybean imports have become part of the narrative in recent years.

[2]Before 2000, Chinese processing enterprises owned 90 percent of the soybean oil capacity in China. By 2010, Chinese state-owned enterprises accounted for about 30 percent of the soybean oil-pressing capacity, Chinese private enterprises 32 percent and foreign enterprises about 37 percent.

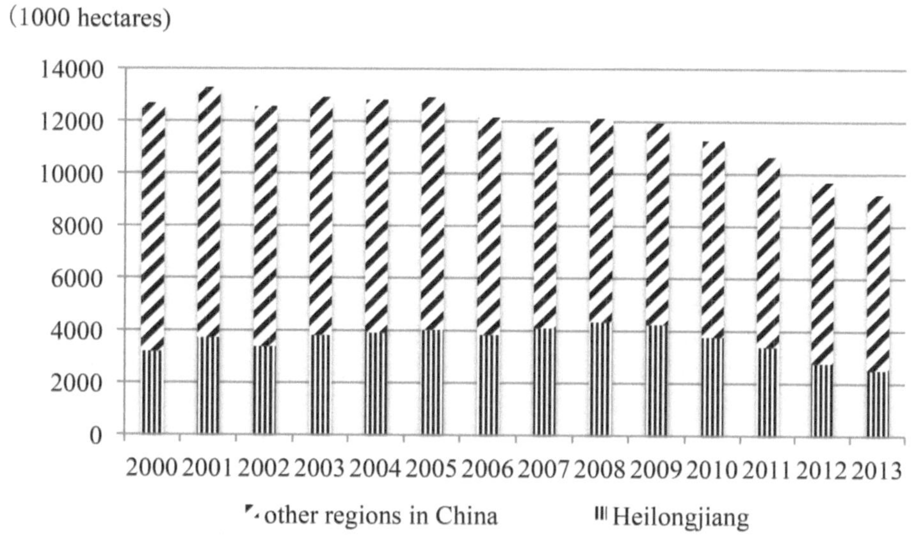

Figure 1. Soybean planting area in China (2000–2013).
Source: Data for planting area of soybean are drawn from China's Statistical Year Book (2001–2014, tables 12-14, 13-15, 12-13, 13-13, 12-8), edited by Zhongguo guojia tongjiju (2002–2015).

will discuss how engaged scholars and activists regard and localize the notion of food sovereignty in China in relation to, but also beyond, the soybean crisis. Observations in China that 'South America produces soybeans, China buys soybeans, and the US sells soybeans' also point to the global division of labor in the current neoliberal food regime that needs further examination (Friedmann 2009; McMichael 2009).

Narratives of modernization, liberalization and the economic logic of China's soybean import

There are at least three narratives about China's food imports, interpreting the soybean import as manifesting an inevitable process of modernization or an economic logic. One narrative, premised on the modernization thesis, explains China's food imports as an inevitable modernization effect. It underlies Lester Brown's 1994 alarmist prediction of a coming food deficit in China and its dire consequence for the world at large. In this essay, later turned into a monograph, Brown uses the modernization experience of Taiwan, South Korea and Japan as templates to outline China's future: a rise in the income pushes a more protein-rich diet; loss of farmland to urbanization and industrialization leads to diminished food production capacity; these factors are compounded by an exceedingly large population that is still growing (Brown 1994, 1995). Pointing to the reductions in grain-producing capacity in Taiwan, South Korea and Japan, Brown predicted that China will no longer able to feed itself and its deficit will more than engulf the supply on the global food market. Brown states, 'The bottom line is that when China turns to world markets on an ongoing basis, its food scarcity will become the world's scarcity', and 'it will – in one way or another – take the rest of the world with it' (Brown 1994, 10, 19). In Brown's framework, food deficits in East Asia are only a function of the modernization effect; thus, China on the path of modernization is facing an irresistible destiny.

Brown's view of the modernization effect has made invisible how geopolitical-economic conditions, particularly US hegemony, helped create food deficits in the region. He assumes national governments as sovereign and autonomous actors and food deficits as resulting from internal dynamics within each society. Taking Taiwan as an example, several factors contributed to its structural adjustment during 1982–1991. In the earlier period, the Taiwanese government implemented a grain-procurement policy and had thus borne a fiscal burden. US merchants, through the US government, pressured the Taiwan government to limit Taiwan's export of rice. The confluence of both factors pushed Taiwan to abandon its earlier policies. In the period of structural adjustment demanded by the US, Taiwan abandoned its policy of self-sufficiency, opened its market and liberalized its agricultural trade. It was under these conditions that the Taiwan government discouraged domestic rice production and provided subsidies for 20 percent of its land to remain fallow (Cai 2010). Brown's narrative of South Korea also ignores how its agriculture since the beginning of the Cold War was inserted into 'a food regime centered on both the United States and global corporation' (Yoon, Song, and Lee 2013, 57). In the early 1950s, when US food aid arrived in South Korea, it pushed grain prices so low that many farmers were forced out of agricultural production. With many leaving for the city and some even burning their harvests, the base of agricultural production became violently eroded. In 2011, four corporations – Cargill, ADM, Bunge, and Louis Dreyfus – were responsible for 57 percent of South Korea's grain imports (Yoon, Song, and Lee 2013, 58).

Brown's modernization-as-destiny outlook also has its echo in China. Similarly to Brown, Chinese economist Mao Yushi also cites the cases of South Korea and Japan as exemplary for China, but as successful cases of more liberalization. Mao wondered aloud that if these two countries do just fine with food self-sufficiency rates at 40 percent and 60 percent, respectively, why can China not similarly embrace economic integration, by allowing more food imports from the global market? Mao regards economic integration as destiny. He thus argues against policies of state intervention and national self-sufficiency. The difference between Brown and Mao is that while Mao celebrates this destiny, Brown is pessimistic about its planetary consequence. Known as an arch-neoliberal figure in China, Mao famously argues that all trade that is both equal and voluntary is beneficial and that importing food is equivalent to importing land and water, both of which are in short supply in China (Mao 2008). Thus, Mao shares a similar narrative to Brown, but has an opposite view on the outcome.

The benefit of importing food, particularly soybeans, is further elaborated by the well-known scholar Philip Huang and his co-author Gao Yuan (Huang and Gao 2014). Huang and Gao, who support a Chayanovian vision for agriculture, are not bound to the modernization thesis. Thus, their pragmatic arguments regarding soybean imports may seem convincing to many in China. However, Huang and Gao treat the economic logic of soybean imports as both inevitable and rational, and in this way their narrative falls into the logic of modernization. In Huang and Gao's view, China's soybean import fits Huang's thesis about China's invisible revolution in agriculture. By 'invisible revolution', Huang means that a sizable portion of Chinese farming households are switching from an 'old agriculture' of low-value grains and soybeans, whose production is land intensive, to a 'new agriculture' of high-value greenhouse vegetables, fruits and animal farming, which is both labor- and input intensive (Huang 2010). Huang and Gao argue that because soybean is among the most land intensive, but least productive, least value-generating crops, it is the economic advantage of the new agriculture that woos farmers away from soybeans.

In the view of Huang and Gao, it is also economic logic at work that US soybeans outcompete Chinese soybeans, because the US has used the technological package of glyphosate-based herbicides and glyphosate-resistant GM soy (created and marketed by Monsanto as RoundUp and RoundUp-Ready seeds) to reduce labor demand and production costs.

Table 1. Percentages of old and new agricultures' planting in China's total.

Year	Planting of vegetables (new agriculture, %)	Planting of soybeans (old agriculture, %)	Planting of other grains (old agriculture, %)
1995	6.3	5.4	59.6
2000	9.7	6.0	54.6
2005	11.4	6.2	52.7
2009	11.6	5.8	55.7
2010	11.8	5.3	55.9
2011	12.1	4.9	56.1
2012	12.5	4.4	56.7
2013	12.7	4.1	57.0

Source: Data for percentages of different crops are drawn from *China Rural Statistical Yearbook* (1996, table 6–3; 2001, table 7–3; 2006, table 7–3; 2010–2014, table 7–3), edited by Zhongguo guojia tongjiju (1997, 2002, 2007, 2011–2015).

Following this, they argue that it is not only rational for Chinese farming households to give up on soybeans for new agriculture, but also advantageous for the Chinese state to facilitate the import of land-intensive soybeans and to save China's scarce farmland for new agriculture. They surmise that had China grown the soybeans that it imported in 2009, it could have used 15–20 percent of its farmland for other grains and high-value agricultural products. They suggest that those who show concern about global corporate domination in the soybean trade forget that the Chinese state 'is one of the few powers in the world that can possibly counter-balance transnational capital and can possibly struggle for an alternative path of development' (Huang and Gao 2014, 179–80, 185). As will be shown below, this characterization of the Chinese state has been challenged by critical narratives.

There are several problems with Huang and Gao's argument about economic logic, and with their regard for the role of the state. First, their argument that the economic advantage of new agriculture has brought about the soybean decline is a feasible argument in the abstract, but is not supported by the national data. Below, we reproduce and update Huang and Gao's national data on the planting areas of old and new agriculture (Huang and Gao 2014). Table 1 shows that the most rapid expansion of the new agriculture (vegetables) – from 6.3 to 11.4 percent of the total planting in China – took place during 1995–2005, but soybean also experienced a modest growth from 5.4 to 6.2 percent in the same period. Thus, the presumed inverse relationship between the new agriculture and soybeans as the old agriculture does not hold water.

Second, the decline of soybean planting after 2008 did not give rise to 'new agriculture' in the case of Heilongjiang Province, China's main soybean-producing province. There, soybeans are mainly displaced by similarly 'old agriculture' corn and, sometimes, rice (Figure 2). More importantly, the displacement of soybeans by corn and rice cannot be attributed to a simple economic logic of the market, because the Chinese government regulates the prices of soybeans, corn and rice. The China Soybean Industry Association (CSIA) argues that the difference between soybean and corn prices plays an important role in driving farming households from soybeans to corn (CSIA 2014, 32).[3] When the ratio of soybean price to corn price grew to the high of 2.77 in 2007, 2008 saw the largest soybean spring planting area in Heilongjiang. When the ratio dropped to 2.25 in 2009 and further

[3]It was not until 2009 that there appeared a strong and obvious correspondence between the soybean/corn price differential and the respective decline and increase in planting areas of these two crops.

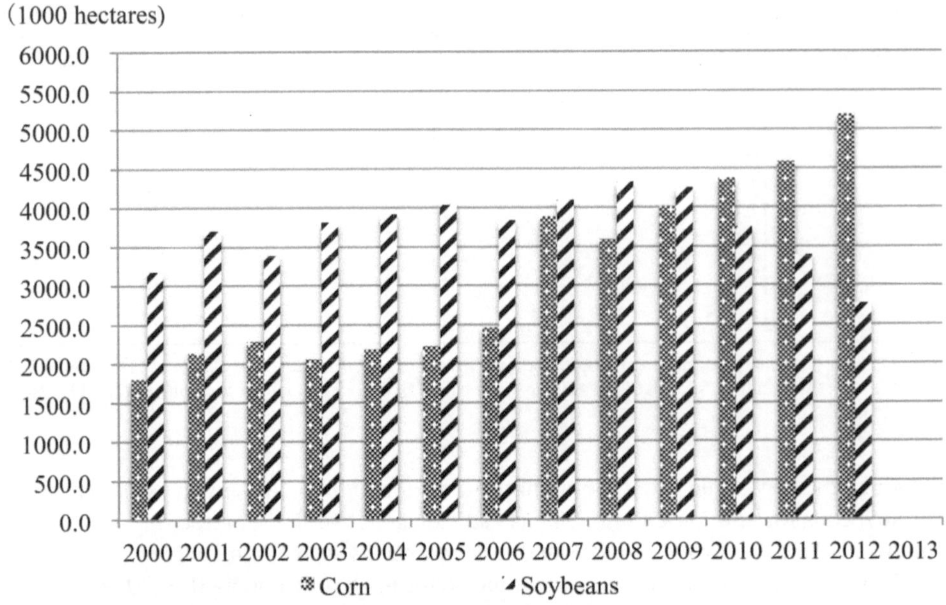

Figure 2. Planting area of soybeans and corn in Heilongjiang province (2000–2012).
Source: Data for planting area of corn and soybeans in Heilongjiang province are drawn from China Rural Statistical Yearbook (2001–2014, tables 12-14, 13-15, 12-13, 13-13, 12-8), edited by Zhongguo guojia tongjiju (2002–2015).

deteriorated in the following years, the planting of soybeans in the province experienced a continual decline, while that of corn climbed. Indeed, the soybean/corn price ratio in China has often been lower than that in the US (Figure 3). The changes in the government-regulated price differential have contributed to the relative low income derived from growing soybeans and thus has pushed famers to grow corn (Zhang and Li 2005, 26).[4]

The comparative income derived from growing rice, corn and soybeans in China (Figure 4) demonstrates that since 2009, the income from planting rice and corn increased at a faster pace than that from soybeans. The growing gap since 2009 has pushed farmers away from soybeans. The conversion to successive planting of corn has ecological consequences, as it has destroyed the local practice of yearly rotation between soybean and corn, and caused a stress and deterioration in the soil fertility. However, not all farmers have switched to corn. In Heilongjiang, rural households have been accustomed to growing soybeans and relying on bean stalks as their winter fuel. Switching to corn requires greater expenditures even while it may bring more income. Our field research finds that poorer farmers tend to stick to soybeans, both because they cannot cover the cost of making the crop switch and because, not being able to afford purchased fuel, they need dried soybean stalks as their winter fuel.

Thus, the question is why the Chinese government allows the prices of rice and corn to increase at a faster pace than that for soybeans. An official of the CSIA argues, in our interview with him in 2015, that the state has made a choice to 'abandon soybean and protect

[4]In the US, it is the opposite: the US government's greater subsidy for soybeans than corn enabled greater soybean expansion.

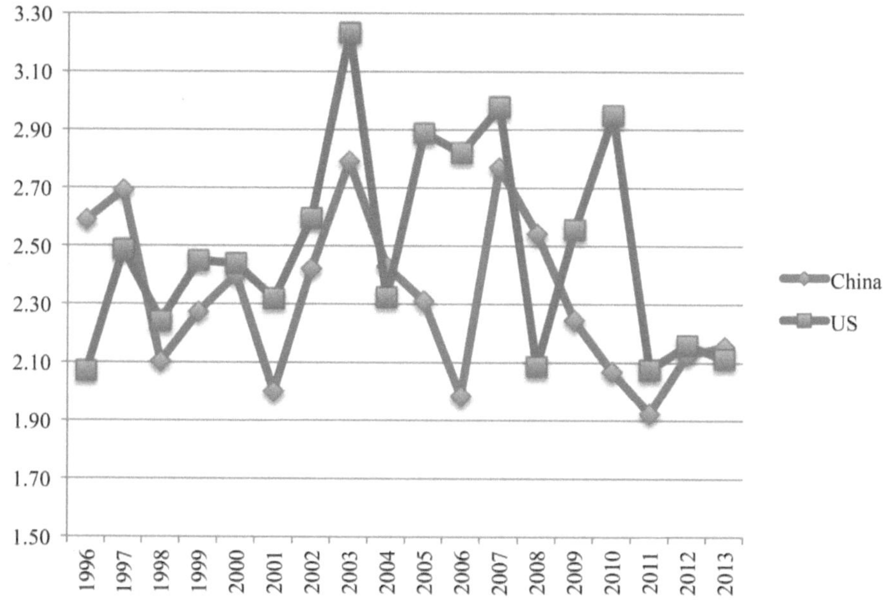

Figure 3. Price ratio of soybean and corn in China and the US (1996–2013).
Source: Data for corn price in China are drawn from National Agricultural Production Cost and Revenue Information Summary (2008, table 11811; 2014, table 1-8-1), edited by Guojia fazhan he gaige weiyuanhui jiagesi (2009, 2014), and those for soybean price are drawn from National Agricultural Production Cost and Revenue Information Summary (2008, table 1-9-1; 2014, table 1-9-1), edited by Guojia fazhan he gaige weiyuanhui jiagesi (2009, 2014). Data for soybean and corn price for the US are drawn from the United States Department of Agriculture (2006, 2013).

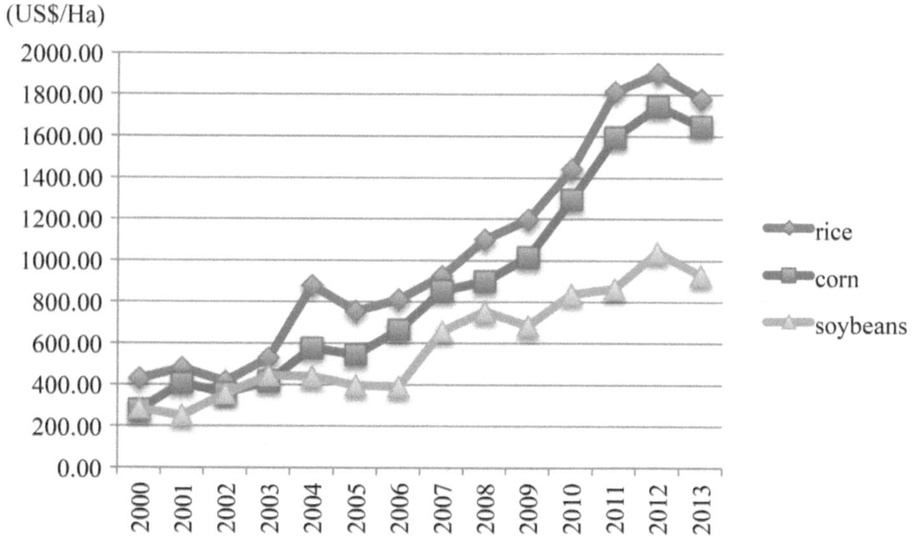

Figure 4. Comparison of incomes from three crops in China.
Source: National Agricultural Production Cost and Revenue Information Summary (2008, tables 1-2-1, 1-8-1, 1-9-1; 2014, tables 1-2-1, 1-8-1, 1-9-1), edited by Guojia fazhan he gaige weiyuanhui jiagesi (2009, 2014).

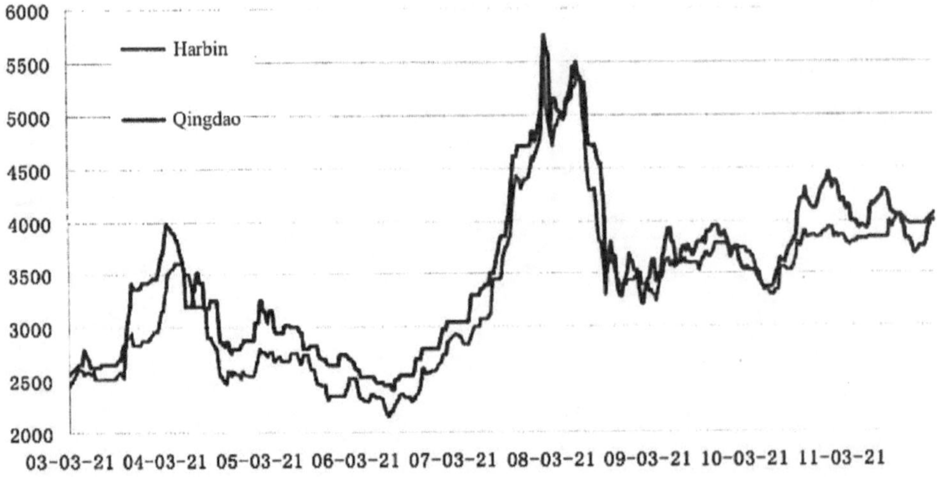

Figure 5. The prices of domestic (Harbin) and imported (Qingdao) soybeans (2003–2011).
Source: This figure is quoted from the China Soybean Industry Association (CSIA 2014, 56).

grains' (弃豆保粮). Indeed, in 1996, before China's entry into the WTO, the Chinese government temporarily slashed the import-quota tariff from 114 to 3 percent. When joining the WTO in 2001, China agreed to abolish multiple-schedule tariffs and adopt a single-schedule tariff of 3 percent for soybeans. We should add that while the Chinese government can more or less regulate the prices of corn and rice, it has lost its capacity to regulate the price of soybeans, even when it has tried to, because soybean is now the most liberalized crop.

The liberalization of soybean is observed by a soybean industry insider, who noted in an interview with us that the domestic market for soybeans is already part of the international market.[5] Indeed, Figure 5 shows that the price of China's domestic soybeans for 2003–2011 was not independent of the price of imported soybeans, but corresponded very closely to it.

Under the overwhelming dominance of imported soybeans, the Chinese government's price regulation for domestic soybeans has only a limited function. Following the social outcry about the foreign takeover of the Chinese soybean industry, the Chinese government took some steps to protect traditional soybeans in the Northeast provinces, including Heilongjiang. One key measure is for the government to purchase soybeans from farmers at a designated price. When the government announced its first purchase plan on 20 October 2008, it pushed the market price up only for about two weeks before it fell again. When the government announced its second purchase plan in December, it barely made any dent in a market dominated by imported soybeans (CSIA 2014, 33). The soybean planting area has continued its fall without interruption since 2009 (Figures 1 and 2; Nanfang Dushi Bao [The Southern Metropolis] 2010).

In short, one cannot use an 'invisible agricultural revolution' to explain the fall in soybean production in Heilongjiang, because farmers did not switch to the 'new

[5]Because general social consciousness critical of GM soybeans is rising, the Chinese market is beginning to distinguish domestic soybeans as 'food soybeans' from imported soybeans as 'feed soybeans'. However, soybean merchants, not soybean farmers, have so far benefited from the higher market price for domestic soybeans that consumers are willing to pay.

agriculture'. And what has pushed farmers from soybeans to corn is not a pure economic logic, but the price differential between soybean and corn. What is at play is that the Chinese government can effectively regulate the price of corn and rice, but is unable to intervene in the price of soybeans.

If the price for China's domestic soybean were to gain any autonomy from that of imported GM soybeans, there would have to be some kind of delinking between domestic and imported soybeans. The organizer of Heihe Soybean Association in Heilongjiang argues that domestic soybeans ought to be treated differently from imported GM soybeans: 'Whereas Europeans and Japanese recognize the distinction of our non GM soybeans and pay a higher price, why are our beans in China not treated differently from imported GM soybeans?'[6] Although the law in China requires GM foods to be labeled, in practice labeling is not strictly enforced. Existing labels tend to be inadequate to inform consumers, not to speak of GM soybeans that often escape labeling when they are illegally mixed with non-GM foods (Nanfang Ribao [Nanfang Daily] 2014; Zhongguo guangbo wang [China National Radio Net] 2014). A lack of government enforcement on GM labeling has made it more difficult to delink domestic non-GM and imported GM soybeans.

More recently, the Ministry of Agriculture's intervention has made the delinking more difficult. In 2014, the Ministry of Agriculture ordered China's Central Television to censor 'misguiding' advertisements that acclaim non-GM foods as healthier and safer. Aiming to pacify public anxiety and discontent over GM foods, an official publication called for a 'rational approach' and informed the public that 'all GM foods on the market are safe' (Xinjingbao [The Beijing News] 2015). These efforts by government agencies in China indicate that supra-economic forces are at play in establishing a discursive equivalence in the use value between GM and non-GM foods, even when domestic producers and consumers expect otherwise. To understand the massive import of GM soybeans requires us to make these forces visible. Indeed, an argument for GM soybean import based on a comparison of the cost of production already assumes equivalence in the use value between two different kinds of soybeans.

This is not to deny that there is economic logic in China's import of soybeans. For example, CSIA's deputy director Liu Denggao acknowledged that oil content is a little higher in the imported GM soybeans (about 18.5 percent) than in soybeans domestically grown in China (around 16.5 percent), making GM soybeans attractive to oil-extraction enterprises (Guo 2012). Yet this economic reductionism – eliding all other differences and making only one quality visible and valuable – already relies on the supra-economic power that enforces a comparability between GM and non-GM soybeans. Thus, the economic logic does not function as a pure and single factor, because it always exists in the context of political choices. Huang and Gao (2014) assume that the use of glyphosate herbicide has made the US soybean cheaper and more competitive, and that this economic logic explains and justifies the massive import of GM soybeans. A comparison of the costs of production in the US, Brazil and China does indicate that Brazil has the lowest cost of soybean production and China has the highest, per 50 kg. The cost of soybean production in China was about the same as that in the US until 2003. It grew somewhat faster in 2006, and much faster than that in either the US or Brazil after 2009 (Figure 6). This suggests that the key to understanding the divergence is less about what happened with production in the US or Brazil and more about what happened with production in China. As the

[6]Liu Min, presentation at workshop 'Food Sovereignty and Protection of the Chinese Soybean Industry,' 11–12 October 2014, Beijing. See also Chen (2014).

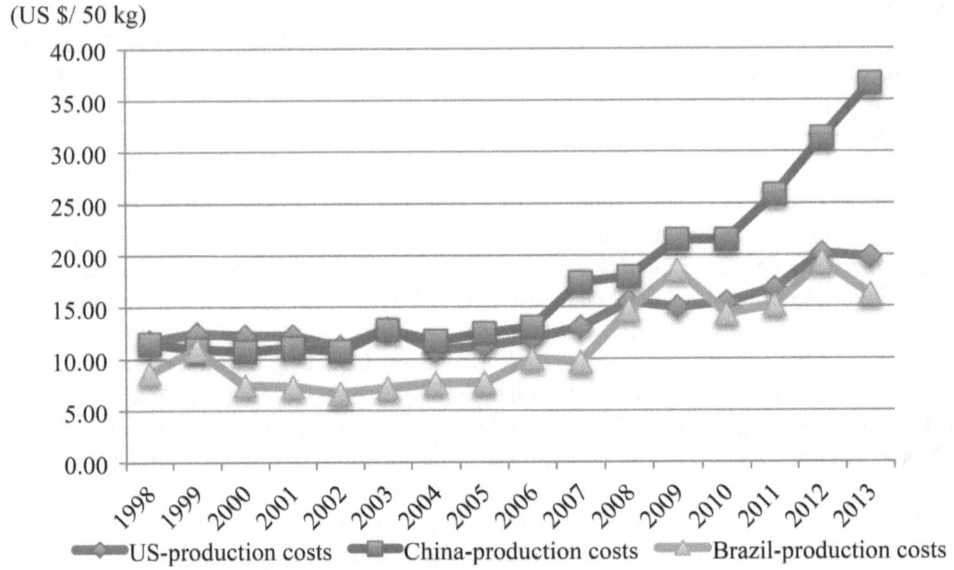

Figure 6. Comparison of the costs of soybean production: US, Brazil and China.
Source: The US data are drawn from the official website of the United State Department of Agriculture. The Chinese data are drawn from the National Agricultural Production Cost and Revenue Information Summary (2005, table 1-9-1; 2008, table 1-9-1; 2014, table 1-9-1), edited by Guojia fazhan he gaige weiyuanhui jiagesi (2006, 2009, 2014). The Brazilian data are drawn from CONAB/SUINF (2015).

glyphosate weed killer has been used in the US since the introduction of the GM soybeans in the 1990s, this constant factor cannot explain the divergence since 2009.

A comparison of the costs of soybean production in the US and China for the period of 2006 to 2013 shows that every cost has increased for both countries, but more so in China than in the US (Table 2). During 2006–2013, the material and service cost (M&S) is the largest cost in soybean production in both China and the US, about 40 percent of China's total and about 55–60 percent of the US total. From 2006 to 2013, this cost has increased 115 percent in China and 70 percent in the US. Opportunity costs of unpaid

Table 2. Costs of soybean production: China and the US.

China	Material and service costs	Hired labor	Opportunity cost of unpaid labor	Land rental	Opportunity cost of land	Total costs
2006	5.28	0.38	3.62	0.52	3.19	13.07
2013	11.35	1.03	10.72	3.14	9.77	36.61
Increase	114.93%	175.78%	195.75%	507.70%	205.84%	180.12%
US	**Material and service costs**	**Hired labor**	**Opportunity cost of unpaid labor**	**Land rental**	**Opportunity cost of land (rental rate)**	**Total costs**
2006	6.60	0.08	0.65	N/A	3.70	11.93
2013	11.23	0.13	0.75	N/A	6.53	19.82
Increase	70.20%	69.38%	15.11%	N/A	76.63%	66.16%

N/A: not applicable. Source: *National Agricultural Production Cost and Revenue Information Summary* (2008, table 1-9-1; 2014, table 1-9-1), edited by Guojia fazhan he gaige weiyuanhui jiagesi (2009, 2014). United States Department of Agriculture (2006, 2013).

labor and land, the next two most weighty costs in China, are each about 23–30 percent of the total and have increased by 196 and 206 percent, respectively. Among the costs for soybean production in China, land rental has had the most remarkable increase.[7]

However, soybean is not the only crop that has seen a dramatic rise in its cost of production in China. As Table 3 shows, from 2006 to 2013, three major crops – rice, wheat and corn – have also experienced increases in each and every cost of production, including M&S, opportunity cost of unpaid labor (OCUL), hired labor, land rental, and opportunity cost of land (OCL). Among the costs compared, soybean production has the highest increase in M&S. While soybean's total cost of production has the highest increase of 180 percent, other crops are close behind in their respective increases.

Yet soybean alone has experienced the most drastic decline in its planting area. This brings us back to the earlier analysis in the paper: that the Chinese government's liberalization of soybean has undermined its capacity for intervention. The narratives examined so far have seen China's massive soybean import as an inevitable outcome of the logic of modernization. As will be shown below, critics and activists in China have challenged these narratives that present soybean import as both inevitable and apolitical.

Debates and activism

Unlike the narratives that construct China's soybean import as a destiny or a win for China, critics point out multiple risks and dangers. These critics do not see economic logic as independent, but regard it as intertwined with larger issues, whose stakes go far beyond economic logic. These critics and activists are mainly academics, writers, journalists, activist citizens and a few military commentators. They emerged after November 2009, when the Ministry of Agriculture granted safety approval for two GM rice varieties and one GM corn. It sent a shock wave through much of the society (Guanyu 2010). Around 2004, when Greenpeace was investigating GM soybean imports' impact in China, its effort did not arouse much public attention, in large part because the public then had barely heard about GM foods. However, the earlier effort by Greenpeace laid a foundation for later mobilizations.[8] The 2009 approval was the first one ever granted in China for GM crops and involved rice, the most popular staple food.

Following that, activism has opened up debate about national security, the reach of corporate power, the relationship between science and commercial interest, discursive power, complicity between government agencies and corporations, etc. These challenges to the more mainstream narratives might be grouped into several counter-narratives that cover different, but overlapping, fronts. While each of these narratives has its own focus, they

---

[7]As farmland is not a commodity in China, agricultural producers in China who wish to expand their production may rent land from those who have land-use rights. This practice is referred to as 'land transfer' (tudi liuzhuan) in China. For farmland institutions in China, see Ye (2015). Using available statistics, we calculate that the rate of rural farmland transfer in Heilongjiang is about 37 percent. The total farmland transfer in Heilongjiang for 2014 is 65 million mu; the total village farmland in Heilongjiang is 177.45 million mu. The rate of land transfer is 65 million/177.45 = 37 percent. The actual rate of self-cultivating farmland is likely lower than 63 percent, because some intra-village land transfer may not be captured in government statistics on land transfer. See Heilongjiangsheng tongjiju (Heilongjiang province bureau of statistics) and Guojia tongjiju heilongjiang diaocha zongdui (NBS survey office in heilongjiang province) (2013, table 12-28); Zhongguo guojia tongjiju (National Bureau of Statistics of China) (2013, table 13-3), and Zhenxing dongbei (Revitalization of Northeast China) (2014).

[8]Interview with a former staff member of Greenpeace's China Office, 21 July 2014.

Table 3. Costs compared: rice, wheat, corn and soybean in China.

Costs and increase (USD/50 kg)		Rice	Wheat	Corn	Soybean
M&S	2006	3.67	4.11	2.79	5.36
	2013	8.02	9.00	5.95	11.95
Increase	118.58%	118.84%	113.30%	122.93%	
OCUL	2006	2.36	2.07	2.08	3.62
	2013	7.37	7.19	7.06	10.72
Increase	212.89%	248.05%	239.19%	195.75%	
Hired labor	2006	0.32	0.07	0.14	0.38
	2013	1.00	0.22	0.47	1.03
Increase	211.11%	236.62%	240.02%	175.78%	
Land rental	2006	0.17	0.04	0.08	0.52
	2013	0.76	0.30	0.34	3.14
Increase	337.67%	619.16%	316.33%	507.70%	
OCL	2006	0.93	0.93	1.01	3.19
	2013	2.55	3.02	2.91	9.77
Increase	174.56%	223.86%	189.95%	205.84%	
Total	2006	7.45	7.22	6.10	13.07
	2013	19.70	19.73	16.74	36.61
Increase	164.47%	173.38%	174.54%	180.11%	

Note: In the table, M&S = materials and service costs; OCUL = opportunity cost of unpaid labor; OCL = opportunity cost of land.
Source: National Agricultural Production Cost and Revenue Information Summary (2008, tables 1-2-1, 1-7-1,1-8-1, 1-9-1; 2014, tables 1-2-1, 1-7-1, 1-8-1, 1-9-1), edited by Guojia fazhan he gaige weiyuanhui jiagesi (2009, 2014).

form a united front, not necessarily by design, but because these foci are interconnected by social logic.

Some critics make use of the Chinese government's white paper stance on food security to challenge China's massive soybean import. They focus on how the corporate power of US-based transnational agribusiness corrodes China's national food security. Even though Brazil and Argentina are also major soybean exporters to China, Chinese critics perceive South American producers as sharing a similar marginal position in the global soybean complex (Zhou 2014). These critics' concern and challenges are sometimes accommodated by government media. Some go where the wind and waves appear the highest: the topic of GM safety and the right of consumer choice. Still others extend the critique to the mode of capital-driven agricultural development, the role of the market and even the nature of the Chinese state. It might be the first time in reform-era China that one issue has enabled such a wide variety of topics to be debated. Activism has taken place on the internet and TV, in print media and conference halls and on university campuses, as well as in public parks and streets. In this context, the case of soybean imports no longer stands as an isolated instance, but serves as the warning for what would happen if China were to lose control of its own food system. Critics have sounded the alarm that the same misfortune may happen to corn in China (Guan, Duanfang, and Cai 2010).

Responding to the soybean imports, critics have embraced the state-sponsored notion of food security, but have also gone much beyond it, as will be shown below. Chinese government and public understanding of 'food security' grew out of the experience of being subjected to embargo during the Cold War. This legacy has shaped the notion of food security in China as a production-based concept. The Chinese government's 1996 white paper on food security in China – perhaps a response to Lester Brown's 1994 'Who will feed China' – evaluates China's ability for self-reliance, understood as its own productive capacity in meeting its domestic needs (The State Council Information Office of PRC 1996). This take on food security is different from how it is more conventionally framed (Xiao 2004). The FAO's definition of food security centers around access to food, which can be supplied by 'the level of production, stock levels and net trade' (FAO 2008).

While all critiques are concerned with national food security, one counter-narrative takes this as its primary concern and has gone beyond the 1996 white paper's focus on China's productive capacity. An exemplary piece is a short essay published in *Global Times*, a newspaper under *People's Daily*, the main official newspaper in China. Authored by Peng Guangqian, a military commentator, it questions the promotion of GMO for staple foods in China and challenges some of the government's policies. Peng fired a series of eight questions: Why should China import GM foods when there is a lack of consensus about GM food safety? Given evidence showing the harms of GM foods, why should we ignore these warnings and what evidence do we have for rushed imports? In terms of research funding by Chinese government agencies, why is GM seed technology favored over conventional seed research by a ratio of 166:1? While the US has blocked China's access to high-level technology, why does it make an exception for GM technology? When some government departments do nothing to stop GM rice and corn from being illegally planted in China, whose interest do they guard? Is there a hidden chain of interlinked interests between those promoters of GM technology in China and their US background? When China surrenders its food supply to the West, what is the bottom line for China's food security? If GM technology were weaponized, does China have its own biosecurity plan (Peng 2013)? All eight questions are critical of Chinese government agencies' existing policies and funding priority, willful lack of government oversight and even principles.

Peng's questions are not out of the blue, but are a synthesis and representation of the main concerns that have already been circulating on the internet for a while. For example, Greenpeace, in its 2009 report, and a group of Beijing academics in 2010, in a petition letter, warned that China does not own the core technology for the approved GM varieties and such a decision would surrender Chinese rural producers' livelihoods and China's food sovereignty to transnational corporations. (Greenpeace 2009).[9] Peng is also not alone in questioning interests and intentions. Cai Weidong, an activist writer, raised the following question at a workshop organized by the Ministry of Environmental Protection: 'Is the fundamental mission of agriculture to feed the country's people or allow corporations to make profit? If there is conflict of interests, how do we ensure that the public voice is not suppressed?' (Mu et al. 2011).

Concerns about China's soybean crisis have also presented Chinese critics with an opportunity to understand national food security in the global food system. Thus the notion of national food security has been expanded to include not only the capacity to produce, but also the capacity to 'speak'. Activists and critics point out that China has little 'discursive power' (话语权) in the global market. The term initially appeared in Chinese in 1994 in the field of literary and cultural criticism, but moved after a few years to a variety of social fields. In 2002, an article discussed the problem of Chinese peasants' lack of discursive power in the context of the WTO (Fu and Zhang 2002). The soybean crisis has made China's lack of discursive power a more concrete experience. Critics often point out that China has no discursive power – that is, it has little influence on the pricing of soybeans, despite China's massive import volume. With soybean

[9]See also 'Guanyu zanhuan tuiguang zhuanjiyin zhuliang de huyushu' [A Petition on Postponing the Propagation of GM Staple Foods] (2010).

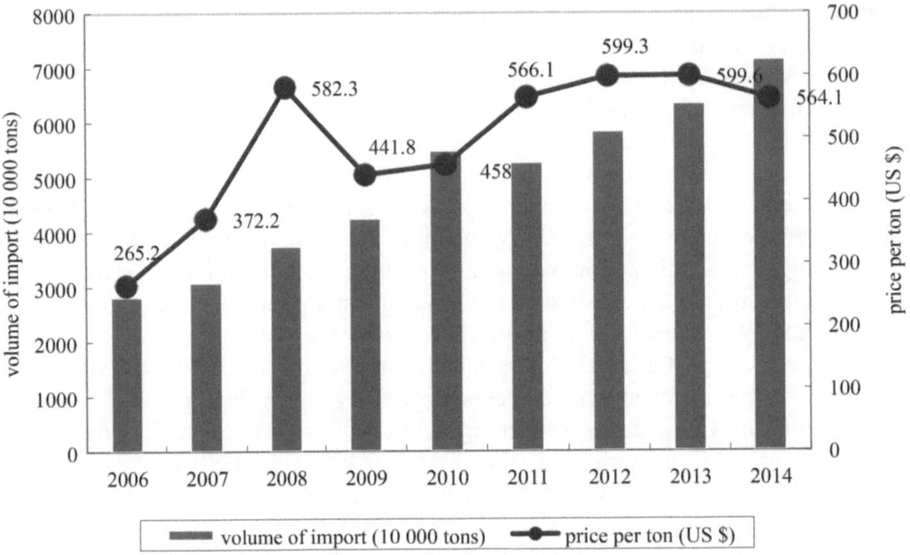

Figure 7. Tonnage and price of imported soybeans (2006–2014).
Source: The data for the volume of soybean import and the price are drawn from Guomin jingji he shehui fazhan tongji gongbao (Statistical bulletin for national economic and social development), edited by Zhongguo guojia tongjiju (National Bureau of Statics of China) (2007–2015).

imports doubling, from 28.24 million tons in 2006 to 63.4 million tons in 2013, the price of soybean imports increased 126 percent (Figure 7), whereas the cost of soybean production in the US in this period increased 60 percent (Table 2). This illustrates what has become popularly known in China about its lack of discursive power on the globalized market: 'whatever China sells on the international market becomes cheap; whatever China buys from the international market becomes dear'. The Chinese public do not see it as an economic phenomenon of the 'big country effect' on supply and demand, but as a complex political-economic issue that involves the capacity for organization. For example, a critic points out that unregulated competition among domestic exporters and a lack of organization has led to a 'dirt cheap' price for China's rare earth export, whereas the organization of oil producers such in the form of OPEC (Organization of the Petroleum Exporting Countries) has enabled more 'discursive power' of these resource exporters (Langtou 2015).

Emerging as a social keyword in China at the turn of the century, the term 'discursive power' in China has its own disposition, dispersion and diversification that is not found in what discourse and power mean in Foucault-inspired post-structural connotations.[10] 'Discursive power' in Chinese usage, particularly in the context of the food imports, is more lodged in the political-economic context. If the Foucauldian notion of discourse and power are without a center, 'discursive power' in the Chinese usage often points to a center of power. If the Foucauldian notion of discourse and power is about the persistent presence – 'always already' – of discourse and power, 'discursive power' in the Chinese usage often points to a lack of discursive power of a specific social group or entity.

[10]This observation is based on archival research via CNKI (China knowledge resource integrated database), a full-text Chinese journal/magazine article database.

Discussions of 'discursive power' in China are thus often about their uneven and unequal distribution, and the very discussion of the topic often has a critical-interventionist edge.

For critics of China's soybean imports, where discursive power concentrates is indicated by the price of soybeans being determined at the Chicago Board of Trade (CBOT), and the US Department of Agriculture – with its vast apparatus for agricultural statistical collection – being the central source of information that feeds the CBOT. Critics have also come to see that this discursive power does not just reside in the global market, but works within China to impinge on its policy making. Many activists have targeted Mao Yushi, winner of the US-based neoconservative Cato Institute's 2012 'Milton Friedman Prize for Advancing Liberty' and a founding member of the think tank Unirule Institute in China. For example, a blogger entitled his warning about China's food security 'The US Achieved Its Goal by Funding Mao Yushi: Food Prices in China are Now Up to those in the US' (Laojianghu kan shijie 2012). Critics have also called for an indictment of Li Jiayang, China's Deputy Minister of Agriculture, who is suspected of having retained his position on DuPont's biotechnology advisory panel for some time after he had been promoted to Deputy Minister (Zhiyanliao 2012).

A counter-narrative has also developed along with concerns about GMO safety, GMO regulation and consumer choice. This counter-narrative has mainly grown in the very struggle against pro-GM forces, which include government agencies in the agriculture sector, agricultural research institutions in China and transnational corporations. In the context of discursive struggles, the Ministry of Agriculture has attempted to settle public debates by claiming that officially approved GM foods are as safe as traditional foods (Jinghua shibao [Jinghua News] 2013), but this claim has further spurred the debates. As one observer put it, 'questioning began in the print media, but grew and expanded on the internet' (Yin 2013).

In these debates, science is a site of a fierce contestation over who has scientific authority and where science is located. Pro-GM forces have derided critics and activists mostly having humanities and social science backgrounds and dismissed their critiques on the ground of their science illiteracy. Science illiteracy is also seen as the root cause of public unease about GM safety. Contesting this hubris, a young activist pointed out malign practices by corporations and government agencies, and fired back: 'transnational corporations that are actively promoting GM technology do not have a scientific attitude; China's scientists working on GM do not have a scientific attitude; the Ministry of Agriculture also does not have a scientific and responsible attitude' (Mu 2014). This implies a redefinition of science as an attitude or an approach that anyone can acquire, instead of an object for expert possession.

The institutional location of science in labs and academic papers is also openly contested. When activists present counter-evidence from fields, they are told by pro-GM forces that their evidence is not scientific and that they need to cite papers in academic journals. A senior activist has responded, 'Do not ask us for papers. I invite you to rural areas for field inspection, to search for evidence!' (Mu 2011). What is being critiqued is the monopoly of an institutionalized science that excludes a science that is open to the field, to everyday observation and to experiences of farmers and concerned citizens (Marglin 1996; Gu 2015).[11] In order to respond to pro-GM publicity, critics and activists in China

[11] In 'Farmers, seedsmen, and scientists: systems of agriculture and systems of knowledge', Stephen Marglin (1996) usefully compared different knowledge systems, which he refers to as episteme and technic.

have collected and disseminated a great amount of information and have built transnational alliances with critical researchers and writers abroad. If the overall pursuit of modernization in China has put science and scientists on a holy pedestal, these debates have posed a serious challenge to its holiness.[12] The status of science in society, and its relationship with the state, capital, power and institutions have all been questioned (Tian 2014).

In the context of growing commodification of knowledge and knowledge institutions in China, these debates have called into question the interest of knowledge production, perhaps the most serious critique of knowledge production since the end of the Cultural Revolution in China. What has been critiqued includes complicity between governments and bio-tech agribusiness, and collusion between research institutions and corporations in countries of liberal democracy and in China. Apart from struggles on the internet, anti-GM activities have included, for example, protesting in front of the Ministry of Agriculture, challenging pro-GM scientists in their public lectures, producing videos for public dissemination, disseminating leaflets on the street, etc.[13]

Debates about GM safety necessarily entail the question of consumer choice. In August 2012, the exposure of the 'Golden Rice' scandal shocked the public who learned that researchers from US-based Tufts University tested GM golden rice on 25 primary school children in China in 2008 (GM Watch 2014). While pro-GM elites have defended the test as harmless and characterized the incident as an issue of research ethics, netizens (users of the internet) in China were outraged about a lack of government oversight and the complicity of Chinese researchers in the incident. Some regarded it as a 'national shame' and rhetorically asked why 'Chinese elites have no human compassion' (Huasheng 2012). In September 2013, Fang Zhouzi, a pro-GM opinion leader and popular science writer, vowed to 'create conditions for Chinese to eat GM food every day' and pulled a publicity stunt by inviting volunteers to taste GM corn (Nanfang Zhoumo [The Southern Weekend] 2014). That was one of a series of organized 'tasting' events (Nanfang Zhoumo 2013). Cui Yongyuan, a well-known media figure, then challenged this act on a microblog – a Chinese equivalent of Twitter – by raising the issue of consumers' right of choice. The ensuing battle between the two on microblog lasted for months, drawing nearly 20 million followers, with only 4.63 percent expressing support for pro-GM Fang Zhouzi (Guanchazhe wang [Social Watch] 2013; Fenghuangwang [Phoenix Net] 2014).[14] In anti-GM activism, 'consumer choice' seldom appears in its single dimension, but is often connected with inequality among consumers. For example, activists have criticized the Ministry of Agriculture for using non-GM cooking oil in its own kindergarten while preaching about how safe GM foods are to the public (Renmin wang [People's net] 2013).

The most critical narrative concerns not just policies about GM food imports or consumer rights of choice, but the dominant paradigm of development in China and the world. For these critics, China's rising food imports are not a march of inevitable logic, but the very symptom of the dominant paradigm's failing in sustainability. They argue that global capitalism

[12]The politics of science, knowledge, and expertise was subjected to interrogation during the years of the Cultural Revolution (1966–1976) in China (Wei and Brock 2013).

[13]This video is titled 'zhuanjiyin de qianshi jinsheng' [GMO's foregoing and present life] (2011).

[14]In pro-GM arguments, 'international standards', particularly US policies and regulations of GM foods, are frequently cited authoritative references. To contribute to the debate, Cui Yongyuan self-financed the production of a documentary on GMOs in the US, based on his interviews with scientists, farmers, activists and consumers in that country. The video has attracted wide attention. See 'Xiao Cui Kaocha zhuanjiyin' (2013).

generates both overconsumption and overproduction, and that overconsumption in China – the growing use of imported soybean as animal feed for a fat-rich and protein-rich diet – is driven by a relentless pursuit of a gross domestic product (GDP)-oriented development (Wen 2009). The so-called market demand for soybean imports needs critical deconstruction, instead of being satisfied. Consumerism and overconsumption – part of the current development paradigm – do not facilitate but occlude consumer rights, and contribute to China's stress over food security (Yan and Chen 2013a, 2013b). Going against the grain of the narratives found in Brown, Mao and Huang, critics and activists critique and reject China's soybean imports and the larger development paradigm with it.

Even the seemingly indisputable advantage of US productivity is turned on its head. Wen Tiejun, dean of the School for Agriculture and Rural Development at Renmin University and director of a civil organization called the Center for Rural Construction, has provided a staunch and in-depth critique of modernization. While the mainstream often laments the low labor productivity of agriculture in China compared to US agriculture, Wen reverses the verdict by highlighting China's higher land productivity. He argues that even with more than twice the subsidy, the capital-driven agriculture in the US only produces over 300 million tons of food on 193 million hectares, but Chinese agriculture produces over 500 million tons of food on 100 million hectares. What are being contrasted and therefore debated are two different modes of agriculture. Wen associates many of the current ills in Chinese agriculture, such as the overuse of chemical inputs, with China's blind copying of Stalinist and Fordist agriculture a few decades ago (Wen 2013). Wen thus locates the problem with China's food security not in presumed low productivity; instead, he argues that food wastage and over-consumption in China are more of a problem (Wen 2014).

Within this general rubric against capital-driven agriculture, the critical narrative is not entirely homogenous. Some in China take a pro-peasant stance that is similar to that of food sovereignty activists elsewhere. There are also differences among critics in China about what the 'small peasant' refers to. Wen and his network of rural advocates draw on the legacy of the earlier reformist rural construction movement before the Communist-led land reform and inherit the earlier view of small peasants as an undifferentiated whole.[15] For those in the food sovereignty network in China, the legacy to be valued is the socialist rural collectives of the Mao period. These critics and activists are thus critical not only of capital-driven agriculture, but also of the market orientation for agricultural development (Renmin shiwu zhuquan wang [Food Sovereignty Network] 2013).

Also critiquing the dominant paradigm of development, including the Green Revolution, Gu Xiulin, a professor and senior activist against GM technology, takes issue with W. Arthur Lewis and Theodore Schultz, two Nobel Laureate development economists. She turns their concern on its head and argues that the key question is not whether agriculture can be a source of economic growth – that is, generating enough profit – but whether the market-oriented path of agricultural development can possibly allow for a long-term sustainable development of agriculture. To eliminate small peasants is to eliminate the foundation of sustainable agriculture (Gu 2014, 77).

Interestingly, in Gu's view, the collective rural economy developed in the Mao era is a constructive innovation of 'a modern organizational form that can best inherit advantages of traditional agriculture' and can fend off capital's invasion of agriculture (Gu 2014, 74). For Gu, Lester Brown's question 'Who will feed China?' assumes a resource determinism and a

[15]For a detailed analysis, see Yan and Chen (2013a, 2013b).

market determinism. Dismissing these determinisms, Gu considers the real question to be whether China can find a secure, sustainable model for agriculture that fits China's own conditions (Gu 2014, 80). Gu sees the capital-driven and market-oriented paradigm of agriculture development as the threat to China's food and agriculture security. In her view, the ills of the Green Revolution and massive soybean imports are part and parcel of that paradigm. Gu shares with other critics the same concern that what has happened to soybeans in China may be replayed for corn.

Among those who critique capital-driven agriculture, the more outspoken activists clearly articulate the choice of paradigm to be socialism or capitalism. A senior activist proposes that the debate cannot just remain about one specific issue or another, but has to be about one principal issue: whether the market should play a determinative role in development. Some of these critics use parks and streets or the internet as public spaces of activism, and place anti-GM information sheets alongside their literature on Mao Zedong. On the internet, websites linked with 'red songs' – songs from the Chinese revolution and the Mao era – are very critical of GM technology. All critics see the GM issue as a political issue in multiple dimensions. However, facing the possibility of official censorship, some argue that depoliticizing anti-GM activism would make it appealing to the broadest spectrum of people.

Spurred by the Chinese government's approval of GM rice, and China's soybean crisis, debates in China have moved beyond the government's definition of food security. In the general context of social concerns over food security and food safety, the social impact of the soybean crisis becomes a context for an emerging food sovereignty advocacy in China (Yan and Chen 2013a, 2013b; Yan 2014; Zhou 2014). Advocates hold that food sovereignty has dual subjects, including the state and people who are producers and consumers, with the subjects sometimes in tension with each other, as demonstrated in the debates about the soybean crisis.[16] For advocates of food sovereignty, two sources provide legitimacy for this alternative concept in China. First, the Chinese government is one of the 58 governments that accepted the UN-sponsored Global Report on the *International assessment of agriculture knowledge, science and technology for development* (IAASTD). The IAASTD report introduces food sovereignty as representing a paradigm shift from the earlier concept of food security (IAASTD 2009, 111). Second, food sovereignty was developed by peasant organizations in Latin America and La Via Campesina collectively (Boyer 2010), and hence has its origin in developing countries, among which China still claims its place.

Sharing the above-mentioned critiques about the current paradigm of development, advocates link food sovereignty with China's socialist legacy. They argue that food – because of its irreplaceable place in national security, ecological security and livelihood security – should be a public or semi-public good, instead of being a commodity, and hence the system that facilitates its production and circulation should be a social and public institution. Supporters of food sovereignty criticize China's post-1978 rural reform because it has fragmented the peasantry in China and commodified most means of production. Class differentiation and agrarian capitalism are emerging in rural China (Zhang and Donaldson 2008; Yan and Chen 2015). The way forward to food sovereignty is neither capital-driven agriculture nor market-determined economy, as pursued by the government's policy, but cooperative and collective farming producing food as a public good (Renmin shiwu zhuquan wang [Food Sovereignty Network] 2013).

---

[16]Zhou Li argues that food sovereignty should have triple subjects, including nature. See Zhou (2014).

Conclusion

In *Who will feed China*, Lester Brown regards the modernization logic as inevitable for China and warns the world of the implications (1995). As this essay demonstrates, the modernization logic has not been universally accepted in China, and it is increasingly questioned and challenged. The case of the Chinese soybean crisis and debate also demonstrates that what is being questioned is not only domestic politics over soybean production and imports, but a division of labor in the global food regime such as 'South America produces soybeans, China buys soybeans, and the US sells soybeans'. Whether China will fulfill Brown's modernization destiny, and how food production and consumption in China affect the current food regime in formation, is open to further examination and debate.

The Chinese public has not taken China's massive soybean imports for granted. Along with the massive imports, soybean farming in China has experienced unprecedented decline. While it is perceived as a win or a necessity by some, it has generated an acute sense of crisis and intense debates and social activism. Concerns over the soybean crisis and the massive imports of GM soybeans have propelled many to re-examine the dynamics of globalization, the logic of modernization, the role of the state, the politics of science and knowledge production, and the reach of corporate power in everyday life. China's soybean crisis is also a context for some to take on food sovereignty as an alternative concept through which to articulate an ecological socialism. Through critical knowledge and image dissemination about ecological and public health devastation in South America, particularly Argentina, Chinese critics have come to see the massive soybean imports as a lose-lose business for publics in both the country of consumption and the countries of production. Argentina in particular stands out because there criticism and opposition to soybean production have focused most strongly on agro-toxic impacts to human health, while in Brazil it is land concentration and deforestation that are highlighted. In the wide internet circulation in China about Argentinian suffering in GM soybean planting areas, the Argentinian experience stands as a shared calamity facing developing countries or humanity in globalization.[17] In their critiques and protests against the power of transnational corporations that dominate in the soybean trade and GM technology, Chinese critics have been drawing from and participating in a global solidarity of new knowledge production and circulation.

Acknowledgements

The authors wish to thank the two anonymous reviewers for their comments, and Gustavo Oliveira for providing and interpreting the data for soybean production in Brazil. The research for this essay is part of the project 'Rural China in Globalization: the Soybean Crisis and Its Everyday Impact' and is supported by Hong Kong Research and Grants Council for funding (project No. B-Q39N).

Disclosure statement

No potential conflict of interest was reported by the authors.

[17]For example, the article by an anonymous author 'Agenting yuku wulei: diyige bei zhuanjiyin huihuai de guojia' [Argentina cries with no tears: the first country ruined by GM technology]' (2012) is widely circulated. Other articles are well circulated in China, such as The Seattle Times (2013).

References

'Agenting yuku wulei: diyige bei zhuanjiyin huihuai de guojia' [Argentina cries with no tears: the first country ruined by GM technology']. 2012, March 23. http://bbs.tianya.cn/post-worldlook-451980-1.shtml.

BBC. 2015. Five ways China's economic crisis will affect Africa, Aug. 27. http://www.bbc.com/news/world-africa-34060934.

Boyer, J. 2010. Food security, food sovereignty, and local challenges for transnational agrarian movements: the Honduras case. *Journal of Peasant Studies* 37, no. 2: 319–351.

Brown, L. 1994. "Who will feed China?" World Watch 7(5): 10–19.

Brown, L. 1995. *Who will feed China?: Wake up call for a small planet*. New York: W. W. Norton.

Brown, L. 2014. Can the World Feed China? Feb. 25. http://www.earth-policy.org/plan_b_updates/2014/update121.

Cai, Peihui. 2010. Youcaihuatian de beihou: nongye jiegou yu liangshi anquan [Behind the field of rape flowers: agricultural structure and grain security]. *Xin shi zhe* 120.

Chen, Hangying. 2014. Dadou de kuibai qiaoxiang le baowei zhongguo shiwu zhuquan de jinzhong [The Collapse of Soybeans Sounds the Warning for Protecting Food Sovereignty in China], *Pengpai*, Oct. 24. http://www.thepaper.cn/newsDetail_forward_1272721.

China Radio International. 2014. Guanzhu liangshi anquan xilie baodao zhi yi [Serial report concerning grain security], http://gb.cri.cn/42071/2014/03/19/6871s4470171.htm

China Soybean Industry Association (CSIA). 2014. Dadou chanye jiance yujing baogao. Document for Internal Circulation.

CONAB/SUINF. 2015. Estimated Cost of Production for Soybeans. National Supply Company of Brazil (CONAB), Agribusiness Information Department (SUINF).

Earth Policy. 2014. Who will Feed China? Sept. 24. http://www.earth-policy.org/blog/who_will_feed_china.

FAO. 2008. An Introduction to the Basic Concepts of Food Security, www.fao.org/docrep/013/al936e/al936e00.pdf

Fenghuangwang (Phoenix Net). 2014. Cui Yongyuan, Fang Zhouzi dazhan zhuanjiyin: Kexue qu na'er le? [Cui Yongyuan and Fang Zhouzi battle over GMO: where is science?]. January 13. http://news.ifeng.com/opinion/wangping/zhuanjiyin/.

Friedmann, H. 2009. Feeding the empire. *Socialist Register* 41: 125–142.

Fu, Shundong, and Zhang Xiangyu. 2002. Nongmin de huayuquan: Yingdui jiaru shimao de chongji (Peasants' Discursive Power: Responding to the Impact of the WTO). *Shehui (Society)* 7: 29–32.

Global Post. 2014. Good News: US export to China soar, setting a new record. Jan. 21, http://www.globalpost.com/dispatch/news/regions/asia-pacific/china/140120/good-news-us-exports-china-soar-setting-new-record.

GM Watch. 2014. Gold rice paper to be retracted amidst scandal. July 18, http://www.gmwatch.org/index.php/news/archive/2014/15536-gm-golden-rice-paper-to-be-retracted-amid-ethics-scandal.

Greenpeace. 2009. Shui shi zhongguo zhuanjiyin shuidao de zhenzheng zhuren [Who are the real owners of China's GM rice], Feb. 25, http://www.greenpeace.org/china/zh/publications/reports/food-agriculture/2009/3045095/

Gu, Xiulin. 2014. Qiantan yinxiang liangshi anquan he nongye anquan de jige genben wenti [A Preliminary Discussion of Some Fundamental Problems Affecting Food Security and Agriculture Security]. 2: 72–83. (accessed here: http://www.cefc-ngo.co/uploadfiles/file/20141114154734845O.pdf)

Gu, Xiulin. 2015. Zhuiwen zhuanjiyin fengxian [Pursuing the question about the safety of genetically modified foods]. Fenghuangwang, January 18, http://news.ifeng.com/exclusive/lecture/special/zhuanjiyin2015/guxiulin.shtml

Guanchazhe wang (Social Watch). 2013. Cui Yongyuan Fang Zhouzi jiu zhuanjiyin shipin wenti weibo wu luan jizhan [The five rounds of heated battles between Cui Yongyuan and Fang Zhouzi regarding the GMO foods], September 10. http://www.guancha.cn/Media/2013_09_10_171478.shtml.

Guan, Kejiang, Tao Duanfang, and Wei Cai. 2010. Kuaguo liangshang zuoyou quanqiu liangjia [Transnational agribusiness controls global food prices], April 26, http://www.globalview.cn/ReadNews.asp?NewsID=21083

'Guanyu zanhuan tuiguang zhuanjiyin zhuliang de huyushu' [A Petition on Postponing the Propagation of GM Staple Foods]. March 10, 2010. http://shc2000.sjtu.edu.cn/20110220/zanhuan.htm

Guo, Yanchun. 2012. Guochan dadou chanyelian chu zai bengkui de bianyuan [Domestic soybean industry is on the brink of collapse]. Zhongguo shangbao (China Business Newspaper), July 4, http://www.caas.cn/nykjxx/fxyc/64560.shtml

Guojia fazhan he gaige weiyuanhui jiagesi [National Development and Reform Commission of the People's Republic of China] eds. 2006, 2009, 2014. *National Agricultural Production Cost and Revenue Information Summary* [Quanguo nongchanpin chengben shouyi ziliao huibian]. Beijing: Zhongguo tongji chubanshe.

Heilongjiangsheng tongjiju [Heilongjiang province bureau of statistics] and Guojia tongjiju heilongjiang diaocha zongdui [NBS survey office in heilongjiang province]. 2013. *Heilongjiang Statistical Yearbook [Heilongjiang tongji nianjian]*. Beijing: China Statistics Press.

Huang, Zongzhi. (Huang C.C. Philip). 2010. *Zhongguo de yinxing nongye geming [China's Invisible Agricultural Revolution]*. Beijing: Falü chubanshe.

Huasheng, Zaixian. 2012. 'Huangjin dami' shijian zheshe keyan liangzhi sangshi daijin [The 'Golden Rice' issue reflecting the consciencelesss of scientific research]. http://www.zznews.gov.cn/news/2012/1207/94185.shtml

Huang, CC Philip, and Gao Yuan. 2014. Dadou shengchan he jinkou de jingji luoji [The economic logic of soybean production and import]. *Kaifang shidai (Open Times)* 1: 176–188.

International Assessment of Agriculture Knowledge, Science and Technology for Development (IAASTD). 2009. Agriculture at a crossroads. http://www.unep.org/dewa/agassessment/reports/IAASTD/EN/Agriculture%20at%20a%20Crossroads_Global%20Report%20(English).pdf.

Jinghua shibao (Jinghua News). 2013. Nongyebu cheng zhuanjiyin shipin yu chuantong shipin yiyang anquan [MOA claiming GM food as safe as traditional food]. September 2. http://politics.people.com.cn/n/2013/0902/c1001-22769100.html.

Langtou Langtou. 2015. Weishenme hui chuxian mai shenme shenme pianyi, mai shenme shenme gui de guai xia xiang? [Why there is a strange phenomenon of 'Whatever [China] buys becomes cheap, whatever [China] sells becomes dear?'] http://club.kdnet.net/dispbbs.asp?boardid=3&id=10875669

Laojianghu kan shijie. 2012. Meiguo zanzhu Mao Yushi de mudi decheng: zhongguo de liangshi jiage dei kan meiguolao lianse [The US has achieved her goal sponsoring Mao Yushi's work: China's grain price slavishly depended on the US]. August 11. http://blog.ifeng.com/article/19310673.html

Mao, Yushi. 2008. Mao Yushi fangtan lu [Interviews with Mao Yushi]. http://business.sohu.com/20081030/n260342560_6.shtml.

Marglin, S.A. 1996. Farmers, seedsmen, and scientists: Systems of agriculture and systems of knowledge. In *Decolonizing Knowledge: From development to Dialogue*, ed. Frédérique Apffel-Marglin and Stephen A. Marglin, 185–247. Oxford: Clarendon Press.

McMichael, P. 2009. A food regime genealogy. *Journal of Peasant Studies* 36, no. 1: 139–169.

Ministry of Agriculture. 2009. Dadou chanye fazhan jizhi chuangxin shidian gongzuo fangan [Plans for piloting institutional innovation for the development of the soybean industry]. Printed in Zhongguo dadou chanye fazhan jizhi chuangxin gaoceng lunan: ziliao huibian [Advanced forum for institutional innovations for the development of China's soybean industry: a collection of documents], complied by China's Soybean Industry Association, Beijing, 157–165.

Mu, Chuan, Jin Wei, Hei Juyi, and Gu Xiulin. 2011. 11 ci jilie jiaofeng: shengwu anquan guoji luntan shang de zhuanjiyin [11 vigorous arguments: GM issues on international forum of biosafety], May 11. http://www.hswh.org.cn/wzzx/llyd/zz/2013-05-01/1477.html.

Nanfang Dushi Bao [The Southern Metropolis]. 2010. Guojia butie zhichi nandi jinkou dou yajing, dongbei guochan dou jiagong zai xian tinggongchao [It is hard with state subsidies to counterbalance the pressure from imported soybeans, Northeast domestic soybean processing shutting down]. Feb. 2nd. http://gcontent.oeeee.com/1/41/1415db70fe9ddb11/Blog/528/3e2df4.html

Nanfang ribao [Nanfang Daily]. 2014. Cui Yongyuan: xiwang mingque zhuanjiyin chengfen de biaoshi [For GM Labelling, Cui Yongyuan Hopes for Clear Definition]. March 3. http://news.xinhuanet.com/fortune/2014-03/03/c_126212022.htm

Nanfang Zhoumo [Southern Weekend]. 2013. Qian ren shichi zhuanjiyin dami: yi chang yansu de kepu huodong? [Thousands people tasting GM rice: is it a serious popularization of science?]. 26 July. http://www.infzm.com/content/92818

Nanfang Zhoumo. 2014. Ruguo shi chang dubo, ta ya dui le ma? [If it were a gamble, is he backing the right horse?]. 23 January. http://www.infzm.com/content/97730

Oliveria, G. de L.T. 2015. The geopolitics of Brazilian soybeans. *The Journal of Peasant Studies* 1–25. doi:10.1080/03066150.2014.992337.

Oliveria, G. de L.T., and Schneider Mindi. 2014. The politics of flexing soybeans in China and Brazil. Transnational Institute (TNI), Think Piece Series on Flex Crops and Commodities, No. 3.

Peng, Guangqian. 2013. Bawen zhuliang zhuanjiyin hua [Eight questions on staple grains going GM]. *Global Times*, August 21. http://opinion.huanqiu.com/opinion_china/2013-08/4266709.html.

Renmin shiwu zhuquan wang [Food Sovereignty Network]. 2013. Renmin shiwu zhuquan xuanyan [A declaration of the food sovereignty] http://www.shiwuzq.com/food/website/declaration/declaration.html.

Renmin wang [People's net]. 2013. Zhuanjiyin shiyongyou anquan xing zai zao zhiyi: nongyebu youeryuan jinyong [GM cooking oil is doubted again, Ministry of Agriculture's kindergarten forbids it]. Dec. 3, http://finance.sina.com.cn/chanjing/cyxw/20131203/015717506176.shtml.

The Seattle Times. 2013. Associated Press investigation: 'Argentines blame birth defects, cancer, on agrochemicals for biotech crops'. http://seattletimes.com/html/nationworld/2022094305_apxargentinaagrochemicals.html?syndication=rss.

The State Council Information Office of the PRC. 1996. Zhongguo de liangshi wenti [Grain issues in China]. http://www.scio.gov.cn/zfbps/ndhf/1996/Document/307978/307978.htm.

Tian, Song. 2014. Jingti kexuejia [Beware of scientists]. *Dushu* 4: 105–114.

United States Department of Agriculture. 2006, 2013. http://www.ers.usda.gov/data-products/oil-crops-yearbook.aspx

Wang, Shaoguang. 2013. Dadou de gushi: ziben ruhe weiji renlei anquan [The Story of Soybeans: How Capital Endangers Human Security]. *Kaifang Shidai [Open Times]* 3: 87–108.

Wang, Kejing, and Li Fushan. 2000. Woguo yesheng dadou (G. soja) zhongzhi ziyuan jiqi zhongzhi chuangxin liyong [The germplasm resources of China's wild soybean G. soja and their developing utilization]. *Zhongguo Nongye Keji Daobao [Journal of China's Agricultural Science and Technology]* 2, no. 6: 69–72.

Wei, Chunjuan Nancy, and Darryl E. Brock, eds. 2013. *Mr. Science and Chairman Mao's Cultural Revolution: Science and Technology in Modern China*. New York: Lexington Books.

Wen, Jiajun. 2009. Gao fuli, di xiaofei: tongxiang shengtai wenming zhi lu [High welfare, low consumption: a path to ecological civilization]. *Lu Ye* 3: 110–116. http://www.snzg.cn/article/2011/0216/article_22165.html

Wen, Tiejun. 2013. *Zunzhong lishi yu kexue zhengwei [Respect history and scientific falsification], foreword for Shengtai nongchang jishi [A documentary of an ecological farm]*. Beijing: Zhongguo kexue jishu chubanshe. http://blog.sciencenet.cn/blog-475-698624.html

Wen, Tiejun. 2014. Langfei xing xiaofei hen pubian, liangshi anquan nan zhuize [Wasteful consumption is universal, accountability for food safety is difficult to make], Feb. 26 http://www.caogen.com/blog/infor_detail/57677.html.

Xiao, Junyan. 2004. 'Liangshi anquan' haishi 'shiwu anquan': jianping zhongguo shiwu anquan wenti ['Grain security' or 'food security': Comments on China's food security issues]. *Diaocha yanjiu baogao* [Survey and research report]: 1–23.

'Xiao Cui kaocha zhuanjiyin' [Xiao Cui investigating GMOs]. 2013. http://www.youtube.com/watch?v=IbrfJ0pTEHk.

Xinjingbao [The Beijing News]. 2015. Nongyebu: Shangshi zhuanjiyin shipin dou anquan [Ministry of Agriculture of the PRC: All GM food on the market are safe] 29 Aug. http://epaper.bjnews.com.cn/html/2015-08/29/content_595937.htm.

Yan, Hairong, et al. 2014. Dadou chanye weiju yu baowei shiwu zhuquan [China's soybean crisis and food sovereignty]. *Jingji Daokan [Economic Herald]* 12: 16–21.

Yan, Hairong, and Chen Yiyuan. 2013a. Cong dadou weiji kan shiwu zhuquan [The soybean crisis from the perspective of food sovereignty]. *Nanfeng chuang [Southern Review]* 19: 36–40.

Yan, Hairong, and Chen Yiyuan. 2013b. Debating rural cooperative movements in China, the past and the present. *Journal of Peasant Studies* 40, no. 6: 955–981.

Yan, Hairong, and Chen Yiyuan. 2015. Agrarian Capitalization without Capitalism?: Capitalist Dynamics from Above and Below in China. *Journal of Agrarian Change* 15, no. 3: 366–391.

Ye, Jingzhong. 2015. Land transfer and the pursuit of agricultural modernization in China. *Journal of Agrarian Change* 15, no. 3: 314–337.

Yin, Shuaijun. 2013. Jueding mingyun de yinian [The fate-deciding year]. Sept 25 http://www.haijiangzx.com/2013/attention_0925/130175.html.

Yoon, Byeong-Seon, Song Won-Kyu, and Lee Hae-jin. 2013. The struggle for food sovereignty in South Korea. *Monthly Review* 65, no. 1: 56. http://monthlyreview.org/2013/05/01/the-struggle-for-food-sovereignty-in-south-korea/.

Zang, Yunpeng. 2013. *Zhongguo nongye zhenxiang [The Truth about Chinese Agriculture]*. Beijing: Beijing daxue chubanshe.

Zhang, Xiaoshan. 2012. Zhongguo de liangshi wenti ji qi duice [China's food issues and resolutions]. http://www.globalview.cn/ReadNews.asp?NewsID=30425.

Zhang, Huipeng. 2014. How to Protect Chinese Soybeans. Paper presented at the workshop of Food Sovereignty and the Protection of Chinese Soybean Industry, Oct. 11–12, in Beijing, China.

Zhang, Shengyong, and Li Congguang. 2005. Zhongguo dadou de bijiao youshi ji zhong mei dadou chengben xiaoyi de jingjixue fenxi [The comparative advantage of Chinese soybean and an economic analysis of the cost and benefits of soybean production in the US and China]. *Zhongguo nongcun faxian [China Rural Survey]* 1: 18–26.

Zhang, Q. Forest, and J.A. Donaldson. 2008. The rise of agrarian capitalism with Chinese characteristics: Agricultural modernization, agribusiness and collective land rights. *The China Journal* 60: 25–47.

Zhao, Gang. 2009. Dadou chanye de wenti pojie he tizhi jizhi chuangxin [The Solution to the Problems of the Soybean Industry and Institutional Innovation]. *2009 nian zhongguo dadou chanye fazhan jizhi chuangxin gaoceng luntan ziliao huibian*, 116–212. Beijing: Zhongguo dadou chanye xiehui.

Zhenxing dongbei [Revitalization of Northeast China]. 2014. Heilongjiang: duo cuoju tuidong tudi liuzhuan guimo jingying cu liang zengchan [Heilongjiang: Promote land transfer in multiple ways, enlarge the scale of farm operation, increase grain output]. http://www.chinaneast.gov.cn/2014-11/03/c_133761666_2.htm.

Zhiyanliao. 2012. Yifa tanhe Li Jiayang de renming [Indict Li Jiayang's appointment according to law], 22nd Jan, http://zhiyanle.blog.hexun.com.tw/72380393_d.html.

Zhongguo guangbo wang [China National Radio Net]. 2014. Duoming lvshi qisu zhuanjiyin shiyongyou biaoshi buqing, yiyou jiu jian li'an [Many lawyers sue over unclear labeling of GM cooking oil, nine cases already established]. Sept. 20. http://china.cnr.cn/ygxw/201409/t20140920_516472443.shtml.

Zhongguo guojia tongjiju [National Bureau of Statics of China] eds. 1997, 2002–2015. *China Rural Statistical Yearbook [Zhongguo nongcun tongji nianjian]*. Beijing: China Statistics Press.

Zhongguo guojia tongjiju (National Bureau of Statistics of China) eds. 2002–2015. *Zhongguo tongji nianjian 2001–2014 [China's Statistical Year Book 2001–2014]*. Zhongguo tongji chubanshe [China Statistics Press].

Zhongguo guojia tongjiju [National Bureau of Statics of China] eds. *Guomin jingji he shehui fazhan tongji gongbao [Statistical bulletin for national economic and social development]*. 2007–2015. Available from the official website of the National Bureau of Statics of China.

Zhou, Li. 2014. Nongqi he xiaonong de jingzheng: zai tan dadou weiji [Competition between agribusiness and small peasants: Further discussion of soybean crisis], *Jingji Daokan* (Economic Herald). (12), http://www.shiwuzq.com/food/rights/producer/2015/0126/1277.html.

'Zhuanjiyin de qianshi jinsheng' [GMO's Foregoing and Present Life], 2011. http://v.youku.com/v_show/id_XMjQ1NjQwNTQ0.html.

Different farming styles behind the homogenous soy production in southern Brazil

Bert Vander Vennet, Sergio Schneider and Joost Dessein

Soy is often perceived as a typical example of a homogenous capitalist agricultural model that is responsible for ecological damage and social conflicts. But this monolithic perception of soy production can be challenged: more than 30 percent of the soy producers in Rio Grande do Sul (Brazil) are family farmers. In this contribution, we study soy production, the soy producers and their institutional environment from an actor-oriented perspective. We have uncovered different farming styles behind soy production: the colonial farmer, the niche farmer and the entrepreneurial farmer. The farming styles differ from each other not only in the farming system, but also in attitudes (for example, towards the forest). We found that the institutional environment and the technology are mainly focused on the entrepreneurial farmer. However, also, non-governmental organizations (NGOs) neglect the existence of small-scale soy producers. These results have several repercussions for further analysis of this problematic crop and how it can shift towards a more sustainable agricultural production model as small-scale farmers might produce soy more sustainably.

Introduction

In the 1950s, soy was introduced to Brazil in a small town: Santa Rosa, located in the southern Brazilian region, Rio Grande do Sul. It was the start of an agricultural success story. In a period of 30 years, the area cultivated expanded extremely quickly from Santa Rosa towards the central-west and the northeast regions of Brazil. Various drivers spurred this vast and rapid expansion. One was the existing infrastructure for cereal production, which could be easily adapted to the soy crop; another was geopolitical events such as the foundation of the Common Agricultural Policy in Europe, which substantially accelerated the European demand for imported soy (Brum 1987; McMichael 2009). Research efforts coming from EMBRAPA (Brazilian Enterprise for Agricultural Research), the federal agricultural research institute of Brazil, made it possible to modify the soy crop from a temperate-climate crop into a tropical crop (Brown and Purcell 2005). The Economist (2010) calls it 'the miracle of the Cerrado'.[1] Soy

[1]The Cerrado is the region south of the Amazon, which used to be a vast area without potential for agricultural production. The region, which represents 23 percent of Brazilian land, is an ancient biome characterized by rich biodiversity (Ratter et al. 1997).

was one of the main drivers of the Green Revolution in Brazil, and this process is still going on today in the Cerrado region.

Despite its successes, such as impressive yield increases, and its positive impact on the Brazilian national trade balance, the modernization of soy production is often regarded as a classical example of an unsustainable agricultural model. Ratter, Ribeiro, and Bridgewater (1997) stated that 35 percent of the Cerrado has been transformed into an anthropic landscape. As of 2005, 50 percent had been transformed (Klink and Machado 2005). In addition to this landscape transformation, soy production is also held responsible for huge ecological problems, such as vast monocultures, deforestation and soil erosion (Pengue 2005, 2009; Hecht 2005; Brown et al. 2005; Morton et al. 2006; Steward 2007; Hecht and Mann 2008; Gudynas 2008; Altieri 2009; Walker et al. 2009, and others). The production of soy in huge monocultures is necessarily accompanied by a rising pesticide use to compensate for the increased vulnerability to pests, which, in turn, results in a reduction in biodiversity (Belo et al. 2012; Porto and Soares 2012). A number of social problems are also linked to soy production, such as the increasing number of farmers who can no longer maintain their farms (Gudynas 2008), and farmers who face difficulties with farm succession.

In the western world, environmental organizations (e.g. Genetic Modification Watch (GMWatch), Carbontradewatch, La Soja Mata, Agricultures Network, Fairfood, World Rainforest Movement (WRM), Action for Solidarity Environment Equality and Diversity (ASEED)) and the feed concentrate industry often regard soy as one monolithic production model. Wield, Chataway, and Bolo (2010), quoting James (2008), state that the nature and the geography of the crops indicate that the soy production in Brazil and Argentina is done by capitalist farmers, including corporate farming enterprises. Furthermore, soy production is often linked with the problem of land-grabbing by foreign companies and the strong effect of international soy prices on Brazilian land prices (Altieri and Pengue 2006; Li 2011, Lambin and Meyfroidt 2011; Borras *et al.* 2010; Oliveira 2013). Wolford (2008) also describes how large-scale farming could benefit from better access to credit and technological assistance, as these farms 'deserve' committed government support to develop modern agriculture that will feed the world in the twenty-first century. The Economist (2010) states that 'big is beautiful' and that mainly the large-scale, modern farmers will be able to feed the world instead of the 'hobby farmers'.

Regardless of the favorable or unfavorable regard for large-scale soy production, the above-mentioned trend in the literature seems to agree that soy production is a synonym for capitalistic, large-scale agriculture. This monolithic perception of one specific capitalistic industrial production model heavily affects the discussion about sustainable agriculture, because many of the current discussions about sustainable agriculture are tightly linked with scale. Large-scale farms are seen as fostering huge monocultures which negatively affect the ecological system (e.g., Weis 2010 and Woodhouse 2010). Furthermore, the strong concentration of the genetically modified (GM) crops (with soy as one of the top three GM commodity crops) has generally been considered a main driver for homogenization.

However, Wield, Chataway, and Bolo (2010) have already suggested that only little research has been done on the class differentiation of farmers cultivating GM, glyphosate-resistant soy. Schneider (2009) and Schneider and Niederle (2010) showed that family farms in the southern region of Brazil are a diverse group with varying agricultural practices. Up to now, very few studies have focused specifically on diversity in the case of soy producers. In reality, soy is produced in diverse ways, ranging from large-scale to small-scale production, with GM, conventional and organic soy all being grown

(Schlesinger, Nunes, and Carneiro 2008; Cassol 2010). In Rio Grande do Sul, family farms[2] are responsible for 36 percent of the state's soy production (de França et al. 2009). These diverse soy production practices do not automatically lead to different outcomes regarding sustainability. As Brown and Purcell (2005) demonstrate in the case of rubber and coffee production in Brazil, it is not the scale itself but the specific political, economic and ecological details of an agricultural project that are important in the sustainability outcome. Therefore, the question arises: Do these diverse soy production practices fit within a global, corporate food regime (Friedmann 1987)? If so, how? Why would small-scale producers still continue with soy production if they have to compete with large-scale producers? And is their production technique as standardized as is generally assumed in the literature?

Van der Ploeg (1993) suggests that specific farming practices are linked with a specific patterning of resources in a goal-oriented way, where different logics shape specific farming practices. We hypothesize that the producers adhering to different farming practices have different attitudes towards soy crop production, with different ecological positions and different future development paths. This contribution, which is based on a qualitative comparative case study approach, uses the concept of farming styles (van der Ploeg 1993). Van der Ploeg (1993, 241) defines a farming style as

> a specific way of organizing the farm enterprise: farmer practice and development are shaped by cultural repertoire, which in turn are tested, affirmed or adjusted through practice. Therefore a style of farming is a concrete form of praxis, a particular unity of thinking and doing, of theory and practice.

He continues, 'a farming style is a specific pattern for tying together land, labor, cattle, machines, networks, knowledge, expectations and activities: this is done in a goal oriented knowledgeable and coherent way' (van der Ploeg 2010, 4). This concept, which is a specific way to organize farms based on theory and practice, is used to explore the soybean production system as an embedded entity of a larger global food regime (e.g. Friedmann 1987 and McMichael 2005). We visited two regions in the southern region of Brazil to gather information using semi-structured interviews with farmers, bank employees, extension workers, teachers, scientists and consumers. This research topic is operationalized in two research questions: (1) Can we distinguish different farming styles with respect to Brazilian soy production? (2) How do these differences in farming styles influence soy production, and what is their relation to technology use?

To answer these questions, we have to know in which ways family farms have become and can still become integrated and embedded into the current food regime. The case of soy commodity production in Brazil sheds light on this question. Furthermore, this paper represents a contribution to a broader discussion on agrarian and rural development. We mainly use the perspective of policies and initiatives that need to be addressed. We address the need to take such a diverse realm of actors in the whole agribusiness chain into account and explore how different actors become integrated in the social division of labor in the global food regime.

We will first discuss the soy crop embedded in the food regime, the concept of farming styles and the characteristics of the case studies. Then we will describe the context and history of soy production in Brazil. After that, we discuss the social reality of the soy

[2]In Brazil, a family farm is defined as a farm with an area smaller than 4 fiscal units (in Rio Grande do Sul, between 56 and 140 ha, depending on the municipality) and it should be mainly managed with family labor (Planalto 2006).

crop in Brazil, the three farming styles, how their visions about good farming and their future aspirations differ and how these are influenced by differing values. Finally, we discuss how the institutional environment and technologies have a different impact on each farming style.

Soy production system becomes embedded in a food regime

The soy production system, and more particularly the genetically modified organism (GMO) soy production system, is strongly embedded in a global food regime (Pechlaner and Otero 2008). Inspired by a global political economy perspective, Friedmann (1982, 248) first conceptualized a food regime as 'a stable set of complementary state policies whose implicit coordination creates specific prices relative to other prices, a specific pattern of specialization, and resulting patterns of consumption and trade', resulting in a process of 'accumulation and class formation on a world scale' based upon agricultural and food production, processing and commercialization. Several food regimes can be distinguished. The first food regime, from 1870 to the 1930s, is characterized by the combined import of colonial tropical products with basic grain and livestock imports coming from European colonies. These imports provided the emerging European industrial classes with low-cost food. The second food regime (1950s–1970s) was characterized by the adoption of the model of national agro-industrialisation by the 'development states', expressed by the Green Revolution technologies. 'Agribusiness elaborated transnational linkages between national farm sectors, which were subdivided into a series of specialized agricultures linked by global supply chains (e.g. the transnational animal protein complex linking grain/carbohydrate, soy/protein, and lot-feeding)' (McMichael 2009, 141). From the 1980s until now, we are in the third food regime, a corporate food regime (McMichael 2005).

Pechlaner and Otero (2008, 352) argue that

> the inter-relationship between regulatory change and genetic engineering (GE) are integral to the emerging third food regime: national neoregulation and supranational regulation around important aspects of the technology (such as intellectual property rights (IPR)) provide the means for biotechnology's ascendancy as a central technology for capitalist agriculture (Mascarenhas and Busch 2006), while the technology itself provides a means for further corporate concentration and integration of the food regime.

Pechlaner and Otero (2008, 352) wonder in addition 'whether it is inevitable that neoregulation and the introduction of corporate agri-biotechnology will, in fact, widely reorganise agricultural production relations in the somewhat homogenous manner suggested by food regimists'. Or, in other words, does the food regime standardize the farming practices of the soy production system, or does the soy production system, with its many practices, challenge this claim of homogenization?

The current food regime is often linked with its disastrous effect on sustainability (Oosterveer and Sonnenfeld 2012; Buttel 2006). For example, Weis (2010) mentions the biophysical limits of the capitalist industrial agriculture and argues for rebuilding it into a biodiverse food system. Moore (2010) even questions whether the socio-ecological crisis of the current food system could be the end of the capitalistic system, and wonders whether the coming transformation will be able to incorporate new agro-ecological conditions. Many studies about soy focus on environmental degradation: for example, Pengue (2005, 2009) and Altieri (2009) state that soy production not only causes landscape transformation but also causes tolerance of weeds to glyphosate, nutrient depletion, soil-structure degradation, potential desertification and loss of species. Hecht (2005) links soy production with a biotic loss of the anthropogenic landscapes. Brown et al. (2005) link soy production with deforestation, as do Morton et al. (2006), Steward (2007), Hecht and Mann (2008) and Walker et al. (2009). Gudynas (2008) especially focuses on how peasants and rural dwellers are being increasingly marginalized by soy crop production.

Our study contrasts with those that treat soy as one homogenous production system associated with large-scale capitalist agriculture. We have explored the different attitudes, practices and interfaces that shape the various farming styles existing in the soy production system. The concept of farming styles has been elaborated and tested by different researchers such as Roep, van der Ploeg, and Leeuwis (1991), Commandeur (2003), Vanclay (1994), Vanclay et al. (2006), Howden and Vanclay (2000) and de Bruin (1997). Some critics point out the heuristic character of farming styles, the influence of the specific research questions and the false suggestion of farming styles as fixed identities.

However, as Vanclay et al. (2006) state, the usefulness of the farming style concept depends greatly on the research questions and the aim of the research. Given our research question, we rely on farming styles as a key concept to examine how farmers operate within the soy production system according to their proper values and knowledge, and how this results in a specific soy-farming style based on specific farming practices, knowledge, future aspirations and visions of good farming and the environment. This discussion has a two-step structure: first, we explore the different materializations of the soy farm practices and, second, we exemplify the different world views behind these practices.

The farming styles are identified by individual identification of several elements as mentioned by van der Ploeg (2010): these include land, labor, knowledge, expectations and others. Farmers respond either differently or equally to some of these elements. For example, labor can be divided into (1) reduction of labor and increased use of machinery, (2) focusing labor on the production of high value products, or (3) increasing the labor use in order to reduce the fixed costs. We also allowed other aspects to emerge. For each element the farmers gave different responses, and, after this exercise was done for each item and each farmer, different clusters could be made. Each cluster is made up of a consistent logic behind the several elements that were shared by a group of farmers – a specific pattern behind the combination of the elements. This logical combination of the constructions shared by a group of farmers resulted in a specific farming style. The number of clusters depends on the explanatory power of the different clusters, and their internal consistency. To better understand the clusters, we apply an actor-oriented approach that treats soy as a crop that can be produced differently according to the values that inform a particular production style (Long 1985, 2001).

Empirically, we explore the data using the case study approach as described by Eisenhard (1989) and Eisenhard and Graebner (2007), with semi-structured interviews as the most important data-gathering technique. The interviews were complemented with informal conversations, as well as secondary data sources such as newspapers and policy documents. To gain a thorough understanding of the different farming styles, we opted for a sample of maximum contrast instead of a statistically representative sample. Based on discussions in the scientific literature (as mentioned above) and the reports of NGOs (such as Via Campesina, Movimento dos Trabalhadores Rurais Sem Terra (MST), Wervel, Aseed, Oxfam) about sustainable farming, we distinguished the following criteria for farmer selection: large-scale vs. small-scale, organic vs. conventional vs. transgenic; specialized or mixed (with other activities such as milk, vegetables and off-farm activities). Moreover, we added the farmers who stopped (organic) soy production and shifted to another

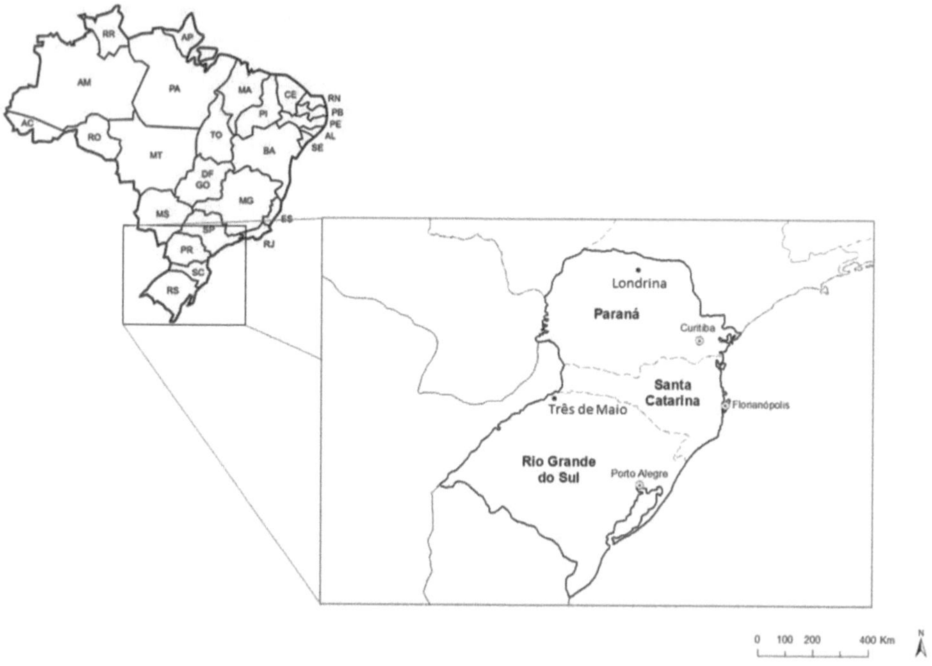

Figure 1. Map of Brazil, with Três de Maio in Rio Grande do Sul and Londrina in Paraná.
Source: Jetdicas 2012 and Santiago 2012.

production system in their search for alternatives in order to sustain their farm. When combining these criteria, and relying on insights from extension workers, farmers and other practitioners, we used multiple entry points to start a snowball sampling until the total number of farmers interviewed was sufficient to saturate the database with relevant information (Glaser and Straus 1967). Several discussions with agricultural specialists and peers, and a confrontation with the secondary data sources, allowed for a triangulation of the data.

We have selected two municipalities in our case study area (Figure 1): Três de Maio and Londrina. Três de Maio is a municipality in the northern area of Rio Grande do Sul. This community was founded by European immigrants in 1915 and is characterized by its rather small-scale family farming (15 ha is a common farm size). Soy production is very important in this municipality, which has also a cooperative called Cotrimaio, with 13,104 members in total (including members from other municipalities). Londrina, located in the northern area of Paraná, was founded in 1920. The region is characterized by family farms and large-scale farms producing mainly feed crops such as soy, wheat and corn, as well as coffee and oranges. The region of Londrina is more densely populated, with more small towns in the neighborhood than is the case for Três de Maio. Twenty farmers were interviewed in Três de Maio. Their farm size varied from 4 to 400 ha. The farmers were all male except two, and were between 28 and 53 years old. In Londrina, 11 farmers were interviewed, with a farm size varying from 12.5 to 1500 ha.[3] The farmers here were all male, aged

[3]A farm of 1500 ha is a big farm in South Brazil; however, in the Center-West region (Mato Grosso, for instance), this would be a small farm. But because of a lack of contacts and limited time, we didn't choose this region as a case study. As such, the huge farms of more than 4000 ha are not included in our study. We agree that the difference in size between farms of 1500 and 4000 is considerable, but we

25–75. The interviews took from 30 minutes up to 2 hours. The farmers' interviews were complemented with interviews with other people involved in the soy market: employees from the cooperative Cotrimaio (the director and a technician), a private soy processor in Londrina, an administrator of the soy cooperative, an employee of the Bank of Brazil, 15 consumers at the farmers market in Três de Maio and four experts (an agronomy professor, two scientists at Instituto Agronômico do Paraná or Agronomic Institute of Paraná (IPAR)'s department of socio-economy (the governmental agricultural research institute of Paraná) and the director of the Setrem agronomy school).[4]

Context and history of soy production in the southern region of Brazil

In 2010, Brazil produced 70 million tons of soybeans on 24.7 million ha. Thirty million tons of those soybeans were exported (MAPA 2011). Of the 70 million tons in total, Rio Grande do Sul accounted for 12 percent and Paraná for 20 percent (MAPA 2011). In 2006, 195,000 farmers produced soy on 6.8 million ha in the southern region (including the state of Santa Catarina), which is an average of 35 ha of soy cultivation per farmer (Zanon et al. 2010). This is much less compared to the farmers of the central-east area (including Mato Grosso), where 500 ha of soy was cultivated per farm (Zanon et al. 2010). In 2009, Brazil exported 28.6 million tons of soy, which was worth USD 11.4 billion (FAOSTAT 2012). This represents 11 percent of the total Brazilian export value (Lamucci 2011). Soy is therefore a strategically important commodity for Brazil. For 40 years, this commodity product has been heavily promoted by the government for the same reasons as today: the acquisition of foreign currencies (Brum 1987 and Lamucci 2011).

We will describe the historical evolution of the policies regarding soy and their effects on the local level in our two research areas, Rio Grande do Sul and Londrina.

Soy has been grown in Rio Grande do Sul for nearly a century. In that time, its production has undergone a significant structural and functional change. The present regional diversity of farming styles is influenced by different events in history. In the nineteenth century, European immigrants arrived in Rio Grande do Sul and gradually deforested the Atlantic rainforest, or *mata*, to become the owners of small plots of land. The activities of these small-scale family farms were mainly structured around the production of lard pigs (*banha*), together with subsistence crops. From 1930 on, soy was used to feed the pigs (Shurtleff and Aoyagi 2009). In addition to these activities, the farmers in this region started to produce wheat, which was highly promoted by the government because this reduced Brazil's need to import wheat. This brought the construction of infrastructure such as silos and other machinery. Cooperatives were founded, which had a very important influence on this process of wheat production and infrastructure construction.

In the 1970s, three interrelated events led most farmers to shift to the system of soy/wheat production and abandon the polyculture of meat and crop production. First,

stress that there is not so much difference between the farms in terms of what concerns technology, because the machinery and commercialization system are quite similar.

[4]The reason to choose the two regions was based on the maximum contrast method. Within these regions, the two specific municipalities were chosen because of the availability of contacts. As we started in Três de Maio, we had already a lot of data regarding some farming styles; as such, we had a faster data saturation in Londrina. The research was more complicated in Londrina because of the larger distances and more complicated relations between our local connections and the extension workers and the farmers. The land distribution is more skewed in Londrina compared with Três de Maio; as such, we expected – correctly – to find more large-scale soy producers in Londrina.

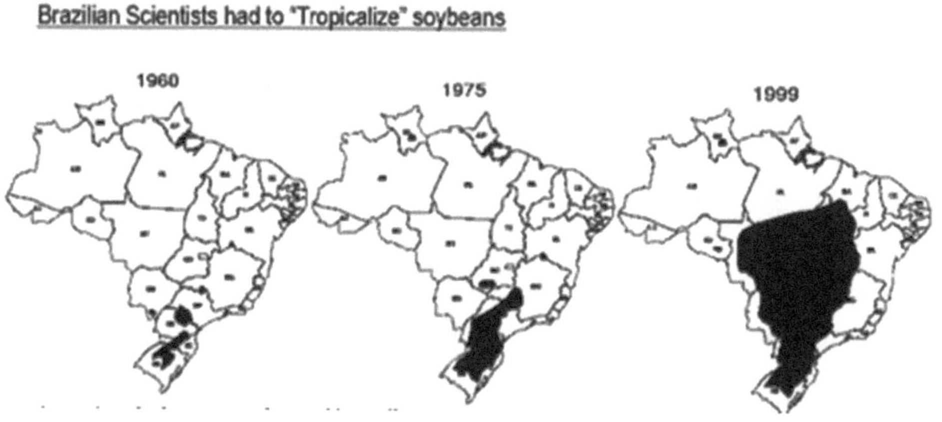

Figure 2. Soy expansion in Brazil.
Source: Embrapa, Soy/Londrina/Paraná; slide by Maurício Antonio Lopes 2011.

Europe was organizing its 'Never Hungry' campaign after the Second World War (this was the foundation of the Common Agricultural Policy). Milk and pig production increased drastically. As a result, many farms shifted to an industrial farming model that uncoupled feed production from pig and cattle farming. This led to heavy reliance on external protein sources such as soy. The price of soy, one of the main ingredients in pig and dairy concentrates, rose dramatically. Second, simultaneously with the European policy shift, Brazil started the so-called Green Revolution, an agricultural modernization project. Supported by the Rockefeller Foundation, the subsistence family agriculture in Rio Grande do Sul was transformed into a high input/high output agricultural system. Knowledge centers such as EMBRAPA were founded in order to intensify and facilitate this process. Third, the existing infrastructure for wheat production only needed small adjustments in order to respond rapidly to the increasing soy demand in the world market (Brum 1987). This increase in scale forced farmers to make one of three choices: (1) abandon farm activities and move to the city to live in often impoverished conditions (Brum 1987), (2) sell their farm to buy land in the cheaper central-west region to produce soy on a much larger scale (Mato Grosso do Sul, Mato Grosso, Goiás; Figure 2; Schlesinger, Nunes, and Carneiro 2008[5]) or (3) successfully take over the other farms that were left by the two previous groups of farmers. The phenomenon of land purchase in the central-west area is still going on and is partially responsible for the deforestation of the Amazon.[6] Figure 2 illustrates how soy was initially grown in the South region and how the soy production area expanded very rapidly towards the central-west region. EMBRAPA played an important role in the soy expansion towards the central-west area because it actively selected soybeans that would

[5]The process of either quitting farming or buying cheap land instead of instituting land reform is also described as *modernização conservadora,* or conservative modernization. This was the modernization process during the Brazilian dictatorship without any land reform (Wolford 2008; Da Silva 1982).

[6]There is an intense debate about this process. We find different explanatory hypotheses. The oldest one says that some families living in very bad conditions at the frontier of the Amazon are claiming property rights of a plot of forest. They cut the trees (illegally), sell the timber and light the rest on fire. After that, the land is sold to the cattle rangers who use the land for a maximum of 5 years. The land is then sold in overexploited conditions to the soy farmers. But, recently, another explanation has been proposed: the indirect cause is the deforestation of the savannah region (the Cerrado) by soy producers, which drives the cattle rangers into the forest (Arima et al. 2011).

grow in the acid tropical soils of that region. As such, soy, which normally grows in temperate climates, was 'tropicalized'.

In the period between the 1950s and the 1980s, cooperatives were adjusting and adapting their infrastructure to soy/wheat production. However, this process ended abruptly during the economic crisis of the 1980s. In this period, many cooperatives had financial trouble and thus had to start the 1990s under impoverished conditions. In the 1990s, the national productivism of the 1970s shifted to a multinational productivism, in which many cooperatives were bought up by companies such as the four major commodity companies (ADM, Bunge, Cargill and Dreyfus; Benetti 2004).[7] It was also in Rio Grande do Sul in the 1990s that the first GM soy beans were imported and cultivated illegally, which forced the federal government in 2003 to legalize GM crops in Brazil.

The region around Londrina (Paraná) has a slightly different development path than the soy production areas of Rio Grande do Sul, although the region was affected by the same international events as described above. Londrina was characterized by pristine forests inhabited by indigenous people, not explored by European colonists. This changed in1930, when the Northern Paraná Land Company bought the property rights and started to divide the region into different parcels. People from São Paulo, Minas Gerais and Japan migrated into this area and began the processes of deforestation and transformation into agricultural land. The region was especially famous for its coffee production. However, this production suffered a heavy crisis during the 1960s, which caused many of the coffee producers to move towards soy production, as this crisis coincided with the Green Revolution as described above.

Culturally, the farmers in Londrina still feel a significant connection to coffee production, a very labor-intensive crop which had a completely different way of farm organization than did cereal production. However, because of the increasing labor costs, coffee production has become more and more complicated and farmers are gradually moving towards soy production. The region consists of 80 percent small-scale family farms, responsible for 50 percent of the land use. The other 50 percent belongs to large landowners (*fazenderos*). In this sense, the soy production in Paraná differs from that in Rio Grande do Sul because of the more skewed land distribution in the former area.

The stories behind the numbers: the social reality of soy production

Brazilian farmers are squeezed between rising input prices, increasing cost of living and decreasing output prices. They have developed three strategies to secure their income and to keep farming: (1) upscaling, (2) diversification and (3) alternative agricultural strategies (direct selling, organic agriculture).

Although we focused on soy producers, we found that the majority of the farmers visited (95 percent) had a diversified production on their farms in addition to soy production. Referring to these as *colonial activities*,[8] this practice coincides with the local notion of 'a good farmer', someone who is autonomous ('good farmers don't buy food

[7]It is fundamental to understand that the cooperatives were "capitalist cooperatives" or large enterprises (which has nothing to do with the original idea of cooperatives or even with cooperation), which had a key role in the commercialization of soy. And this was also before the rise of multinational productivism.

[8]In Portuguese: *produtos coloniais* (referring to the European peasant farmers who went to colonize the region of Rio Grande do Sul); they are characterized as home-produced products, with no or low use of agrochemicals, such as a mixture of fruit trees and vegetables, free range chickens called *frango*

in the supermarket' (Interview, colonial farmer, Três de Maio, 17 May 2011)). Furthermore, this concept refers to 'those who conquest the *mata* (the Atlantic rainforest) and changed this unexploited natural richness into a fertile land for Brazil's food production' (Interview, colonial farmer, Três de Maio, 12 May 2011). The colonial style is a *Gaucho* style and defines the cultural identity within the region of Rio Grande do Sul. This region was also the country of origin of the *Gauchos*, settlers who went to the north and central-western area of Brazil to exploit the pristine forests in order to produce soy. They are following in their ancestors' footsteps by transforming *unused* land into fertile fields that give food to the people.

In some cases, animal production was also found on the farm (mostly with smaller producers, i.e., niche farmers and colonial farmers; see below). The respondents framed the production of animals for cooperatives or integrators as *diversification*, as it is considered a new and promising strategy that ensures a secure monthly income. Soy is more risky for the following reasons: the major investment is at the start of the season (inputs, prepare the land, seeds); the yields can drop (because of weather events and new diseases); and prices can fall at the moment that the farmer needs to sell his crop. The diversification can be split up into milk production, meat cows, and industrial poultry and pork production. Industrial poultry and pork production is mostly on contract for meat processors and is directly sold to them. Milk production is linked with the milk processors, a private company or cooperatives. Meat cattle production is done on a rather extensive scale and is characterized by a more informal, small-scale trading system based on intermediaries.

The forest, as a part of the land, is generally seen as very important. Most interviewed farmers (colonial farmers and niche farmers; see below) were very concerned with the preservation of the forest (70 percent of the interviewed farmers). This became clear in the landscape: the fields were always attached to pristine or secondary forest: 'The forest is important for the preservation of the biodiversity, as a climate regulator, because it is beautiful, and because it is my duty to maintain it' (Interview, colonial farmer, Três de Maio, 19 May 2011). Older farmers talked about the period that the government and *Banco do Brasil* were stimulating the farmers to deforest. They refer to that period by saying 'this was the way things were' (Interview, colonial farmer, Três de Maio, 16 May 2011), but they are now glad about the reforestation.

Três de Maio faces a general labor shortage because the young people are massively migrating to the cities. In Londrina, farmers suffer from a shortage of labor as well, but the underlying dynamics differ. This region is more densely populated; thus, the competition with jobs in the city is much higher. This makes the labor shortage economic rather than geographical. This labor shortage, together with a profit squeeze (see below) influences many of the farmers' decisions. A farmer reported: 'I would shift to organic agriculture, because it's much healthier, but I cannot do it on my own. My children are now working in the city for a much better income and they are living in better conditions than I do' (Interview, colonial farmer, Três de Maio, 13 May 2011). This results in a general shift to less labor-intensive cropping techniques. Milk production is also limited because of this labor shortage.

A huge variety in land size (4–1500 ha) and differences in land ownership status (renting vs. owning land) determine how the farmers will combine the production factors of land and labor to respond to the cost/profit squeeze. The farmers rent extra land to optimize the

---

caipira ('farmers' chickens') and the local breeds for the swine production that are not as vulnerable to diseases as are the breeds from the integrators.

capacity of the machinery they have bought. Rented land, however, is seen as a mere production factor, which plays no part in the culturally laden concept of 'good farmer' as discussed above: rented land is only used to maximize production. A farmer does not have to be a caring steward for his rented land, as it is merely a resource to be exploited.[9]

The above-mentioned production activities are linked with different marketing channels: soy and other staples are often sold to cooperatives and private processors. Chicken and pork production is marketed on contract with large companies, while meat cows are sold via more traditional marketing channels. Colonial products are either produced strictly for home consumption or sold in farmers' markets. Milk is marketed via the cooperatives, private companies or public procurement.

Three farming styles: surprising diversity hidden behind the monolithic homogeneity of soy

Based on the different responses to the above-described elements and the functional links between the elements, we can distinguish three farming styles. The 'colonial farmer' is connected with the commodity crops, but still has a strong relation with the forest and his colonial activities. The 'niche farmer' searches for alternative markets, such as organic farming and farmers' markets. Finally, the 'entrepreneur' has a productivist approach to his farm and the future[10] (see Table 1 for more detail, and Figures 3 and 4 for illustrative pictures).

These three farming styles result from a governmental initiative, the Brazilian Green Revolution of the 1970s. The most typical style still present is the 'colonial farmer' farming style: these farmers are strongly connected with the technological package that was delivered by the institutions of the Green Revolution, while still maintaining other activities as colonial products. This farming style is increasingly endangered because of the profit squeeze originating from increasing input prices and increasing competition with large-scale soy producers. In reaction to this profit squeeze, we observed two reactions. The niche farmer searches for other (autonomous) modes of production, and the entrepreneur seeks to scale up production, which is a logical consequence of the Green Revolution model that has now been taken over by multinational companies.

The colonial farmer

The colonial farming style is the dominant style in the region of Três de Maio and Londrina (the highest number of farmers). Farmers in this style produce commodity products (soy and wheat, and poultry, cows and pigs) with the help of intensified mechanization, high external-input use and a high dependency on external knowledge. The common marketing channels are used to sell their commodity products: cooperatives, private processors and also futures markets. The colonial products are strictly for home consumption. However, farmers are complaining about the ever-more-challenging profit squeeze, which forces them to shift towards diversification. A few farmers are getting an additional income from non-farming activities. These farmers continue farming 'as our parents did', although their children have often

[9]The difference in attitude towards owned land and rented land decreases when talking to the industrial soy producers, as they already have the maximization strategy on their own land (see the description of the different farming styles below).

[10]The farming styles as such are a conceptualization of the farming practices and their related world views. However, for the sake of simplicity, we will refer in this paper to, for example, the colonial farmer, while meaning 'the farmers within the farming style called the "colonial farmer"'.

Table 1. Summary of the characteristics of the different farming styles.

	Colonial farmer	Niche farmer	Entrepreneur
Income strategy	Improve production results per ha, production of milk and/ or pigs for the integrators	Search farmers' markets, deliver for the school food program	Scale enlargement, improve production results
Main activity	Soy/wheat or pigs and dairy cows	Colonial products, organic products	Soy/wheat
Forest	Important functional element in the farm	Important functional element in the farm	No functional element in the farm
Colonial products	Functional element in the farm, but only for home consumption	Core business of the farm	No function, sometimes not present at all
Materialization of the production logic	Partly monoculture of wheat/soy or integration, partly forest and agro-ecological crops	Partly forest, mainly agro-ecological crops, sometimes organic, sometimes wheat and soy	Mainly monoculture of wheat/soy, small pieces of forest, agro-ecological crops insignificant
Technologies	Older technologies (tractors older than 1990, absence of no-tillage tools), or new technologies used for production for integrators	Older technologies (tractors older than 1990), new technologies to comply with food legislation for processed food	New technologies (recently bought combines and tractors, tools for no-tillage agriculture), fit for large-scale activities
Labor	Maximize labor use by diversification: labor-intensive milk and pig production, next to soy production	Maximize labor use by own marketing, ecological production, colonial products, pastry	Maximize labor use by scale enlargement

Figure 3. Monoculture. The field of an entrepreneur.
Source: the author.

Figure 4. Diversity. The field of a colonial farmer or a niche farmer.
Source: the author.

moved to the city. This makes farm succession a big problem. The colonial farmers see their farm as a whole: an organic entity with different activities that *must* be present on the farm, such as soy, the forest and some colonial activities. About the forest, they said: 'it is important, that's a reason enough, you just have to preserve the forest' (Interview, colonial farmer, Três de Maio, 19 May 2011). This gives the forest an unquestionable importance. After asking why it is important, a farmer responded 'because it is, because you don't need to deforest everything and it wouldn't solve my problems anyway' (Interview, colonial farmer, Três de Maio, 16 May 2011). Colonial products are important 'because a good farmer produces his own food' (Interview, colonial farmer, Londrina, 4 July 2011). This seems quite surprising, as the size of the farms ranged from 6 to 24 ha, which means that this group is having many different activities on a rather small plot of land, but this way of production is inherently linked with their farming style.

Colonial farmers produce soy in the same way as entrepreneurs do, but with smaller and less up-to-date technologies on much smaller cultivated surfaces. However, they use also no-tillage techniques, usually by hiring a contractor, as in most cases they do not own the appropriate machinery. They also carefully follow the advice of technical assistants and apply the same pesticides, varieties and inputs as the entrepreneurs.

The niche farmer

Unlike the colonial farmer and the entrepreneur, the niche farmer is characterized by farmers who try to search new markets for their products in order to escape the increasing input prices and decreasing output prices. They find them in short marketing chains such as farmers' markets. They criticize other agricultural models and search actively for new ways to produce, based on alternative information channels. In this sense, the niche farmer is very innovative. This group has one thing in common: the amazing diversity in strategies to search for autonomy and to increase income.

In the case of the organic farmer, soy is but one of the many farm products that is sold directly to consumers. The soy variety grown is fit for direct human consumption; thus, it becomes part of a larger cluster of farming activities that integrate the production of colonial as well as commodity products. Other niche farmers use conventional farming techniques, but focus strongly on direct sales. In Três de Maio, a group of farmers has organized a farmers' market in order to get a better price for their products. Three major groups of products are sold here: (1) fresh products (fruits, vegetables), (2) processed products (honey, meat and cheese), and (3) bread and pastries. The fruit and vegetable producers (category 1) often also have conventional soy production, an additional activity that increases their income up to 20 percent. Their main activity is selling fresh products at farmers' markets. The bread and pastry producers have a more conventional soy and wheat production. The wheat is used to make the bread and pastries. This specific niche farming style therefore has few differences from the colonial farmer style, but it makes it possible for these farmers to maintain their property and increase their income by seeking direct contact with consumers.

Most of them used to be traditional soy farmers, but they have moved towards other products (organic products, colonial products) or selling methods (farmers' markets, farm shops). The farm scale of the niche farmers is comparable to the farm scale of the colonial farmers. And, like the colonial farmers, the niche farmer's farm is seen as one organic entity, where the forest takes an important place, together with the colonial products. In this style, farm succession is also problematic, for the same reasons. We can conclude that the niche farmer, in contrast with the colonial farmer, gets his income especially from activities other than soy. Soy might be present, but the main source of income is derived from fresh products, processed products, bread and pastries or organic products (which can include organic soy as well).

The entrepreneur

The entrepreneur farming style consists of specialized (larger scale) soy producers, who are intensely interlinked with the market. Their market-oriented discourse sees land as a capital factor, an input in their economic activity. In addition to soy and cereals, they also produce colonial products, but these are mostly seen as a relic from the past to be maintained, though without an economic function and only a small identity function, in the sense that colonial products are not a necessary condition to be a good farmer. In Londrina, there are farms (although only 20 percent) where the focus goes strictly to soy and cereal production without extra colonial activities, and where food is bought in the supermarket.

Focusing merely on soy and cereal production, the entrepreneur invests in new machines and tries to intensify this technology use in order to increase labor productivity. A good farmer uses state-of-the-art technology, which means modern tractors and planting and harvesting equipment. Most machinery is bought on credit, a necessary condition in order to increase his competitiveness vis-à-vis other farms. As machines allow him to cultivate more land, the modernization of the farm is accompanied with land acquirement (bought or rented) from farmers who have stopped farming or moved to town.

The entrepreneur's attitude towards the forest is quite specific: the forest is not seen as an economic resource and is even perceived as an obstruction to farm development. He doesn't pay special attention to the forest, as 'a forest doesn't give you an income'. Hence, very little forest is preserved in the landscape: 'Why should we care about this? How much forest do you see in the city? Maybe they can oblige the people in the city to buy forests. We need to produce, and for this, we need land' (Interview, entrepreneur, Londrina, 7 July 2011). A forest law is in force that requires farmers to keep 20 percent of their property forested. As a way around this law, some entrepreneurs buy plots of already-existing pristine forest that are disconnected from their farms.

Future aspirations and visions of good farming as values behind the materialization of practices

Future aspirations

In general, most colonial farmers talked quite positively about modernization processes such as mechanization and transgenic technology: 'It's much easier, we can produce more with less labor. Soy production used to be a very tough activity, because the seeds had to be planted by hand, in the sun. Yes, I'm glad' (Interview, colonial farmer, Três de Maio, 13 May 2011). Still, they were complaining that it was harder to get the same income as before because of increasing costs. The recent need to use pesticides serves as an illustration. Up until a few years ago, farmers did not need to use pesticides, but the yield has been very low recently when no pesticides are used: 'I don't know what's the cause, but the pressure from diseases is increasing. Maybe it's because of the changing climate? Whatever the cause is, we need to use those expensive pesticides to save our crop' (Interview, colonial farmer, Londrina, 8 July 2011). The increasing costs, together with the absence of farm succession, make these farmers continue farming like they used to do: maximizing production results by carefully listening to the technical advisors, while hoping that good results will overcome the profit squeeze: 'You have to follow the latest evolutions! You have to increase your production, and thanks to the help of the extension workers, this is possible' (Interview, colonial farmer, Três de Maio, 25 May 2011).

Most of the niche farmers are able to maintain their income in a better way than the colonial farmers. They are optimistic that farming will still be possible in the future, still with long working days, but with more options:

> The adaptation towards the new system was quite difficult, but it's better now, we have a better income. The consumers like our fresh and healthy food without pesticides. But it is still a risky business and we hope that the consumers will continue to buy our food.
>
> (Interview, niche farmer, Três de Maio, 20 May 2011)

This is a very different attitude compared to that of the other farmers in this region. Three years ago, the Brazilian government started a special program for family farms (farms smaller than 80 ha), where, for example, a special credit provision for this farm type is arranged. Schools are obliged to buy at least 30 percent of the products for their school food from family farms. This has given an enormous boost to the family farms to produce colonial products, which makes them perceive the future more positively. However, the farmers in the niche farming style do not have farm successors either, because their children don't believe in their production model:

> They think that the farm is too small to continue, maybe it's true, I don't know. Farming is a hard job. They just want a job in the city with a good income, work from nine to five. I can't force them to stay on the farm.
>
> (Interview, niche farmer, Londrina, 7 July 2011)

Entrepreneurs perceive the future as successful and promising, because access to credit is much easier than in the past. During the period of hyperinflation, it was very expensive to invest in machines or land. The current access to credit enables them to invest in and improve their farm. The future for soy is also seen as promising: 'Everybody needs soy, look at China. We will be able to keep on selling our soy' (Interview, entrepreneur, Londrina, 13 July 2011). The future developments will bring an increasing upscaling, which is translated as a continuous investment in new technologies and obtaining new land. However, sometimes the new environmental laws are seen as a threat for their future. For example, the forest regulations (which oblige the farmers to maintain 20 percent of their land as forest) curtail their development. The new demands are seen as unrealistic, and they propose a revision. This farming style has no problem with succession. This means that, in contrast with the other farming styles, the entrepreneur sees the continuation of his farm as guaranteed.

Vision of a good farm

When the colonial farmers were asked about the qualities of a good farmer, they responded that the colonial activities and the forest form an inherent part of their farming activities; it is an expression of their identity of a *colonial* farmer. The farmer should also be familiar with the current evolution and be prepared for new technologies in order to be able to continue his farm: 'If you don't evolve, you will lose' (Interview, colonial farmer, Três de Maio, 16 May 2011). It is seen as important that a farmer should try to produce more on the same amount of land. This is done by choosing the right technologies and the right varieties. When talking about the production of soy and cereals, the colonial farmer is moving towards an entrepreneurial model, i.e., increasing production by intensifying technology use. For this reason, the no-tillage techniques provided by the extension workers were rapidly adapted in the region – because 'a good farmer needs to take care of the soil by using no-tillage techniques to avoid erosion and to apply fertilizers in the right manner' (Interview, colonial farmer, Três de Maio, 25 May 2011).

Niche farmers say that a good farmer should have different activities, be able to manage the farm and be willing to work hard: 'If you don't love the work, you will never be able to be a good farmer' (Interview, niche farmer, Três de Maio, 23 May 2011). It became clear that these niche farmers, when asked about their property, automatically referred to their forest, as was the case for the colonial farmers. The niche farmer also views the forest as an integrated part of their farm. The organic farmers strongly stress the ecological production of food as an important aspect of good farming practice: 'A farmer should care about the environment, produce crops without damaging the ecosystem and without poisoning our neighbors. The consumers have the right to have healthy food and that's what our task is' (Interview, niche farmer, Três de Maio, 20 May 2011).

The entrepreneur's description of a good farmer seems to come very close to that of the colonial farmer: you have to be a good manager, have good technologies, follow the current evolution, choose the best products with the help of technical advisors and avoid soil erosion. The big difference between the entrepreneur and the colonial farmer is in the attitude towards colonial products and the forest. While the colonial farmer focuses on production results for the production of soy and cereal crops, he sees how colonial production and forestry are also necessary to fulfill the image of a good farmer. Entrepreneurs pay much less attention to these *side* activities. Several entrepreneurs also focus on having a *clean farm*: preventive elimination of weeds, in order to have a clean field without weeds – even when these plants would not harm the crops: 'Cleaning the weeds off my property, that's my free-time activity. It's nice to have a clean farm' (Interview, entrepreneur, Londrina, 30 June 2011). This reflects a strict separation between nature and agriculture.

Different farming styles and their perspectives on soy production and its accompanying technologies

In this section, we explore how soy production and the accompanying technological package are approached in the different farming styles. Soy production is strongly institutionalized, and is embedded in a network of the cooperatives, knowledge institutions and technologies that are intensively linked with the crop. However, the increasing use of technology in soy cultivation (no-tillage, GMOs, pesticides), as facilitated by the accompanying institutions, has consequences for the different farming styles.

The relation between the attitudes, worldviews and ways of production, and the relation with the forest, allowed us to distinguish three soy farming styles. In all three farming styles, however, *the practices of soy production itself* were surprisingly homogenous. Even in farms where organic soy was produced, the production of conventional or even transgenic soy was also a part of the farming activity. Although this appears contradictory at first glance, it can be explained by the maximization of the production of premium products (such as vegetables, fruits, processed food, organic products) because farmers use their scarce labor for these products which use less land. The rest of the land is used for labor-extensive cultivation of soy and cereals. Moreover, most soy farmers will keep producing soy, because of its cultural embeddedness:

> Soy is a tradition, my father used to grow it; it's easy to grow and the soil and the climate are suitable for it. It's like an inherent activity connected with my farm. I wouldn't know what else to produce.
>
> (Interview, colonial farmer, Três de Maio, 23 May 2011)

Some even use soy as a local exchange unit: farmers use soy to pay for inputs, like pesticides or contract work, or even machines and land.

However, some farmers have left soy production: 'Soy production is for the bigger producers, I'm just too small. It is too complicated' (Interview, niche farmer, Três de Maio, 20 May 2011). In these cases, they shifted from soy to milk or pig production, which made it possible to earn an income on a smaller plot of land. Others were thinking of leaving soy production, but only on the condition that there would be a feasible alternative to soy.

Despite this homogeneity in soy production practices, the different farming styles point to the farmers' attitudes towards integration of their production within the environment; the forest and colonial products (i.e., products grown without pesticides) are important aspects for the colonial farmers and the niche farmers. The functional integration of these diversified agricultural practices principles causes these two groups to have a co-evolving attitude with the ecological environment. In other words, agriculture and nature are not mutually exclusive in these farming styles, but are constitutive of each other. This is entirely different for the entrepreneurs, who do not see the forest and colonial activities as conditions necessary for farming. The entrepreneurs treat forest and colonial activities as separate from farming. To comply with the forest regulations, their best option is to buy existing forests without any relation to, or integration into, their farms.

Technology and knowledge

Technology is very important for all of the farmers because 'You need to move forward, or you lose' (Interview, colonial farmer, Três de Maio, 17 May 2011). This coincides with the market-oriented attitude of the colonial farmers and entrepreneurs regarding cereal, soy and milk production. For these commodity activities, the farmers stress the importance of following the new trends. Technical assistants have a very important role regarding this issue. Most colonial farmers and entrepreneurs told us that a good farmer follows the advice of the technicians, and listens carefully to the advice of the extension services in order to be able to follow the latest trends. This gives the extension services a powerful position. Most of the soy buyers (*agropequarias*), who are also tied to agrochemical factories, also give technical advice. Due to their very biased advice, the colonial farmers were not introduced to the new marketing possibilities such as farmers' markets and the school food program. This voluntary withholding of information causes a discrepancy between the view of the colonial farmers about the consumers ('consumers only want clean products without any spot' (Interview, colonial farmer, Três de Maio, 25 May 2011)) and an existing, real demand from the consumers at the farmers' markets ('I buy here because I want healthy food without pesticides; it's more natural, it's healthier, it's fresh, it's cheaper, it's family agriculture' (Interview, consumers, Três de Maio, 20 May 2011)).

The colonial farmers in general see technology as useful, as it reduces the workload. It is good as long as it improves the production results. However, the cost/price squeeze brings colonial farmers into an inconvenient position, as the technologies needed to maintain their production and to cope with the lack of labor force are too expensive for these smaller farms. For the colonial activities, almost no external technologies are used. The colonial farmers rely strongly on the extension services for information to improve their production results. This information is strongly focused on technical results, which keeps them uninformed about alternative farming strategies. In case of diversification, the processors are still working on milk quality:

> Thanks to the technical assistants from the processors, my milk quality has improved a lot. The occurrence of mastitis decreased drastically, just by applying some simple techniques. But I have to, because the processors oblige me to improve the results continuously. But it's worth the effort as they pay a lot of money.
>
> (Interview, colonial farmer, Três de Maio, 16 May 2011)

The niche farmers in general perceive that technology is helpful. The technologies used to produce colonial products for the farmers' markets were fit for the scale of the activities and conform to the legislation. However, in the case of the organic farmers, the technologies were hard to access and, also, organic varieties of several crops were harder to find. Niche farmers are mostly searching for information in an autonomous way because the relevant information, owned by the mainstream extension services, is generally difficult to access. Information about organic farming was lacking. However, in the case of Três de Maio, one technical assistant gave advice to the farmers in the niche farming style about specific issues concerning direct selling strategies and compliance with the regulations.

The entrepreneurs use technologies that are appropriate for their farm size and make it possible to scale up production. The technologies induce further scale enlargement; they are mostly over-dimensioned. The farmers in this group were very proud of their modern machines. The entrepreneurs are tightly linked with the extension services. Given their dependency on a single crop, market or production risks might have a higher impact on their farm results compared to the other styles. This makes them highly dependent on the extension services for advice, whenever an event occurs that could threaten yields.

Soy and the different farming styles – final remarks

Our analysis has shown that the farming styles, as defined by van der Ploeg (1993), do not lead to different soy production techniques. The homogeneity of production techniques is a result of the strong exogenous character of the soy, where most of the knowledge, technologies and markets are neither locally produced nor created. Soy production has always been strongly embedded within an institutional environment that fits within the Green Revolution paradigm and that provides the necessary *technological package* to the farmers. The soy production system is indeed embedded in a global food regime that is reorganizing the practices of soy production in a homogenous way. However, our results clearly show that despite the implication that all soy production systems are homogenous, the farms where the soy is produced do have important differences. The discovery of different farming styles behind an apparently homogenous soy production has major repercussions on several levels: it affects the accompanying institutions, the NGOs and the scientific community.

Soy and cereal production is embedded in the current third food regime, which is exemplified by modern technology, such as tractors, agrochemicals, improved seeds, transgenic crops and the accompanying knowledge coming from the extension workers. The colonial activities are rather disconnected from this technology and research-driven practices, as they are based on traditional varieties without treatments using agrochemicals. This disconnection from the food regime is simultaneously a connection with their own endogenous resources and territorial capital, as these products are 'not for the market, but for us, and they are not toxic' (Interview, colonial farmer, Três de Maio, 25 May 2011). Because the local breeds are endogenous, they are adapted to the local conditions and resistant to local pests.

The institutional environment of the global food regime provides the same technological package to the entrepreneurs and the colonial farmers: farming practices such as no-tillage combined with GM soybeans and the use of pesticides. This results in several frictions between the different farming styles and the surrounding institutions. Although the focus remains on improving production results, the question remains whether the increasing influence of technology on soy production really brings solutions fit for the colonial farmers. Most of the new soy-related technologies induce scale enlargement, which drives the colonial farmers towards an entrepreneurial farming style. This will affect the position of the colonial products and the forest within their production system, as these are especially related to owned land, and much less to rented land. Consumers show increasing interest in colonial products, which could give colonial farmers new opportunities to maintain their farming style. However, knowledge is structured in such a way that a large portion of the colonial farmers are not informed about other promising alternatives. The niche farmers still feel a lack of interest in their farming style from the surrounding institutions. In this sense, soy production is strongly embedded and linked with the global food regime, with its dominant technology-driven paradigm that leads to upscaling, while alternative agricultural activities are largely ignored by the main actors and institutions within this regime. This causes an ongoing process of concentration accompanied by vertical integration. The concentration can be subtly recognized in the different dynamics in farm succession between the entrepreneur farming style and the two other farming styles. The entrepreneurs' better access to technical assistance that is fit for their farm dimensions lends an extra impulse for the entrepreneur farming style. We can therefore expect a further tendency of scale enlargement in the region, at the cost of the other farming styles. This further concentration will also increase dependency on technical assistance, which will drive the entrepreneur farms further in the direction of the input industry. This increasingly blurs the boundary between independent farmers and vertical integration. Our findings therefore support the conclusions of many articles about the increasing concentration and vertical integration of the soy crop.

But the entrepreneur farmers are not the only ones cultivating soy in Brazil. The colonial farmer and the niche farmer both produce soy but they have a totally different relation to their natural environment. It seems that they are more adaptable and resistant than usually assumed by authors who state that large scale will make small scale disappear. Apparently, many farmers still see soy production as an important aspect of their farm, so we can assume that they will continue to produce it. However these farming styles are mainly neglected by the institutional environment. Despite the fact that Brazil started a program especially designed for family farms (Programa Nacional de Fortalecimento da Agricultura Familiar (PRONAF)), it seems that the institutions do not follow this recognition. The fact that small-scale soy producers are largely neglected by the NGOs is a pity. The NGOs' framing of large-scale soy production as unsustainable should therefore be nuanced, especially if they consider the colonial farmers and the niche farmers to be important potential partners for the strengthening of family farming and agro-ecology. By addressing soy production as inherently unsustainable, the discourse of international NGOs is not likely to touch these farmers. In this sense, we should move the discussion from a monolithic perception of soy production as a homogenous activity towards a model that embraces the discussion of farming styles. Each crop can be produced by a different farming style, which will ultimately affect the sustainability of an agricultural production system.

Wield, Chataway and Bolo (2010) suggested that we need to take the stratification of the soy producers into account. In this study, we can convincingly argue that this is indeed a necessary condition to understand the reality of the soy farmers. Our findings may nuance Woodhouse (2010), who treats the sustainability problem by questioning whether large- or small-scale farmers produce more per hectare. Instead, we should shift the question toward *which set of values behind farming practices actually results in sustainable agricultural production?* Even in the case that the homogenous soy crop, as such, heavily affects the environment, farmers can dramatically influence how this environmental degradation might occur. This research raises new questions: can there be sustainable production of soy in a diversified way, and how might this be organized? What is the potential of the colonial farmer and the niche farmer to avert the process of increasing concentration and vertical integration? Can a change in the organization of the accompanying institutions construct another soy production system with the help of the colonial farmers? Or should the colonial farmer try to valorize his colonial products instead of producing soy? These questions have the power to shift and extend the debate about sustainable agriculture.

Acknowledgements

We gratefully acknowledge a scholarship grant from ILVO, the Institute for Agriculture and Fisheries Research, to accomplish this research.

References

Altieri, M.A. 2009. The ecological impacts of large-scale agrofuel monoculture production systems in the Americas. *Bulletin of Science, Technology, Society* 29, no. 3: 236–44.

Altieri, M.A., and W. Pengue. 2006. GM soybean: Latin America's new colonizer. *Seedling*, January: 13–7.

Arima, E.Y., P. Richards, R. Walker, and M.M. Caldas. 2011. Statistical confirmation of indirect land use change in the Brazilian Amazon. *Environmental Research Letters* doi:10.1088/1748-9326/6/2/024010

Belo, M.S.S.P., W. Pignati, E.F.G.C. Dores, J.C. Moreira, and F. Peres. 2012. Uso de agrotóxicos na produção de soja do Estado do Mato Grosso: um estudo preliminar de riscos ocupacionais e ambientais. *Revista Brasileira Saúde Ocupacional* 37, no. 125: 78–88.

Benetti, M.D. 2004. *Globalização e Desnacionalização do Agronegócio Brasileiro no Pós 1990.* Porto Alegre: FEE.

Borras, S.M. Jr, P. McMichael, and I. Scoones. 2010. The politics of biofuels, land and agrarian change: Editors' introduction. *The Journal of Peasant Studies* 37, no. 4: 575–92.

Brown, J.C., M. Koeppe, B. Coles, and K. Price. 2005. Soybean production and conversion of tropical forest in the Brazilian Amazon: The case of Vilhena. Rondônia. *Ambio* 34: 462–9.

Brown, J.C., and M. Purcell. 2005. There's nothing inherent about scale: Political ecology, the local trap, and the politics of development in the Brazilian Amazon. *Geoforum* 36, no. 5: 607–24.

de Bruin, R. 1997. Dynamiek en Duurzaamheid. Beschouwingen over bedrijfsstijlen, bestuur en beleid. Thesis (PhD) Wageningen: WUR.

Brum, A.J. 1987. *Modernização do agricultura trigo e soja* . Petrópolis, RJ: Vozes.

Buttel, F.H. 2006. Sustaining the unsustainable: Agro-food systems and the environment. In *Handbook of rural studies*, eds. P. Cloke, T. Marsden and P.H. Mooney, 213–30. London: Sage Publications Ltda.

Cassol, A. 2010. As instituições tácitas e formais e as novas formas de interação social no meio rural. Thesis (Msc) Porto Alegre: UFRGS.

Commandeur, M. 2003. Styles of pig farming : a techno-sociological inquiry of processes and constructions in Twente and The Achterhoek. Thesis (PhD) Wageningen: WUR.

Da Silva, J.F.G. 1982. *A modernização dolorosa: estrutura agrária, fronteira agrícola e trabalhadores rurais no Brasil.* Rio De Janeiro: Zahar Editores.

Eisenhard, K.M. 1989. Building theories from case study research. Academy of Management Review 14: 532–50.

Eisenhard, K.M., and M.E. Graebner. 2007. Theory building from cases: Opportunities and challenges. *Academy of Management Journal* 50: 25–32.

FAOSTAT. http://faostat.fao.org/ (accessed June 6, 2012).

de França, G.C., M.E. Del Grossi, and V.P. de Azevedo Marques. 2009. *O censo agropecuário 2006 e a agricultura familiar no Brasil* . Brasíla: MDA.

Friedmann, H. 1982. The political economy of food: The rise and fall of the postwar international food order. *American Journal of Sociology* 88: S248–86.

Friedmann, H. 1987. Family farms and international food regimes. In *Peasants and peasant societies*, ed. T. Shanin, 258–76. Oxford: Basil Blackwell.

Glaser, B.G., and A.L. Strauss. 1967. *The discovery of grounded theory: Strategies for qualitative research* (4th, 2009 ed.). Rutgers: Transaction Publishers.

Gudynas, E. 2008. The new bonfire of vanities: Soybean cultivation and globalization in South America. *Developement* 51, no. 4: 512–8.

Hecht, S.B. 2005. Soybeans, development and conservation on the Amazon Frontier. *Development and Change* 36, no. 2: 375–404.

Hecht, S.B., and C. Mann. 2008, January 10th. How Brazil outfarmed the American farmer. *Fortune*, January: 92–105.

Howden, P., and F. Vanclay. 2000. Mythologization of farming styles in Australian Broadacre cropping. *Rural Sociology* 65, no. 2: 295–310.

James, C. 2008. *The Global Status of Commercialized Biotech/GM Crops. ISAAA Briefs No. 39.* Ithaca, New York: ISAAA.

Jetdicas. 2012. Map of Brazil. http://www.jetdicas.com/

Klink, C.A., and R.B. Machado. 2005. Conservation of the Brazilian Cerrado. *Conservation Biology* 19, no. 3: 707–13.

Lambin, E.F., and P. Meyfroidt. 2011. Global land use change, economic globalization, and the looming land scarcity. *Proceedings of the National Academy of Sciences* 108, no. 9: 3465–72.

Lamucci, S. 2011. Cinco commodities garantem 43% da exportação do Brasil. *Valor Economico*, http://www.valoronline.com.br/impresso/vale/1902/367747/cincocommodities-garantem-43-da-exportacao-do-brasil (accessed 25 May 2013).

Li, T.M. 2011. Centering labor in the land grab debate. *The Journal of Peasant Studies* 38, no. 2: 281–98.

Long, N. 1985. Creating space for change: A perspective on the sociology of development. *Sociologia Ruralis* 24, no. 3–4: 168–84.

Long, N. 2001. *Development sociology: Actor perspectives*. London: Routledge.

Lopes, M.A. 2011. Agricultural innovation and challenges for promotion of knowledge and information flows in agrifood systems in Brazil. http://www.oecd.org/tad/agricultural-policies/48218869.pdf (accessed 24 December 2014).

MAPA (Ministério da Agricultura Pecuária a Abastecimento) - Assoria do Gestão Estratégica. 2011. *Brasil Projeções do Agronegócio 2010/2011 a 2020/2021* . Brasília: Ministério da Agricultura Pecuária a Abastecimento.

Mascarenhas, M., and Busch, L. 2006. Seeds of change: Intellectual property rights, genetically modified soybeans and seed saving in the United States. *Sociologia Ruralis* 46: 122–138. Cited in Pechlaner, G., and G. Otero. 2008. The third food regime: Neoliberal globalism and agricultural biotechnology in North America. *Sociologia Ruralis* 48, no. 4: 351–71.

McMichael, P. 2005. Global development and the corporate food regime. In *New directions in the sociology of global development (Research in Rural Sociology and Development)*, eds. F.H. Buttel and P. McMichael, 265–99. 11.

McMichael, P. 2009. A food regime genealogy. *Journal of Peasant Studies* 36, no. 1: 139–69.

Moore, J. 2010. The end of the road? Agricultural revolutions in the capitalist world-ecology, 1450–2010. *Journal of Agrarian Change* 10, no. 3: 389–413.

Morton, D.C., R.S. DeFries, Y.E. Shimabukuro, L.O. Anderson, F.D. Espirito-Santo, R. Freitas, et al. 2006. Cropland expansion changes deforestation dynamics in the southern Brazilian Amazon. *Proceedings of the National Academy of Sciences of the United States of America* 103: 14637–41.

Oliveira, G. 2013. Land regularization in Brazil and the global land grab. *Development and Change* 44, no. 2: 261–83.

Oosterveer, P., and D.A. Sonnenfeld. 2012. *Food, globalization and sustainability*. London/ New York: Earthscan.

Pechlaner, G., and G. Otero. 2008. The third food regime: Neoliberal globalism and agricultural biotechnology in North America. *Sociologia Ruralis* 48, no. 4: 351–71.

Pengue, W.A. 2005. Transgenic crops in Argentina: The ecological and social debt. *Bulletin of Science, Technology & Society* 25, no. 4: 314–22.

Pengue, W.A. 2009. Cuestiones económico-ambientales de las transformaciones agrícolas. *Problemas del Desarrollo, revista latinamericana de economía* 40, no. 157: 137–61.

Planalto. 2006. Lei 11326. http://www.planalto.gov.br/ccivil_03/_ato2004-2006/2006/lei/l11326.htm (accessed 21 December 2014).

van der Ploeg, J.D. 1993. Rural sociology and the new agrarian question. A perspective from The Netherlands. *Sociologia Ruralis* 33: 240–60.

van der Ploeg, J.D. 2010. Farming styles research: the state of the art. *Historicising Farming Styles*, Melk, Austria. 1–15.

Porto, M.F., and W.L. Soares. 2012. Modelo de desenvolvimento, agrotóxicos e saúde: um panorama da realidade agrícola brasileira e propostas para uma agenda de pesquisa inovadora. *Revista Brasileira Saúde Ocupacional* 37, no. 125: 17–50.

Ratter, J., J. Ribeiro, and S. Bridgewater. 1997. The Brazilian Cerrado vegetation and threats to its biodiversity. *Annals of Botany* 80: 223–30.

Roep, D., J.D. van der Ploeg, and C. Leeuwis. 1991. *Zicht op duurzaamheid en kontinuïteit: bedrijfsstijlen in de achterhoek.* Thesis (PhD) Wageningen: Vakgroep Agrarische Ontwikkelingssociologie Landbouwuniversiteit Wageningen.

Santiago. 2012. Map of Southern Region. http://santiagosiqueira.blogspot.com.br/

Schlesinger, S., S.P. Nunes, and M.S. Carneiro. 2008. *A AGRICULTURA FAMILIAR DA SOJA NA REGIÃO SUL E O MONOCULTIVO NO MARANHÃO*. Rio de Janeiro: FASE.

Schneider, S. 2009. *A Diversidade da Agricultura Familiar* (2nd edition ed.). Porto Alegre: Editor of UFRGS.

Schneider, S., and P. Niederle. 2010. Resistance strategies and diversification of rural livelihoods: The construction of autonomy among Brazilian family farmers. *The Journal of Peasant Studies* 37, no. 2: 379–405.

Shurtleff, W., and A. Aoyagi. 2009. *History of soybeans and soyfoods in South America (1882–2009) Extensively annotated bibliography and sourcebook* . Lafayette: Soyinfo Center.

Steward, C. 2007. From colonization to "environmental soy": A case study of environmental and socio-economic valuation in the Amazon soy frontier. *Agriculture and Human Values*, 24, no. 1: 107–22.

The Economist. 2010, August 26th. The miracle of the cerrado. Brazil has revolutionised its own farms. Can it do the same for others?

Vanclay, F. 1994. Actors and structures in agriculture: Reflections of an Australian in Wageningen. In *"With a rural focus", An edited and refereed collection of papers with a rural focus presented to the annual conference of The Australian Sociological Association Inc.*, ed. F. Vanclay, 89–128. Wagga Wagga: Centre for Rural Social Researchn, Charles Sturt University.

Vanclay, F., P. Howden, L. Mesiti, and S. Glyde. 2006. The social and intellectual construction of farming styles: Testing dutch ideas in Australian agriculture. *Sociologia Ruralis* 46, no. 1: 61–82.

Walker, R., R. DeFries, C. Vera-Diaz, Y. Shimabukuro, and A. Venturieri. 2009. The expansion of intensive agriculture and ranching in Brazilian Amazonia. In *Amazonia and global change*, eds. M. Keller, J.C. Gash, and P.S. Dias, 61–81. Washington, DC: American Geophysical Union.

Weis, T. 2010. The accelerating biophysical contradictions of industrial capitalist agriculture. *Journal of Agrarian Change* 10, no. 3: 315–41.

Wield, D., J. Chataway, and M. Bolo. 2010. Issues in the political economy of agricultural biotechnology. *Journal of Agrarian Change* 10, no. 3: 342–66.

Wolford, W. 2008. Environmental justice and the construction of scale in Brazilian agriculture. *Society & Natural Resources* 21, no. 7: 641–55.

Woodhouse, P. 2010. Beyond industrial agriculture? Some questions about farm size, productivity and sustainability. *Journal of Agrarian Change* 10, no. 3: 437–53.

Zanon, R.S., M.S. Saes, L.J. Corrar, and M.A. Macedo. 2010. Produção de soja no Brasil: principais determinantes do tamanho das propriedades. *48° Congresso SOBER, Sociedade Brasileira de Economia, Administração e Sociologia Rural* . Campo Grande, MS, Brazil. 1–20.

Soybean agri-food systems dynamics and the diversity of farming styles on the agricultural frontier in Mato Grosso, Brazil

Mateo Mier y Terán Giménez Cacho

Soybean production in South America has become a symbol of commodity crops produced on a large scale for the agribusiness geared to global markets. However, in practice, there are diverse styles of farming, ranging from the very large scale to the small scale, associated with different social relations of production. This diversity of farm types is often overlooked within the focus on the stereotypical large-scale farm. Any understanding of the social dynamics of relations of production and farming practices, and so insights into longer term trajectories of agrarian change, must also go beyond a simplistic dichotomous vision of large-scale versus small-scale farming.

Introduction

Soybean production in South America has become a symbol of commodity crop production on large-scale, highly mechanized, labour saving, chemically based farms, associated with corporate agribusinesses geared to a global market. The political use of this symbol, by advocates of its further expansion as well as by most groups criticizing its negative environmental and socio-economic implications, has led to the construction of a homogenous stereotype of soybean producers; particularly for Mato Grosso (MT) state, Brazil, where the scales of production are among the largest in the world. Moreover, the implications of the rapid expansion of soy production over various ecosystems, including the Cerrado-savannah and Amazon Forest in Brazil, and the *Chaco* and Atlantic rainforests in Argentina, Paraguay and Bolivia (Gudynas 2006) are highly contested (Azevedo and Pasquis 2009). In the case of Brazil, soybean is associated by some with an economic development making the country an agricultural power and the granary of the world (Pinazza 2007; Economist Intelligence Unit 2010; Monteiro 2010). For others, the expansion of soybean production has implied multiple negative environmental and socio-economic consequences causing ecosystem devastation and the exacerbation of poverty and inequality (Shiki 2000; Fearnside 2001; Rohter 2003; Bertrand 2006; Nepstad, Stickler, and Almeida 2006; Schlesinger and Noronha 2006).

Yet the image used to depict soybean production – whether cast as positive or negative – of large monocrop soybean fields with a row of harvesters in a V formation portrays a particular view of soybean farming in Brazil. However, as this paper argues, a focus on ways of farming soybean shows that there is a diversity of styles of farming, ranging from the mega-large-scale to the small-scale, associated with different socio-technical, economic and ecological relations of production. Any understanding of the social dynamics of relations of production and farming practices, and so insights into the longer term trajectories of agrarian change, must also go beyond this simplistic dichotomous vision of large-scale versus small-scale farming or labour versus landowners (Azevedo and Pasquis 2009; Heredia, Palmeira, and Leite 2010), to a perspective that highlights the contrasts and interactions between a range of farm sizes and structures, each with different labour regimes, production systems and farming styles, situated in an evolving agrarian structure.

Through a detailed analysis of farming practices in Querência, MT, Brazil, this contribution argues that the predominant soybean narratives influencing policies are locked, partly by choice, partly by ignorance, in a narrow conception of existing soybean farming styles in Brazil. This paper argues that a more nuanced understanding is required, differentiating soy farming styles, and that this can contribute to more effective policies for sustainable development. The in-depth examination of five different farms highlights the contrasts in soy production, across a range of farm sizes from 100 ha to 10,000 ha.

The farm cases are explored through the lens of 'farming styles', defined following van der Ploeg (1994; 2008) as units of discourse and practice that express how farmers strategically combine and order components used in the process of agricultural production, and represent particular relations between the economic, social, political, ecological and technological dimensions. 'Taken together, they make up a richly chequered range that extends from different forms of peasant agriculture, via highly complex combinations, to different expressions of entrepreneurial agriculture' (van der Ploeg 2008, cited in van der Ploeg 2010, 5; Marsden et al. 2001; Schmitzberger et al. 2005; Van Averbeke and Mohamed 2006). The focus of analysis in this paper is the 'different expressions of entrepreneurial agriculture'. Studying farming styles enables a nuanced understanding of the relations of commodity producers with the global agri-food system. This contributes to a better understanding of the multi-level definition of farming practices. It is argued that the studied farming styles are influenced by migration patterns, labour regimes, technology, markets and organizations.

This research was motivated by previous research on the socio-economic implications of the introduction of the genetically modified Roundup Ready Soybean and governance dynamics through global value chains in Rio Grande do Sur, in southern Brazil (Mier y Terán 2008), a place where – in contrast to the general stereotype of soy farmers as large-scale farmers – different farming scales and styles were found. MT state was then selected for this research to find out if some of this diversity was replicated in what is considered the main state of large-scale farmers in Brazil (Schlesinger 2008), a migration destination for many farmers from the south.[1] After two months of scoping study on the eight main soy production municipalities in three different areas of MT, Querência was selected.[2] Actor-network mapping, field interviews and participant observation between September 2009 and September 2010, with a follow-up visit in July 2013,[3] provided the empirical data for this contribution. Within Querência, eight case studies were chosen for intensive study using a stratified random selection from a list of soy producers provided by the municipal Rural Union, cross-checked and complemented with a list provided by the

[1]Southern Brazil comprises Rio Grande do Sul, Santa Catarina and Paraná.
[2]Decisive factors for this choice were the high presence of land reform settlements and the location within the Amazon biome.
[3]In this visit, the research conclusions were presented for feedback from soy producers, and specific data presented here were rectified.

municipal secretary of agriculture and environment to assure the identification of small-scale soybean producers not registered in the Rural Union.[4] This followed a broad scoping survey across the whole area, combined with interviews on policy positions, production systems and marketing of soy. The studied farm cases were chosen following the analysis of secondary data on land holding size and distribution from the 1995/1996 Brazilian Institute of Geography and Statistics (IBGE) Census of Agriculture for MT (see next section).[5] Cases of soy producing farms within four size categories – mega large (>10,000 ha), large (1000–10,000), medium (100–1000), and small (<100 ha) – were chosen.[6] Data from five are presented here, excluding one mega-large and two small-scale cases to focus on the middle range.

The next section of the paper introduces the study area. This is followed by a brief overview of the main narratives influencing policy, demonstrating how these ignore patterns of social difference and diversity in agrarian structure. The following section introduces the farm cases studied and the factors identified as influencing agrarian diversity in the area. The conclusion follows, emphasizing how an appreciation of this variation in social relations of production challenges standard policy narratives, as soy can be incorporated in farm systems in dissimilar ways, with different implications for the political economy of agrarian change in the region.

The soy frontier: the Mato Grosso study area

Mato Grosso (MT) – nicknamed 'soylandia' by Hecht and Mann (2008) – is the Brazilian state representing the largest and most technologically intensive soybean plantations in the country. Since the harvest of 1999–2000, it has surpassed the production of the southern states Rio Grande do Sul and Paraná. By 2012/2013, MT had 7.8 million ha under soybean, representing 29 percent of national production (Conab 2013). The expansion of soybean has altered rural landscapes, forming new agro-industrial frontiers and transforming previously colonized areas (Klink and Moreira 2002; Hecht 2005; Brando et al. 2013). In many areas in MT, towns and municipalities were formed rapidly, including Querência, Sapezal, Sorriso and Canarana. These municipalities, with an uneven presence of the soybean agri-food system, form the soybean frontier in MT (Glass, Gomes, and Biondi 2011). Querência, presented here through farm cases, reflects broad similarities with other soybean municipalities: a predominant migration from southern Brazil, private colonization processes, a range of middle scale soy farmers, indigenous territories, land reform settlements (LRSs),[7] both rural workers' and *patronal*[8] unions, environmental initiatives on soybean production, and connectivity to trading and processing infrastructure (cf. Mier and Terán 2014). All of these factors have shaped the heterogeneity of farming styles found in these municipalities.

Querência is located in the Lower Araguaia region in the northeast of MT state. Its recent history has been influenced by the demarcation of the Xingu and Wawi indigenous

[4]Small-scale soybean producers are registered within the Rural Workers Union; however, in the list provided by this union, it was not possible to identify who was producing soybean.
[5]The newest (2006) Census of Agriculture was not yet available at the time the farm cases were chosen.
[6]Further information on the methodology used and an extended study of the eight cases are presented in Mier y Terán (2014).
[7]LRS refers here to the Brazilian government distributed land through land reform programs.
[8]Rural Employer – an official class category of farmers in Brazil.

territories, the land grab during the 1960s that led to the formation of large-scale speculative ranches, the private colonization started in 1985 by the Cooperativa Mista de Canarana (Joint Cooperative of Canarana, COOPERCANA) and more recently the formation of LRSs (Mier y Terán 2014). It was designated a municipality in 1991. As mentioned by other authors (Moreno 1993; Jepson, Brannstrom, and Filippi 2009), land use policies have been determinant in shaping the territories in MT. In the case of Querência, the history of colonization has created socio-cultural diversity, from indigenous people to multiple migrations from various states of Brazil (Sanches and Gasparini 2000).

Table 1 shows the agrarian structure and unequal land distribution in Querência. Thirteen percent of the properties have more than 1000 ha and occupy 90 percent of the area, including the Roncador farm with more than 150,000 ha, and the Tanguro Farm owned by Grupo Amaggi with more than 80,000 ha. In contrast, 65.6 percent of the properties that have less than 100 ha occupy only 3.4 percent of the area. However, the statistics also show the relevance of medium-scale farms with soybean production between 100 ha and 1000 ha. Although these occupy only 18.3 percent of the soybean planted area, they represent 61 percent of the soybean producers. If the farmers with soybean areas between 100 ha and 5000 ha are combined, these represent 83 percent of soybean producers and 51.4 percent of the soybean area plated in 2012/2013.[9] Both of these aspects, unequal land distribution and an important number of medium-scale producers, are also reflected in the aggregate statistics for the state of MT (cf. Castrillon Fernández 2007). This paper focuses on the producers characterized as medium- and large-scale, and so leaves out both mega-large scale and small-scale, extremes which have gained disproportionate attention in policy and research commentary.

The dominance of farms with soybean areas under 5000 ha coincides with the argument that '[t]he evolution of agriculture in the Brazilian Center-west frontier starting in the seventies was a movement of fractioning of large private farms, into smaller but still large farms' (Ferreira Filho and Vian 2013, 29). Certainly the farming styles over these scales, particularly the medium range, have a considerable influence in shaping the landscape and the socio-economic and political dynamics in the soybean municipalities. However, this does not negate the existence of mega-large scale farms that reproduce the historical land concentration in Brazil (Pádua 2004), nor the current trend to legitimize it (Wolford 2008; Sauer and Leite 2012; Oliveira 2013).

Querência has a fairly flat topography, with 80 percent of the area of the municipality considered seasonal evergreen forest (Rossete 2008, 59). It is in the transition region where the Amazon Forest and Cerrado-savannah intersect. This is an important fact when it comes to policy, since the biophysical uniqueness of this region – with differentiated ecological dynamics from those of the core Amazon and Cerrado – is not fully understood and it is often neglected (Rossete 2008; Marimon et al. 2014). It is a region characterized by an 'ecological tension', as the majority of grain and fibre production in MT occurs here (Faria et al. 2009, 19). The production of soybean was started by the first inhabitants of the COOPERCANA Querência Colonization Project in the late 1980s. However, it was in the 2000s that production increased rapidly, at an average rate of 33 percent annually, from 14,400 ha in 2000/2001 to 215,000 ha in 2010/2011 (IBGE 2012), while the initial increase in soybean area in Querência during the 1990s

[9]As a notion of large scale, 2000 ha up to 4000 ha were mentioned in interviews with different soybean producers in Querência as minimum areas for an economically viable soybean farming enterprise.

Table 1. Number of properties (2006) and soybean producers (2012/2013) by scale of production in Querência, Mato Grosso MT.

Total area of property	Number of properties and (%)[a]	Area (ha) and (%)[a]	Soybean planted area	Number of soy producers and (%)[b]	Soybean planted area (ha) and (%)[b]
Small-scale (<100 ha)	400 (65.5)	26,158 (3.4)	Small-scale (<100 ha)	20 (9.7)	1073 (0.4)
Medium-scale (100–1000 ha)	131 (21.5)	49,540 (6.5)	Medium-scale (100–1000 ha)	125 (61)	53,883 (18.3)
Large-scale (>1000 ha)	79 (13)	689,559 (90)	Large-scale (1000–5000 ha)	47 (22.9)	97,316 (33.1)
			Large-scale (5000–10,000 ha)	9 (4.4.)	64.850 (22.1)
			Mega-large scale (>10,000 ha)	4 (2)	76,800 (26.1)
Total	610 (100)	756,257 (100)	Total	205 (100)*	293,922 (100)**

Source: [a]IBGE-FAO (2006).
[b]Municipal Secretary of Agriculture and Environment-Querência (2012–2013).
Notes: *There are 226 producers registered in the source, but 21 are not included due to lack of data on soybean production.
**This corresponds to the total planted area of the available data of 205 producers.

was mostly from converting pasture. In the first half of the 2000s, although cattle production was still an important economic activity, soybean became the main driver of deforestation (Arvor et al. 2010).[10] The 313,240 deforested ha in 2000 rose to 510,000 ha in 2012; that is 17.5 percent to 28.5 percent of the total area of the municipality (INPE 2013). This put Querência on the government blacklist of municipalities with the highest deforestation rates from 2007 until 2011, when they reduced the rate of deforestation and enrolled at least 80 percent of the properties under the Rural Environmental Registry (CAR in Portuguese, MMA 2009). This land-use change would have been more dramatic if it was not for the demarcation of 41 percent of the municipality as indigenous territory, which is maintained largely as native forest by the indigenous population (Nepstad et al. 2006; Rossete 2008; Arvor et al. 2010; Schwartzman et al. 2013). Moreover, the deforestation trend slowed down considerably after 2006, mainly due to the government closing illegal sawmills, using police to identify infringers, and cutting bank loans (Brando et al. 2013). It was in this context that the Brazilian Forest Code (FC), that regulates land use in private properties, went through a controversial reform in 2012, and its impacts are yet to be assessed.[11]

While the initial establishment of soy farming took place under the COOPERCANA colonization project, the major expansion of soy production took place after 2000 when soybean multinational corporations and traders, such as Bunge, Cargill, ADM, Grupo Amaggi and Caramuru started to have a stronger presence in eastern MT (Jepson, Brannstrom, and Filippi 2009). Such investments used and expanded the earlier infrastructure developed by the colonizers, including the network of private input suppliers and grain silos. With the arrival of more capitalized farmers, including corporate investors, soybean became the predominant crop driving the local economy, geared towards exports. Alongside the direct economic activities around soybean in Querência town there has been an increase in the service sector, with the arrival by 2013 of banks, schools, supermarkets, a farmers' market, government institutions, local radio stations, a mobile phone company and other retail shops. This urban–rural centre, as in other soybean production municipalities, has become a political space where disputes about the pathways to sustainable agriculture take place and to a certain degree are shaped (Brondizio, Ostrom, and Young 2009). That is the case with the Y Ikatu Xingu Campaign, that gathered multiple stakeholders, including soybean producers, around the conservation of riparian forest with agroforestry principles (Durigan, Guerin, and Costa 2013), the CAR and the Forest Code, to mention a few.

Querência has transitioned from an agricultural frontier to an agroindustrial frontier, and recently to a municipality with corporate input providers and grain trading. The town has become pivotal for further expansion of the soybean agri-food system in the Araguaia region, particularly to the north of Querência, where historically there are more smaller scale producers (Barrozo 2007). Meanwhile the cattle ranching and new policies to

[10]Interviews with soybean producers, Querência-MT, 2010, and Rodrigo Junqueira from Instituto Socioambiental (ISA in Portuguese), Canarana, Brazil, 8 June 2010. See also the Kurtises' case below. The extent to which soybean will be the direct cause of future deforestation in Querência is contested (Brando et al. 2013).

[11]For the cases discussed in the time of research, the FC required landowners to leave 80 or 35 percent of the native vegetation on their rural properties as Legal Reserve (LR) and additional Areas of Permanent Preservation (APPs) to conserve riparian forest. Among other changes, the 2012 reform has exempted properties with up to four fiscal modules (360 ha in Querência) the restoring of native vegetation of LR, and has given amnesty to irregular use of protected areas until 22 July 2008 (Law 12,651 of 25 May 2012, cited in Crove and Clemente 2014; see also Soares-Filho et al. 2014).

promote biofuel from soy link the municipality with Amazon-wide dynamics (Walker et al. 2009; Walker 2011). The town is also host to a variety of organizations linked to soybean production, from producer groupings to input providers. The configuration of these is attuned to the structure of production and its politics, with the most powerful actors associated with large and mega-large production. Indeed the rural workers' union in the municipality was formed in 1991, collapsed in 1999 and was only reformed in 2006, with the increasing presence of land reform settlers, led by members of the National Confederation of Agricultural Workers (CONTAG in Portuguese). Moreover, the lack of organizations specific to a range of underrepresented medium-scale farmers is also evident. Although the soybean farming styles in Querência are embedded in the global dynamics of the soybean agri-food system, leading to a homogenization of agricultural practices and a concentration of wealth, the differences in styles are relevant to understanding the local agricultural and agrarian dynamics.

Policy narratives: framing soy

In Brazil, several narratives frame policy debates around the development and sustainability of soybean agri-food systems. These derive from a variety of different sources, both within Brazil and internationally. The most influential ones focus on the consequences of large-scale – often mega-large – mechanized soy production, or simply 'agribusiness'. The discursive antagonism between these narratives is paralleled by the existence of two distinct ministries dealing with agriculture: the Ministry of Agriculture, Livestock and Food Supply (MAPA in Portuguese) and the Ministry of Agrarian Development (MDA in Portuguese). This has led to an institutional differentiation of agricultural producers and traders according to their links to two broad categories, agribusiness and family farming. However, these are disputed categories and do not delineate policies precisely (Medeiros 2001; Heredia, Palmeira, and Leite 2010).

The first, the agribusiness narrative, is promulgated by leading firms and by the national and local state. It highlights the successes of transforming what was deemed unproductive beef ranching savannah and forest land into high-value export agriculture. In the words of Carlo Lovatelli (2009, 8), president of the Brazilian Vegetable Oil Industry Association (ABIOVE in Portuguese), 'Brazil has ample resources in arable land suitable for food production. When you factor in the climate, sunlight, rainfall and entrepreneurial spirit of the tropical farmer, success is inevitable'. For the Chief Executive Officer (CEO) of the Mato Grosso Soy and Corn Growers Association (APROSOJA), Marcelo Monteiro (2010, 12) 'Brazil's agriculture is considered a reference and an example on how to feed the world […] Brazilian soybean production will continue to be competitive and will certainly contribute to meet the world demand increase'. These statements illustrate how soy is seen as a symbol of Brazil's wider economic transformation, and how agribusiness is viewed as the basis for economic growth and global penetration of agricultural markets.

The second, the environmental narrative, focuses on socio-environmental consequences of soy expansion. This narrative is promoted by environmental organizations and some parts of the Brazilian state, including those associated with agrarian reform, environmental protection and climate change. It focuses on the destruction of natural habitats, particularly the Amazon Forest (Fearnside 2001; Klink and Machado 2005; Greenpeace 2006; The Dutch Soy Coalition 2006). Certainly, forest loss due to agricultural expansion has been massive in Brazil including in MT, but there are often particular stereotypes that are deployed by this narrative that do not match realities. These include the construction of 'the Amazon' and the 'rainforest', both global signifiers for environmental protection, particularly in combating climate change. With Brazil keen to preserve its image not only as a major agribusiness power, but also as a country concerned with environmental protection – not least around the Rio+20 conference in 2012 – major attempts to offset environmental destruction have been instituted, including strict controls on forest removal and requirements for reserving forest areas on farms (DeFries et al. 2013).

These two narratives have given place to efforts to generate more effective environmental management and responsible business investment regimes, around a third narrative centred on 'responsible', 'sustainable' soy production (Dros 2004; van Berkum and Bindraban 2008). Multi-stakeholder groupings, such as the Round Table on Responsible Soy (RTRS) and the Soybean Moratorium, are intended to generate a strong, inclusive narrative arguing that large-scale soy production can be good for both business and the environment if certain safeguards are applied (WWF 2004; The Nature Conservancy 2012). Such efforts, involving some of the major soy companies operating in Mato Grosso as well as the Brazilian state and mainstream environmental non-governmental organizations (NGOs), again concentrate on large-scale soy farming, and try to show the more positive sides of soy expansion.

Whether coming from an agribusiness standpoint or from environmental campaigners, or a 'responsible sustainability' compromise between the two, these predominant narratives focus on the consequences of the expansion of large-scale soybean farming. This is unquestionably an important phenomenon, as the data above for Querência show. However, 61 percent of soybean producers still cultivate 18.3 percent of the land under soy areas between 100 ha and 1000 ha. In the context of Querência, these are categorized as small- and medium-scale farms, and it is commonly mentioned that 'less than 500 ha is not viable'.[12] What happens on these farms, and their impact on production, labour and environment, is barely considered in these predominant narratives that have latched on to an 'agribusiness promise' or 'environmental disaster' style of narrative, or increasingly a 'sustainable compromise' version that tries to balance the two.[13]

The large to medium farming group, excluded from the most influential narratives, represents around a third of farms in Querência, cultivating around half of the soy area (also in other soybean production municipalities in MT; Castrillon Fernández 2007). While highly diverse, as the case studies introduced in the next section show, these are poorly understood soy production styles, with little space in the policy debate which has evolved largely around very large-scale production, or the counter narrative around small-scale agroecological or family farming alternatives incompatible with industrial soy production. The lack of advocacy groups, governmental representation and forms of organization for this 'hidden' group of soy producers is striking. The following section offers an insight into five case studies of farms with soybean areas between 100 ha and 4000 ha.

Case studies: different farming styles

The following five case studies explore the farming styles of medium to large farms.[14] These summaries are based on in-depth analysis of each farming operation, involving

[12]Interview with soybean producer in Querência, 18 February 2010.

[13]Another narrative focuses on small-scale production, and an agroecological alternative to corporate farming (Mier y Terán 2014). As Table 1 shows, 65 percent of producers own less than 100 ha, although they contribute only 0.4 percent of the farmed soy area. This paper does not focus on this group, nor on the agroecological alternative narrative, and instead concentrates on the large- and medium-scale farming range.

[14]Names used for the farmers are pseudonyms.

Table 2. Characteristics of soybean farm cases, Querência, 2010.

Characteristics	Range for selection (ha soybean area)				
	(1000–10,000)		(100–1000)		(<100)
Family name	Kurtis	Oshemback	Durero	Graciano	Prestig
Age (him, her)	55, 48	46, 37	46, 42	46, 38	36, 35
Arrival in Querência	2001	1999	1988	1986	1999
Area of soybean planted in 2010 (ha)	3400	1400	835	160	90
Forest area owned (ha)	23,400	4800	133.9	20	24
Total land owned (ha)	30,000	7000	372	180	60
Other crops	Maize	Rice, and fruit and vegetable garden	Palm tree, teak, and fruit and vegetable garden	Sorghum for no-till agriculture	Diversified
Cattle and area of pasture in 2010	5000 heads on 4500 ha	5 heads on 800 ha	Few for self-consumption	none	50–60 on 120 ha
No. of permanent and temporary workers	18–22	6	5	1	None
Equipment owned	Eight large-machines	Six large machines	Four large machines	Three medium machines	Three medium (old) machines
Sells to	Soybean to Louis Dreyfus; maize to local traders.	Soybean to *Condominio*, Bunge, and Louis Dreyfus; rice to local trader	Soybean to *Condominio,* Cargill, and Bunge; palmito to Luana agroindustry	Soybean to ADM and local trader	Soybean to Cargill, and Caramurú; rice to local trader
Union membership	*Patronal* – Rural Union	*Patronal* – Rural Union	*Patronal* – Rural Union	Rural Workers' Union	Rural Workers' Union

extended visits and discussions with those involved, including input vendors, extension agents, soy traders, union leaders, politicians and NGO members. These farming cases are of course not representative in any statistical sense, but offer an insight into the range of experiences and contexts in this 'middle' range of soybean farming operations. The section ranges from the largest of the five to the smallest, highlighting differences in agronomic practices, farm organization, labour regimes and interactions with input suppliers and service providers, and through the analysis of life histories shows how the different farms are located in patterns of migration and colonization. Their main characteristics are presented in Table 2.

Igor Kurtis (b. 1955) and Carol Kurtis (b. 1962)

Igor Kurtis migrated from the US to Paraná, Brazil in 1973 at age 18 to work on a farm. He later married Carol from Goias and they had five children. They acquired a farm and built up their wealth with support from Igor's family, who also farm soy in the US. In 2000, Igor, Carol and three of their children bought a large farm in Querência. They sold their well-established farm with a silo and 3000 ha of soybean in Goias to buy a much bigger area. By 2010, they had cleared 26 percent of the 10,000 ha of forest they had acquired, and produced 3400 ha of soy, including 800 ha on a neighbour's plot. They also acquired a large cattle ranch (20,000 ha) in the neighbouring municipality, where they had 5000 heads on 4500 ha of pasture and 14,500 ha of forest. According to Igor, 'we plant maize after soybean in the same season to use the long rains and intensify production'.[15] During the soybean harvest season, they had 12 temporary and nine permanent workers who would move between the soybean farm and the cattle ranch. The soybean farm was managed with two planters, three combine harvesters, a new elevated sprayer, two tractors, a grain cart, a motorcycle for the manager, a truck and other equipment. The farm also had a 500-ton silo with additional infrastructure to dry the harvest and store it to the standards required by the processing industry. It is precisely the silo that gives the Kurtises some leeway to capture more value from the soy while negotiating the price with the corporate traders – including the capacity to sell soy outside Querência for better deals.

In terms of farming styles, the Kurtises represent a large-scale, capitalized farming operation, linked closely with international corporations and reliant on them for finance and marketing. However, they differentiate themselves from the large corporate players in important ways. Coming as colonists, they see themselves as part of the wider group of migrants, mostly from southern Brazil, carving out a new agricultural frontier. The Kurtises' trajectory is one of risk-taking by investing in land in regions where it 'was still cheap'.[16] This is an example of the increase in soybean production through land expansion, the characteristic pattern during the 2000s. The Kurtises' soybean farm is located at the agricultural frontier of the municipality. Most neighbouring plots got converted from native forest into soybean *fazendas* (farms) in that decade, leaving aside the compulsory forest and riparian reserves). Other plots are still large private forest holdings awaiting land use change. 'I wanted to plant 50 percent with soybean, but now I have to reforest 6 percent to have the 80 percent of forest legal reserve', Igor explained.[17] The style of farming practised, in contrast with smaller scale production, involves a labour structure

[15]Interview with Igor, Querência-MT, 9 February 2010.
[16]Interview with Igor, 2010.
[17]Interview with Igor, 2010.

with a farm manager, specialized division of labour and working time recording. They intensify their production with short-cycle soybean varieties and maize for a double harvest in a year, a recent approach in the region to intensify production. Moreover, they have an extended network of relationships with input providers that starts in Querência, as well as access to financial resources with international banks, such as Rabobank, the machinery corporation John Deere and the grain multinational traders Louis Dreyfus, Bunge and Cargill, a critical source of financing increases on scales of production.

Antonio Oshemback (b. 1964) and Lumina Oshemback (b. 1973)

Antonio and Lumina arrived in Querência together in 1999, following two of Antonio's brothers. They moved from Rio Grande do Sul to MT, knowing that they wanted to set up a soybean farm. They had been soybean producers in the south since 1992. He remembers, 'soybean yields were around 32–35 sacks per hectare; now the average is 60 sacks per hectare […] in the south we did no-tillage for soybeans using wheat, here we use millet'.[18] When migrating to Querência, they aimed to reproduce their lifestyle as soybean producers: 'we arrived with plans to plant soybeans', Antonio said.[19] With the help of their parents they bought two plots: a 10-hectare *chácara*[20] where they live at the edge of the town, and 3000 ha of forest 12 km from the city. The size of their *chácara* allows them to have cows, an orchard and a vegetable garden, giving them a high degree of self-sufficiency for most of the year, including the food for two permanent workers hired to produce soybeans. Their 3000 ha farm is, in Antonio's words, a '*fazenda caprichada*' (well organized farm),[21] with 1300 ha of soybeans managed with large machinery, no-till agriculture, preserving the riparian area and forest reserve as required by the Forest Code.[22] In 2008, they bought in the neighbouring municipality of Riberão Cascalheira a second farm, 4000 ha located on the former 90,000-ha *fazenda* Maria Teresa.[23] Antonio considered that their only child, a 14-year-old boy, would 'have somewhere to come back to when he finishes technical [agronomic] school' and would manage the property.[24] By 2010, between their two farms, Antonio and Lumina farmed a total of 1400 ha of soybeans, but were planning to expand this further into their new 4000 ha of degraded pasture and forest lands.

The Oshembacks represent farmers who have been able to accumulate capital through soybean production and increase their scale of production by buying a farm within Querência. Their livelihood strategy is to adapt to socio-political and economic pressures to become entrepreneurial farmers. Moreover, their case illustrates the blurring of boundaries between small-scale producers and large-scale producers. Antonio considers himself a medium-scale producer, and Lumina very much associates with the small-scale production she undertakes at their *chácara*. They do not have their own silo but in 2010 acquired a

[18]One sack equals 60 kg of grain.
[19]Interview with Antonio Oshemback, 18 February2010, Querência-MT.
[20]The *chácaras* date back to the private colonization project by the cooperative COOPERCANA, planned as small plots in order to have a belt of small-scale farmers around the urban centre that would supply food. Interview with Querência's urban planner Edio Schwantes, Xavantina-MT, 4 May 2010.
[21]Interview with Antonio, 2010.
[22]According to Antonio, their farm was located within the *Cerrado* and had been registered to be able to deforest 60 percent of the land. Interview with Antonio Oshemback, 20 July 2013, Querência-MT.
[23]A *Fazenda* which had been squatted by large-scale and small-scale farmers and where the situation was not yet legally recognized.
[24]Interview with Antonio Oshemback, 20 July 2013.

second harvester, a new John Deere that 'collects 55 sacks [3300 kilos] of soybeans per minute, just about a hectare per minute', Antonio's worker said.[25] However, Antonio thinks that with the amount of soybeans they produce on 1400 ha and the land on the new farm, they are moving towards becoming large-scale producers. This has brought other changes. One is a change in the administration of the farm, an activity he considers a burden. This means for him that 'they [media, banks and government] want us to say that we are not farmers, but rural entrepreneurs'; he adds, 'I'm managing the farm as a company: the registration is like a company'.[26] Their house has an office with a computer, where Lumina takes care of the accounting and other paperwork of the business, in addition to milking cows, tending the vegetable garden, making cheese and cooking.. In many respects, their farming operation is in transition, and they are also torn between the need to expand as a commercial business, while also wishing to continue their association with the land as practising farmers.

Hector Durero (b. 1964) and Rocio Durero (b. 1968)

Hector and Rocio Durero were among the first people to come to the Querência colonization project started by COOPERCANA in 1985. They arrived in 1987 from Santa Catarina, in southern Brazil, with one of Hector's brothers, Omar, and his family, to work on 750 ha of land that their father had bought. Hector's and Omar's families lived on a shared farm for seven years. Afterwards, Hector, Rocio and their two teenage children moved to the city, 'mainly to be closer to the schools', Rocio commented.[27] Today, having split the property between families, Hector and Rocio's farm is 372 ha, where they still have the homestead of the farm. In addition, they rent a neighbouring plot of 400 ha and another one of 200 ha a few kilometres away. By 2010, they had planted a total of 820 ha of soybeans. Also on their own farm, they had eight ha of *popunha* (peach palm, *Bactris gasipaes*), a vegetable garden and cattle for self-provision, and – next to their required legal reserve and riparian forest – a plantation of trees including neem, eucalyptus, clove and cinnamon. Hector has four workers: three permanent ones that live on the *fazenda*, including a farm manager and a cook. They also have a temporary worker who comes every year for the soybean harvest months before moving to São Paulo for the orange harvest. They have all worked on the farm for eight years and come from Santa Catarina.

The Dureros' farming style does not differ greatly from that of a corporate farm in terms of the use of machinery and improved seeds. However, the scale and relations of production differentiate their farming style. Their network of input providers is bound to Querência, a relatively locally focused network of economic interactions. In the case of purchasing agro-chemicals and the decisions regarding applying *defensivos* (herbicides and pesticides), they have an arrangement with the intermediary SinAgro, a regional agricultural inputs provider with a semi-outsourced service. As Hector explained, 'it is more expensive, but it is a fixed price, a percentage of the harvest', so the company also has a stake in the productivity. This creates a particular socio-technical network that involves sharing responsibilities between producers and input suppliers. The local character of their network is also reflected in their relations with soybean traders, as they have been part of creating an association of 40–70 producers who have owned a silo collectively since 1993. By 2013, it had

[25] Interview with rural worker Raphael, 18 February 2010, Querência-MT.
[26] Interview with Antonio Oshemback, 20 July 2013.
[27] Interview with Rocio Durero, 19 May 2010, Querência-MT.

become a cooperative, doubling the silo capacity and starting to purchase fertilizers for its members. In 2010, limited by the association capacity, the Dureros stored and sold through it 10–12 percent of their harvest, and sold the rest through corporate traders such as Cargill. The history of this association represents a tendency that is rarely depicted in the standard narratives of the large soy producers, and is crucial to understand medium-scale farmers' strategies.

Lorenzo Graciano (b. 1964) and Monic Graciano (b. 1972)

Lorenzo and Monic migrated separately from the south of Brazil to Querência, where they met. She arrived in 1986 with her parents, and he visited his brother in 1990 and stayed. By 2010, they were living with their two children, 13 and 15 years old, in a 70-m^2 flat in Querência town. The case of the Gracianos differs from all other cases in that their livelihoods are not completely based on farming. She is a teacher and he works in a government institution in charge of supervising livestock breeding. The production of soy is a side activity that Lorenzo started in 2002; in his words, 'I realised I had some savings and could invest it in getting land to plant soybeans'.[28] By 2009/2010 he had 160 ha of soybeans on three plots at a LRS. They farmed with a temporary worker, two old tractors and a harvester bought second-hand from his brother, using the usual package of high yield seeds, fertilizers and agrochemicals, managing the soil with no-till farming and precision agriculture. He obtained soybean planting services from the same providers as other larger and middle-scale soybean producers in Querência. This all situates Lorenzo as a capitalized farmer, even though he was planting an area commonly considered, by agronomists and soy producers of the area, too small to be economically viable.[29] His accumulation of capital derived partly from his salary, but also because he accessed a five-year loan from *Banco do Brasil* (State National Bank) for *custeio* (production costs), and a federal government loan exclusively for family farmers at subsidized rates from the National Program to Strengthen Family Farming (PRONAF-*custeio*),[30] which he used to pay for the limestone soil acidity correction.[31]

Lorenzo's soybean farming practices do not differ radically from other middle and large-scale producers, as he uses technology similarly. However, details indicate differences in practice that shape and are shaped by the scale of production and the degree of capital used. For example, he does not apply fertilizers according to precision agriculture due to the high cost of the service and machinery; and he plants long-cycle varieties (140 days) instead of early maturing varieties (90 days), given that the latter 'are less productive',[32] are used to manage different harvesting times in larger areas and are generally used to introduce maize as second crop – a recent intensification practice – instead of the sorghum used for no-till agriculture. Moreover, the employment conditions at Lorenzo's

[28]Interview with Lorenzo, 15 February 2010, Querência.

[29]Although there are different opinions among producers, 500 ha is often mentioned as the minimum planted area to have an economically viable soybean farm in Querência-MT.

[30]Brazilian legislation (Decree no. 84.685 of 6 May 1980) indicates that for Querência, the limit of four modules that farmers can have to be classified as family farming can be of 80 ha each (maximum 320 ha total).

[31]These loans were registered for use on another plot outside the LRS. This allowed him to avoid the government environmental embargo on public funding for LRS in Querência, meant to enforce compliance with the environmental licensing and restrictions on deforestation (ISA and ICV 2006).

[32]Interview with Lorenzo, 2010.

farm neglect health and safety regulations: for example, a hammock for temporary sleeping in the same wooden shack as the diesel barrels. Unlike most soybean farmers in Querência, Lorenzo is a member of the Rural Workers Union – rather than the Rural Union – and, as mentioned, accesses public credit directed to small-scale family farming. While he is officially registered as belonging to the category of *Agricultura Familiar* (family farming), and his volume of production is considered small-scale in the context of Querência, his socio-economic status is that of a medium-scale soybean producer and his social relations are with large-scale soybean producers, including his brother. The farming styles expressed in the Gracianos' case differ from that of families who actually live on the LRS, as in the next case.

Fernando Frank Prestig (b. 1974), Camila Prestig (b. 1975) and family

Fernando, Camila and their extended family migrated to MT from southern Brazil. Fernando moved in 1999 to work as a labourer on an 18000-ha farm, where 'the forest needed to be cleared' for soybean production.[33] By 2005, they cultivated 2500 ha of soybean. Fernando became the farm manager, and his family, including his wife Camila, his brother and parents, joined him. In 2005, the whole family decided to move to the neighbouring LRS, where they had acquired plots for each family – one of 60 ha, three of 90 ha and one of 120 ha. The core activity on their plots, according to Fernando's brother Uber, 'is cattle breeding, [and] we all have productive fruit and vegetable gardens'.[34] The parents take care of other livestock on their plot, including chickens, pigs, goats and dairy cows, which feed all the family members. By 2013, Fernando and his family were also planting 200 ha of soybean and about 50 ha of rice distributed on more than eight rented plots. All these plots reserved the required forest area.

Fernando and Camila's farming style is one of diversified production, sustained by strong family ties that make their farming enterprise a collective endeavour. Their life histories reflect a trajectory from rural workers to land reform settlers and soybean farmers, one that is not acknowledged in the narratives of expansion of soybean production. Fernando and Camila's farming shows how soybean production can increase via mechanized small-scale production, including within the LRSs. Fernando mentioned that 'there are those who say that settlements are not meant to have soybean, but I don't know if it is politics or what'.[35] In agreement, Uber expressed that 'soybeans bring progress. I don't know why the government is trying to stop it'.[36] Fernando was not sure if there was a formal prohibition on producing soybean in the LRS; nevertheless, despite this uncertainty he wanted to plant more.[37]

Explaining agrarian diversity

These five case study soy farms are broadly representative of farms in the large- to medium-scale range found in Querência. Each incorporates soy into their farming operations and

[33]Interview with Fernando Prestig, 26 May 2010, LRS, Querência-MT.
[34]Interview with Uber Prestig, 25 May 2010, LRS, Querência-MT.
[35]Interview with Fernando, 2010.
[36]Interview with Uber, 2010.
[37]There is no legal restriction on planting soybean in Querência's LRS, but policies geared towards LRS do not encourage it. On the contrary, within the narrative of the Ministry of Rural Development (MDA), soybean production is often framed as a crop for agribusiness, as opposed to family farming.

livelihoods in different ways. Between them there are huge differences, and a clear pattern of differentiation is observed. This section asks: what are the factors that explain this pattern of agrarian diversity, and the variation in farming styles seen? And, in turn: why is it important to take notice of the dynamics of change on and between farms between the mega-large and the small scale? Five features are suggested as critical to understand the diversity: the migration histories of the farmers, the farm labour regimes, the types of technology used, the markets engaged with and the organizations within which farmers operate.

As all the cases show, the history of expansion of soybean in Brazil is associated with the migration of farmers from southern states. These farmers had previous knowledge of the use of machinery, agrochemicals, improved seeds and markets, particularly since their home areas already produced soybean in a mechanized way (Castrillon Fernández 2007; Ferreira Filho and Vian 2013). However, the heterogeneity within this group is often overlooked. First, there were large differences in access to capital and in farming knowledge and experience at their time of arrival to Querência. These aspects are certainly present in most areas in MT where soybean became a predominant crop in the past decades (Jepson 2003; Castrillon Fernández 2007). The first affected their farming strategies in relation to access to land, crops produced and the scale of production in which they engaged on arrival. The second refers to the micro-practices and knowledge they brought with them of food production for self-consumption and for making processed products, such as bread, cheese and cold meats – products that in the southern states are today associated with a niche market of family farmers (Schneider and Niederle 2010). Some arrived with limited capital and had to move between jobs and farms before settling to establish their soy farms; others arrived with capital and were immediately able to acquire large areas of land. Some started out as workers and only transitioned to farming soy through assistance from the land reform programme, or through the accumulation of assets from farm work, starting out with small plots, and expanding by renting, leasing and purchasing (Mier y Terán 2014). All of them combine soy production with other activities, including livestock ranching, but also small-scale vegetable gardening and dairy, often with family members. Despite the size of some farms, the styles of farming often combine large and small-scale features, and there are no cases that conform in all aspects to the standard image of the large-scale monoculture of soy.

Soy farming evolved in the area through the process of migration and colonization (Jepson 2003). Many early migrants did not have large capital and had to clear land with limited resources (Falabretti 2010). Only later did the richer, capitalized farmers arrive.[38] The role of soy, as part of wider farming systems and crop mixes, changed too. Many of the early migrants did not consider soybean central in their farming strategies, and those who started to plant soybean by the early 1990s also had livestock and vegetable gardens. COOPERCANA promoted crop diversification such as the native fruit tree *piqui* (*Caryocar brasiliense*), rubber trees and apiculture. Today all these activities, even if relatively marginal, form part of the diversification alternatives of small- and medium-scale farmers in Querência. It was the more capitalized farmers, as in the case of the Kurtises, who arrived later and created larger farms centred exclusively around soybean production. These farmers benefited from the emergence of a more dynamic private land market and corporate investment in the region, after the dismantling of COOPERCANA in 1994. This led to the fragmentation of large properties that were previously used for

[38] A similar process has been noticed in the neighbouring state of Rondonia (cf. Browder et al. 2008).

extensive cattle ranching or forest. In sum, the contrasting levels of capital and difference in time of arrival shaped the way soybean has been used, and its uneven consequences.

As the case studies show, increases in the scale of production involve the employment of more workers per farm and different labour management, in particular the division of labour and specialization. Yet in many of the cases, including that of large, mechanized farms, there are relatively few workers and the farm owner's family takes an active role in farming. This includes daily visits to farm or supervise farming, cooking and doing the farm accounts, often with clear gender divisions of labour. Indeed, the persistence of aspects of family farming reflects a resistance to the pressures to move to a 'business' and 'managerial' model. In the words of Antonio Oshemback, 'we are family farmers'.[39] Although they do not fit the official Brazilian, *Agricultura Familiar* category, as they employ permanent workers, in his view, the farm is run around and by the family. It is also representative that, unlike other producers on a similar scale, the Kurtises – who have reached an enormous land accumulation of 30,000 ha, with 3400 ha of soy in 2010 (see Table 2) – go daily to the *fazenda*, supervise production decisions and manage the labour and accounting between husband and wife, rather than settling for a more corporate structure that is decoupled from the family and is managed from a distance with a fully specialized structure of employees.

Labour laws and health and safety regulations are changing the labour regimes and affecting farmers in different ways. The vigilance of workers' rights and inspection of compliance of regulations has become a pressure to improve working conditions towards a managerial corporate model (Nepstad and Stickler 2008). In this process, non-corporate soybean farmers have been unwilling and/or unable to comply with what often are seen by soybean producers as absurd legislative interventions (cf. Azevedo and Pasquis 2009). For example, employers resent the requirement to restrict rural workers to specialized jobs, preventing flexibility to deploy them on other activities outside the main farming season. Hector Durero considers that 'the government obstructs, […] it over-complicates. Imagine we can be fined because a door is a few centimetres narrower than what is officially required.'[40] Most soy producers recognize the importance of regulations (Azevedo and Pasquis 2009), such as in the use of agrochemicals and in relation to salaries and working conditions, yet these farmers point to the regulations as being driven by a concern with the large, corporate farms, undermining their ability to respond flexibly.[41]

The same applies to environmental legislation. Again, legitimate concerns about deforestation have resulted in strict limits on the percentage of land that can be cleared. For a very large farm, retaining 80 percent as forest still allows soy production over thousands of hectares. But such a restriction on a medium-scale farm may have a major impact on the economics of production.[42] Lorenzo Graciano, with a critical view of environmental pressures, asked rhetorically: 'tell me which other country has such restrictive laws? Where else is riparian forest so well preserved? And how can a farmer make a living from 20 percent of the farm?'[43] He argues that 'people in the cities don't think about where the food comes from […] they want us to leave the forest standing but they want to continue eating high-quality steaks'.[44] Lorenzo's vision differs from that

[39]Interview with Antonio Oshemback, 18 February 2010, Querência-MT.
[40]Interview with Hector Durero, 20May 2010, Querência-MT.
[41]This was raised as a collective concern at a soybean producers meeting at APROSOJA's offices in Querência, 2013.
[42]These regulatory pressures have reduced with the 2012 FC reforms (see note 11).
[43]Interview with Lorenzo, 2010.
[44]Interview with Lorenzo, 2010.

of soybean corporations that recently adopted a socially and environmentally responsible narrative and are pursuing the compensation for conservation in the carbon credit market. This suggests that initiatives promoting sustainable soybean agriculture may have to consider differences in soy farming styles and be more inclusive rather than replicate the existent wealth concentration patterns.

Soybean farming at medium and large scales is often assumed to be very similar in terms of its technological and market characteristics. Roessing and Lazzarotto (2004, 36) comment that 'the differences between the technological levels implemented in properties of 1000 ha and of 10,000 ha are minimal'. Yet the cases show that there is enormous variation, linked to particular technological–market trajectories of intensification. These cases are indicative of four potential trends: first is the rapid expansion and high-capital intensification seen in the Kurtises' farm; second is the careful expansion and intensification with integration of crop–livestock production seen with the Oshembacks; third is a pattern of moderate diversification as with the Dureros; and fourth is the incremental expansion and multifunctional diversification, as with the Prestigs.[45] Certainly, soy production has a high degree of standardization. But are farming practices and farming styles to be defined only by the soy plot? Or are all the practices taking place on the farms – including forest preservation, fixing old tractors, and producing and selling other crops – also relevant to understand farming practices and the related agricultural and agrarian dynamics?

In general terms, the farms with more than 100 ha of soybean practise no-till farming or semi-till farming, and use multiple seed varieties – including conventional and transgenic seeds, with different growth cycles. They use a combination of machinery – owned, shared or hired – aspects of precision agriculture, and a fairly standard suite of agrochemicals. However, there are important differences in the practices of intensification. Larger-scale farmers, like the Kurtises, who started with high levels of capital and increased their production quickly, focus on increasing the productivity of a monocrop, twice a year, with highly mechanized operations. They are for the time being on the verge of consolidating as an agribusiness, but for a range of reasons (notably difficulties with loan finance), they have found it difficult to maintain. Hence, they retain characteristics of a more family-oriented operation, particularly in the use of family labour and scale-enlargement strategies that include considering property division for inheritance.

In contrast, large-scale and medium-scale farming operations, such as those of the Oshembacks and Dureros, have taken a precautionary approach to technology, and increase their scale of production and capital accumulation more slowly. These different practices derive from their farming trajectories and may be a base for a different approach to land use intensification (cf. Nepstad and Stickler 2008; Arvor et al. 2010). These farmers have combined a family economy based on resources from the land (e.g. food production for household consumption) and a cautious involvement in the commodity market based on external inputs (e.g. a gradual increase of soybean-planted area). Differences in their life histories have contributed to variations in their intensification strategies. While the Oshembacks bought more land – a cattle ranch with degraded pasture – and are adopting integrated crop-livestock, production, the Dureros – after the division of their farm between brothers due to a divorce – have rented degraded land, are improving the soil fertility with soybean and have diversified their production with 10 ha of peach palm to be sold to a regional agroindustry.

[45]These are not fully distinct trajectories, as there are aspects in common, but the underlined differences are relevant for the way the socio-economic and ecological landscape is being shaped.

In the case of farms with soybean areas around 100 ha, where the farmers live on the land, such as the Prestigs, the norm is that soybean is part of diversified production. The limited access to capital shapes farming practices. They must rely on old machinery, and must reduce the cost of production with non-recommended practices, such as the use of stronger and cheaper agrochemicals, and the elimination of planting sorghum for no-till agriculture. Given that the production of soybean is not only a cause of displacement of small-scale farming (Barrozo 2007) but is also increasing in family farms in Querência and neighbouring municipalities,[46] it is relevant to recognize the different relation to technology, and the multifunctional use of soybean, e.g. as a local currency, to feed livestock, and for recuperation of degraded soil.

All soy produced in Querência is exported by corporate traders.[47] However, most farmers in Querência rely on local input and service providers and grain traders to varying degrees. More capitalized large-scale farmers, like the Kurtises, tend to extend their networks to other municipalities, while other, slightly less capitalized farmers rely on local providers and share services, such as agronomic advice or the collective ownership of a silo. Farmers with smaller scale production, like the Prestigs, have reduced interaction with local actors and struggle to receive equal treatment, for example in prices of inputs or of soy. These differentiated relations with input providers and traders not only reproduce disadvantages for small-scale producers (Gudynas 2006; The Dutch Soy Coalition 2006), but also form a dense network of relationships which differentiate middle-scale from corporate farmers.

The institutional framework for soy farming – in terms of input supply, advice and marketing – is heavily influenced by the dominant corporate actors. In this context, paraphrasing Wolford (2008, 651), increasingly, bigger is better; a single notion of 'large-scale agriculture is privileged in the Brazilian Cerrado [and soybean agricultural frontiers] by this belief that large[r] scale is "naturally" more efficient, more productive and more appropriate for the country than small[er] scale'. While those with large operations like the Kurtises are able to benefit from this, others compete and lose. Access to inputs, services and markets are shaped for ever-larger scales, and alternative arrangements, including cooperative and sharing institutions, have to be worked out.

Brazil's dichotomous division of agriculture institutions, policies and rural unions between large-scale corporate agribusiness and small-scale 'family farming' or rural workers is historically embedded in Brazil's dualistic structure of agriculture and regional differences (Goodman, Sorj, and Wilkinson 1985; Medeiros 2001; Barrozo 2007; Schneider, Shiki, and Belik 2010). However, within Querência, this has the consequence that medium- and many large-scale farmers are not taken account of, and so have no distinct voice in policy debates. Patterns of differentiation within the fast-evolving soy farming areas are such that these classic divisions, as reflected in forms of organization in unions, producer groups, and indeed political formations, do not represent the diverse class interests of those in the 'middle' group of farmers focused on in this paper.

---

[46]Confirmed in an interview with Eleandro Mariani Ribeiro, former president of the rural workers union (2008–2011), and municipal secretary of agriculture and environment (2012–present), 18 July 2013, Querência-MT. He mentioned the arrival of a biodiesel company established in the region that was signing contracts with small-scale farmers. I was also able to talk with land reform settlers that entered the contract scheme (see also Bernardes and Aracri 2011).

[47]Projects to increase local use are marginal, since there are no processing industries to convert it to livestock feed, and the few direct uses, such as human soymilk consumption or small-scale processing by farmers, are still minor.

Conclusion

The policy narratives that dominate the debate about soy farming in Brazil focus on very large corporate agribusiness, whether as critiques from environmental or social standpoints or as a celebration of Brazil as a global agricultural power. Such narratives mask the complexity and agrarian diversity observed in soy farming areas, such as Querência in Mato Grosso state. This paper focuses on large- and medium-scale farms cultivating between 100 and 5000 ha of soy. These intensify along different pathways, with much diversity in labour regimes, patterns of technology adoption and market engagement.

As argued by Schlesinger and Noronha (2006), the increase of corporate farming exercises pressure to homogenize production and concentrate wealth in fewer and larger farms. However, rather than a linear trajectory towards a single type of mega large-scale soybean production that continues expanding, the existing differences in farming styles reflect a more complex dynamic. In broad terms, some capitalized farmers will continue to increase their scale and intensify production, while complying or not with environmental land-use regulations; others will intensify their production in the same area through diversification; for example with integration of agriculture, livestock and forestry, or production of crops for the national market; soybean production in family farming will still have a role as part of diversified production systems; and, finally, some farmers may opt to get out of farming completely, selling their land. The diversity provides potential solutions to unforeseeable future challenges and needs more attention in both the academic and policy spheres.

Moreover, none of the cases conforms to the stereotypical pattern of a large-scale monoculture-based, corporate agribusiness. Despite their scale, they are often 'family farms', managed by in-migrants to the area. They are thus associated culturally, socially and sometimes still economically with farming systems elsewhere, and bring these attributes with them to the agricultural frontier in Mato Grosso. Their farming styles – encompassing the range of technical, market, ecological and social relationships that make up a farming operation – are thus different from both smallholder family farms and large-scale agribusiness enterprises. Yet, despite such farms representing a majority of soy producers in Querência (see Table 1), their distinct characteristics have not been a central focus for policy discussions and decisions. Some see them as entrepreneurial farmers in 'transition' in an evolutionary trajectory from small-scale farming to a somehow inevitable incorporation into large-scale agribusiness (van der Ploeg 2008). Yet, as the currently dominant pattern of production, such farms, in all their diversity, need greater attention. For, with their diversified patterns of production, combining soy with other activities, and their socially embedded characteristics as (large) family farms, with personal commitments to the land and the farming operations, they are unlikely to disappear as quickly as some imply. Instead, this diverse grouping is in fact the core of the Brazilian soy farming story. This variety forms part of what Thompson and Scoones (2009, 386) describe as 'diverse "rural worlds" and their potential pathways to sustainability through agriculture'. Understanding the future of soy in Brazil needs greater attention to this group of farmers, labourers, farming practices, market relations and socio-political organizations.

Acknowledgements

I would like to thank Ian Scoones and John Thompson for their supervision during my PhD, and for the invaluable comments that they provided on earlier versions of this paper. I would also like to thank João Carlos Barrozo, Kees Jansen and Evan Killick, for their earlier feedback on this research, and my two anonymous reviewers.

Funding

This work was supported by the Mexican National Council of Science and Technology (CONACYT) [grant no. 228833].

References

Arvor, D., M.S.P. Meirelles, R. Vargas, L.A. Skorupa, E.C.C. Fidalgo, V. Dubreuil, I. Herlin, and J. Berroir. 2010. Monitoring Land Use Changes Around the Indigenous Lands of the Xingu Basin in Mato Grosso, Brazil. In *Geoscience and Remote Sensing Symposium (IGARSS), 2010 IEEE International*, 3190–93. IEEE.

Azevedo, A.A., and R. Pasquis. 2009. Produtores de soja do Mato Grosso: que agenda ambiental construir? *Projeto Diálogos.* Brazil: CDS/UnB, CIRAD, ICV, IPAM e WWF-Brasil.

Barrozo, J.C. 2007. Incertezas no Araguaia: a enxada enfrenta o trator. In *Política, Ambiente E diversidade cultural*, ed. Vitale Joanoni Neto, 36–47. VI Seminario de ICHS-UFMT. Cuiabá: EdUFMT.

Bernardes, J.A., and L.A. dos Santos Aracri, eds. 2011. *Novas fronteiras do biodiesel na Amazônia: limites e desafios da incorporação da pequena produção agrícola.* Rio de Janeiro: Arquimedes Ed/CNPQ/UFRJ.

Bertrand, J.-P. 2006. The brazilian soybean is competitive, but at what cost to the environment? *INRA-MONA Ivry France* 4 (September).

Brando, P.M., M.T. Coe, R. DeFries, and A.A. Azevedo. 2013. Ecology, economy and management of an agroindustrial frontier landscape in the southeast Amazon. *Philosophical Transactions of the Royal Society B: Biological Sciences* 368, no. 1619: 20120152.

Brondizio, E.S., E. Ostrom, and y.O.R. Young. 2009. Connectivity and the governance of multilevel social-ecological systems: the role of social capital. *Annual Review of Environment and Resources* 34: 253–78.

Browder, J.O., M.A. Pedlowski, R. Walker, R.H. Wynne, P.M. Summers, A. Abad, N. Becerra-Cordoba, and J. Mil-Homens. 2008. Revisiting theories of frontier expansion in the Brazilian Amazon: a survey of the colonist farming population in Rondônia's post-frontier, 1992–2002. *World Development* 36, no. 8: 1469–92.

Castrillon Fernández, A.J. 2007. Do cerrado à Amazônia : as estruturas sociais da economia da soja em Mato Grosso. PhD thesis, Universidade Federal do Rio Grande do Sul.

Companhia Nacional de Abastecimento. Historical series for farming season 1976/77 and 2012/2013 for planted area, productivity and production. Soybean. Accessed December 5 2013. http://www.conab.gov.br

Covre, J., and F. Clemente. 2014. New brazilian forest code: changes and prospects. EAAE 2014 Congress 'Agri - Food and Rural Innovations for Healthier Societies', Slovenia.

DeFries, R., M. Herold, L. Verchot, M. N. Macedo, and Yosio Shimabukuro. 2013. Export-oriented deforestation in Mato Grosso: harbinger or exception for other tropical forests? *Philosophical Transactions of the Royal Society B: Biological Sciences* 368, no. 1619: 20120173.

Dros, J.M. 2004. Managing the soy boom: two scenarios of soy production expansion in South America. Report commissioned by WWF Forest Conversion Initiative. Amsterdam: AIDEnvironment. http://assets.panda.org/downloads/managingthesoyboomenglish_nbvt.pdf.

Durigan, G., N. Guerin, and J.N. Martorano Neves da Costa. 2013. Ecological restoration of Xingu basin headwaters: motivations, engagement, challenges and perspectives. *Philosophical Transactions of the Royal Society B: Biological Sciences* 368, no. 1619: 20120165.

Economist Intelligence Unit. 2010. The global power of Brazilian agribussiness. The Economist-Accenture.

Falabretti, E. 2010. Homenagiando familias que ajudaram a construir a historia de nossa ciudade. *Conexão Xingu Araguaia*, January 1, 20 edition, sec. Editorial.

Faria, A.M. de Melo, et al. 2009. *Análise do Mapeamento e das Políticas para Arranjos Produtivos Locais no Norte e Nordeste do Brasil e dos Impactos dos Grandes Projetos Federais em Estados Nordestinos Selecionados.* Projeto pesquisa. Cuiabá-MT, Brasil: Ministério da Educação; Faculdade de Economia, Universidade Federal de Mato Grosso.

Fearnside, P.M. 2001. Soybean cultivation as a threat to the environment in Brazil. *Environmental Conservation* 28, no. 01: 23–38.

Ferreira Filho, J.B. de Souza, and C.E. de Freitas Vian. 2013. The Brazilian Experience With the Occupation of the Cerrados: The Dynamics of Large Farms x Small Farms. Working paper N° 5 Series Estudios Rurales. Rimisp – Centro Latinoamericano para el Desarrollo Rural.

Glass, V., M. Gomes, and A. Biondi. 2011. *Valuing the amazon. Impacts of soy and soy export infrastructure in the Brazilian Amazon region.* Centro de Monitoramento de Agrocombustíveis, Repórter Brasil.

Goodman, D., B. Sorj, and J. Wilkinson. 1985. Agroindústria, políticas públicas e estruturas sociais rurais: análises recentes sobre a agricultura brasileira. *Revista de Economía Política* 5, no. 4: 31–55.

Greenpeace. 2006. *Eating Up The Amazon.* Greenpeace.

Gudynas, E. 2006. El debate sobre la soja en el cono sur. Nuevas tendencias y opciones de acción. Presentation at Controle Social do Mercado da Soja: Ações e Estratégias, Articulação Soja Brasil / Coalizão Soja Holanda, São Paulo, Brazil.

Hecht, S.B. 2005. Soybeans, development and conservation on the Amazon frontier. *Development and Change* 36, no. 2: 375–404.

Hecht, S.B., and C.C. Mann. 2008. How Brazil outfarmed the American farmer. *Fortune*, January 19.

Heredia, B., M. Palmeira, and S. Pereira Leite. 2010. Sociedade e economia do "agronegócio" no Brasil. *Revista Brasileira de Ciências Sociais* 25(octubre): 159–76.

IBGE. 2012. SIDRA - Automatic Data Recovery System. http://www.sidra.ibge.gov.br. (accessed December 2, 2012).

IBGE-FAO. 2006. Censo Agropecuário 2006 - Agricultura Familiar - FAO/INCRA. *Sistema IBGE de Recuperação Automática - SIDRA.* http://www.sidra.ibge.gov.br.

Instituto Nacional de Pesquisas Espaciais - INPE. PRODES - Desflorestamento nos municipios. Accessed: December 6 2013. http://www.dpi.inpe.br/prodesdigital/prodesmunicipal.php.

ISA and ICV. 2006. *Sistema de licenciamento ambiental em propiedades rurais do estado de Mato Grosso: análise de sua implementação.* Série Estudos 7. Brasília: Ministério do Meio Ambiente.

Jepson, W.E. 2003. A political ecology of land-use and land-cover in the central Brazilian Savanna since 1970. PhD thesis, Los Angeles, USA: University of California.

Jepson, W.E., C. Brannstrom, and A. Filippi. 2009. Access regimes and regional land change in the Brazilian Cerrado, 1972–2002. *Annals of the Association of American Geographers* 100, no. 1: 87–111.

Klink, C.A., and R.B. Machado. 2005. Conservation of the Brazilian Cerrado. *Conservation Biology* 19, no. 3: 707–13.

Klink, C.A., and A.G. Moreira. 2002. Past and current human occupation and land use. In *The Cerrados of Brazil: Ecology and natural history of a neotropical savanna*, ed. P.S. Oliveira and R.J. Marquis, 69–88. New York: Columbia University Press.

Lovatelli, C. 2009. The success case of soy agribusiness in Brazil. Presented at the UK Farmers Club. ABIOVE - Associação Brasileira das Indústrias de Óleos Vegetais.

Marimon, B.S., et al. 2014. Disequilibrium and hyperdynamic tree turnover at the forest–cerrado transition zone in southern Amazonia. *Plant Ecology & Diversity* 7, no. 1–2: 281–92.

Marsden, T., J. Banks, H. Renting, and J.D.V.D. Ploeg. 2001. The road towards sustainable rural development: issues of theory, policy and research practice. *Journal of Environmental Policy & Planning* 3, no. 2: 75–83.

Medeiros, L.S. de 2001. "Sem terra", "assentados", "agricultores familiares": considerações sobre os conflitos sociais e as formas de organização dos trabalhadores rurais brasileiros. In N. Giarracca, comp. *Una nueva ruralidad en América Latina?* Buenos Aires: CLACSO, Consejo Latinoamericano de Ciencias Sociales. 103–128.

Mier y Terán Giménez Cacho, M. 2008. Global value chains and biotechnology: governing small farmers' use of genetically modified soybean in Rio Grande Do Sul, Brazil. MPhil thesis, Oxford: QEH, University of Oxford.

Mier y Terán Giménez Cacho, M. 2014. The political ecology of soybean farming systems in Mato Grosso, Brazil. A cross-scale analysis of farming styles in Querência-MT. PhD thesis, UK: IDS, University of Sussex.

MMA. 2009. *Plano de ação para prevenção e controle do desmatamento da amazônia legal. Operação Arco Verde.* Brasilia, Brazil: Ministério do Meio Ambiente.

Monteiro, M.D. 2010. *APROSOJA - Agriculture overview.* APROSOJA www.aprosoja.com.br.

Moreno, G. 1993. Os (des) caminhos da apropriaçâo capitalista da terra em Mato Grosso. PhD thesis, São Paulo, Brazil: Departamento de Geografia, Universidade de São Paulo.

Municipal Secretary of Agriculture and Environment -Querência. various years. *Relação dos produtores de Querência-MT*. Excel document. Querência-MT, Brasil.

Nepstad, D., S. Schwartzman, B. Bamberger, M. Santilli, D. Ray, P. Schlesinger, P. Lefebvre, et al. 2006. Inhibition of Amazon deforestation and fire by parks and indigenous lands. *Conservation Biology* 20, no. 1: 65–73.

Nepstad, D.C., and C.M. Stickler. 2008. Managing the tropical agriculture revolution. *Journal of Sustainable Forestry* 27, no. 1–2: 43–56.

Nepstad, D.C., C.M. Stickler, and O.T. Almeida. 2006. Globalization of the Amazon soy and beef industries: Opportunities for conservation. *Conservation Biology* 20, no. 6: 1595–1603.

Oliveira, G.de L.T. 2013. Land regularization in Brazil and the global land grab. *Development and Change* 44, no. 2: 261–83.

Pádua, José Augusto. 2004. *Agricultura sustentavel*. Brazil: Manuscript for O Estado do Nosso Meio Ambiente: Uma Visão da Sociedade Civil.

Pinazza, Luiz Antonio (coord). 2007. *Cadeia produtiva da soja*. Série Agronegócios. Brasilia, Brazil: Ministério da Agricultura, Pecuária e Abastecimento; Instituto Interamericano de Cooperação para a Agricultura.

Ploeg, J.D.V.D. 1994. Styles of farming: An introductory note on concepts and methodology. In *Born from within. Practice and perspectives of endogenous rural development*, ed. J.D.V.D. Ploeg and A. Long, 7–30. Assen, The Netherlands: Van Gorcum.

Ploeg, J.D.V.D. 2008. *The new peasantries. Struggles for autonomy and sustainability in an era of empire and globalization*. London, UK: Earthscan.

Ploeg, J.D.V.D. 2010. Farming styles research: the state of the art. Keynote lecture for the Workshop on Historicising Farming Styles. Melk, Austria.

Roessing, A.C., and J.J. Lazzarotto. 2004. *Criação de empregos pelo complexo agroindustrial da soja*. Documentos 233. Londrina: Embrapa Soja.

Rohter, L. 2003. Relentless foe of the Amazon jungle - soybeans. *NYTimes*, septiembre 17.

Rossete, A.N. 2008. Zoneamento ambiental do município de Querência – MT. PhD Thesis, São Carlos: Universidad Federal de São Carlos.

Sanches, R., and R. Gasparini. 2000. O Xingu na mira da soja. *Instituto Socioambiental - ISA*. www.socioambiental.org/esp/soja/8.shtm

Sauer, S., and S. Pereira Leite. 2012. Agrarian structure, foreign investment in land, and land prices in Brazil. *Journal of Peasant Studies* 39, no. 3–4: 873–898.

Schlesinger, S. 2008. *O grão que cresceu demais – a soja e seus impactos sobre a sociedade e o meio ambiente*. Rio de Janeiro, Bazil: FASE.

Schlesinger, S., and S. Noronha. 2006. *O Brasil está nu! O avanço da monocultura da soja, o grão que cresceu demais*. Brasilia, Brazil: Federação de Órgãos para Assistência Social e Educacional - FASE.

Schmitzberger, I., Th Wrbka, B. Steurer, G. Aschenbrenner, J. Peterseil, and H.G. Zechmeister. 2005. How farming styles influence biodiversity maintenance in Austrian agricultural landscapes. *Agriculture, Ecosystems & Environment* 108, no. 3: 274–90.

Schneider, S., and P.A. Niederle. 2010. Resistance strategies and diversification of rural livelihoods: the construction of autonomy among Brazilian family farmers. *Journal of Peasant Studies* 37, no. 2: 379–405.

Schneider, S., S. Shiki, and W. Belik. 2010. Rural development in Brazil: overcoming inequalities and building new markets. *Revista de Economia Agraria* LXV, no. 2: 225–59.

Schwartzman, S., A. Villas Boas, K. Yukari Ono, M. Gesteira Fonseca, J. Doblas, B. Zimmerman, P. Junqueira, et al. 2013. The natural and social history of the indigenous lands and protected areas corridor of the Xingu River Basin. *Philosophical Transactions of the Royal Society B: Biological Sciences* 368, no. 1619: 20120164.

Shiki, S. 2000. Globalização do domínio dos cerrados: sustentabilidade do sistema agroalimentar sob regulação privada. Projeto Brasil Sustentável e Democrático. In *O futuro do cerrado: degradação versus sustentabilidade e controle social*. Série Cadernos Temáticos 2. Rio de Janeiro: FASE.

Soares-Filho, B., R. Rajão, M. Macedo, A. Carneiro, W. Costa, M. Coe, H. Rodrigues, and A. Alencar. 2014. Cracking Brazil's forest code. *Science* 344, no. 6182: 363–64.

The Dutch Soy Coalition, and AIDEnvironment. 2006. *Soy – Big Business, Big Responsibility. Addressing the social- and environmental impact of soy value chain*. http://www.bothends.nl/index.php?page=6&documentId=12

The Nature Conservancy. 2012. *Good agricultural practices and socio-environmental certification in Brazil. Towards sustainability*. Brazil: The Nature Conservancy.

Thompson, J., and I. Scoones. 2009. Addressing the dynamics of agri-food systems: an emerging agenda for social science research. *Environmental Science & Policy* 12, no. 4: 386–97.

Van Averbeke, W., and S.S. Mohamed. 2006. Smallholder farming styles and development policy in South Africa: The case of Dzindi Irrigation Scheme. *Agrekon* 45, no. 2: 0–0.

Van Berkum, S., and P.S. Bindraban. 2008. *Towards sustainable soy. An assessment of opportunities and risks for soybean production based on a case study Brazil*. The Hague: LEI Wageningen UR.

Walker, R. 2011. The impact of Brazilian biofuel production on Amazônia. *Annals of the Association of American Geographers* 101, no. 4: 929–38.

Walker, R., J. Browder, E. Arima, C. Simmons, R. Pereira, M. Caldas, R. Shirota, and S. de Zen. 2009. Ranching and the new global range: Amazonia in the 21st century. *Geoforum*, 40, no. 5: 732–45.

Wolford, W. 2008. Environmental justice and the construction of scale in Brazilian agriculture. *Society & Natural Resources: An International Journal* 21, no. 7: 641–655.

WWF. 2004. Mato Grosso a hot spot of soy expansion – and deforestation. *Forest Conversion News*, febrero, 3 edición.

Farming is easy, becoming Brazilian is hard: North American soy farmers' social values of production, work and land in Soylandia

Andrew Ofstehage

Since the late 1980s, North American farmers have been migrating to Brazil to produce soybeans and escape a general farm crisis in the United States. This paper analyzes their work, values, social relations and relations with the land in order to understand transnational farming and agrarian change from the perspective of transnational farmers. North Americans' migration to Brazil and soy production in Brazil can inform our understanding of the mechanisms of the soy boom and unpack the relative significance of social values at play in intensive, technified and financialized agriculture. It also provides an evocative perspective of the soy boom as it engages with issues of transnationalism, crisis, migration and change in business and farming practices. Using ethnographic data, this paper explores the intimate and emerging realities of agrarian change by detailing four elements of transnational farming – migration, farm management, land use and work – through the narration of three farmers' career histories. These cases address the transformation of social values of work, land and social relations through the processes of migration and agrarian change. Farmers' work, it is found, emerges out of an entanglement of regulations, expertise, meanings of work and land, worker relations and the political economy of Brazil and the United States.

Introduction

Sitting in the shade on his Western Bahian soybean farm, John Hanson[1] told me that his 'family has always farmed, [we] came from Ireland sometime in the 1800s and probably just always farmed'. The Hanson family is one of dozens[2] of migrant US soy farm families in Western Bahia, Tocantins, Roraima, Goiás and Mato Grosso, Brazil. North American farmers came to Brazil in two waves, one stretching from 1988 to 1994 and another from 2000 to 2004. Generally, the first cleared native Cerrado to develop agricultural land, while the second purchased cleared and developed land.[3] In Bahia, Brazilian and North Americans produce cotton, soy and corn, and several own cotton gins. North American narratives of migration and soy production in Brazil demonstrate how relative cultural

[1]All names in this paper are pseudonyms, in accordance with informed consent agreements.
[2]Estimated at 30 to 40 families, based on interviews with farmers and policy makers in Brazil.
[3]In addition to this group of Midwestern family farmers, a colony of 70 Mennonite families has farmed soybeans near Rio Verde, Goiás, since migrating from the United States in the late 1960s (Mello and Silva 2011).

outsiders confront, adapt and co-opt the practices and values of soy farming in Brazil. These narratives add a transnational perspective on the motivations behind and in front of the international spread of the Brazil model of farming. These contributions are of particular interest in a moment of expansion of the Brazil Model of farming within Brazil and its spread, particularly to Sub-Saharan Africa. The phenomenon of US soy producers in Brazil provides an evocative perspective of the soy boom as it engages with issues of transnationalism, crisis, migration and change in business and farming practices.

An analysis of their migration bridges work that situates soy production in the political and economic histories of production (Hecht 2005; Oliveira 2013; Wolford 2008) with work that reveals the transformations of social relations and social practices within the field of soy production (Gudynas 2008; Peine 2010). I engage with both sets of work via a framework of social values of work to demonstrate the interconnected processes of farm crisis in the United States and the ways that North Americans transform, preserve and defend social relations, practices and values of production. I ask: what values, meanings, and practices of work, land, and production emerge through transnational soy commodity production? By taking seriously the social values and social context surrounding these global economic processes, I connect market dynamics with social values of soy production and link production dynamics in Brazil, primarily in the Cerrado, with agrarian transformation in North America.

Soybeans in the Cerrado

Soy production in the Cerrado of Mapitoba (Maranhão, Piauí, Tocantins and Bahia), a soy frontier region where the majority of US farmers have settled, is the product of government policy to incentivize agricultural expansion and intensification. The Brazilian State has mobilized soy to implement colonization and modernization projects in the Cerrado, replicating similar nationalist projects carried out in Amazônia (Hecht and Cockburn 2011) and Bolivia (Hecht 2005; Valdivia 2010).

State policies in Amazônia give legal title of unregistered lands to landless workers to reduce land grabbing and foreignization in that region. At the same time, the state incentivizes large-scale production in the Cerrado by implementing lower required Legal Reserves (20–35 percent in the Cerrado, depending on the state, compared to 80 percent in the Amazon ecosystem) and improving transportation infrastructure in the Cerrado region (Oliveira 2013). The greater exertion of state control over the Amazon, paired with the opening up of the Cerrado through policy and infrastructural incentives, 'actually result in greater [gross] deforestation and land concentration [in the Cerrado]' (Oliveira 2013, 279). State-sponsored research of soil amendment techniques and seed varieties, along with credit provisions, has incentivized large-scale soy production. This technocentric state support has increase production in the Cerrado while worsening distributional and procedural inequity (Wolford 2008).

A Great Transformation?

The Brazil Model of soy production, the assemblage of business and farming practices that make up the Brazilian agricultural system, is celebrated for providing work, modernizing agriculture in Brazil, increasing production and feeding the growing world population, all with limited direct governmental support (compared to US crop subsidies and European support for multifunctionalism), and little deforestation (Economist 2010). The characteristics of the model are no-till production, use of genetically modified organisms and a shift from managers of land to managers of labor;[4] it also constitutes what Gudynas calls the 'Great Transformation of rural South America' (Gudynas 2008, 514). Recalling Polanyi's observations of the commodification and dehumanization of twentieth-century capitalism (Polanyi 2001), Gudynas frames the transformation of farm organization and agricultural work in Soylandia as a disembedding of production from social life and social relations.

The transformation of soy production in South America (see also Reboratti 2010 and Teubal 2006) entails a shift from integrated agricultural systems to monocultures, partial commodification to strong commodification, and local-based to export-oriented production. The farm itself is managed as a hierarchically organized and highly capitalized business (Gudynas 2008) in which the 'classic image of poor farmers and rich ranchers is replaced by one of rural managers, most of them with university-level education, living away from the farm, and specialized in business management' (Gudynas 2008, 515).

The transformation enables large-scale soybean producers to see themselves as pioneers and heroes of market-oriented export agriculture and look to the market as legitimation of their work. Peine writes that soy farmers are

> thought to be the unwitting victims of the corporate food regime, [but] they are inexorably incorporated into and therefore dependent on it...they become its agents, promoting its extension as a way to preserve their own livelihoods even as the total value of what they produce is expropriated by the companies that rule the regime. (2009, 59–60)

Gudynas' analysis, supported and expanded by the work of Peine, deftly lays clear the stakes of agrarian transformation and the articulations of this change at the level of farm practice and business organization. However, while work, land and production are increasingly commoditized through the proliferation of the Brazil Model, farmers continue to articulate distinct values and meanings of production. US farmers conform to Gudynas' characterization in many aspects, but the transformation of their values and practices is neither homogenous nor complete. Neither are they able to impose North American values and practices unchallenged. Here I echo recent critiques of Polanyi's 'Great Transformation' (Mitchell 2008) which argue that the economy cannot be disembedded from social relations and cultural meanings. Farmers' practices and values of soy production emerge out of constant negotiation between themselves and Brazilian farmers, workers, governmental policies and soy agro-ecosystems in the Cerrado. I conceptualize the reproduction, transformation, and emergence of these different values and practices of soy production using David Graeber's 'anthropological theory of value'.

Farming as work and work as socially valued action

Midwestern farmers in Brazil possess highly technified machinery, operate within market-driven institutions, and engage in a global soy commodity market. It may be attractive to view them as 'capital personified', their souls being the soul of capital (Marx 2012, 257), or to propose that, like Wall Street bankers, their values of work directly reflect the demands of the market (Ho 2009). However, research on large-scale producers throughout Latin America (Adams 2008; Adams 2010; Bobrow-Strain 2007; Eakin, Bausch, and Sweeney 2014; Hoelle 2012; Mackey 2011; Valdivia 2010) indicates that landowners

[4]See Goldfarb and Zoomers (2013) for an extensive analysis of the expansion of this technology frontier in Argentina.

hold ethical and social values of work aloft as they defend their own work, often balancing agrarian and industrial regimes of value, other times focusing more completely on 'neoliberal' values of production. US farm families also struggle to uphold non-monetary agrarian dimensions of success, including daily work autonomy, opportunities for achievement and reward, and spiritual connections to nature and to agrarian work (Barlett 1993). Farmwork is subject to both capitalist and agrarian regimes of value. I propose that US farmers' work in Brazil entails a transformation toward a set of values and practices associated with the Brazil Model of farming, but also the reproduction of farmers' own identity, culture and work. I argue that the social values of soy production emerge out of economic and agro-ecological conditions and as well as contestations over the relative importance of action, or work.

Work, or action, is the subject of value realization and contestation; it is also a product of that negotiation (Graeber 2001). Value, as the importance of action, is subject to legitimation as actors defend their know-how and their work, thus entailing a process of 'people-making' in two senses (Graeber 2013). On one hand, actions establish one's status as a legitimate social actor, be it a good farmer, an upstanding community member or an innovative entrepreneur. On the other, actions, apparent in everyday practices (Certeau 2011) and everyday ethics (Brodwin 2012) of agriculture, re-make farmers as they tie emerging actions, such as managing labor or speaking Portuguese, to articulations of the self and community. Farmwork emerges out of the material realities of production – political economy and agro-ecology – but also out of social imaginations of what constitutes good and legitimate work.

This notion of value and action is useful in understanding the integration of the US farming community into Brazilian worlds of soy production. One can observe how the practices of the Brazil Model (i.e. hiring labor, use of genetically modified organisms and no-tillage) become valued as efficient, rational and economistic across diverse groups of producers. It also allows one to ask how farmers from different cultural or socio-economic classes re-affirm their identity and difference through practices and discourse on practices (see for example Ofstehage 2011, 2012). This is visible in how Brazilian and North American farmers adopt practices associated with the Brazil Model and then use these practices to define themselves as economically progressive farmers. It becomes apparent in how North Americans use practices that separate them from Brazilians to self-identify as bringing American values or know-how to Brazil.

Writing on elite landholders in Chiapas, Bobrow-Strain states that landowners' calculations of costs and benefits of defending their estates against Zapatistas 'were not the disembodied rational calculations of *Homo economicus*. Rather, they were painful struggles over identity in which uncertain landowners grappled with the upending of the once reliable categories of nature, race, development, good government, and masculinity' (2007, 208). This experience of embodied struggles, reflecting non-capitalocentric behavior (Gibson-Graham 2006; Miller 2001; Mitchell 2008), defies liberal, rational calculations and emerges in other studies of elite landholders.

Ranchers in Acre, Brazil, for example, connect their migration from southern Brazil to family legacies of migration from Europe and to nationalist discourses on taming, colonizing and developing Amazonia. They also defend their work as contributing to national food security (Hoelle 2012). Brazilian landholders and soy producers in Bolivia also make claims on their work as development or improvement (see Li 2007); in the words of one farmer, 'We came with money, we put in technology, it was really a win–win relationship. We all win, the Brazilians won, Bolivia won, and the Bolivian producer won' (Mackey 2011, 20). Soy producers in Bolivia juxtapose themselves against the Morales government as moral compasses of capitalism (Valdivia 2010). Gaúcho farmers in Santarem, Brazil, explain their move to the Center-west as a response to high land prices in the South and increasing land concentration in Mato Grosso, but also refer to their roles as missionaries of modernity and highlight family histories of migration and farming (Adams 2010). Each of these cases demonstrates both economic and non-economic objectives. Profit sits alongside pursuit of individual and collective desires, pleasure, and freedom as objectives for landowners in Latin America.

Methods

I conducted 19 interviews with US soy producers who operate farms in Brazil, with the US Foreign Agriculture Service and with local crop advisors in Brazil. I conducted several daylong farm tours and attended get-togethers of the US community in Brazil, and recorded career history interviews and interviews centered on their everyday farming and social lives in Brazil. On farm visits, I recorded agricultural landscapes and practices that distinguished farmers' practices in Brazil – ranging from the mundane (soccer fields on the farm) to the more agriculturally significant (high-tech machinery, security fences and housing for workers). Research participants were identified through a literature review of newspaper articles on North American farmers in Brazil prior to research, and snowball sampling after arrival in Brazil.[5]

I present three farmers' career histories as integrated narratives in order to show the inter-relatedness of socio-economic processes in Brazil and the United States and to show the gradual transformation being done on farmers, even as they themselves transform practices and the land in Brazil. These career histories, chosen for their ability to represent difference and sameness among the research group, narrate personal moments in actors' lives that are indicative of both sameness among North American soy producers in Brazil and difference in farm practices, social relations and social values. In concert, the career histories tell a situated story of agrarian change (Colloredo-Mansfeld 2009). In presenting integrated realities, I hope to establish also the difficulty of interpreting any one of these farmers as a purely liberal subject. Each is caught up in multiple projects, some collective and some more personal. Two of the career histories follow North Americans in Brazil's Cerrado and a third career history follows a US farmer in Roraima.

I divide the remainder of this paper into two sections. First, I review the literature on a general farm crisis in the United States and an expanding commodity frontier in the MaPiToBa region of Brazil in order to show the inter-connectedness of these processes. Second, I analyze the social values of work that are produced and reproduced through transnational soy production.

Twenty-first-century US farm crisis

I conceptualize structural aspects of production alongside actors' social worlds of production through crisis. Crisis is a dynamic process in which 'particular sets of social relations, political connections and economic practices' link up with development interventions and forces of political economic change to construct post-crisis realities (Fraser, Fisher, and Arce 2013, 68). Crisis frames agrarian change in relation to both the 'big

[5] I presented myself primarily as a researcher interested in the role and everyday lives of US farmers in Brazil.

battalions' of capital, state and land (Scott 1985), and improvisation, social values and community-making. It integrates cultural, ecological and economic phenomena and re-affirms that 'economic crises are ecological crises are cultural crises' (Escobar 2008, 14).

Studies on large-scale soy producers in South America reflect the dual process of crisis and frontier expansion. Southern Brazilian Gaúcho soy farmers in the Northern Brazilian city of Santarem, for example, explain their internal migration as a response to shortage of land in the South and highly consolidated farms in Mato Grosso (Adams 2010). Likewise, Gaúcho cattle ranchers in rural Acre, Brazil, explain their migration as a search for opportunity to expand land acreage, but also to continue ranching and migration legacies (Hoelle 2012). I place the work and migration of North Americans in a similar context – crisis is not deterministic of their migration, but an understanding of their migration is impossible without an understanding of the social, ecological and economic conditions of farming in the United States.

The US farm crisis of the 1980s, or 'farm financial stress', precipitated from heavy debt loads, rapid declines in the value of farm land and other agricultural assets, low prices for many of the most important US agricultural commodities, and a somewhat heightened pace of voluntary and involuntary liquidation of assets (Buttel 1989). Farmers changed management styles, became part-time workers and engaged with government farm support, together entailing a clash of industrial and agrarian values (Barlett 1993; Dudley 2002).[6] The agrarian crisis in the United States has continued into the twenty-first century in the form of higher farmland values, farm corporatization, farm consolidation and changing agricultural practices. These processes, in turn, have narrowed profit margins, increased dependency on finance markets, shifted expertise from agro-ecological knowledge to knowledge of high finance and reduced labor requirements (MacDonald, Korb, and Hoppe 2013).

Inaccessibility of land

Two factors have contributed to the decreasing accessibility of land for farmers, especially young farmers in the United States – rising farmland prices and consolidation of farmland under fewer entities. Since the farm crisis of the 1980s, land prices have risen from around USD 1000/acre to nearly USD 2000/acre (Nickerson et al. 2012). This average masks even higher farmland values in highly productive regions. Even poor-quality land in Indiana, as of the summer of 2013, sold for an average of USD 5750/acre (+14.7 percent from 2012), while average-quality land sold for USD 7446/acre, (+17.1 percent) and top-quality land sold for USD 9177/acre (+19.1 percent) (Dobbins and Cook 2013). Iowan farmland is similarly expensive, with the state average rising to USD 8296 in 2010. Iowan agricultural producers cited high commodity prices, low interest rates, cash/credit availability and lack of land availability as key factors in the rising cost of land (Duffy 2013).

As the price of farmland has increased, so has farm consolidation. Despite an increasing number of small farms, the average size of the US farm has increased from 589 acres per farm in 1982 to 1105 acres per farm in 2007. United States Department of Agriculture (USDA) researchers credit changing technology, changes in the organization of farms, and government policy for the changing make-up of US farms (MacDonald, Korb, and Hoppe 2013). Family owned and operated farm businesses dominate the agricultural

[6]My own father experienced this crisis on our South Dakota farm as pressure to consolidate and narrow his production to a corn/soybean rotation, and as a community trauma. This experience and the general situation of my family's farm was shared with research subjects.

landscape of the US, but the size of the farms comes with a shift toward reliance on hired labor, rented land and contracted services (MacDonald, Korb, and Hoppe 2013). In order to manage this new agrarian model, families have formalized their businesses as limited-liability companies (LLCs) and family or non-family corporations (O'Donoghue et al. 2011).[7] As of 2007, 43 percent of US farm production originated on farms organized as partnerships or corporations, an increase of 9 percent from 1982 (O'Donoghue et al. 2011).

Displacement of farmwork by technology

The 1980s farm crisis changed the value and meaning of labor in which 'well-tended fields, hard work, and performing tasks 'in a timely manner' ceased to evoke praise (Barlett 1993, 238). With the introduction and widespread adoption of biotechnology,[8] no-tillage cultivation and reduced chemical pesticide use, productivity has risen despite reductions in the use of labor and, as Midwestern farmers claim, 'made good farmers lazy and lazy farmers good'.[9] Since 1982, operator labor has fallen by 40 percent and hired labor has fallen by 30 percent (MacDonald, Korb, and Hoppe 2013). The operator and spouse labor accounts for three fifths of labor on small farms, one half on medium-sized farms, and 11 percent on 'large farms – the remainder being hired labor' (MacDonald, Korb, and Hoppe 2013). Forty percent of US farmers worked off-farm for at least 200 days per working year; indeed, many rely on this labor for financial survival (MacDonald, Korb, and Hoppe 2013).

Increasing farmland prices have benefited landholders, new forms of business organization protect families (and investors) against risk, and new production practices allow producers to reduce labor and pursue part-time work; however, all of these processes raise barriers for young farmers to enter into production agriculture. Young farmers confront a post-1980s agricultural landscape in which land is inaccessible, technology displaces labor, and the outlook for small-scale family farms is dire. This landscape interconnects with the political economy (see Oliveira 2013) of Brazil to co-construct the migration of North American soy producers to Bahia and the emerging soy frontier of Mapitoba.

The soy frontier of Mapitoba

Restrictions to soy production in Amazônia and encouragement from public policies and infrastructural development have transformed the Brazilian Cerrado into a 'hotspot' of deforestation and soy production (Hecht 2005); it also serves as a haven for certain US farmers. In the soy frontier region of Mapitoba, low-biomass forests are being rapidly converted to soy production. Investors and soy producers have flocked to Mapitoba to seek out a large area of available land, low land prices compared to comparable agricultural production regions of the United States or Brazil, a reduced presence of environmentalists compared to Amazônia and a favorable infrastructural situation (Mondardo 2010). In Western Bahia, the site of most non-Mennonite US-owned soy farms, agricultural production increased from approximately 2.05 km^2 in 1979 to 1615 km^2 in 1986, then to 5743 km^2 in 2000 and to 7259 km^2 in 2005, covering 55 percent of total land in that year (Brannstrom

[7]LLCs and corporations reduce economic risk by protecting non-business assets from liability, and generally reduce tax burden.

[8]Ninety percent of corn planted, 90 percent of cotton planted and 93 percent of soybeans planted in 2013 were genetically modified (USDA, National Agricultural Statistics Service 2013).

[9]Suggesting a leveling process and a shift away from agro-ecological expertise and in-field work.

2009). The rapid influx of capital and 'agents of modernization' transformed the region from an isolated region economically based on low-density cattle ranching and shifting cultivation to one based on capitalist agriculture, provoking conflicts between capitalist farmers and peasant families (Diniz 1984). Mirroring Jason Moore's 'commodity frontier' analysis (Moore 2010a, 2010b), US farmers flee crisis in the United States; they help to consolidate a regional process of soy expansion and development in Mapitoba; they also, perhaps unknowingly, participate in a historical transformation of a region and the continuation of social conflict. In the following section, I analyze issues of migration, farm management, land use, and work and worker relations via three farmers' career histories. Narratives on migration illustrate the presence of crisis and hope in migration decisions. An analysis of their Brazilian farm management indicates US farmers turn to financial capital and the process of locating farm property. The final two sections on land use and workers illustrate the uneven spread of farming and business practices associated with the Brazil Model as US farmers 'become Brazilian', as well as farmers' legitimization of these practices.

Migrating to Brazil: sleepless nights and National Geographic

Jacob and Paul Miller: the entrepreneurs

Jacob Miller's family is descended from a German farmer who migrated to northwest Iowa in the nineteenth century; the family still owns the century farm. Jacob's father, Paul, inherited the homestead in the early 1970s, during the inflationary period, and continued to operate the farm until the 1990s. He now rents out the farm and manages his financial services and consulting business. Interested in other investment options, Paul took a Brazilian agricultural tour in 2001 and was promptly amazed by the extensive availability of flat land, although he reported that the soil was degraded and had a poor soil profile and was 'nothing but Cerrado'. Jacob, then still in university, reported sleepless nights, saying,

> I remember I couldn't sleep the first three nights 'cause I was so excited about it and you come from Iowa and you know, these 80-acre[10] fields and you come down here [Bahia] and you see everyone has these 1000-acre fields and you didn't even know stuff like that existed and you'd go home and explain it to people and they'd either think you're exaggerating or you know they don't believe you.

They witnessed the clearing of Cerrado and pasture, 'not much besides 10-foot trees and grassland', and were told that the site was near to a limestone quarry that would be useful for making the poor soils fertile. The two Iowans found the scope and scale impressive enough for them to suggest that someday Bahia could look like their home acreage. The tour operator informed them of the relatively cheap but 'unskilled' labor force in the region, and the family began to seriously consider purchasing land in Brazil.

Kurt and Caleb Carter: the missionaries

The Carter family came to Brazil in 1972 after Caleb Carter sold the Indiana farm to become a missionary in São Paulo and Paraná. His son Kurt stayed in Brazil until 1977 when, drawn

[10]In this paper I use both 'acres' and 'hectares' to describe area; my use of each term is determined by how each farmer referred to land, whether using the US measure (acres) or the Brazilian measure (hectares). For reference, a hectare is equal to 2.56 acres.

by an interest in farming, he moved back to Indiana to work on his grandfather's farm and attend university. After 10 years in Indiana, Caleb asked Kurt to leave the Indiana farm for a few years to help out with the expanding missionary organization. Kurt departed Indiana with his wife and then-10-month-old son Brad and worked in the missionary's accounting department. This work provided experience with Brazilian business practices and international money transfers that would later prove invaluable. After seven years working for the missionary, the young family moved to central Brazil to fill a perceived 'spiritual burden for farmers in central Brazil [Mato Grosso], [where] there's no church, or bible study or anything' and also as a means of re-connecting with agricultural production. According to Kurt's son Brad Carter, and current manager on the family farm, 'my parents made the decision from that standpoint to go to central Brazil, and agriculture was a little bit of a secondary question. They had to do something to be there in the community'.

John and David Hanson: the manager and the farmer

David Hanson grew up on a large family farm operated by his father, uncle and grandfather, and he 'never had any doubt he was going to farm'. He was studying business at a state university when, drawn by press coverage on the soy boom and personal connections to tour operators, his father and uncle went on a short agricultural tour in Brazil's Cerrado. Months later, they went on a second, longer tour. David's father and uncle saw Brazil as a means of acquiring more land to expand and maintain the family farm as well as provide more on-farm work opportunities to the wider family. David, on the other hand, saw it as a means of starting to farm straight out of college without heavy capital investment to buy land in the Midwest or waiting to take a more central role in the family farm. He also wanted to farm without depending on gifts of land and machinery from the family.

First impressions of Brazil and expressions of crisis

The first experience for most North American farmers in Brazil regularly begins with an agricultural tour intended to confirm the whispers of farming opportunities in Brazil. Farmers' tour experience highlight is often the sight of felling trees and extensive, productive land. The sight of felling trees simultaneously reminds them of the clearing of native prairies in the Midwest and highlights the transition from wasteland to productivity occurring in Brazil. The main objective for purchasing land in Bahia is mixed. Some do so to sustain the family farm, others to capture a seemingly ripe and easy profit, and, in the Carters' exceptional case, to conduct missionary work. For most farmers, Brazil serves as a solution to the rising costs of production in the United States and the increasing difficulty of finding land for farm expansion. There is also a sense of adventurism – one farmer even reported that it felt like he was living in the pages of National Geographic.

A sense of crisis is present in all three narratives – the Millers' crisis of farm profitability in Iowa, the Carters' spiritual crisis and early transition from agriculture and the Hansons' crisis of land scarcity. As we see in the Millers' and Hansons' narratives, the migration of US producers to Brazil cannot be disentangled from conditions of opportunity and threat in the United States. The Millers respond directly to the unprofitability of farming in the Midwest, while the Hansons respond to a more complex situation in which it is not only profitability, but also a way of life that is threatened.

Farm management: accumulating land and courting financiers

The Millers

In 2001, the family pulled together 'a few million [dollars]' through a private placement off ing, a non-public funding stream, and the family purchased land from another North Americ in Western Bahia. A private placement offering requires that investors must have a pri relationship with the company; thus, the family attracted interest from business partnei nearby farmers and other colleagues. They also raised capital by selling Iowa land USD 3100/acre and purchasing Bahian land at USD 800/acre. David Miller's son Jaco packed his bags and, without speaking Portuguese or having much farming experience o his own, relocated to Brazil. The Miller family purchased land from a fellow Americai near Luis Eduardo Magalhães, Bahia, who then managed the farm. It became operational ir 2004. After some management issues in the early years, they turned the business over to Jacob. Today, the farm employs 90 farmworkers, 60 workers for the cotton gin, 10 office employees, an agronomist and an operations manager. They cultivate 4000 hectares of soybeans and 3000 hectares of cotton, and own 9000 hectares of Cerrado in Legal Reserve. Looking ahead, the family has taken on an international funder as their major investor. The long-term plans of the funder are to package the farm with agricultural production investments in Argentina and South Africa to be sold as private shares on the stock exchange. Jacob and Paul are currently sitting on the farm's executive board with plans for Jacob to remain in his position and possibly translate his experience into work as an analyst or consultant.

The Carters

As a way of financing this mission, and re-engaging with the agrarian life, Kurt courted like-minded investors, driven by missionary objectives as well as capital accumulation, to begin a soybean farm in Mato Grosso – some investors are also missionaries, and many are farmers. The farm grew slowly and the family witnessed, and implicitly contributed to, the tremendous growth of the region, going from having 'nothing, no bank or phone lines' to producing 1,000,000 metric tons of soy annually. The family continued to do missionary work and grow the farm until 2001, when they perceived an economic and spiritual opportunity in the North of Brazil. Land prices were rising in Mato Grosso while prices in Roraima remained low. Roraima also had extensive infrastructure and expected market opportunities with Venezuela. They sold the Mato Grosso farm for 40 percent over their expected market price. However, the move came with difficulties. Land prices in Roraima at the time that the family decided to move (2001) were 40 reals, then increased to 100 reals by 2003, and then rose by another 800 reals in the next 18 months. To make this turn of events worse, poor weather conditions in Roraima placed a stress on farming, as I will discuss below.

The Hansons

The Hanson family purchased 4400 acres in 2001, not including Legal Reserve, for a tenth of the price of Midwest farmland. Unlike most US farmers in Brazil, the Hansons self-funded their move to Brazil using capital from their Midwest farm. This purchased land was cleared, but not yet completely 'developed'. At the time, David spoke no Portuguese, and had relatively poor knowledge of the local agro-ecological conditions. Through the years he has learned the language informally and developed a cropping rotation that is less focused on soy and more dedicated to cotton – a rotation that more closely resembles

that of Brazilian producers in the region. At the time of research (2012), the farm had 30,000 acres of production land, 4000 of which was cleared by the business; the remainder was cleared prior to being purchased. These 30,000 acres are in a rotation of corn, soy and cotton, with the relative mix of the three crops changing year to year with the changes in commodity markets. In addition to their own land, they manage a sizeable acreage for a US investment firm.

Mirrors of Midwest farm corporatization

Financing for US farms in Brazil often come through personal contacts, much in the way that investors for ethanol plants in the Midwest are often local farmers. For farmers who distrust the stock market or other investment funds, ethanol plants and international production groups organized into limited liability companies offer an expected return on investment, but also a measure of legibility as a less abstract, more personal investment. Nearly all US soy producers in Brazil form corporate bodies such as LLCs or corporations, or are moving towards that model. This reflects the nature of large-scale production in Brazil as much as the changing landscape of family farm corporatization in the United States (MacDonald, Korb, and Hoppe 2013). Farm investments also mirror global trends of farm financialization (Fairbairn 2014) in which financial motives, markets, actors and institutions take on a larger role in farm production and capitalization (Epstein 2005). Through the own-operate model of farm financialization, land is both a productive asset and a speculative one, meaning that while farmers do intend to profit annually from sales of their harvest, they also expect to sell their developed land at a profit (Fairbairn 2014).

Uses and visions of the land: farming is easy

The Millers

Portuguese is not among the challenges of farming in Brazil for Jacob Miller; he has acquired more than a working knowledge of the language. Neither is it the actual process of farming; he hired consultants and learned more appropriate agricultural techniques (e.g. implementing no-till farming systems on some of his fields and adopting a cropping rotation that better suits the soils and the crops). Rather, the most difficult element it is the process of making land 'productive'. 'The clearing', he said, 'is really the easy part, it's the development that's the hard part, it takes a lot of time, it's an expensive process'. Once workers clear the land, one adds lime, gypsum, and phosphorus to increase the pH level of the soil; thereafter, one continues to build the soil with mineral amendments and no-tillage farming. Development of the land, he said, 'takes years, it's not even a matter of just adding fertilizer, it takes years to build up the fertility and soils and, and build 'em up to the point where you produce optimal yields so we kinda learned about that the hard way'.

The Millers faced challenges in both disease pressure management and producing cotton. Although fungicides are rarely used in Iowa, in Bahia they typically apply insecticides and fungicides to soybean five times annually. For cotton they applied 17 rounds of insecticides and fungicides, using 41 different insecticides. Cotton production was entirely new to the Millers and required a large capital investment for machinery and for a cotton gin, although not all farmers in Bahia own their own gin.

The Carters

Of the North Americans I interviewed, the Carters had the greatest difficulties with the actual production process. In their first year in Roraima, floods prevented most of their planting and the little that they could plant was seeded too late to fully mature. Adding to this difficulty, a 15-day drought during a key moment in the maturation process prevented seeds from fully filling out. For what was planted, they harvested 60 percent of the expected yield. The year after, the family again experienced a series of untimely dry periods and heavy rain periods, which prevented planting of some acres, forced them to plant later than hoped, or forced them to replant hundreds of acres. At the time of research they were expecting 50 percent of expected yield. This experience has led the Carters to consider moving to a more corporate style of farm organization and to consider merger options in order to reduce personal financial risk.

The Hansons

When asked about the region prior to their settlement, David responded that before they came there was nothing: it was unproductive, had no decent roads and no real infrastructure. Part of the effect of their conversion from Cerrado to farmland was to contribute to the agrarian and economic development of the region. The Hansons developed the land by adding large amounts of lime and gypsum the first year and then supplementing with phosphate, potash, magnesium, nitrogen and lime thereafter.

Although production does not even register in David's list of top challenges of producing soy in Brazil, the regulations on land use do (along with uncertainty, lying, capital, weather, theft and culture). Among the most intrusive, according to David and other farmers, is that all cotton and soybean plants must be removed from agricultural fields by September 1 of every year; the cost of infraction is a USD 15,000 fine. This law is meant to reduce disease, pest and fungal pressure on crops by removing hosts from the agro-ecology, but David sees this as a prominent example of government intrusion on farmers' day-to-day operations.[11] Seen as less intrusive but no less costly is the required 20 percent Cerrado Legal Reserve in Bahia, intended to preserve native Cerrado land.

From waste to value

North Americans' productivist agrarian visions of Cerrado land reflect this statement below, which in turn reflects the research and work of Philip Warnken (1999):

> The Cerrado area has been defined as a wasteland with stunted twisted trees. *The Cerrados are not rainforests*. The soils of the Cerrado are highly acidic, saturated with aluminum, deficient in phosphorous [sic] and have low water-holding capacity .... It is clear that *there is a potential for large increases in crop production in the Cerrados*. (McVey, Baumel, and Wisner 2000, emphasis added)

North Americans in Brazil repeat the narrative of bringing unproductive land into the productive economic sphere, a narrative that is common among soy producers in Bahia and other soy frontier regions of Brazil and Bolivia. They also repeat the narrative that is

[11]Much like farmers' resistance to the United States Environmental Protection Agency (EPA) control of air pollution and non-point source pollution in the US Midwest (Nixon 2014).

alluded to in the quote above, that the Cerrado is neither nature worth preserving (as the Amazon is), nor fertile land demanding to produce. US farmers in the Brazilian Cerrado re-configure meanings of the land as modern or traditional, mirroring agrarian change elsewhere in Brazil (Sauer 2012), and echoing agro-industrialist discourse in the soy frontier of lowland Bolivia where low-biomass forests are less valued (despite high ecological value) and are 'easily, and relatively cheaply, cleared with heavy machinery' (Hecht 2005, 278). The construction of the Cerrado and other 'non-productive' lands mirrors statements by Brazilian immigrant soy producers in Paraguay (Albuquerque 2005) and reflects what Gidwany and Reddy refer to as the corollary of value – waste (2011). Opening up production frontiers, settling where there was 'nothing' and juxtaposing the Cerrado against Amazonia constructs non-productive land as waste – as the other of capitalist value. This in turn lends legitimacy and even heroism to large-scale soy production as well as forging a link to European ancestors who opened up the plains by homesteading.

The Cerrado is a wasteland, neither productive nor pristine, which, through hard work, can be made productive. The experience does differ between those who purchased native Cerrado land and those who purchased cleared or developed land. Those who cleared native land often expressed special pride in that action. This development of land is also frequently tied to the so-called development of Luis Eduardo Magalhães as seen in the towns' growth, the expansion of infrastructure and lowered unemployment. The town has grown from little more than a truck stop in the early 1990s to the 'Capital of Agribusiness' and a clean and orderly refuge for Southern Brazilian gaúchos in Bahia (Brannstrom and Brandão 2012). The Cerrado itself is frequently seen as something that does not require protection. It is something that should be excluded from discussion of wildlife conservation, despite the rich biodiversity that is present.

Although the 'development' of land can take years and farmers are under threat of crop pests, drought and flooding, few farmers include production practices among the most difficult aspects of producing soybeans in Brazil. As I will show in the following section, the nature of agricultural work shifts from work in the field to work in the office. Farm work by US farmers is generally limited to decisions on which seed variety to plant, whether to use no-tillage or when to spray a field – and these decisions are made with support from farm managers and agronomists. Farmers have experience with soy production and general farm management, but depend on local expertise, at least in their early years in Brazil, for producing cotton and responding to higher pest pressure in Brazil. While North Americans perceive the land as a wasteland in need of development, they are able to outsource the management of the land to local agronomists and farm managers, leaving the management of workers as the primary difficulty.

Work and workers: becoming Brazilian is hard

The Millers

Jacob's biggest difficulties are labor regulations and managing labor. Workers have 'no skill and no training' and, according to him, they are coming from Southern Bahia and some have backgrounds in small-scale agriculture, but little experience with large-scale soy production or high technology machinery. On labor laws, he said,

> it's not very, it's not fair for the employers, it's not practical ... I can understand it to a certain extent 'cause I think, there is a huge index, a difference between rich and poor and the rich do, or have a history of exploiting the poor in Brazil, but I don't know, it's just. But it's unfair to those that are trying to do what they can.

Partially for this reason, the pushback from workers and labor laws, Jacob has dismissed previous concern for providing work and training for Brazilians; several farmers claimed an intent to bring jobs and opportunities to Brazilian workers until finding out that workers in Brazil had the backing of relatively strong labor laws and the farmworker federation. Recalling Mauss' theory of the gift, the apparent rejection of their job opportunities as gifts, farmers were somewhat offended and felt liberated from responsibility (Mauss 2000). Jacob perceives his role and the role of North Americans in general in the development of Brazil by saying, 'I think we've added, just, added to the community just from the fact that we pay our bills on time and conduct ourselves in a serious business-like manner [compared to Brazilian farmers].' The statement is at once less grandiose than the Santarem Gaúchos' claim to be 'missionaries of modernity' (Adams 2010) and less interested.

The Carters

The Carters chose to invest in long-term employee relationships. The family pays over market price for labor with the understanding that they expect long-term employment. They employ a foreman who manages the farm, one or two operators and seasonal help, a huge difference in total numbers compared to the typical Brazilian farmer. Brad, the youngest active member of the family, also mentioned that receiving payment in sacks of soy, what other North Americans often deride as bartering, is a way of engaging with the local farming community, and rather than a sign of backward business practice, an indication of real farming experience in Brazil. The Millers employ what Brad calls objective-driven management, meaning that they intend and are required by investors to produce an economic profit, but that their management decisions are driven by an interest in treating workers well and being stewards of the land.

The Hansons

The Hansons' farm employs 11 managers and 165 workers; the work crew is cut to 14 workers in the off-season. The workers labor in teams such as soil fertility, harvesting and public relations, live in on-farm housing and typically leave the farm once every two weeks for three days. The workers represent one of David's greatest frustrations. David explains that Bahia isn't a traditional agricultural region and so workers aren't used to the odd hours and the realities of agrarian production. In his eyes, they are uneducated, untrained and often lazy. Compounding this perceived shortcoming of Brazilian agriculture are the labor laws. Labor laws, which regulate firing of workers, housing requirements and workdays, are seen as both intrusive and as superfluous. In response, David has worked around labor by adopting new technologies to replace labor force and reduce the risk of offending labor laws. He has also hired a public relations (PR) firm and a team of lawyers. David compares the current situation with the situation prior to the arrival of large-scale agriculture, saying: 'when we came down it was commonplace for employees … to live in plastic tents, you know, out in the Cerrado … and now you can get a fine if you don't have the right brand of fire retardant'.

For David and his family, the farm has many different objectives. For David, the role of Americans is to bring some degree of rationality or business principles to Brazil. The main difference between North Americans and Brazilians, he says, is that Brazilians have a higher level of debt and more of a penchant to barter. What David brings is an understanding of the market and business principles. The immediate objective of the farm, however, is to function as a business and to make an economic profit as well as serve as a means for him to stay involved in agriculture. For the older generation, the farm expansion provides an opportunity for nieces and nephews, sons and daughters to find on-farm employment and to continue to strengthen the family farm. One of David's struggles in this regard is to encourage the older generation to see the value of working in the office serving business needs rather than driving a tractor in the field. For David, farming is not about sitting in a tractor; it's a business to be managed.

Irrational regulations and lazy workers

Section 31.23.5.1 of Brazil's regulatory laws of agricultural work is the target of many North American farmers' contempt: 'Worker housing must: have beds with mattresses, separated by at least one meter; when bunk beds are used must be limited to two beds in height, and have at least 110 centimeters in height between bunk beds' (Moraes 2013, translation by author). For North Americans like David and Jacob, rules like these are both overly constrictive and illogical, opinions that are repeated by Brazilian farmers in the regions. Others, like Brad Carter, see these restrictions as well as restrictions on firing/hiring of employees as forms of protection that are difficult to deal with, but valid. Despite many complaints about Brazilian regulations and the relative strength of the union of rural workers, few US farmers participate in negotiations between farmers and rural worker unions. Most avoid political involvement altogether. For example, while many Brazilian and some US farmers supported a national farmers' protest in June 2005, several US farmers openly derided the protest during interviews.

The mixed feelings of workers can be summed up by one farmer's statement that landless workers are thieves who try to get something for nothing, but at the same time admitted that were he in their situation he would participate in land occupations. The portrayal of farmworkers, especially Baianos, as quick to take advantage is widespread among both US and Brazilian farmers. Several farmers claimed an intent to bring jobs and opportunities to Brazilian workers, until finding out that workers in Brazil had the backing of relatively strong labor laws and the farmworker federation. However, not all farmers expressed such aversion towards Brazilian workers. Almost universally, interviewees reported that becoming a farm manager and getting used to working with Brazilians was the most difficult part of farming in Brazil. One reported that he thought 'learning plant names, insect names, learning to farm here' would be hard, but 'that's easy. Crop scouts, farmworkers know everything here. They tell you all you need to know, [they] know how to operate machinery'. The difficult part is becoming a manager.

In becoming a manager of labor rather than a manager of land, most North Americans are taking steps towards the Brazilian model of farm management as elaborated by Gudynas. This role allows many North Americans to claim that farming is easy in Brazil, it's getting used to Brazilian culture and regulations that are difficult. Farmers sit in tractors only rarely, and then only for agricultural tours or ethnographic interviews. Most of their time is spent in an office ordering delivery of inputs, securing sale contracts and meeting with farm managers. In the Hanson family, this has led to conflicting ideas of what a good farmer should do, the older generation preferring to be in a tractor demonstrating their hard work and technical skill, the younger generation demonstrating their business savvy from the office. Several younger farmers mentioned that this management style is one element of Brazilian agriculture that they would implement on farms in the Midwest. However, as we seen in all three narratives, there are elements of other visions of agrarian development that persist. The older generation on the Hanson farm still long for time in the tractor, the Miller family proposes that they can introduce US models of accounting into Brazil and the Carters seek out objective-driven worker relations.

Conclusion

Large-scale farmers struggle between a degree of autonomy (or freedom) and power. I use the concept of value as the importance of action to connect narrative of the soy boom as political phenomenon and cultural phenomenon by proposing that their migration to Brazil is a decision based on a confluence of material realities of soy production in Brazil, crisis in the United States, and personal and collective values and identities of work. In addition to this, new values are emerging as farmers engage with local forms of production, local actors, Brazilian culture and Brazilian legal regulations, and as they take on new roles as transnational farmers and as managers of labor.

While I was researching in Western Bahia, interviewees frequently turned my questions on agricultural practice and landscape transformation around on me to say that farming is easy in Brazil, it's getting used to Brazil that's difficult; or, as one North American caricatured his fellow migrants in an echo of a long-standing Brazilian joke, 'they say Brazil is great except for the Brazilians'. Farmers point to stories of North Americans who came to Brazil as fully competent farmers, but quickly went bankrupt due to worker conflicts, misunderstanding of labor or environmental regulations, or simply because they failed to adjust to the way things were done in Brazil. Patience, they said, is the key trait of a successful farmer in Brazil, next to rationality. The North Americans' recognition of the difficult process of becoming good Brazilian farmers, dismissive as it may be, reveals the importance of understanding the lived-in realities of the soy boom.

North American farmers bring with them collective and personal meanings of land, production and good farming. These meanings and imaginaries are mobilized to legitimize, even celebrate, their work in Brazil. Their very presence in Brazil is rarely explained by economistic language itself, but also through pursuit of good work (be it evangelical, entrepreneurial or agricultural) and the survival of family farming legacies. In looking at these non-economistic forms, globalization, through its articulation in the soy boom, can become less abstract and possibly less indisputable (Cravey 2003).

This ethnographic data finds that North American farmers in Brazil, like their elite landholding counterparts in Chiapas, Bolivia and Santarem, are working for more than economic profit, and their work is unfolding in relation with (not in domination over) land, workers, political economy and crisis. North American farmers remain neither fully autonomous in their engagement with soy production nor powerless against the forces of agrarian change; rather, they are semi-autonomous as their work emerges out of an entanglement of regulations, expertise, meanings of work and land, worker relations and political economy. For North Americans, then, producing soybeans in Brazil has many faces – it is the end of the US family farm and its continuation; it is an escape from the conditions of production in the US and their reproduction; it is production for the market and the making of people.

Acknowledgements

The author would like to thank Rudolf Colloredo-Mansfeld, Arturo Escobar, Gabriela Valdivia, Elizabeth Havice and Caela O'Connell for comments, and Susanna Hecht and Gustavo Oliveira for organizing this collection. Thanks also to Jun Borras and two anonymous reviewers for their comments. This research would not have been possible without funding from the Tinker Foundation and the Halperin Memorial Fund.

References

Adams, R. 2008. Large-scale mechanized soybean farmers in Amazônia: New ways of experiencing land. *Culture and Agriculture* 30: 32–7.

Adams, R. 2010. *Elite Landowners In Santarém: Ranchers, Gaúchos And The Arrival Of Soybeans In The Amazon*. PhD diss, Indiana University.

Albuquerque, J.L.C. 2005. Campesinos Paraguayos y Brasiguayos en la frontera este del Paraguay. In *Enclave sojero: merma de soberania y pobreza*, 149–82. Asunción: Centro de estúdios rurales interdisciplinares.

Barlett, P.F. 1993. *American dreams, rural realities: Family farms in Crisis*. Chapel Hill: The University of North Carolina Press.

Bobrow-Strain, A. 2007. *Intimate enemies: Landowners, power, and violence in Chiapas*. Durham, NC: Duke University Press Books.

Brannstrom, C. 2009. South America's neoliberal agricultural frontiers: Places of environmental sacrifice or conservation opportunity. *AMBIO: A Journal of the Human Environment* 38, no. 3: 141–9.

Brannstrom, C., and P.R.B. Brandão. 2012. Two hundred hectares of good business: Brazilian agriculture in a themed space. *Geographical Review* 102, no. 4: 465–85.

Brodwin, P. 2012. *Everyday ethics: Voices from the front line of community psychiatry*. Berkeley: University of California Press.

Buttel, F.H. 1989. The US farm crisis and the restructuring of American agriculture: Domestic and international dimensions. In *The International Farm Crisis*, 46–83. New York: St. Martin's Press.

Certeau, M. de. 2011. *The practice of everyday life*. Berkeley: University of California Press.

Colloredo-Mansfeld, R. 2009. *Fighting like a community: Andean civil society in an Era of Indian uprisings*. Chicago: University of Chicago Press.

Cravey, A.J. 2003. Toque una Ranchera, Por Favor. *Antipode* 35, no. 3: 603–21.

Diniz, J.A.F. 1984. Modernização e conflicto na fronteira ocidental do Nordeste. *Revista GeoNordeste* 1, no. 1: 12–20.

Dobbins, C., and K. Cook. 2013. *Purdue Agricultural Economics Report (August 2013)*. West Lafayette, IN: Purdue University. https://www.agecon.purdue.edu/extension/pubs/paer/pdf/PAER8_2013.pdf (accessed September 16, 2013).

Dudley, K.M. 2002. *Debt and dispossession: Farm loss in America's heartland*. Chicago: University of Chicago Press.

Duffy, M. 2013. *2012 Farmland Value Survey* (No. File C2-70). Ames, IA: Iowa State University. http://www.extension.iastate.edu/agdm/wholefarm/pdf/c2-70.pdf (accessed September 1, 2013).

Eakin, H., J.C. Bausch, and S. Sweeney. 2014. Agrarian winners of neoliberal reform: The 'Maize Boom'of Sinaloa, Mexico. *Journal of Agrarian Change* 14, no. 1: 26–51.

Economist. 2010. The Miracle of the Cerrado. *The Economist*, 08/28/2010 Print Edition.

Epstein, G. 2005. Introduction: Financialization and the world economy. In *Financialization and the world economy*, ed. G. Epstein, 3–16. Northampton: Edward Elgar.

Escobar, A. 2008. *Territories of difference: Place, movements, life, redes*. Durham: Duke University Press.

Fairbairn, M. 2014. 'Like gold with yield': Evolving intersections between farmland and finance. *Journal of Peasant Studies* 41, no. 5: 777–95.

Fraser, J., E. Fisher, and A. Arce. 2013. Reframing 'crisis' in fair trade coffee production: Trajectories of agrarian change in Nicaragua. *Journal of Agrarian Change* 14, no. 1: 52–73.

Gibson-Graham, J.K. 2006. *A postcapitalist politics*. Minneapolis: University of Minnesota Press.

Gidwani, V., and R.N. Reddy. 2011. The afterlives of 'waste': Notes from India for a minor history of capitalist surplus. *Antipode* 43, no. 5: 1625–58.

Goldfarb, L., and A. Zoomers. 2013. The drivers behind the rapid expansion of genetically modified soya production into the Chaco region of Argentina. In *Biofuels - Economy, environment and sustainability*, ed. Z. Fang. Rijeka, Croatia: InTech: 73–95.

Graeber, D. 2001. *Toward an anthropological theory of value: The false coin of our own dreams*. New York: Palgrave Macmillan.

Graeber, D. 2013. It is value that brings universes into being. *HAU: Journal of Ethnographic Theory* 3, no. 2: 219–43.

Gudynas, E. 2008. The new bonfire of vanities: Soybean cultivation and globalization in South America. *Development* 51, no. 4: 512–8.

Hecht, S.B. 2005. Soybeans, Development and conservation on the Amazon Frontier. *Development and Change* 36, no. 2: 375–404.

Hecht, S.B., and A. Cockburn. 2011. *The fate of the forest: Developers, destroyers, and defenders of the Amazon*. Chicago: University Of Chicago Press.

Ho, K. 2009. *Liquidated: An ethnography of wall street*. Durham: Duke University Press.

Hoelle, J. 2012. Black hats and smooth hands: Elite status, environmentalism, and work among the ranchers of Acre, Brazil. *Anthropology of Work Review* 33, no. 2: 60–72.

Li, T.M. 2007. *The will to improve: Governmentality, development, and the practice of politics*. Durham: Duke University Press Books.

MacDonald, J.M., P. Korb, and R.A. Hoppe. 2013. *Economic Research Report Number 152: Farm Size and the Organization of U.S. Crop Farming* (Economic Research Report No. 152). USDA Economic Research Service. http://www.ers.usda.gov/publications/err-economic-research-report/err152/report-summary.aspx#.UjSslWR4bfU (accessed December 10, 2013).

Mackey, L. 2011. Legitimating Foreignization in Bolivia: Brazilian Agriculture and the Relations of Conflict and Consent in Santa Cruz, Paper presented at the International Conferernce on Global Land Grabbing. 6–8 April, Brighton: university of Sussex. http://www.iss.nl/fileadmin/ASSETS/iss/Documents/Conference_papers/LDPI/23_Lee_Mackey.pdf (accessed September 20, 2011).

Marx, K. 2012. *Capital, volume one: A critique of political economy*. Mineola, NY: Courier Dover Publications.

Mauss, M. 2000. *The gift: The form and reason for exchange in Archaic societies*. New York: W. W. Norton and Company.

McVey, M., P. Baumel, and B. Wisner. 2000. *Brazilian soybeans - What is the potential?* Ames, IA: Iowa State University. http://www.extension.iastate.edu/agdm/articles/others/McVOct00.html (accessed November 1, 2012).

Mello, H.A.B. de, and S.D.S. Silva. 2011. Revisitando Contextos Bilíngues em Goiás. *Dossiê Imigrantes no.* 10: 11–39.

Miller, D. 2001. Turning Callon the right way up. *Economy and Society* 31, no. 2: 218–33.

Mitchell, T. 2008. Rethinking economy. *Geoforum* 39, no. 3: 1116–21.

Mondardo, M.L. 2010. A 'territorialização' do agronegócio globalizado em Barreiras - BA: migração sulista, reestruturação produtiva e contradições sócio-territoriais. *Revista NERA* 13, no. 17: 112–30.

Moore, J.W. 2010a. 'Amsterdam is standing on Norway'Part I: The alchemy of capital, empire and nature in the diaspora of silver, 1545–1648. *Journal of Agrarian Change* 10, no. 1: 33–68.

Moore, J.W. 2010b. 'Amsterdam is standing on Norway'Part II: The global North Atlantic in the ecological revolution of the long seventeenth century. *Journal of Agrarian Change* 10, no. 2: 188–227.

Moraes, G. 2013. *Legislação de Segurança e Saúde no Trabalho 2013 - 10a Edição*. Gerenciamento Verde Editora, Brasilia, Brazil.

Nickerson, C.J., M. Morehart, T. Kuethe, J. Beckman, J. Ifft, and R. Williams. 2012. *Trends in US farmland values and ownership*. USDA Economic Research Service. http://www.ers.usda.gov/media/377487/eib92_2_.pdf (accessed December 10, 2012).

Nixon, R. 2014. E.P.A.'s Proposed Rules on Water Worry Farmers. *New York Times*. 05/12/2014. http://www.nytimes.com/2014/03/13/us/politics/environmental-protection-agency-water-rules.html (accessed October 29, 2014).

O'Donoghue, E.J., R.A. Hoppe, D.E. Banker, R. Ebel, K. Fuglie, P. Korb, M. Livingtston, C. Nickerson, and C. Sandretto. 2011. *Economic information nulletin number 88: The changing organization of US farming*. Washington, DC: USDA Economic Research Service.

Ofstehage, A. 2011. Nusta Juira's gift of Quinoa: Peasants, trademarks, and intermediaries in the transformation of a Bolivian commodity economy. *Anthropology of Work Review* 32, no. 2: 103–14.

Ofstehage, A. 2012. The construction of an alternative quinoa economy: Balancing solidarity, household needs, and profit in San Agustín, Bolivia. *Agriculture and Human Values* 29, no. 4: 441–54.

Oliveira, G. de L.T. 2013. Land regularization in Brazil and the global land grab. *Development and Change* 44, no. 2: 261–83.

Peine, E. 2009. *The Private State of Agribusiness: Brazilian Soy on the Frontier of a New Food Regime*. PhD diss. Cornell University.

Peine, E. 2010. Corporate mobilization on the Soybean frontier of Mato Grosso, Brazil. In *Contesting development: Critical struggles for social change*, ed. P. McMichael, 132–46. New York: Routledge.

Polanyi, K. 2001. *The great transformation*. 2nd ed. Boston: Beacon Press.

Reboratti, C. 2010. Un mar de soja: la nueva agricultura en Argentina y sus consecuencias. *Revista de Geografía Norte Grande* no. 45: 63–76.

Sauer, S. 2012. Land and territory: Meanings of land between modernity and tradition. *Agrarian South: Journal of Political Economy* 1, no. 1: 85–107.

Scott, J.C. 1985. *Weapons of the weak: Everyday forms of peasant resistance*. Bethany, CT: Yale University Press.

Teubal, M. 2006. Expansión del Modelo Sojero en la Argentina: De la Produccion de Alimentos a los Commodities. *Realidad Economica, 220*(mayo/junio), 71–96.

USDA, National Agricultural Statistics Service. 2013. *Acreage (June 2013)*. Washington, DC: USDAo. http://www.usda.gov/nass/PUBS/TODAYRPT/acrg0612.pdf (accessed July 15, 2013).

Valdivia, G. 2010. Agrarian capitalism and struggles over hegemony in the bolivian lowlands. *Latin American Perspectives* 37, no. 4: 67–87.

Warnken, P.F. 1999. *The development and growth of the soybean industry in Brazil*. Ames: Iowa State University Press.

Wolford, W. 2008. Environmental justice and agricultural development in the Brazilian Cerrado. In *Environmental Justice in Latin America: Problems, Promise, and Practice*, ed. David Carruthers, 213–38. Cambridge, MA: MIT Press.

Green for gold: social and ecological tradeoffs influencing the sustainability of the Brazilian soy industry

Rachael D. Garrett and Lisa L. Rausch

In this study we assess the sustainability of the Brazilian soy industry over the past 40 years in comparison to alternative land uses. We conclude that Brazilian soy production performs as well as or better than sugarcane or cattle production in a number of areas, including macroeconomic contributions, local economic development and land use efficiency, though it involves similar tradeoffs between growth and equity, and food production and conservation. While there is no evidence that soy has reduced food security in Brazil, tax redistributions and value-added activities from soy remain limited, particularly in comparison to sugarcane production. Emerging environmental governance measures have helped to reduce the land cover impacts from soy; however, little effort has been taken to minimize the impacts of intensification.

1. Introduction

In the last 50 years, Brazil has emerged as one of the largest agricultural exporters in the world, playing a major role in the global supply of beef, poultry, soy, sugar, oranges and coffee. The expansion and intensification of Brazilian agriculture has generated substantial economic benefits for Brazil (Awokuse 2009). Unlike most other economic sectors in Brazil, agriculture has a positive trade balance, with exports exceeding USD 95 billion in 2012 (IMEA 2013). Nevertheless, these benefits have come at a huge environmental and social cost (Martinelli et al. 2010; Lapola et al. 2013). Over the last 40 years, 18 percent of the forest in the Legal Amazon region, 50 percent of the native vegetation in the Cerrado, Pampas and Caatinga, and 88 percent of the native vegetation in the Atlantic Forest has been cleared, primarily for agriculture (INPE 2014; Ferreira et al. 2012; Ribeiro et al. 2009). Agriculture now occupies 30 percent of the total land area of the country (Sparovek et al. 2010). Much of this agricultural land is controlled by large farms, particularly in sectors oriented toward export markets, such as sugarcane, soy and rice (Martinelli et al. 2010). 'Family farms',[1] which typically produce a diverse set of agricultural products for domestic consumption, occupy only 25 percent of the area (Lapola et al. 2013).

Amid these broader trends of increasing export-oriented agriculture in Brazil, soy production has increased faster than any other crop. Between 1970 and 2010, soy yields

[1] 'Family farmers', alternatively referred to as small-holders, are defined here as farmers with up to four fiscal modules (20 to 400 hectares depending on the county) that rely almost entirely on household labor for production, following Brazilian Law number 11,326 (MDS 2014).

doubled and planted area increased by 22 million hectares, contributing to a 10-fold increase in soy production (FAO 2013). While much of the recent literature has focused on the impacts of Brazilian soy production on land cover change, land concentration and social conflict (Soares-Filho et al. 2006; Vera Diaz, Kaufmann, and Nepstad 2009; Richards et al. 2012; DeFries et al. 2013; Morton et al. 2006; Macedo et al. 2012; Steward 2007; Baletti 2014; Fearnside 2007, 2001; Brannstrom 2009; Sparovek et al. 2010; Lapola et al. 2013), we provide a more nuanced assessment of the sustainability of the Brazilian soy sector by considering a broader suite of ecological and social impacts and tradeoffs.

Sustainable development is defined as development that leaves future generations with as many opportunities as current generations have, if not more (Serageldin 1996). It implies that current generations use resources in ways that increase the well-being of the current generation and the *potential* well-being of all generations that follow (Arrow et al. 2012). Applying this definition to soy production in Brazil, a 'sustainable soy sector' is one that meets the needs of current generations (by providing food, improving access to healthcare and education, and creating broadly distributed income and employment opportunities), without compromising future generations' opportunities to meet whichever needs they may have. This means, chiefly, that soy production must not deplete the natural capital stock on which all current and future generations' well-being depend (Daly 1991). However, this definition of sustainability creates challenges for policy design because it does not tell us exactly how much food, income and employment the soy sector should generate, or how much environmental degradation is possible without irreparably damaging the natural capital stock and endangering the well-being of future generations. Nor does it tell us anything about path dependency and what sorts of policy and development pathways are feasible given past trajectories.

For this reason, it is easier to view the goal of Brazilian soy sustainability, or any form of sustainability, for that matter, as a moving target – to achieve greater potential well-being *relative to* historical land use and development pathways, echoing the approach of Matson (2011) for the wheat sector in Mexico. In practice, this means that soy production is becoming *more* sustainable if it creates: (1) higher local incomes, (2) more jobs, (3) improvements in local infrastructure, (4) fewer greenhouse gas emissions, (5) less conversion or more restoration of native vegetation, particularly around streams, and (6) improved soil quality *relative to* previous production practices. It is becoming *less* sustainable when it: (1) increases income inequality, (2) exposes people to new health risks, (3) marginalizes/excludes small farmers and (4) reduces biodiversity. This multi-criteria definition follows the efforts of Loos et al. (2014) to broaden the discussion of 'sustainable agriculture systems' beyond a limited set of environmental criteria.

To understand the historical impacts of Brazilian soy we begin this paper by providing a brief overview of soy development and expansion in the region. We then analyze a wide range of secondary data and existing literature to identify the relative impacts of the soy sector on current well-being versus other potential land use activities. In describing impacts on current generations' well-being (i.e. inclusive development) in Brazil, we focus on the contributions of soy production to the economy of Brazil (macroeconomic conditions), national and household food security, access to goods and services in rural areas, and rural equity. In assessing hypothetical impacts on future generations' well-being (i.e. sustainability) we focus on the soy sector's influence on natural capital within Brazil and globally through land cover change, greenhouse gas emissions and pollution. We conclude by discussing the potential of current social and environmental governance mechanisms to improve the sustainability of the soy sector, including federal policies and industry and civil society-led market mechanisms.

Table 1. Variables examined in this analysis.

Impacts	Scale	Variable
Macroeconomic conditions	National	Total exports and share of agricultural exports
		Tax revenue
		GDP from agriculture
		Agricultural loans
Food security and quality	National	Direct food security measures
		Changes in domestic staple production
		Protein production per hectare of soy versus other crops
Access to goods and services	National	Local infrastructure development and urbanization
	County	HDI-total, HDI-income, HDI-longevity, HDI-literacy (2003)
Equity	National	Foreign ownership of land
		Employment in soy compared to other crops
		GINI coefficient for land compared to other crops
	County	Theil index of income inequality (2000)
	Case studies	Exclusion of existing land users
		Displacement
Land cover change	Biome	Historical and total land cover change
		Soy area per biome
		Carbon debt from land conversion to soy
		Watershed impacts of land cover change on soy farms
Input intensification	Biome	Fertilizer used on average soy farm in Amazon and Cerrado
		N and P emissions from soy fields in Cerrado and Amazon

2. Methods

We examine the influence of the Brazilian soy sector on intergenerational well-being using a variety of socio-economic and ecological variables at four different scales: national, biome, county and community. While the use of multiple scales in this study limits tradeoff analysis, it is necessary since many data are not available, quantifiable or relevant at all scales. Details on variables examined at each scale are presented in Table 1. Below is a short synthesis of our approach:

- Macro-economic impacts are analyzed at the national scale, the only scale relevant for these impacts.
- Food security impacts are analyzed at the national scale, since no data are available at the local level for all regions.
- Natural capital/ecological impacts are analyzed at the biome scale, since this is the scale at which the use of agricultural inputs and impacts from land cover change from agricultural expansion vary the most. We also include data on land cover and watershed impacts from the state of Mato Grosso, which spans multiple biomes, since this is the largest and most well studied soybean production region in the country.
- Household well-being and equity impacts are analyzed at the municipal scale, since data is available for almost all counties (n = 5539) through secondary sources. We applied an analysis of variance (ANOVA) test followed by the post-hoc Tukey test to examine whether mean levels of well-being are different for counties that are 'soy dominant' in comparison to counties that contain other dominant land uses. A county is classified as soy dominant (SOY) if in 2000 it had more than 50 percent of its crop area planted in soy, sugar dominant (SUGAR) if it had more than 50

Table 2. Methods for classifying dominant land uses.

Class	Soy	Sugar	Cattle	Other AG	Non AG
Criteria	>50% of crop area in soy	>50% of crop area in sugar	>31,000 head cattle, <50% of crop area in soy or sugar	>10% of area in crops <50% of crop area in sugar and soy <31,000 head cattle	<10% of area in crops <31,000 head cattle
No. of cases	337	373	1167	1663	2099

Note: Where 'soy' = soy dominant counties; 'sugar' = sugar dominant counties; 'cattle' = cattle dominant counties; 'other AG' = counties dominant in other forms of agriculture; and 'non AG' = non-agricultural counties.
Source: (IBGE 2013a, 2013b).

percent of its crop area planted in sugarcane, cattle dominant if it had above-average cattle herd size but less than 50 percent of its crop area planted in soy or sugarcane (CATTLE), and other agricultural (OTHER) if more than 10 percent of the county area contains cropland but cattle herd size is below average and not more than 50 percent of its crop area is planted in soy or sugar. Finally a county is characterized as non-agricultural (NON-AG) if less than 10 percent of the county area is comprised of cropland and the county contains a below-average cattle herd (Table 2). Average household well-being in each county is represented by the Human Development Index (HDI) from 2003, which includes three components: income, literacy and longevity. The HDI values range from 0 to 1, signifying that the population in a county achieving a score of 0.8 will have higher average per-capita income, years of education and life expectancy than the population in a county achieving a score of 0.5. The HDI measure represents only *average* income, literacy and longevity in each county, not the *distribution* of income, education or health. Therefore, to examine equity, we use the Theil Index of income inequality from 2000. The Theil values also range from 0 to 1, signifying that the population in a county achieving a score of 0.8 will have higher income inequality than the population in a county achieving a score of 0.5. Unfortunately, there are no data on the variance of health and education levels within each county. The following controls are included in the statistical analysis since they may also influence HDI and Theil levels: lagged transportation costs to São Paulo (1995), the date the county was established, population (1996) and the percentage of the population that is urban (1996). Distance to São Paulo, date of establishment and percentage of the population that is urban are included to reflect the history of the county and/or the composition of the economy, in particular the presence of service and manufacturing opportunities. For example, an urban municipality near São Paulo that was established 150 years ago may have better health services than a remote, recently established county. Population is included as a control because it may influence the resources available within a county, i.e. health and education infrastructure may be more abundant in a county with a high population. Descriptive statistics and data sources for all non-land use variables are presented in Table 3.

- Local development and equity outcomes were also examined at the community level through first and second author observations from over five years of fieldwork in soybean-producing regions in Brazil, and by reviewing previously published field studies, which examine more complex social outcomes, such as investments in local infrastructure, agro-urbanization, displacement, exclusion and rising costs of living.

Table 3. Distribution of HDI and Theil by dominant land use.

Variable	Soy Mean (SD)	Sugar Mean (SD)	Cattle Mean (SD)	Other AG Mean (SD)	Non AG Mean (SD)
HDI-total	0.77 (0.04)	0.72 (0.09)	0.72 (0.06)	0.71 (0.09)	0.67 (0.08)
HDI-income	0.68 (0.06)	0.63 (0.10)	0.63 (0.08)	0.61 (0.09)	0.57 (0.10)
HDI-literacy	0.86 (0.05)	0.79 (0.10)	0.80 (0.07)	0.78 (0.09)	0.76 (0.09)
HDI-longevity	0.77 (0.05)	0.73 (0.08)	0.72 (0.06)	0.72 (0.09)	0.69 (0.08)
Theil	0.56 (0.12)	0.47 (0.10)	0.56 (0.11)	0.50 (0.10)	0.52 (0.10)
Transport cost to São Paulo (index)	1166 (474)	1296 (1031)	1642 (1194)	1396 (896)	1903 (1350)
Population (1000 people)	20.3 (41.6)	31.5 (65.7)	40.8 (111.7)	15.5 (24.4)	41.6 (289.2)
% of population that is urban	58 (23)	74 (18)	62 (22)	50 (22)	55 (24)

Note: Where 'soy' = soy dominant counties; 'sugar' = sugar dominant counties; 'cattle' = cattle dominant counties; 'other AG' = counties dominant in other forms of agriculture; and 'non AG' = non-agricultural counties.
Sources: IPEA. Total and urban population are from the demographic census (http://www.sidra.ibge.gov.br).

One of the main limitations of our quantitative analysis is that our indicators of well-being are not available for more recent years, nor do we possess a time series of these data. Instead, we must infer the relationship between soybean production, human development and equity from spatial variations in land use, HDI and income inequality. Nevertheless, this analysis is helpful for understanding how 'soy counties' compare to 'sugar and cattle counties', since evaluations of the sustainability of soy production frequently lack comparisons with other land use alternatives. Without this type of information, it is impossible to understand whether another land-use strategy might be capable of providing more benefits to society than soy, with fewer ecological impacts.

3. A brief history of soy production in Brazil

Soy production in Brazil began in the South, in the Atlantic Forest and Pampas ecosystems (Figure 1), where the naturally fertile red latosols soils require fewer inputs than the soil of other regions in Brazil. There, soy was double-cropped with wheat, a commodity that was heavily subsidized by the federal government in the 1960s and 1970s to reduce dependence on imports (Faminow and Hillman 1987). The varied topography and intergenerational land fragmentation limited the size of farms in this region. While soy became one of the larger land uses in this region, it never fully displaced livestock, which continues to be the largest land use in the region (Table 4; IBGE 2006).

Soy production advanced into the Cerrado savanna ecosystem in the Center–West region of Brazil during the 1960s and 1970s as part of a larger government process to modernize and increase the legibility of this 'empty space' (Oliveira 2013; Hecht 2005; Scott 1998; Hecht and Mann 2008). The high-modernism ideology of the period included an unwavering confidence in the potential of science and technology to achieve social progress (Scott 1998), which led the government to invest heavily in agricultural research and development. Plant geneticists and agronomists from the Brazilian Agricultural Research Corporation (Embrapa) were sent to the United States for training and charged with the task of developing new agricultural technologies for the heretofore un-arable landscape (Hecht and Mann 2008; Oliveira 2013). This federal investment led to the development of so-called miracle soy cultivars capable of tolerating the metal-heavy, nutrient-poor, acidic soils of the Center–West, with high levels of biological nitrogen fixation. (Spehar

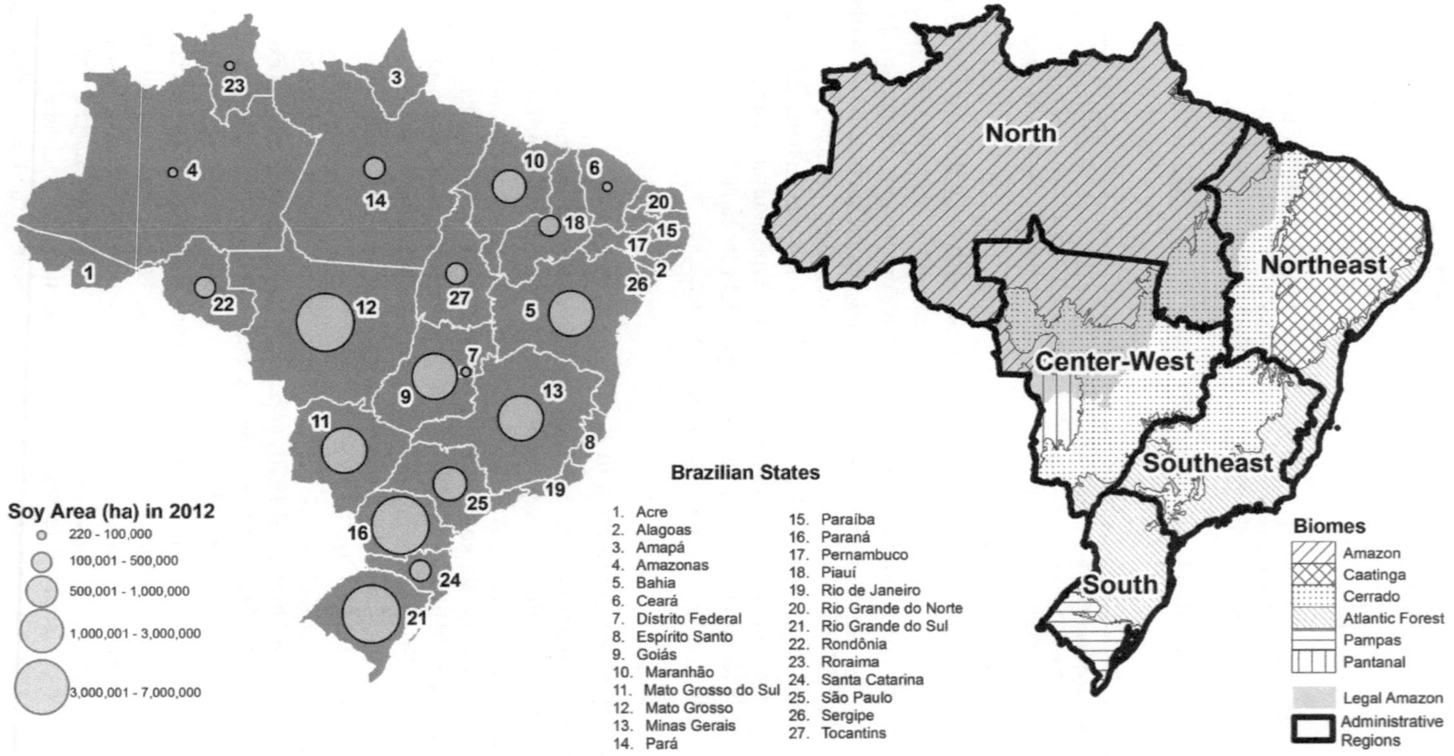

Figure 1. (a) Soy area by state in Brazil; (b) major geographical regions and ecological biomes in Brazil. Source: (IBGE 2013a).

Table 4. Major land uses by region (1980–2006).

		% of agricultural[a] area in each land use			
Region	Land use	1980	1985	1995	2006
North	Permanent crops	1	1	1	3
	Annual crops	3	3	2	4
	(Soy)	<1	<1	<1	1
	Pastures	19	33	42	48
Northeast	Permanent crops	5	5	3	5
	Annual crops	11	11	10	15
	(Soy)	<1	<1	1	2
	Pastures	39	38	41	40
Southeast	Permanent crops	5	5	5	7
	Annual crops	12	13	11	17
	(Soy)	1	1	1	2
	Pastures	59	58	59	51
South	Permanent crops	3	2	1	4
	Annual crops	28	28	26	33
	(Soy)	13	13	11	17
	Pastures	44	45	47	38
Center–West	Permanent crops	<1	<1	<1	1
	Annual crops	5	7	6	11
	(Soy)	1	2	3	7
	Pastures	60	60	58	56

Source: (IBGE 2006).
[a]Agricultural area includes pastures and forest reserves.

1995; Sousa and Busch 1998; Luna and Klein 2006; Alves, Boddey, and Urquiaga 2003). Embrapa also facilitated the transfer of no-till/direct planting technology, which was critical to the long-term sustainability of soy production in the region by reducing erosion, enhancing soil permeability for water infiltration, and increasing organic material in the soil (Gasques and Bastos 2010).

In contrast to the predominantly state-led colonization programs in the Amazon, families moved to the Cerrado during the late 1970s and 1980s, through a combination of state efforts, spontaneous migration and organized private colonization (Rausch 2014). When state support was not available, private colonization firms helped migrants navigate the complicated government bureaucracy, volatile markets and lack of infrastructure (Jepson 2006a). Despite harsh initial conditions, competition and collaboration between multiple settler groups accelerated the development of the region (Rausch 2014). Soy production in the Cerrado soon became one the most lucrative rural land uses in Brazil, particularly since the relatively flat terrain allows for large farms and economies of scale. Soy farms in the region now average 1000–2000 hectares in size, but sometimes exceed 10,000 hectares, much larger than the national average soy farm size of roughly 130 hectares (IBGE 2006; Garrett, Lambin, and Naylor 2013a). Given the low natural fertility of the Cerrado soils, soy production requires large applications of potassium, phosphorus and lime, but is now capable of achieving soy yields averaging 3 tons per hectare there (IBGE, 2013a), rivaling, if not exceeding, productivity in the United States (Ash 2012).

Soy expansion in the Amazon forest ecosystem in the North of Brazil did not proceed in earnest until the 1990s. By the time soy first arrived in the region, many of the 'soy-suitable'[2] regions of the Amazon, particularly in Pará and Rondônia, were already settled by previous colonization projects (Brown et al. 2005; Steward 2007). In these regions, particularly post-2006, soy expansion occurred mainly on areas that had been previously cleared for small-holder production and pastures (Rudorff et al. 2011; Brown et al. 2005; Steward 2007). Consequently, the expansion of soy in these regions was fraught with cases of resistance, conflict, displacement and indirect land-use change, typified by the highly publicized soy-producing regions of Paragominas and Santarém (Steward 2007; Gardner et al. 2013). In contrast, soy expansion in Southern Amazonia (Northern Mato Grosso and Southern Rondônia) was implicated in a large amount of direct deforestation of primary forest in the early 2000s (Morton et al. 2006), but fewer reported social impacts.

In addition to investments in agricultural research and development, the government promoted the expansion and modernization of Brazilian agriculture through supportive monetary and agricultural policies. The Bank of Brazil and the National Rural Credit System were established to provide cheap credit to farmers, with interest rates fixed below the rate of inflation (Warnken 1999; Sousa and Busch 1998). In the Cerrado, the government created the Superintendency of the Development of the Center–West (SUDECO) to implement fiscal incentives for migration, the Development Program for the Cerrado (POLOCENTRO) to fund agricultural research and extension, improvements to roads and electricity infrastructure, and the Japanese–Brazilian Cooperative Program for Cerrado Development (PRODECER) to subsidize the construction and purchase of silos and farm equipment (Jepson, Brannstrom, and Filippi 2010; Fearnside 2001; Jepson 2006a). More broadly, the government also provided minimum price supports for commodities such as soy, corn and wheat (Damico and Nassar 2007), but high global prices have rendered these price supports largely unnecessary in recent years.

The private sector took the leading role in the development of the soy industry during the 1990s as a result of changes in macroeconomic, trade and agricultural policies (Jepson 2006a). The Brazilian government lowered import tariffs, exempted soy exports from the interstate movement tax, joined MERCOSUR (a regional free trade agreement),[3] and stabilized the national currency (Flaskerud 2003; Chaddad 2006; Luna and Klein 2006). These changes made Brazilian soy more competitive in international markets and helped Brazil double its agricultural exports between 1990 and 2005 (Chaddad 2006). Brazil's liberalized markets, macroeconomic stability and investments in infrastructure continued to attract foreign investment to the Brazilian soy industry throughout the 2000s (Luna and Klein 2006; Jank, Leme, and Nassar 2001). As foreign investment by multinational food companies increased, the soy sector became more vertically coordinated and internationally integrated. Multinational agribusinesses connected distant consumers to the far reaches of the Brazilian Cerrado through advanced logistics services (Jank, Leme, and Nassar 2001). They also helped reduce producers' risks from price instability by offering futures contracts and improved access to storage facilities (Vera-Diaz et al. 2008).

Throughout the 1990s and 2000s, the soy industry became increasingly concentrated through mergers and acquisitions, displacing domestic processing, transportation and

[2]Defined as regions with both adequate biophysical conditions (rainfall greater than 1200 mm and less than <2200 mm per year, a distinct dry season to help control pests and weeds, flat topography, soils with adequate drainage) and good transportation networks (access to paved roads, federal highways and or major waterways).

[3]MERCOSUR is regional trade agreement between Argentina, Brazil, Paraguay and Uruguay, and associate members Chile and Bolivia, established in 1991 (Belik and Santos 2002; Polaquini, Souza, and Gebara 2006).

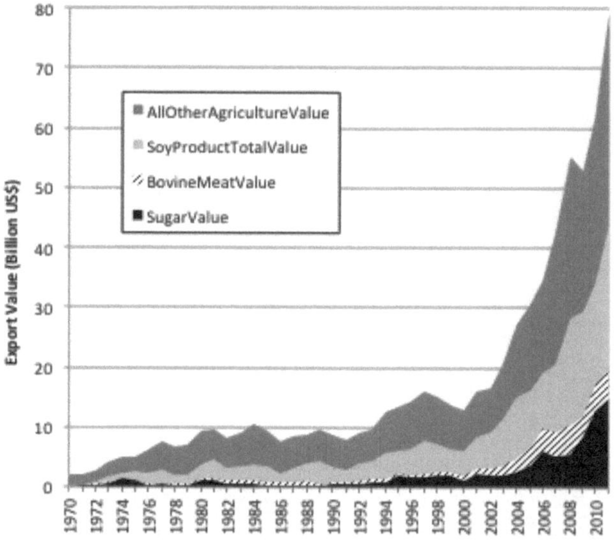

Figure 2. The value of soybean, oil and meal exports in comparison to the value of all other agricultural exports in Brazil (1970–2010).
Source: (FAO 2013).

retailing firms. A small number of soy firms, including Archer Daniels Midland (ADM), Bunge, Cargill, Louis Dreyfus and Maggi Group now control a majority of the soy market in Brazil (Jank, Leme, and Nassar 2001; Goldsmith and Hirsch 2006). Cooperatives also played a critical role in the development of soy production in the Cerrado by facilitating land acquisition and tenure, increasing producers' access to inputs, providing technical assistance and helping producers market their grain for higher prices (Jepson 2006b; Garrett, Lambin, and Naylor 2013b, 2013a; Jepson, Brannstrom, and Filippi 2010). Infrastructure remains the largest challenge inhibiting soybean profitability in the Center–West of Brazil, where limited road and rail networks pose extremely high costs for the industry. In Northern Mato Grosso (one of the largest soy producing regions in the country), transportation costs account for 25 percent of the total price of soy (Vera Diaz, Kaufmann, and Nepstad 2009).

4. Impacts on current generations' well-being

4.1. *Macroeconomic conditions*

The value of soy exports (unprocessed beans and processed oil and meal) has increased exponentially since 1970, reaching USD 28 billion in 2013, which is roughly equal to the value of all other agricultural exports combined (MDIC 2013; Figure 2). The export revenues from soy production have contributed to an increase in the value of the Brazilian currency, which increases consumers' purchasing power in world markets. On the other hand, the appreciation of the Brazilian Real reduces the competitiveness of Brazilian manufactured goods in world markets, which may decrease employment opportunities in the manufacturing sector (Gaulard 2012).

Soy production requires a large number of supporting services and thus has a fairly complex supply chain (Figure 3), including input provision, financing, marketing, distribution and processing (Garrett, Lambin, and Naylor 2013b). In 2007, the soy supply

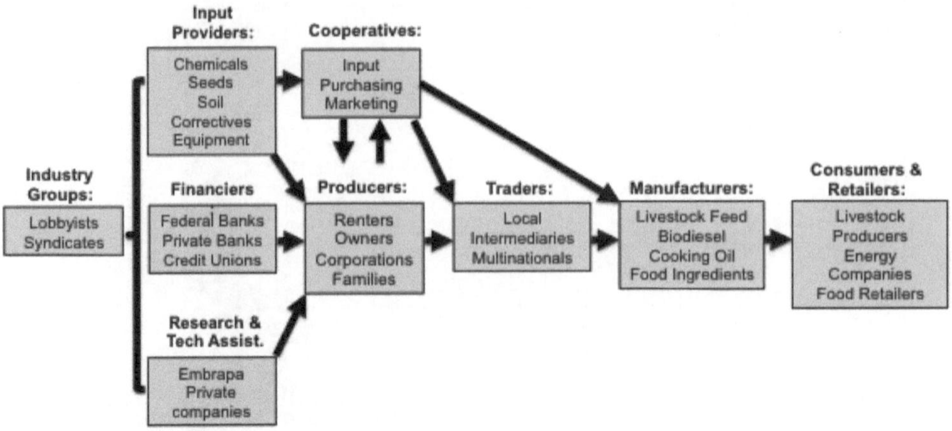

Figure 3. Typical Brazilian soy supply chain.

chain generated USD 17.2 billion toward the Brazilian gross domestic product (0.7 percent of the national GDP; Barros, Fachinello, and Silva 2011). Roughly 45 percent of this GDP came directly from soy production, 40 percent from services, 8 percent from processing and 7 percent from inputs. Still, an increase in the level of domestic soy processing could further amplify the revenues obtained from Brazilian soy production given that more than 50 percent of all the soy produced in Brazil is exported without any processing (FAO 2013).

Even without the addition of more value-added activities, soy production and associated agribusiness generate substantial tax revenue for the country. The tax contribution from domestic sales of whole soybeans alone was USD 1 billion in 2004[4] (Lazzarotto and Roessing 2004). A portion of the tax revenues generated by soy production and processing (through the federal tax programs Fundo de Assistência ao Trabalhador Rural (FUNRURAL), Programa de Integração Social (PIS), and Seguro Acidente de Trabalho (SAT)) are used to support welfare programs, such as conditional cash transfers, social security, unemployment insurance and local rural development.[5] On the other hand, since the passage of the federal Kandir Law in 1996 (Law no. 87), soy destined for export is exempt from the interstate movement tax (ICMS), a major source of tax revenue for states. The Kandir Law was part of the suite of policies developed during the 1990s to increase foreign investment and exports. States' loss of this revenue is theoretically compensated by redistributions of federal funds to the states and municipalities. However, state governments claim that the redistributions fall short of the ICMS losses, though the benefits of greater investment brought by lower taxation on the production of commodities destined for export likely outweigh any such shortfalls (Varsano 2013). Finally, municipal property taxes, which are based on land values, can be channeled directly into improving local services and infrastructure in soy-producing regions. Nevertheless, the overall tax rate on soy production, like that on other Brazilian farming activities, is fairly low, and there are no taxes on soybean, meal or oil exports.

---

[4]No later estimates were available that we could locate.

[5]The rates and names of these taxes are not the same in every state and can be substituted by other tax programs. For producers, they also differ depending on whether the farm is run as a corporation or a family.

Contributions to national export revenues, GDP and taxes must be evaluated alongside contributions that the soy industry receives from the federal government. In contrast to the United States and Europe, Brazil provides little direct support to farmers through subsidies, payments or insurance (OECD 2010). However, Brazilian soy farmers benefit substantially from state-supported/subsidized agricultural loans at below-market interest rates. Agricultural loans to cover soy production costs, value added activities and commercialization totaled more than USD 7 billion in 2012 (using an exchange rate of 1 BR Real = 0.55 USD; BCB 2012). The total value of agricultural loans and credit subsidies for soy production exceeded every other crop between 1969–1990 (Helfand 2001). The soy industry also benefits greatly from federal investment in new roads and pavement of existing roads in key production regions, such as Mato Grosso and Pará (Soares-Filho et al. 2004), but these investments have not been quantified.

4.2. *Food security and quality*

Food security is defined as access by *all people at all times* to enough food for an active, healthy life (World Bank 1986). The expansion of soy production in Brazil – a commodity crop with few direct food uses – raises questions about tradeoffs between global and local food security, as well as between physical and economic food security. According to Brazil's Survey of Demography and Children and Women's Health, 38 percent of Brazilian families reported some form of food insecurity in 2006, with higher levels of food insecurity in the North and Northeast, particularly in rural areas (PNDS 2006). Among the food insecure, *economic* food insecurity (the inability to purchase desired foods even when they are physically available) featured more prominently than physical food insecurity. In particular, approximately 90 percent of food-insecure families reported that they were worried that their food would run out before they received more money, could not afford to purchase balanced meals and relied on low-cost foods for children (Reis 2012). Thus, poverty tends to be a far more important driver of food security than physical food scarcity in Brazil.

Approximately 50 percent of the soy produced in Brazil is exported, but this leaves more than 30 million tons of soybeans for domestic use, which can be converted into oil and meal to be consumed directly in cooking and indirectly via pork and poultry production, or else exported at a higher value (FAO 2013). Per capita consumption of poultry and pork in Brazil have grown by 280 percent and 56 percent over the last three decades (1980–2010),[6] meaning that Brazilians are now indirectly consuming more soy than ever before in the form of animal proteins.

Nevertheless, the use of soy as a livestock feed brings up the issue of food quality versus quantity, and the land intensity of food production. While processed meats provide fat and protein, more diversified farming systems produce foods with a wider range of nutrients. Producing domestic staples, fruits and vegetables, alongside export crops like soy and cotton, can increase farmers' autonomy from volatile commodity markets, resilience to climate change and weather shocks, and sovereignty over the means of production (Altieri and Toledo 2011; Schneider and Niederle 2010). In terms of domestic food supply, soy has not reduced the availability of rice, beans and manioc in the country, since production of these crops has actually increased over the last 20 years through a reorganization of cropping patterns and intensification of production (Brown, Rausch, and Luz

[6]The second author calculated these figures using data from FAOSTAT Food Balance Sheet and Brazilian Population Census data (IBGE 2010; FAO 2013).

2013; Lapola et al. 2013). In many regions, soy is double-cropped with another product – corn in the Center–West and wheat, rice and beans in the South. Soy production has replaced (or displaced) some pastures for grazing cattle over the past decade (Rathmann, Szklo, and Schaeffer 2010; Walker 2011), but total pasture area has not increased and the national cattle herd has continued to grow through intensification on the existing land base (IBGE 2013b; Soler, Verburg, and Alves 2014; Martha Jr, Alves, and Contini 2012).

The use of soy as feed in confined livestock operations is argued to be a land-intensive source of protein (Brown et al 2014), meaning that it takes a relatively large amount of land to produce a unit of food. Yet it is still more land efficient, on average, than open-range, grass-fed cattle production in Brazil. Given a meal-to-oil ratio of 80:20 and soy and conversion efficiencies of 7:1 for cattle meat, 4:1 for pork and 2:1 for chicken (in a confined system; Brown 2006), a soy yield of 3000 kg per hectare will produce 343 kg of cattle meat, 600 kg of pork or 1200 kg of poultry per hectare if used in confined livestock operations. In contrast, cattle production, which occupies the largest area of any agricultural sector in Brazil, roughly 200 million hectares, produces on average only one head and 250 kg of cattle meat per hectare (Walker, Patel, and Kalif 2013). Furthermore, soy production is technically more land efficient in producing *nutritious* food than sugarcane cultivation, since sugar offers little nutritional quality and is often deleterious to human health (Brownell et al. 2009).

4.3. *Access to services*

Soy production is a capital-intensive process that facilitates improvements in rural infrastructure and service provision since farmers require adequate access to dynamic input markets, dryers and silos, and transportation systems to compete in the global economy. Soy farmers and agribusiness owners sometimes build new roads and finance new health and education facilities (RDG personal observations in Mato Grosso and Pará). When farm owners actually live in the same county where their farm is located, they spend money locally on goods and services, which can promote developments in infrastructure that benefit all members of the local community (de Souza and Ravache 2009). 'Agro-cities' emerge in these nascent soy regions as new businesses are established to sell non-agricultural goods and services to farm and agribusiness employees, leading to new employment opportunities outside of the agricultural sector (Da Silva and del Grossi 2001; Alves Sobrinho 2009; Jonasson and Helfand 2010). The wage level in these non-farm jobs may be lower than in surrounding regions due to competition in the labor market from increased in-migration (Jonasson and Helfand 2010). However, these agro-cities often offer a variety of health and educational services, restaurants and consumer goods, including upscale gyms, vocational colleges, federal university branches and luxury condominiums patronized by soy producers and soy sector employees, as well as individuals not directly involved in the soy sector (authors' personal observations; Hecht and Mann 2008).

Statistical analysis of land use and the HDI supports these observations of broader local development alongside the soy sector. 'Soy-dominant' counties have significantly higher HDI (across all categories of the index – income, literacy and longevity) than 'sugar- or cattle-dominant' counties and counties that do not contain high levels of cropland as a proportion of the total area (Table 5). These results are consistent with the results of previous quantitative studies by Weinhold, Killick, and Reis (2013) in Pará and VanWey et al. (2013) in Mato Grosso linking soy production to HDI. However,

Table 5. Results of Tukey test for HDI and Theil by dominant land use.

HDI-total	Non AG <other AG <sugar, cattle <soy
HDI-tncome	Non AG <other AG <sugar, cattle <soy
HDI-literacy	Non AG <sugar, cattle, other AG <soy
HDI-longevity	Non AG <cattle, otherAg <soy nonAG <cattle <sugar <soy
Theil	Sugar <non AG <other AG <cattle, soy

Note: ‘<’ signifies that the mean value for the Human Development Indicator (HDI) and Theil in a particular land use group to the left of the ‘<’ is significantly less (p <0.05) than the mean value of the land use group to the right of the ‘<’. ‘,’ denotes no significant difference between groups. Where ‘soy’ = soy dominant counties; ‘sugar’ = sugar dominant counties; ‘cattle’ = cattle dominant counties; ‘other AG’ = counties dominant in other forms of agriculture; and ‘non AG’ = non-agricultural counties.

our results also add new fodder to the discussion of the well-being impacts of soy production because they compare the relationship of soy production on HDI with that of other land uses. This implies that soy production may generate higher on-farm wages and/or off-farm employment opportunities than sugar, cattle or other land uses in Brazil, and/or that soy production tends to result in more local investment in infrastructure to provide goods and services than other land uses. These results are particularly interesting given the findings of Martinelli et al. (2011), which show that sugar counties in São Paulo that have high levels of sugar production and a mill to refine the sugar do better than counties with high levels of cattle production or sugar production but no mill. They highlight the important additive role played by value-added processing activities in influencing local well-being improvements through higher incomes and employment.

The HDI analyses relating specific crop area to HDI differ greatly from the cross-sectional studies relating deforestation levels and forest cover to HDI (Rodrigues et al. 2009) and suggest that soy agro-cities differ in important ways from the dangerous frontier towns associated with extractive boom and bust cycles in the Amazon (see Schmink and Wood 1992). Among these differences is that soy production in the Amazon does not typically take place on deforestation frontiers, but instead follows some other activity, such as ranching, for which the forest is cleared (Morton et al. 2006). Successful soy production also depends on making long-term investments in infrastructure and soil quality, as well as being sufficiently established on the land to diminish the risk of making a considerable investment in planting but having to leave or being unable to harvest the crop. Soy agro-cities also differ substantially from the planned agropolis/agrovila system established by the National Institute of Colonization and Agrarian reform (INCRA; discussed in Goodland and Irwin 1975) in that development is sustained by local market forces rather than inconsistent federal infusions.

4.4. *Equity*

Soy production occurs on more than 171,000 individual farms across Brazil (IBGE 2006). Many ‘family farmers’ participate in the production of soy (Table 6); however, the distribution of these soy lands, in terms of land area per owner, is more unequal than that of cattle ranching and staple crop production (rice, beans and manioc), but less unequal than that of sugarcane production (Martinelli et al. 2010; Lapola et al. 2013). The concentration of land in the hands of fewer people inherently exacerbates rural income inequality if there are few off-farm employment opportunities. The soybean industry is even more concentrated at the soy trading and processing nodes of the supply chain.ADM, Bunge, Cargill and Louis

Table 6. Percentage of farmers that produce soy that fall into the legal category of 'family farmer' (Law 11,326).

Region	Soy	Sugar	Livestock	Horticulture
	(% of farmers)			
Brazil	71	72	78	87
North	32	84	79	90
Northeast	8	82	84	90
Southeast	44	61	72	84
South	77	83	82	88
Center–West	31	67	66	78
	(% of area)			
Brazil	10	9	20	35
North	2	25	23	37
Northeast	0	13	33	45
Southeast	5	7	24	38
South	26	19	27	47
Center–West	2	4	9	17

Source: IBGE 2006.

Table 7. Rural population, farm size and agricultural employment by region.

Region	Rural population (1000)	Avg farm size (ha)	Ranching	Sugar	Soy	Cereal	All Agriculture
			(% of rural population employed)				
Brazil	31,948	141	15	2	1	6	52
North	6178	1104	9	<1	<1	1	27
Northeast	8612	1761	21	3	<1	13	89
Southeast	6902	401	15	4	<1	2	48
South	4791	69	17	2	7	7	61
Center–West	1549	840	38	4	4	3	65

Source: IBGE (2006, 2010). Farm size calculated from a distribution of farms in various sizes. Employment proportion calculated as % of 2010 rural population data.

Dreyfuss now trade a majority of genetically modified (GM) soy produced in Brazil, having acquired many of the domestic trading firms (Jank, Leme, and Nassar 2001). Brazilian soy giant Grupo Andre Maggi is the largest trader of certified non-GM soy (Garrett, Rueda, and Lambin 2013), but there are numerous small producers and traders participating in the non-GM soy chain as well.

Like most mechanized, agro-industrial systems, specialized soy production requires very little on-farm labor, particularly in comparison to ranching or other crops (Table 7). In the South, where the farms are much smaller, soy production employs a larger proportion of the rural population. In the Amazon and the central and eastern Cerrado production zones, however, soy farms tend to be much bigger and employ only a small proportion of the rural population (Table 7; Fearnside 2001; Hall et al. 2009). The low requirement for on-farm labor is simultaneously a blessing and a curse. On the one hand, there are few new employment opportunities for current residents to benefit from the emergence of soybean production. On the other hand, the negligible of role of non-skilled labor in soy production means there are fewer opportunities for the types of indentured and migratory labor that occurred in Brazilian sugarcane production (Chase 1999; Fearnside

2008; Sawyer 2008). Humane working conditions on soy farms (and all other farms) are also incentivized by the 'dirty list of slave labor' published by the President's Secretary of Human Rights. Employers included on this list are barred from accessing public financing, such as subsidized rural credit, which operates as a strong disincentive in the soy industry since soy cultivation depends heavily on subsidized rural credit. Indeed, only 10 out of 583 (<2 percent) of the farms implicated on the list produce soy (Reporterbrasil.org 2014). There are also slave labor prohibitions in the Soy Moratorium and Roundtable on Responsible Soy certification, which are discussed in more detail below.

Most employment opportunities in the soy sector occur off farm, in related agribusiness (transportation, industry and services). However, existing residents are often unable to acquire the new positions created by soy agribusiness since they do not possess adequate skills (Zoomers 2010). New jobs created by soy agribusiness in the Center–West and North are often taken by migrants from the South, leading to conflict between migrants and existing populations or outmigration of the existing rural population.

Another criticism of the soy sector in Brazil is that the revenues generated by soy production are increasingly captured by non-Brazilians, due to growing foreign investment in farmland, termed by some as 'land grabbing'. In the contemporary Brazilian context, though, foreign investment in farmland is considerably more complex than the term land grabbing suggests. While many corporate farms directly export their products to overseas buyers, foreign ownership of land is limited. Foreign non-residents and foreign corporations are not allowed to directly acquire land in Brazil without special federal authorization, although foreign residents can acquire up to three 'undefined exploration modules' (ranging from 5 to 70 hectares in size; Laws 5709 and 6815). As of 2011, roughly 4.5 million hectares (<2 percent of the total agricultural area in Brazil) were owned directly by foreigners (INCRA 2012). Where foreign direct investment does occur, it is more often in the form of foreign–domestic agribusiness partnerships, rather than direct land ownership (Oliveira 2013). Furthermore, foreign investment in land goes two ways, since Brazilian farmers are very active in purchasing land in the agro-industrial soy frontiers of Bolivia and Paraguay. In these regions, Brazilians and other foreigners often control more land than nationals and have been implicated in widespread displacement and evictions (Hecht 2005; Nagel 1999).

Where soy production is profitable and land for expansion is scarce, land prices can rise very quickly (Steward 2007). In some cases, resident small-holders sell their land to soy farmers out of a desire to find new occupations. In other situations, peasants sell to avoid being enclosed by soy farms or because their social networks have been broken by their neighbors' land sales (Steward 2007; Baletti 2014). Land sales often enable families to move to urban areas or invest in new income opportunities, but can also lead to unanticipated livelihood losses when people move to more marginal farming locations or pay higher prices for the new land, or when they are unable to find new employment opportunities (Baletti 2014; Zoomers 2010). In the more extreme cases where tenure rights are not secure, interest in acquiring new land for soy production can lead to the forcible removal of vulnerable populations. Or, if landholders remain, they may be exposed to new economic and biological pressures that reduce local prices and yields. For example, in a recent Ecological Economic Zoning Plan for the BR-163, some lands currently occupied by traditional and indigenous landholders were reassigned as areas designated for soy expansion and extractive industries (Baletti 2012).

Confirming the more qualitative results of previous case studies, we find that counties that are 'soy dominant' have higher levels of income inequality (a higher THEIL index) than counties that are 'sugar dominant' and non-agricultural, although the THEIL index was not statistically different between soy- and cattle-dominant counties (Table 5). These results may be explained by a number of factors. One potential reason is that labor requirements on mechanized soy farms are lower than labor requirements on fruit and horticulture farms that rely on manual harvesting. With respect to sugar production, mechanized soy farming is particularly less labor intensive than sugar production in the northeast where cane burning and manual harvesting remain high (De Moraes 2007; Lehtonen 2012). Another explanation is that most major sugar production regions, especially in the southeast, contain sugar processing mills and biofuel plants, which present additional employment opportunities, typically with higher wages than farm labor (De Moraes 2007). A final possible explanation is that soy production often occurs in areas of change where a new land use activity is expanding rapidly, which attracts migrants who have high hopes of participating in the 'soy economy' but are unable to find high wage employment (RDG personal observations in Mato Grosso). Meanwhile middle-class workers (such as farm managers and agribusiness vendors) often choose not to live permanently in soy-dominant counties located on these frontiers and stay in these areas only during the week, choosing instead to live in the next closest city that has good schools and healthcare (LR personal observations in Mato Grosso). This situation would push the local income data into a more bimodal distribution, increasing the THEIL index, even if the average income level is higher in soy-dominant counties.

5. Impacts on future generations' well-being

5.1. *Land cover change*

The impacts of historical and recent soy expansion on forests and native grasslands are substantial, but difficult to quantify given soy production's tendency to be an underlying, not proximate, driver of deforestation (Barona et al. 2010; Arima et al. 2011). Nevertheless, soy area expansion in the Cerrado and Amazon over the past two decades, and previously in the Atlantic Forest, has undoubtedly contributed to the large-scale conversion of native vegetation and planted pastures to intensive agriculture by adding to the total demand for land in these biomes.

Historical land cover change in the Atlantic forest eco-region, which originally extended from Rio Grande do Sul in South Brazil all the way to Ceará in the Northeast (Ribeiro et al. 2009), was influenced by a diverse set of land uses, including gold mining, logging, sugarcane and coffee production, cattle ranching and silviculture plantations (primarily eucalyptus and pine), as well as production of wheat, beans and other crops. Soybean production was not the primary driver of change in most of the Atlantic forest, but it was an important factor in changing land-use practices and land-cover change in the southern reaches of the biome (the states of Rio Grande do Sul and Paraná, in particular). The direct contribution of soybean expansion to deforestation in these states was likely limited; during the 1970s, when soybean production expanded most rapidly in this region (from just over 2 million hectares in 1970 to over 6 million hectares in 1980), relatively little new farmland (1.9 million hectares) was added there (Kaimowitz and Smith 2001; IBGE 2006). However, there is some evidence that the rapid expansion of soy in the South contributed to the clearing of old-growth forests in this region, such as the *Arucaria* pines of Parana, as well as indirect land-use change in the Cerrado and Amazon when other land uses were displaced (Kaimowitz and Smith 2001).

Similarly, soy production is only one of many land uses pursued in the Amazon biome and by no means the largest proximate source of land-cover change; cattle production and small-holder agricultural production are the primary land uses in the region. There is no exact estimate for how much of the total Amazon has been converted directly for soy production, but the current soy planted area in the Amazon biome is less than 2 million hectares (less than 1 percent of the total area; Rudorff et al. 2011). Direct conversion of Amazon forest for soy has largely stopped since 2006, coinciding with the initiation of the Soy Moratorium (discussed below; Macedo et al. 2012). Most soybean expansion has taken place on areas previously cleared for pasture or, to a lesser extent, other land uses.

Within the Cerrado, however, and the transition area between the Cerrado and Amazon, soy production has contributed to the clearing of savanna and forest vegetation on a much larger scale (Redo, Aide, and Clark 2012). Between 1986 and 1999, 70 percent of the Cerrado was converted for agriculture (mainly soy and cattle production), although a considerable amount of the region also experienced forest regeneration during this period (Jepson 2005, Redo, Aide, and Clark 2012). Still, most soybean expansion has taken place on areas previously cleared, mainly for ranching. While cattle ranching is still the dominant land use in the biome (Lapola et al. 2013), as of 2008, soy production occupied roughly 12 million hectares (12 percent of the area; Smaling et al. 2008).

The ecological impacts of this large-scale transformation of the landscape are significant. The eventual conversion of native forests and savannas to cropland directly reduces plant diversity, and indirectly reduces animal diversity by way of reduced habitat and disruptions to the balance of predator and prey species (Tabarelli et al. 2005; Pardini et al. 2010; Moura et al. 2013; Peres et al. 2010). Landscape simplification, from diverse plant communities to a single land-cover type, i.e. soy, limits food and shelter opportunities for wildlife in comparison to more diverse farming systems (Jose 2009; Fargione, Cooper, and Flaspoher 2009), initiating the 'technological treadmill' of modern agriculture (Ward 1993). The diversity of plant, animal and bacteria species within a landscape influences nutrient and water availability, necessitating the use of synthetic fertilizers, herbicides and pesticides, and sometimes irrigation (Altieri 1999; Goodland and Irwin 1975). The application of toxic pesticides and herbicides to control pests and weeds affects the health and diversity of pollinator and predator species, including birds and bees (Kremen, Williams, and Thorp 2002; Schiesari and Grillitsch 2011), which can reduce yields, increasing reliance on new technologies to overcome the latest human-induced environmental challenge.

Converting forest or savanna to cropland releases substantial amounts of carbon dioxide (CO_2) into the atmosphere, contributing to global warming. For example, each hectare of Amazonian forest stores between 280 and 450 Mg of CO_2 in the soil and above- and below-ground biomass, and each hectare of Cerrado savanna stores between 97 and 170 Mg of CO_2 (PMBC 2013). Consequently, CO_2 emissions from land conversion were the largest source of greenhouse gases in Brazil for much of the last two decades (Cerri et al. 2009). In contrast, a hectare of land planted in soy in the Amazon or Cerrado returns only 0.9 Mg of CO_2 to the soil each year (Fargione et al. 2008). Thus, biodiesel made from soy grown on cropland converted directly from native forest or savanna in Brazil can actually have a higher carbon footprint than petroleum-based fuels (Reijnders and Huijbregts 2008; Fargione et al. 2008). Approximately 9 percent of the emissions from deforestation in Brazilian Amazon between 1990 and 2010 can be attributed to soy that was exported from that region during that period (Karstensen, Peters, and Andrew 2013).

Table 8. Fertilizer usage by biome.

Region/biome	N (nitrogen) – Kilograms	P (phosphorus) – Kilograms	K (potassium) – Kilograms
South/Pampas	4–9	45	40
Center–West/Cerrado	7–10	76–80	68–80
North/Amazon	2–10	36–80	33–80
Northeast/Caatinga	4	39	41
Southeast/Atlantic	9	73	62

Sources: FAO (2004); Raucci et al. (2014); Garrett et al. (2013).

Land conversion also alters watershed hydrology, morphology and water quality. The reduction in forest vegetation reduces evapotranspiration influencing nearby precipitation patterns. Total water export can be three to four times higher in watersheds dominated by soy croplands compared with forest (Hayhoe et al. 2011). The creation of impoundments and the reduction of riparian forest surrounding streams and rivers on soy farms influences water flow and sediment levels and contributes to increased water temperatures in streams (Macedo and Coe 2013; Neill et al. 2013).

5.2. *Input intensification*

Soy production requires high applications of lime, phosphorus and potassium, but very little nitrogen (Alves, Boddey, and Urquiaga 2003), due to inoculation with rhizobia, which live within the soy root system and enable the soy plant to fix nitrogen from the atmosphere (Alves, Boddey, and Urquiaga 2003). In the acidic and nutrient-poor Cerrado and Amazon soils, these applications are particularly high (Table 8; FAO 2004). However, substantial amounts of nitrogen must be applied to soybean areas that are double cropped with corn, ~60 kg/ha (Neill et al. 2013). These amounts are much higher than what is required in ranching, which generally utilizes little nitrogen fertilizer, especially when pastures are mixed with leguminous grasses (Mattos and Uhl 1994; de Andrade et al. 2006) or diversified farming systems, which rely very little on synthetic fertilizers (Altieri and Toledo 2011). Nevertheless, nitrogen applications on soy–corn fields are still lower than sugarcane production, which requires ~80 kg/ha (Martinelli and Filoso 2008).

Nitrous oxides (NO_X) that are released by fertilizer applications present a serious climate problem because they have a larger per-unit impact on global warming than does carbon dioxide (Vitousek et al. 1997). While atmospheric emissions of NO_X are high, nutrient runoff of nitrogen, phosphorus and potassium from soy fields into streams is currently very low in Cerrado and Amazonian soils, since these soils fix and store phosphate and nitrates (Riskin et al. 2013; Neill et al. 2013). On this criterion, soy performs better than sugarcane production in the Atlantic forests of Southeast Brazil, where water pollution has been a common outcome of sugarcane expansion (Martinelli and Filoso 2008).

CO_2 emissions resulting directly from soy production (rather than land conversion) have decreased with the expansion of no-till or low-till agriculture, which reduces the amount of fossil fuel needed for agricultural production by reducing tractor passes, and increases carbon stored in the soil (Lal, Reicosky, and Hanson 2007). When used in conjunction with Roundup Ready soy varieties, this method of farming can also reduce the number and toxicity of herbicides and pesticides applied in soy and corn fields (Qaim and Traxler 2005), although farmers frequently over-apply pesticides due to lack of knowledge about proper levels (LR, conversation with Embrapa official, November 2013). GM crops also present new ecological and economic risks, such as increasing herbicide- and pesticide-specific tolerance in weeds and insects (Gassmann et al. 2011). Weed resistance to the herbicide glyphosate (Round Up) has become increasingly common on soy farms throughout the United States (Eller 2014) and, in 2014, numerous cases of worm resistance to the Bt pesticide in GM corn were reported in Brazil (Diep 2014).

There is substantial evidence that the greenhouse gas and biodiversity impacts of specialized soy systems could be reduced through the use of more diverse (or *multi-functional*) agricultural systems, i.e. systems that include crop, livestock and tree production, particularly in rotation. When livestock are rotated on agricultural fields for a number of years, organic soil matter can recover between harvests. Planting leguminous crops and forages during a rotation period increases the nitrogen content of the soil, reducing the need for synthetic fertilizers (Balbino et al. 2011; Landers 2007). Planting trees and shrubs helps to sequester carbon above and below ground (Jose 2009). Finally, the mosaic landscapes provided by diversified farming systems tend to have lower pest pressure, necessitating fewer chemical inputs, and provide a wider range of habitat to support local fauna in comparison to monocultures (Wilkins 2008; Meehan et al. 2011; Mendenhall et al. 2014).

6. Sustainability governance mechanisms

6.1. *Mechanisms related to improved small-holder participation*

The modernization and expansion of the Brazilian agriculture sector has undoubtedly left behind many of the country's small farmers (Martinelli et al. 2010; Lapola et al. 2013). To promote socially inclusive economic growth in rural areas, the government has created a suite of policies aimed at improving small-holder livelihoods, both through participation in the soy sector and increased productivity in other agricultural areas. To promote social inclusion in all biofuels sectors, the Brazilian government established the National Program for the Production and Use of Biodiesel (PNPB). This program requires grain traders and biodiesel manufacturers to purchase oilseed crops from small-holder farms (as defined above) and provide technical assistance to these farms (Wilkinson and Herrera 2010; Lima, Skutsch, and Costa 2011). The target for small-holder farm participation is 15 percent in the North and Cerrado and 30 percent in the Northeast, South, and Southeast regions of Brazil. However, Lima, Skutsch, and Costa (2011) find that companies tend to avoid settlements where the existing land holdings are very small and the topography prevents large-scale mechanization, limiting the extent to which this program will really benefit the poorest communities. Consequently, total small-farmer participation has fallen far short of stated goals (Wilkinson and Herrera 2010).

The National Program for Strengthening Family Agriculture (PRONAF) provides subsidized rural credit for individual or collective projects that generate income for family farmers and agrarian reform settlers. The program has the lowest interest rates of any source of rural financing in the country (~1 percent), and covers both costs and investments in farm management and value-added activities. In addition to providing credit, the government helps minimize risk for farmers through the Insurance for Family Farmers (SEAF) program. SEAF provides insurance to farmers who adopt certain technologies that conserve natural resources on the farm and reduce their vulnerability to climatic fluctuations. Finally, in 2010 and 2013, the President of Brazil also approved two new laws (Law no. 12,188 and Law no. 5740) to improve and expand technical assistance in rural areas for vulnerable farmers by making the existing technical programs more participatory, multidisciplinary, equitable and culturally appropriate. Although these targeted credit and technical assistance programs may help small farmers improve their existing practices, it is unlikely that they will help small farmers overcome the structural barriers that prevent them from participating in the production of high-value export crops, including a lack of assets, secure land tenure, education and physical access to markets, as well as higher average costs (Page and Slater 2003; Barrett 2008).

6.2. *Mechanisms related to reducing impacts on natural capital*

Brazil has some of the strictest land-use regulations and one of the largest networks of protected areas in the world in the Amazon, as the result of widespread domestic and international mobilization for forest conservation in the late 1970s and 1980s (Hecht 2005). Global interest in reducing deforestation in the Amazon has also led to a suite of novel market mechanisms to dis-incentivize land clearing for soy production. Landowners in the Amazon forest are required by the federal Forest Code (Law 12,651) to preserve 80 percent of their property in a Legal Reserve (LR). Producers that are located in the Legal Amazon, but not the Amazon forest, must conserve 35 percent of their property. However, land-cover requirements are less strict outside of the Legal Amazon (including in the highly threatened Atlantic Forest biome), where 80 percent of the area can be legally cleared. Landowners in all parts of the country also have to conserve steep slopes and the 30–500-meter riparian areas adjacent to waterways (depending on the size of the waterway) as Permanent Protection Areas (PPAs). Historically, the effectiveness of these progressive environmental regulations was substantially undermined by a lack of enforcement; the total LR and PPA deficit in Brazil may be as high as 36–50 million hectares (Sparovek et al. 2010; Soares-Filho et al. 2014). In the most recent revisions of the forest code in 2012, the LR deficit was reduced by roughly 58 percent, meaning that some producers are no longer obliged to reforest as much area as they 'owed' under the past Forest Code. A large part of these deficit reductions came from 'small' producers (ranging from 20 hectares in the South and 440 hectares in the Amazon region) who were given amnesty for illegal deforestation that occurred prior to 2008, though some also came from reductions in PPA widths, which may critically influence water quality in the Amazon in the future (Soares-Filho et al. 2014). At the same time, law enforcement in the Amazon (and, to a lesser extent, the Cerrado) has increased since the mid-2000s. Breaking the law is now beginning to have consequences, if not in terms of fines or jail times for illegal deforesters, at least in terms of the transaction costs of maintaining access to credit and markets.

One of the biggest obstacles to Forest Code enforcement has been a lack of information about property numbers, sizes and locations. As early as 2008, some state governments (Bahia, Mato Grosso, Pará and Rondônia) began to create registries of rural properties and promote environmental licensing to bring them into compliance with land-use regulations and facilitate monitoring. A combination of non-governmental organization (NGO) campaigns and incentives from the state, including a temporary amnesty for previous Forest Code violations in exchange for Cadastro Ambiental Rural (CAR) registration, quickly made CAR registration common among soy properties in Pará and Mato Grosso (SEMA-MT 2014; SEMA-PA 2014; authors' personal observations). The revisions to the Forest Code passed in 2012 established a federal CAR, which will integrate the state CARs into a national cadaster. CAR can now be considered a condition to get access to subsidized rural credit in the Amazonian portion of Mato Grosso (Garrett, Lambin, and Naylor 2013b).

The introduction of demand-side pressures to 'green' soy supply chains has enhanced the effectiveness of federal deforestation policies. To sell to the major soy buyers, farmers in the Amazon biome must meet the requirements of the Soy Moratorium, a voluntary agreement by the soy supply chain to not purchase or finance soy produced in areas deforested after July 2006 (ABIOVE 2010). This agreement has created a 'hybrid' governance scenario where deforestation for soy expansion in the Amazon is prohibited by international market actors employing tools of the state (such as the legal definition of the Amazon biome and the mapping services of the state space agency, INPE; Brown and Koeppe 2013; Brannstrom et al. 2012). There is evidence that most soy producers have either complied with the Soy Moratorium or exited the soy industry (Macedo et al. 2012; Rudorff et al. 2011; Baletti 2014). However, the agreement is unlikely to be maintained permanently (FAO 2014) and its impacts are limited to the Amazon biome, where less than 20 percent of the country's soy is produced.

While most of the above mechanisms focus on punitive incentives for deforestation in specific agricultural industries (Nepstad et al. 2014), Brazil's new 'low-carbon agriculture' policies may offer some positive incentives for conservation at a regional scale. In 2009, Brazil passed its National Climate Change Policy (PNMC), and committed to reducing national greenhouse gas emissions by 36.1 to 38.9 percent by 2020, including a reduction in deforestation rates to 20 percent of the 10-year (1996–2005) average within the same timeframe via specific plans crafted for several major sectors. The plan for the agriculture and livestock sector, the Low-Carbon Agriculture Plan (Plano ABC), was announced in 2010. In addition to supporting research and development in new agricultural technologies, the initiative provides technical assistance and subsidized rural credit lines for activities such as integrated crop livestock forestry practices, no-till planting and pasture recuperation (MAPA 2011). It is estimated that the adoption of integrated crop–livestock–forestry practices on just 4.0 million hectares of existing agricultural land in Brazil has the potential to mitigate 18–22 million tons of CO_2 equivalent (MAPA 2011). Nevertheless, uptake of the Plano ABC among producers has been slow, with only 48 percent of the ABC credit resources being distributed in 2012/2013 (Prado Júnior 2013). Low interest in ABC loans has been attributed to the fact that the benefits and risks of ABC practices have not been adequately explained to producers, producers lack knowledge about how to implement ABC projects, and high levels of bureaucracy limit access to ABC credit (authors' interviews with farmers and rural syndicate lawyers, 2013; Prado Júnior 2013; Mendes 2013).

Another important, but niche, initiative is the Roundtable for Responsible Soy (RTRS), which was established in 2006 as the result of a two-year long multi-stakeholder process. The Standard for Responsible Soy production requires that farms comply with existing environmental and social regulations in the producing country and must not convert native forests or other 'high value conservation areas'[7] to soy fields (although expansion onto land converted prior to 2009 is allowed; RTRS 2013). As of 2013, however, RTRS certified full or partial production from only 30 farms in Brazil, totaling roughly 495,050 hectares and 1,142,100 tons (~1 percent of the country's total production; RTRS 2014). Though RTRS has modified standards to promote the inclusion of smaller-scale farmers, RTRS certification is highly skewed toward large, well-capitalized farmers who can

[7]'High conservation value areas' are critical areas in a landscape, which need to be appropriately managed in order to maintain or enhance the 'outstanding' biological, ecological, social or cultural values present there (RTRS 2013).

afford the costly certification and auditing processes and tap into segregated supply chains (Garrett, Rueda, and Lambin 2013). Critics of RTRS have shown that the standard negotiation process aggravates existing socio-economic inequities in the soy sector because it tends to exclude small-holders, favoring more politically and financially empowered stakeholders in the soy chain, such as Grupo Maggi, in the discourse about what constitutes responsible soy. These entrenched actors are then able to utilize the definition of certification standards to their mutual benefit (Elgert 2012).

7. Discussion

While a large preponderance of the conservation and rural studies literature focuses on the negative impacts of Brazilian soy, there is no denying that the Brazilian soy industry has provided tremendous economic benefits to Brazilian citizens over the last 40 years, particularly when soy farming is coupled with value-added agribusinesses, such as the production of soy foods, livestock and biodiesel. At the federal level, soy production and related agribusiness provide tax revenues that support ambitious social programs and export revenues that raise the value of the Brazilian currency, increasing consumer purchasing power in international markets. At the local level, the soy supply chain stimulates spending and investment in local transport, health and education, infrastructure (which has translated into higher average levels of income, life expectancy and educational attainment) in comparison to counties dominated by other land-use activities.

On the other hand, it is clear that the federal investment in the soy industry has largely excluded the poorest farmers, particularly in the North and Northeast regions, to the benefit of large multinational agribusiness firms, small farmers from the South, and well-capitalized entrepreneurs from the Southeast. In addition, counties with high levels of soybean production tend to have higher income inequality than counties dominated by other land uses. While soy producers do not receive direct subsidies, the Brazilian government has provided the soy industry with substantial financial support, including cheap access to land, subsidized credit with generous repayment terms, relatively low tax rates, tax refunds upon export, and infrastructure investments. Soy production has created these substantial but exclusive economic benefits without reducing domestic food security, since it has not replaced other crops. However, the expansion of soy cropland into native savannas has endangered future generations' well-being by contributing to irreversible climate change and biodiversity loss in ways that are uncertain and immeasurable (Figure 4).

Recent social governance initiatives have attempted to address the problem of small-holder exclusion from the soy supply chain. Targeted agricultural policies, including low-interest credit and improved technical assistance for small-holders, have the potential to provide the rural poor with enhanced opportunities to choose production systems that meet their needs. Nevertheless, the effectiveness of these credit and technical assistance programs remains uncertain.

Meanwhile, new environmental governance measures have successfully reduced deforestation in recent years by encouraging producers to intensify production on the existing land base or expand onto previously cleared pastures. Some critics argue that the greening of the soy supply chain should not be celebrated, since it merely solidifies support for 'neoextractive' activities in Brazil and reinforces the power of the multinational trading companies and NGOs, without addressing the systemic sources of poverty (Baletti 2014). This may be true, but it does not negate the tremendous environmental benefits that have been achieved by these programs in helping to achieve a rapid decrease in deforestation in Brazil over recent years. Furthermore, it is unclear which land-use alternatives

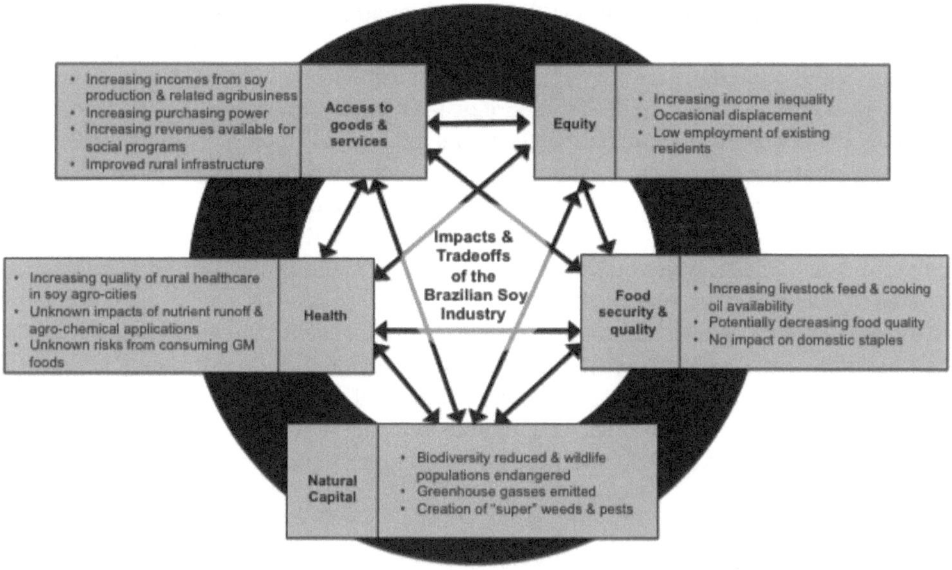

Figure 4. Intra- and intergenerational impacts and tradeoffs of the Brazilian soy industry. The soy industry affects numerous aspects of current and future generations' well-being that are highly interconnected. In some dimensions (access to goods and services, and food security), the modern soy industry is largely helping to promote well-being, particularly for current generations. In other dimensions, the current soy industry is reducing well-being (equity, natural capital and health). To increase the overall sustainability of the industry relative to the past and other land use alternatives, soy production must improve all dimensions represented on the circle, not just one or two.

could replicate the economic benefits and total calorie production of the soy industry without increasing rural inequality and benefiting multinational agribusiness, leaving few options but to improve the existing system.

Current state and private environmental governance mechanisms focus mainly on the land cover impacts of soy expansion, without addressing problems associated with input intensification. They also reflect a 'high forest bias', ignoring entire ecosystems that are threatened by soy production, such as the Cerrado and Chaco biomes (Hecht 2005; Sawyer 2008). In the future, conservation policies should consider more ecological forms of intensification and pay attention to all biomes that are threatened by agricultural expansion if they are going to simultaneously reconcile global demands for increased food production with the protection of the natural capital on which current and future generations depend.

Brazil's ABC plan offers promise in this direction, by promoting more diverse and sustainable agricultural systems in all biomes. The practices incentivized by the program – reducing tillage on existing fields, adopting integrated crop, livestock and forestry practices, and recuperating degraded pastures – can all help to increase the land-use efficiency of agricultural production, while reducing pollution and greenhouse gas emissions. By pursuing the practices advocated in ABC, producers can still tap into major commodity markets and processing opportunities, while reducing their reliance on imported inputs and helping Brazil meet its greenhouse gas reduction commitments, though these benefits will only be realized as the program is made more accessible to farmers.

Despite the large amount of literature referenced above, several significant gaps remain in our understanding of current and future contributions and shortfalls of the soy sector to sustainable development in Brazil. On the socio-economic side, these gaps include better understanding of the impacts of the soy sector on non-farm employment, migration and access to health and education services and mechanisms to include small and vulnerable farmers in soy production. In terms of political economy, there is little understanding regarding processes of cultural homogenization in soy producing regions, impacts on food sovereignty or autonomy, or the influence of soy lobby groups on political power. On the ecological side, there is a huge deficit in our understanding of individual management practices on soy farms and the impacts of lime, fertilizer, herbicide and pesticide applications on climate, water quality and biodiversity. There is also still very little understanding of potential risks associated with transgenic seeds, such as gene transfer and weed and pest resistance, or how these may be changing expansion pathways of soy. Finally, there has been little research on the health risks associated with agrochemical use and runoff in communities adjacent to soy farms. Much of the research on soy production to date has relied on census data and remote sensing, but more in-depth case study research is necessary to understand these more complex socio-economic and ecological processes.

8. Conclusion

The purpose of this paper was to examine the evolving sustainability of the Brazilian soy industry over the past 40 years. In contrast to past studies that focused singularly on the role of soy production in environmental degradation, small-holder displacement and social conflict in isolated case studies, we considered a broader suite of social and ecological impacts of soy production and related agribusiness across Brazil and compared these impacts to alternative rural land uses when possible.

We conclude that the balance between costs and benefits associated with soy production is not uniformly worse or better than that of sugarcane or cattle, particularly in comparison to soy production in neighboring countries. The Brazilian soy industry has contributed to growth in national export revenues at the expense of natural capital stocks. It has also enhanced rural incomes and services, though the cost has been declining rural equity. The soy sector has greatly increased the amount of food produced in Brazil, while also contributing to high carbon emissions and biodiversity loss, and directly generating few rural employment opportunities. Overall, the advancement of mechanized, large-scale agriculture in the Atlantic, Cerrado and Amazon biomes over the past 40 years has come with many social and ecological costs, but it has also generated substantial improvements in well-being throughout Brazil via government revenues for welfare programs, increasing private purchasing power and improvements in rural access to health and education services. Unlike truly 'extractivist' activities that deplete non-renewable resources, such as mining and old-growth logging, the economic benefits of soy production can be reaped and amplified for many years to come, so long as future production proceeds without further deforestation, involves fewer synthetic inputs and is coupled with value-added activities that generate employment.

Recent improvements in environmental governance, new incentives for ecological intensification and the growth in value-added soy industries give us many reasons to be optimistic that some of the tradeoffs between intra- and intergenerational equity and economic growth will be reduced in the future, leading to more benefits across the board. Brazil has already taken many efforts to control deforestation, increase soy processing, reduce land grabbing and include small-holders in biofuels production, especially in comparison to neighboring countries such as Paraguay and Bolivia where soy expansion is still triggering rampant deforestation, and increasing foreign land ownership and small-holder displacement. Additional policies that can help meet these challenges include government investment in loans, technology transfer and marketing assistance for soy production on small farms, support for ecological intensification/low-carbon agriculture through incentives for integrated practices, and support for soy-processing industries that create high-wage employment for rural communities. Voluntary programs that can amplify the social and environmental benefits of these state programs include commitments from soy traders and retailers to purchase only soy that was produced in compliance with existing environmental and labor regulations, as well as a broader set of ecological standards. Our analysis has focused mainly on previously published studies, field observations and secondary data, though we recommend that future researchers undertake coordinated local case studies across the range of impacts identified here to examine whether the national, biome and municipal indicators of soybean sustainability analyzed in this paper hold at finer spatial scales.

Acknowledgements

We are grateful for helpful feedback we received from Eric Lambin and Rosamond Naylor on previous versions of this manuscript, and, in particular, to Susanna Hecht for her extremely helpful comments on the latest version.

Funding

R.D.G. conducted this research partially while funded by the Giorgio Ruffolo Fellowship in the Sustainability Science Program at Harvard University, and partially while funded by the Emmett Interdisciplinary Program in Environment and Resources at Stanford University. Support from Italy's Ministry for Environment, Land and Sea, Richard and Rhoda Goldman, Richard L. Kauffman and Ellen Jewett, and Dan and Rae Emmett is gratefully acknowledged. L.R. is supported by a grant from the Norwegian Government's International Climate and Forest Initiative to the Gibbs Land Use Lab in the Center for Sustainability and the Global Environment at the University of Wisconsin–Madison.

References

ABIOVE. 2010. Soybean moratorium. In *Brazilian association of the vegetable oil industry sustainability*. abiove.com.br/english/ss_moratoria_us.html

Altieri, M.A. 1999. The ecological role of biodiversity in agroecosystems. *Agriculture, Ecosystems & Environment* 74, no. 1: 19–31. doi:10.1016/S0167-8809(99)00028-6

Altieri, M.A., and V.M. Toledo. 2011. The agroecological revolution in Latin America: rescuing nature, ensuring food sovereignty and empowering peasants. *The Journal of Peasant Studies* 38, no. 3: 587–612. doi:10.1080/03066150.2011.582947

Alves, B.J.R., R.M. Boddey, and S. Urquiaga. 2003. The success of BNF in soybean in Brazil. *Plant and Soil* 252, no. 1: 1–9. doi:10.1023/A:1024191913296

Alves Sobrinho, R. 2009. Dinâmica territorial, agronegócio e re-territorialização: o município de Diamantino, MT [Territorial dynamics, agribusiness, and re-territorialization: the municipality of Diamantino, MT]. In *Novas Territorialidades Nas Cidades Mato-Grossenses* [New Territorialities in Mato Grosso Cities], ed. S.R. Romancini, 155–174. Cuiabá, Brazil: Universidade Federal do Mato Grosso.

de Andrade, C.M.S., R. Garcia, J.F. Valentim, and O.G. Pereira. 2006. Grazing management strategies for massai grass-forage peanut pastures: Definition of sward targets and carrying capacity. *Revista Brasileira de Zootecnia* 35: 352–357. doi:10.1590/S1516-35982006000200002

Arima, E.Y., P.D. Richards, R. Walker, and M.M. Caldas. 2011. Statistical confirmation of indirect land use change in the Brazilian Amazon. *Environmental Research Letters* 6, no. 024010: 1–7. doi:10.1088/1748-9326/6/2/024010

Arrow, K.J., P. Dasgupta, L.H. Goulder, K.J. Mumford, and K. Oleson. 2012. Sustainability and the measurement of wealth. *Environment and Development Economics* 17, no. 03: 317–353. doi:10.1017/S1355770X12000137

Ash, M. 2012. Soybeans and oil crops. *United States Department of Agriculture*. http://www.ers.usda.gov/publications/ocs-oil-crops-outlook/ocs-14 h.aspx

Awokuse, T.O. 2009. Does agriculture really matter for economic growth in developing countries? In Paper Presented at the American Agricultural Economics Association Annual Meeting, July 26-28, Milwaukee, WI.

Balbino, L.C., L.A.M. Cordeiro, V. Porfírio-da-Silva, A. Moraes, G.B. Martínez, R.C. Alvarenga, A. N. Kichel, et al. 2011. Evolução tecnológica e arranjos produtivos de sistemas de integração lavoura-pecuária-floresta no Brasil [Evolution of technology and productive arrangements in integrated crop-livestock-forestry systems in Brazil]. *Pesquisa Agropecuária Brasileira* 46, no. 10: 1–12. Embrapa Informação Tecnológica. doi:10.1590/S0100-204X2011001000001

Baletti, B. 2012. Ordenamento territorial: Neo-developmentalism and the struggle for territory in the lower Brazilian Amazon. *The Journal of Peasant Studies* 39, no. 2: 573–598. doi:10.1080/03066150.2012.664139

Baletti, B. 2014. Saving the Amazon? Sustainable soy and the new extractivism. *Environment and Planning A* 46, no. 1: 5–25. doi:10.1068/a45241

Barona, E., N. Ramankutty, G. Hyman, and O.T. Coomes. 2010. The role of pasture and soybean in deforestation of the Brazilian Amazon. *Environmental Research Letters* 5, no. 24002: 1–9. doi:10.1088/1748-9326/5/2/024002

Barrett, C.B. 2008. Smallholder market participation: Concepts and evidence from eastern and southern Africa. *Food Policy* 33, no. 4: 299–317. doi:http://dx.doi.org/10.1016/j.foodpol.2007.10.005

Barros, G.S.C., A.L. Fachinello, and A.F. Silva. 2011. Desenvolvimento metodológico e cálculo do PIB das cadeias produtivas do algodão, cana-de-açúcar, soja, pecuária de corte e leite no Brasil [Development of a methodology to calculate GDP for the cotton, sugarcane, soy, and beef cattle supply chains]. *CEPEA*. Piracicaba, Brasil. http://www.cepea.esalq.usp.br/pibpec/PIB_Cadeias_relatorio2009_10.pdf

Banco Central Do Brasil (BCB). 2012. Annual rural credit statistics. http://www.bcb.gov.br/htms/creditorural/2012/produtCooper.asp?idpai=RELRURAL2012

Belik, W., and R. Rocha dos Santos. 2002. Regional market strategies of supermarkets and food processors in extended MERCOSUR. *Development Policy Review* 20, no. 4: 515–528. doi:10.1111/1467-7679.00187

Brannstrom, C. 2009. South America's neoliberal agricultural frontiers: Places of environmental sacrifice or conservation opportunity. *AMBIO: A Journal of the Human Environment* 38, no. 3: 141–149. doi:10.1579/0044-7447-38.3.141

Brannstrom, C., L. Raush, J.C. Brown, R. Marson, T. de Andrade, A. Miccolis, and R. Andrade. 2012. Compliance and market exclusion in Brazilian agriculture: Analysis and implications for 'soft' governance. *Land Use Policy* 29, no. 2: 357–366. doi:10.1016/j.landusepol.2011.07.006

Brown, J.C. 2006. Feeding seven billion well: Producing protein more efficiently. In *Plan B 2.0: Rescuing a planet under stress and a civilization in trouble*. Washington, DC: Earth Policy Institute.

Brown, J.C., and M. Koeppe. 2013. Debates in the environmentalist community: The soy moratorium and the construction of illegal soybeans in the Brazilian Amazon. In *In Environment and the law in Amazonia: A plurilateral encounter*, 110–126. Sussex: Sussex Academic Press.

Brown, J.C., M. Koeppe, B. Coles, and K.P. Price. 2005. Soybean production and conversion of tropical forest in the Brazilian Amazon: The case of Vilhena, Rondônia. *AMBIO: A Journal of the Human Environment* 34, no. 6: 462–469. doi:10.1579/0044-7447-34.6.462

Brown, J.C., L. Rausch, and V.G. Luz. 2013. Toward a spatial understanding of staple food and non-staple food production in Brazil. *The Professional Geographer* 66, no. 2: 249–259. doi:10.1080/00330124.2013.781492

Brownell, K.D., T. Farley, W.C. Willett, B.M. Popkin, F.J. Chaloupka, J.W. Thompson, and D.S. Ludwig. 2009. The public health and economic benefits of taxing sugar-sweetened beverages. *New England Journal of Medicine* 361, no. 16: 1599–1605. doi:10.1056/NEJMhpr0905723

Cerri, C.E., S.M.F. Maia, M.V. Galdos, B.J. Feigl, and M. Bernoux. 2009. Brazilian greenhouse gas emissions: the importance of agriculture and livestock. *Scientia Agricola* 66, no. 6: 831–843. doi:10.1590/S0103-90162009000600017

Chaddad, F.R. 2006. The evolution of agricultural policies and agribusiness development in Brazil. *Choices: The Magazine of Food, Farm, and Resource Issues* 21, no. 2: 85–90. http://purl.umn.edu/94415

Chase, J. 1999. Trapped workers, urban freedoms and labor control in Brazilian agriculture: Evidence from southwest Goiás. *Journal of Rural Studies* 15, no. 2: 201–220. doi:10.1016/S0743-0167(98)00043-6

Daly, H.E. 1991. *Steady-state economics: With new essays*. Island Press.

Damico, F.S., and A.M. Nassar. 2007. Agricultural expansion and policies in Brazil. In *US agricultural policy and the 2007 farm bill*, 75–96. Stanford: Woods Institute for the Environment at Stanford University.

Da Silva, J.G., and M.E. del Grossi. 2001. Rural nonfarmemployment and incomes in Brazil: Patterns and evolution. *World Development* 29, no. 3: 443–453. doi:10.1016/S0305-750X(00)00103-0

De Moraes, M.A.F.D. 2007. Indicadores do mercado de trabalho do sistema agroindustrial da cana-de-açúcar do Brasil no período 1992-2005 [Labor Market indicators of the sugar cane agroindustrial system in Brazil from 1992-2005]. *Estudos Econômicos* 37, no. 4: 875–902.

DeFries, R.S., M. Herold, L. Verchot, M.N. Macedo, and Y. Shimabukuro. 2013. Export-oriented deforestation in Mato Grosso: Harbinger or exception for other tropical forests?. *Philosophical Transactions of the Royal Society B: Biological Sciences* 368, no. 20120173: 1–8. doi:10.1098/rstb.2012.0173

De Souza, D.B., and R.L. Ravache. 2009. Estruturação do espaço urbano-regional do Centro-Norte Mato-Grossense: Sinop, Sorriso e Lucas do Rio Verde [Structuring of the regional urban space of Central-North Mato Grosso: Sinop, Sorriso and Lucas do Rio Verde]. In *Novas Territorialidades Nas Cidades Mato-Grossense* [New territorialities in mato grosso cities], ed. S.R. Romancini, 199–220. Cuiabá, Brazil: Universidade Federal do Mato Grosso.

Diep, F. 2014, July. Pest In Brazil has evolved resistance against GMO corn. *Popular Science*. http://www.popsci.com/article/science/pest-brazil-has-evolved-resistance-against-gmo-corn

Elgert, L. 2012. Certified discourse? The politics of developing soy certification standards. *Geoforum* 43, no. 2: 295–304. doi:10.1016/j.geoforum.2011.08.008

Eller, D. 2014. 'Superweeds' choke farms. *The Des Moines Register*, June 23.

Faminow, M.D., and J.S. Hillman. 1987. Embargoes and the emergence of Brazil's soyabean industry. *The World Economy* 10, no. 3: 351–366. doi:10.1111/j.1467-9701.1987.tb00105.x

Fargione, J.E., T.R. Cooper, and D.J. Flaspoher. 2009. Bioenergy and wildlife: Threats and opportunities for grassland conservation. *Bioscience* 59, no. 9: 767–777. doi:10.1525/bio.2009.59.9.8

Fargione, J., J. Hill, D. Tilman, S. Polasky, and P. Hawthorne. 2008. Land clearing and the biofuel carbon debt. *Science* 319, no. 5867: 1235–1238. doi:10.1126/science.1152747

Fearnside, P.M. 2001. Soybean cultivation as a threat to the environment in Brazil. *Environmental Conservation* 28, no. 1: 23–38. doi:10.1017/S0376892901000030

Fearnside, P.M. 2007. Brazils Cuiabá-Santarém (BR-163) highway: The environmental cost of paving a soybean corridor through the Amazon. *Environmental Management* 39, no. 5: 601–614. doi:10.1007/s00267-006-0149-2

Fearnside, P.M. 2008. The roles and movements of actors in the deforestation of Brazilian Amazonia. *Ecology and Society* 13, no. 1: 1–23.

Ferreira, J., R. Pardini, J.P. Metzger, C.R. Fonseca, P.S. Pompeu, G. Sparovek, and J. Louzada. 2012. Towards environmentally sustainable agriculture in Brazil: Challenges and opportunities for applied ecological research. *Journal of Applied Ecology* 49: 535–541. doi:10.1111/j.1365-2664.2012.02145.x

Flaskerud, G. 2003. Brazil's production and impact. *North Dakota State University Extension Service* 23092.

Food and Agricultural Organization (FAO). 2004. *Fertilizer use by crop in Brazil.* Food and Agriculture Organization of the United Nations. Rome, Italy. http://ftp.fao.org/agl/agll/docs/fertusebrazil.pdf

Food and Agricultural Organization (FAO). 2013. *FAOSTAT online statistical service*. Food and Agriculture Organization of the United Nations. Rome, Italy. http://faostat.fao.org

Food and Agricultural Organization (FAO). 2014. *Oilseeds, oils, & meals: Market and policy updates – Report No. 57*. Food and Agriculture Organization of the United Nations. Rome, Italy. http://www.fao.org/fileadmin/templates/est/COMM_MARKETS_MONITORING/Oilcrops/Documents/MPPU_Mar__14.pdf

Gardner, T.A., J. Ferreira, J. Barlow, A.C. Lees, L. Parry, I.C.G. Vieira, E. Berenguer, et al. 2013. A social and ecological assessment of tropical land uses at multiple scales: The Sustainable Amazon Network. *Philosophical Transactions of the Royal Society B: Biological Sciences* 368, no. 1619: 20120166-1–11. doi:10.1098/rstb.2012.0166

Garrett, R.D., E.F. Lambin, and R.L. Naylor. 2013a. Land institutions and supply chain configurations as determinants of soybean planted area and yields in Brazil. *Land Use Policy* 31: 385–396.

Garrett, R.D., E.F. Lambin, and R.L. Naylor. 2013b. The new economic geography of land use change: Supply chain configurations and land use in the Brazilian Amazon. *Land Use Policy* 34: 265–275. doi:10.1016/j.landusepol.2013.03.011

Garrett, R.D., X. Rueda, and E.F. Lambin. 2013. Globalization's unexpected impact on soybean production in South America: Linkages between preferences for non-genetically modified crops, eco-certifications, and land use. *Environmental Research Letters* 8: 044055-1–12. doi:10.1088/1748-9326/8/4/044055

Gasques, J.G., and E.T. Bastos. 2010. Produtividade total dos fatores e transformações da agricultura brasileira: análise dos dados dos censos agropecuários [Total factor productivity and transformations of Brazilian agriculture: analysis of the agricultural census data]. In *A agricultura brasileira: desempenho, desafios e perspectivas* [Brazilian agriculture: performance, challenges, and perspectives], 19–44. Brasília: IPEA. http://www.sober.org.br/palestra/15/1188.pdf

Gassmann, A.J., J.L. Petzold-Maxwell, R.S. Keweshan, and M.W. Dunbar. 2011. Field-evolved resistance to Bt maize by western corn rootworm. *PLoS One* 6, no. 7: e22629-1–7. doi:10.1371/journal.pone.0022629

Gaulard, M. 2012. The 'hot money' phenomenon in Brazil. *Revista de Economia Política* 32, no. 3: 367–388. Centro de Economia Política. doi:10.1590/S0101-31572012000300002

Goldsmith, P., and R. Hirsch. 2006. The Brazilian soybean complex. *Choices* 21, no. 2: 97–104. American Agricultural Economics Association.

Goodland, R.J.A., and H.S. Irwin. 1975. *Amazon jungle: Green hell to red desert? An ecological discussion of the environmental impact of the highway construction program in the Amazon basin*. Amsterdam: Elsevier Scientific Publishing Co.

Hall, J., S. Matos, L. Severino, and N. Beltrão. 2009. Brazilian biofuels and social exclusion: Established and concentrated ethanol versus emerging and dispersed biodiesel. *Journal of Cleaner Production* 17, no. 1: S77–S85. doi:10.1016/j.jclepro.2009.01.003

Hayhoe, S.J., C. Neill, S. Porder, R. Mchorney, P. Lefebvre, M.T. Coe, H. Elsenbeer, and A. Krusche. 2011. Conversion to soy on the Amazonian agricultural frontier increases streamflow without affecting stormflow dynamics. *Global Change Biology* 17, no. 5: 1821–1833. doi:10.1111/j.1365-2486.2011.02392.x

Hecht, S.B. 2005. Soybeans, development and conservation on the Amazon frontier. *Development and Change* 36, no. 2: 375–404. doi:10.1111/j.0012-155X.2005.00415.x

Hecht, S.B., and C. Mann. 2008. How Brazil outfarmed the American farmer. *Fortune* 157: 92–105.

Helfand, S.M. 2001. The distribution of subsidized agricultural credit in Brazil: Do interest groups matter?. *Development and Change* 32, no. 3: 465–490. doi:10.1111/1467-7660.00213

Instituto Brasileiro de Geographia e Estatística (IBGE). 2006. Agriculture and livestock census. http://www.sidra.ibge.gov.br

Instituto Brasileiro de Geographia e Estatística (IBGE). 2010. Demography census. http://www.sidra.ibge.gov.br

Instituto Brasileiro de Geographia e Estatística (IBGE). 2013a. Municipal agricultural production survey. http://www.sidra.ibge.gov.br

Instituto Brasileiro de Geographia e Estatística (IBGE). 2013b. Municipal livestock production survey. http://www.sidra.ibge.gov.br

Instituto de Pesquisa Econômica Aplicada (IPEA). 2013. IPEADATA. http://ipeadata.gov.br

Instituto Mato Grossense de Economia e Agropecuária (IMEA). 2013. Agronegócio no Brasil e em Mato Grosso [Agribusiness in Brazil and Mato Grosso]. http://www.imea.com.br/upload/pdf/arquivos/R405_Apresentacao_MT_Portugues_Nova_31_01_2014.pdf

Instituto Nacional de Cidadania e Reforma Agraria (INCRA). 2012. Aquisição e arrendamento de imóveis rurais por estrageiros [Rural land acquisition and rental by foreigners]. http://pt.slideshare.net/SindPFA/aquisio-terras-por-estrangeiros-incra-102012#

Instituto Nacional de Pesquisas Espaciais (INPE). 2014. Projecto PRODES [PRODES project].

Jank, M.S., M.F.P. Leme, and A.M. Nassar. 2001. Concentration and internationalization of Brazilian agribusiness exporters. *The International Food and Agribusiness Management Review* 2, no. 3–4: 359–374. http://www.sciencedirect.com/science/article/pii/S1096750801000325

Jepson, W. 2005. A disappearing biome? Reconsidering land-cover change in the Brazilian savanna. *The Geographical Journal* 171, no. 2: 99–111. doi:10.1111/j.1475-4959.2005.00153.x

Jepson, W. 2006a. Private agricultural colonization on a Brazilian frontier, 1970-1980. *Journal of Historical Geography* 32, no. 4: 839–863. doi:10.1016/j.jhg.2004.12.019

Jepson, W. 2006b. Producing a modern agricultural frontier: Firms and cooperatives in Eastern Mato Grosso, Brazil. *Economic Geography* 82, no. 3: 289–316. doi:10.1111/j.1944-8287.2006.tb00312.x

Jepson, W., C. Brannstrom, and A. Filippi. 2010. Access regimes and regional land change in the Brazilian Cerrado, 1972–2002. *Annals of the Association of American Geographers* 100, no. 1: 87–111. doi:10.1080/00045600903378960

Jonasson, E., and S.M. Helfand. 2010. How important are locational characteristics for rural non-agricultural employment? Lessons from Brazil. *World Development* 38, no. 5: 727–741. doi:10.1016/j.worlddev.2009.11.020

Jose, S. 2009. Agroforestry for ecosystem services and environmental benefits: An overview. *Agroforestry Systems* 76, no. 1: 1–10. doi:10.1007/s10457-009-9229-7

Kaimowitz, D., and J. Smith. 2001. Soybean technology and the loss of natural vegatation in Brazil and Bolivia. In *Agricultural technologies and deforestation*, eds. David Kaimowitz and Arild Angelsen, 195–211. New York: CAB International.

Karstensen, J., G.P. Peters, and R.M. Andrew. 2013. Attribution of CO2 emissions from Brazilian deforestation to consumers between 1990 and 2010. *Environmental Research Letters* 8. doi: doi:10.1088/1748-9326/8/2/024005

Kremen, C., N.M. Williams, and R.W. Thorp. 2002. Crop pollination from native bees at risk from agricultural intensification. *Proceedings of the National Academy of Sciences* 99, no. 26: 16812–16816. doi:10.1073/pnas.262413599

Lal, R., D.C. Reicosky, and J.D. Hanson. 2007. Evolution of the plow over 10,000 years and the rationale for no-till farming. *Soil and Tillage Research* 93, no. 1: 1–12. doi:10.1016/j.still.2006.11.004

Landers, J. 2007. Tropical crop-livestock systems in conservation agriculture: The Brazilian experience. In *Integrated crop management* 5, 1–92. Rome: Food and Agriculture Organization. http://www.fao.org/docrep/010/a1083e/a1083e00.htm

Lapola, D.M., L.A. Martinelli, C.A. Peres, J.P.H.B. Ometto, M.E. Ferreira, C.A. Nobre, A.P.D. Aguiar, et al. 2013. Pervasive transition of the Brazilian land-use system. *Nature Climate Change* 4, no. 1: 27–35. doi:10.1038/nclimate2056

Lazzarotto, J., and A. Roessing. 2004. *Contribuição da Agricultura para Arrecadação Tributária [Tax contributions of agriculture]. Embrapa.* Vol. Documentos. Londrina: Embrapa Soja. http://www.contag.org.br/imagens/f1938embrapa-tributos.pdf

Lehtonen, M. 2012. *Power, social impacts, and certification of ethanol fuel: View from the Northeast of Brazil*, eds. A. Gasparatos and P. Stromberg, 144–170. Cambridge: Cambridge University Press. http://dx.doi.org/10.1017/CBO9780511920899.011

Lima, M., M. Skutsch, and G.M. Costa. 2011. Deforestation and the social impacts of soy for biodiesel: Perspectives of farmers in the South Brazilian Amazon. *Ecology and Society* 16, no. 4: 1–15. http://www.ecologyandsociety.org/vol16/iss4/art4/ES-2011-4366.pdf

Loos, J., D.J. Abson, M.J. Chappell, J. Hanspach, F. Mikulcak, M. Tichit, and J. Fischer. 2014. Putting meaning back into 'Sustainable intensification'. *Frontiers in Ecology and the Environment* 12, no. 6: 356–361. doi:10.1890/130157

Luna, F.V., and H.S. Klein. 2006. *Brazil since 1980.* Cambridge: Cambridge University Press. http://books.google.com/books?hl=en&lr=&id=25r32Amzmd4C&pgis=1

Macedo, M.N., and M.T. Coe. 2013. Land-use-driven stream warming in southeastern Amazonia. *Philosophical Transactions of the Royal Society B: Biological Sciences* 368, no. 1619: 1–9. doi:10.1098/rstb.2012.0153

Macedo, M.N., R.S. Defries, D.C. Morton, C.M. Stickler, G.L. Galford, and Y.E. Shimabukuro. 2012. Decoupling of deforestation and soy production in the southern Amazon during the late 2000s.

Proceedings of the National Academy of Sciences of the United States of America 109, no. 4: 1341–1346. doi:10.1073/pnas.1111374109

Martha, G.B. Jr, E. Alves, and E. Contini. 2012. Land-saving approaches and beef production growth in Brazil. *Agricultural Systems* 110: 173–177. doi:10.1016/j.agsy.2012.03.001

Martinelli, L.A., and S. Filoso. 2008. Expansion of sugarcane ethanol production in Brazil: Environmental and social challenges. *Ecological Applications* 18, no. 4: 885–898. doi:10.1890/07-1813.1

Martinelli, L.A., R. Garrett, S. Ferraz, and R. Naylor. 2011. Sugar and ethanol production as a rural development strategy in Brazil : Evidence from the state of São Paulo. *Agricultural Systems* 104, no. 5: 419–428. doi:10.1016/j.agsy.2011.01.006

Martinelli, L.A., R. Naylor, P.M. Vitousek, and P. Moutinho. 2010. Agriculture in Brazil: Impacts, costs, and opportunities for a sustainable future. *Current Opinion in Environmental Sustainability* 2, no. 5–6: 431–438. doi:10.1016/j.cosust.2010.09.008

Matson, P. 2011. *Seeds of sustainability: Lessons from the birthplace of the green revolution in agriculture*. Washington, DC: Island Press. http://www.amazon.com/Seeds-Sustainability-Birthplace-Revolution-Agriculture/dp/159726525X

Mattos, M.M., and C. Uhl. 1994. Economic and ecological perspectives on ranching in the Eastern Amazon. *World Development* 22, no. 2: 145–158. doi:10.1016/0305-750X(94)90066-3

Ministério da Agricultura, Pecuária e Abastecimento (MAPA). 2011. Plano setorial de mitigação e de adaptação às mudanças climáticas para a consolidação de uma economia de baixa emissão de carbono na agricultura [Sectoral plan for climate mitigation and adaptation and the consolidation of a low carbon economy in agriculture]. Brasilia. http://www.agricultura.gov.br/desenvolvimento-sustentavel/plano-abc

Ministério de Desenvolvimento, Indústria e Comércio Exterior (MDIC). 2013. Foreign trade statistics. http://www.desenvolvimento.gov.br/

Ministério de Desenvolvimento Social (MDS). 2014. Family agriculture. http://www.mds.gov.br/falemds/perguntas-frequentes/bolsa-familia/programas-complementares/beneficiario/agricultura-familiar

Meehan, T.D., B.P. Werling, D.A. Landis, and C. Gratton. 2011. Agricultural landscape simplification and insecticide use in the Midwestern United States. *Proceedings of the National Academy of Sciences of the United States of America* 108, no. 28: 11500–11505. doi:10.1073/pnas.1100751108

Mendenhall, C.D., D.S. Karp, C.F.J. Meyer, E.A. Hadly, and G.C. Daily. 2014. Predicting biodiversity change and averting collapse in agricultural landscapes. *Nature* 509, no. 7499: 213–217. doi:10.1038/nature13139

Mendes, D.H.G. 2013. História E Execução Do Programa Agricultura de Baixo Carbono [History and execution of the low carbon agriculture program]. Universidade de Brasília. http://bdm.unb.br/bitstream/10483/4648/1/2013_EduardoHenriqueGoncalvesMendes.pdf

Morton, D.C., R.S. DeFries, Y.E. Shimabukuro, L.O. Anderson, E. Arai, F. del Bon Espirito-Santo, R. Freitas, and J. Morisette. 2006. Cropland expansion changes deforestation dynamics in the southern Brazilian Amazon. *Proceedings of the National Academy of Sciences* 103, no. 39: 14637–14641. doi:10.1073/pnas.0606377103

Moura, N.G., A.C. Lees, C.B. Andretti, B.J.W. Davis, R.C. Solar, A. Aleixo, J. Barlow, J. Ferreira, and T.A. Gardner. 2013. Avian biodiversity in multiple-use landscapes of the Brazilian Amazon. *Biological Conservation* 167: 339–348. doi:10.1016/j.biocon.2013.08.023

Nagel, B.Y. 1999. The cultural discourse of rural violence and land rights in Paraguay. *Comparative Studies in Society and History* 41, no. 1: 148–181.

Neill, C., M.T. Coe, S.H. Riskin, A.V. Krusche, H. Elsenbeer, and M.N. Macedo, et al. 2013. Watershed responses to Amazon soya bean cropland expansion and intensification. *Philosophical Transactions of the Royal Society B: Biological Sciences* 368, no. 1619: 1–7. doi:10.1098/rstb.2012.0425

Nepstad, D., D. McGrath, C. Stickler, A. Alencar, A. Azevedo, B. Swette, T. Bezerra, M. DiGiano, J. Shimada, and R.S. da Motta, et al. 2014. Slowing Amazon deforestation through public policy and interventions in beef and soy supply chains. *Science* 344, no. 6188: 1118–1123. doi:10.1126/science.1248525

Organization for Economic Cooperation and Development (OECD). 2010. Producer and Consumer Support Estimates Database. http://www.oecd.org

Oliveira, G.L.T. 2013. Land regularization in Brazil and the global land grab. *Development and Change* 44, no. 2: 261–283. doi:10.1111/dech.12009

Page, S., and R. Slater. 2003. Small producer participation in global food systems: Policy opportunities and constraints. *Development Policy Review* 21, no. 5–6: 641–654. doi:10.1111/j.1467-8659.2003.00229.x

Pardini, R., A.A. Bueno, T.A. Gardner, P.I. Prado, and J.P. Metzger. 2010. Beyond the fragmentation threshold hypothesis: Regime shifts in biodiversity across fragmented landscapes. *PloS One* 5, no. 10: e13666-1–10 . doi:10.1371/journal.pone.0013666

Peres, C.A., T.A. Gardner, J. Barlow, J. Zuanon, F. Michalski, A.C. Lees, I.C.G. Vieira, F.M.S. Moreira, and K.J. Feeley. 2010. Biodiversity conservation in human-modified Amazonian forest landscapes. *Biological Conservation* 143, no. 10: 2314–2327. doi:10.1016/j.biocon.2010.01.021

Pesquisa Nacional de Demografia e Saúde (PNDS). 2006. National survey of demography and children and women's health. Ministry of Health. http://bvsms.saude.gov.br/bvs/pnds/seguranca_alimentar.php

Plano Brasileiro de Mudanças Climáticas (PBMC). 2013. Sumário Executivo Do Volume 1 – Base Científica Das Mudanças Climáticas. Rio de Janeiro, Brasil. doi:10.1037/e530172011-002

Polaquini, L.E.M., J.G. Souza, and J.J. Gebara. 2006. Changes in the Brazilian beef cattle production and commercialization system post-Mercosul. *Revista Brasileira de Zootecnia* 35, no. 1: 321–327. doi:10.1590/S1516-35982006000100040

Prado Júnior, S.T. 2013. Agricultura de baixa emissão de carbono: Quem cumpre as decisões? [Low carbon agriculture: Who makes the decisions?]. *ABC Observatório: Fundação Getulio Vargas.* http://www.cetesb.sp.gov.br/userfiles/file/mudancasclimaticas/proclima/file/publicacoes/agropecuaria/portugues/gurgel_abec_quem_cumpre_2013.pdf

Qaim, M., and G. Traxler. 2005. Roundup ready soybeans in Argentina: Farm level and aggregate welfare effects. *Agricultural Economics* 32, no. 1: 73–86. doi:10.1111/j.0169-5150.2005.00006.x

Rathmann, R., A. Szklo, and R. Schaeffer. 2010. Land use competition for production of food and liquid biofuels: An analysis of the arguments in the current debate. *Renewable Energy* 35, no. 1: 14–22. doi:10.1016/j.renene.2009.02.025

Raucci, G.S., C.S. Moreira, P.A. Alves, F.F.C. Mello, L.A. Frazão, C.E.P. Cerri, and C.C. Cerri. 2014. Greenhouse gas assessment of Brazilian soybean production: a case study of Mato Grosso State. *Journal of Cleaner Production* In Press. doi:http://dx.doi.org/10.1016/j.jclepro.2014.02.064

Rausch, L. 2014. Convergent agrarian frontiers in the settlement of Mato Grosso, Brazil. *Historical Geography* 42: 276–297.

Redo, D., T.M. Aide, and M.L. Clark. 2012. Vegetation change in Brazil's dryland ecoregions and the relationship to crop production and environmental factors: Cerrado, Caatinga, and Mato Grosso, 2001–2009. *Journal of Land Use Science* 8: 1–31. doi:10.1080/1747423X.2012.667448

Reijnders, L., and M.A.J. Huijbregts. 2008. Biogenic greenhouse gas emissions linked to the life cycles of biodiesel derived from European rapeseed and Brazilian soybeans. *Journal of Cleaner Production* 16, no. 18: 1943–1948. doi:10.1016/j.jclepro.2008.01.012

Reis, M. 2012. Food insecurity and the relationship between household income and children's health and nutrition in Brazil. *Health Economics* 21, no. 4: 405–427. doi:10.1002/hec.1722

Reporterbrasil.org. 2014. Lista Suja [Dirty list]. *Reporter Brasil.* http://reporterbrasil.org.br/listasuja/resultado.php?busca=soja&submit=Buscar&lingua=pt

Ribeiro, M.C., J.P. Metzger, A.C. Martensen, F.J. Ponzoni, and M.M. Hirota. 2009. The Brazilian Atlantic Forest: How much is left, and how is the remaining forest distributed? Implications for conservation. *Biological Conservation* 142, no. 6: 1141–1153. doi:http://dx.doi.org/10.1016/j.biocon.2009.02.021

Richards, P.D., R.J. Myers, S.M. Swinton, and R.T. Walker. 2012. Exchange rates, soybean supply response, and deforestation in South America. *Global Environmental Change* 22, no. 2: 454–462. doi:10.1016/j.gloenvcha.2012.01.004

Riskin, S.H., S. Porder, C. Neill, A.M.S. Figueira, C. Tubbesing, and N. Mahowald. 2013. The fate of phosphorus fertilizer in Amazon soya bean fields. *Philosophical Transactions of the Royal Society B: Biological Sciences* 368, no. 1619: 20120154-1–11 . doi:10.1098/rstb.2012.0154

Rodrigues, A.S.L., R.M. Ewers, L. Parry, C. Souza, A. Veríssimo, and A. Balmford. 2009. Boom-and-bust development patterns across the Amazon deforestation frontier. *Science* 324, no. 5933: 1435–1437. doi:10.1126/science.1174002

Roundtable on Responsible Soy (RTRS). 2013. High conservation value area (HCVA) assessment and audit guide for Category 2. http://www.responsiblesoy.org

Roundtable on Responsible Soy (RTRS). 2014. About Us>Members. *Roundtable on Responsible Soy.* http://www.responsiblesoy.org

Rudorff, B.F.T., M. Adami, D.A. Aguiar, M.A. Moreira, M.P. Mello, L. Fabiani, D.F. Amaral, and B. M. Pires. 2011. The soy moratorium in the Amazon biome monitored by remote sensing images. *Remote Sensing* 3, no. 1: 185–202. doi:10.3390/rs3010185

Sawyer, D. 2008. Climate change, biofuels and eco-social impacts in the Brazilian Amazon and Cerrado. *Philosophical Transactions of the Royal Society B: Biological Sciences* 363, no. 1498: 1747–1752. doi:10.1098/rstb.2007.0030

Schiesari, L., and B. Grillitsch. 2011. Pesticides meet megadiversity in the expansion of biofuel crops. *Frontiers in Ecology and the Environment* 9, no. 4: 215–221. doi:10.1890/090139

Schmink, M., and C.H. Wood. 1992. *Contested frontiers in Amazonia.* New York: Columbia University Press.

Schneider, S., and P.A. Niederle. 2010. Resistance strategies and diversification of rural livelihoods: The construction of autonomy among Brazilian family farmers. *The Journal of Peasant Studies* 37, no. 2: 379–405. doi:10.1080/03066151003595168

Scott, J.C. 1998. *Seeing like a state: How certain schemes to improve the human condition have failed.* New Haven: Yale University Press.

Secretaria de Estado do Meio Ambiente – Mato Grosso (SEMA-MT). 2014. Licenciamento de Propriedades Rurais: CAR (Licensing of Rural Properities: CAR). http://www.sema.mt.gov.br

Secretaria de Estado do Meio Ambiente – Pará. SEMA-PA. 2014. Licenciamento Ambiental: CAR (Environmental Licensing: CAR). http://www.sema.pa.gov.br

Serageldin, I. 1996. *Sustainability and the wealth of nations: First steps in an ongoing journey.* Washington, DC: World Bank.

Smaling, E.M.A., R. Roscoe, J.P. Lesschen, A.F. Bouwman, and E. Comunello. 2008. From forest to waste: Assessment of the Brazilian soybean chain, using nitrogen as a marker. *Agriculture, Ecosystems & Environment* 128, no. 3: 185–197.

Soares-Filho, B.S., A. Alencar, D.C. Nepstad, G.C. Cerqueira, M.C. Vera Diaz, S. Rivero, L. Solorzano, and E. Voll. 2004. Simulating the response of land-cover changes to road paving and governance along a major Amazon highway: The Santarém-Cuiabá corridor. *Global Change Biology* 10, no. 5: 745–764. doi:10.1111/j.1529-8817.2003.00769.x

Soares-Filho, B.S., D.C. Nepstad, L.M. Curran, G.C. Cerqueira, R.A. Garcia, C.A. Ramos, E. Voll, A. McDonald, P. Lefebvre, and P. Schlesinger. 2006. Modelling conservation in the Amazon basin. *Nature* 440, no. 7083: 520–523. doi:10.1038/nature04389

Soares-Filho, B.S., R. Rajão, M.N. Macedo, A. Carneiro, W. Costa, M. Coe, H. Rodrigues, and A. Alencar. 2014. Cracking Brazil's forest code. *Science* 344, no. 6182: 363–364. doi:10.1126/science.1246663

Soler, L.S., P.H. Verburg, and D.S. Alves. 2014. Evolution of land use in the Brazilian Amazon: From frontier expansion to market chain dynamics. *Land* 3, no. 3: 981–1014.

Sousa, I.S., and L. Busch. 1998. Networks and agricultural development: The case of soybean production and consumption in Brazil. *Rural Sociology* 63, no. 3: 349–371. doi:10.1111/1467-9523.00128

Sparovek, G., G. Berndes, I.L.F. Klug, and A.G.O.P. Barretto. 2010. Brazilian agriculture and environmental legislation: Status and future challenges. *Environmental Science & Technology* 44, no. 16: 14637–14641. doi:10.1021/es1007824

Spehar, C.R. 1995. Impact of strategic genes in soybean on agricultural development in the Brazilian tropical savannahs. *Field Crops Research* 41, no. 3: 141–146.

Steward, C. 2007. From colonization to environmental soy: A case study of environmental and socio-economic valuation in the Amazon soy frontier. *Agriculture and Human Values* 24, no. 1: 107–122. doi:10.1007/s10460-006-9030-4

Tabarelli, M., L.P. Pinto, J.M.C. Silva, M. Hirota, and L. Bede. 2005. Challenges and opportunities for biodiversity conservation in the Brazilian Atlantic Forest. *Conservation Biology* 19, no. 3: 695–700. doi:10.1111/j.1523-1739.2005.00694.x

VanWey, L.K., S. Spera, R. de Sa, D. Mahr, J.F. Mustard, and R. Sa. 2013. Socioeconomic development and agricultural intensification in Mato Grosso. *Philosophical Transactions of the Royal Society B: Biological Sciences* 368, no. 1619: 20120168. doi:10.1098/rstb.2012.0168

Varsano, R. 2013. Fazendo e desfazendo a Lei Kandir [Making and unmaking the Kandir Law]. *Inter-american development bank*. Washington, DC: Inter-American Development Bank. http://54.209.110.9/handle/11319/5774?scope=123456789/6&thumbnail=true&rpp=5&page=3&group_by=none&etal=0

Vera Diaz, M.C., R.K. Kaufmann, and D.C. Nepstad. 2009. The environmental impacts of soybean expansion and infrastructure development in Brazil's Amazon basin. Global Development Environment Institute Working Paper N0. 09-05.

Vera-Diaz, M.C., R.K. Kaufmann, D.C. Nepstad, and P. Schlesinger. 2008. An interdisciplinary model of soybean yield in the Amazon Basin: The climatic, edaphic, and economic determinants. *Ecological Economics* 65, no. 2: 420–431. doi:10.1016/j.ecolecon.2007.07.015

Vitousek, P.M., J.D. Aber, R.W. Howarth, G.E. Likens, P.A. Matson, D.W. Schindler, W.H. Schlesinger, and D.G. Tilman. 1997. Human alteration of the global nitrogen cycle: Sources and consequences. *Ecological Applications* 7, no. 3: 737–750. doi:10.1890/1051-0761(1997)007[0737:HAOTGN]2.0.CO;2

Walker, N.F., S.A. Patel, and K.A.B. Kalif. 2013. From Amazon pasture to the high street: Deforestation and the Brazilian cattle product supply chain. *Mongabay.com Open Access Journal* 6, no. 3: 446–467. http://tropicalconservationscience.mongabay.com/content/v6/TCS-2013_Vol_6(3)_446-467-Walker_et_al.pdf

Walker, R. 2011. The impact of Brazilian biofuel production on Amazônia. *Annals of the Association of American Geographers* 101, no. 4: 37–41. doi:10.1080/00045608.2011.568885

Ward, N. 1993. The agricultural treadmill and the rural environment in the post-productivist era. *Sociologia Ruralis* 33, no. 3–4: 348–364. doi:10.1111/j.1467-9523.1993.tb00969.x

Warnken, P.F. 1999. *The development and growth of the soybean industry in Brazil*. Ames: Iowa State University Press. http://scholar.google.com.br/scholar?q=paul+warnken+1998+soybean+production&hl=pt-BR&as_sdt=0#7

Weinhold, D., E. Killick, and E.J. Reis. 2013. Soybeans, poverty and inequality in the Brazilian Amazon. *World Development* 52: 132–143. doi:10.1016/j.worlddev.2012.11.016

Wilkins, R.J. 2008. Eco-efficient approaches to land management: A case for increased integration of crop and animal production systems. *Philosophical Transactions of the Royal Society B: Biological Sciences* 363, no. 1491: 517–525. doi:10.1098/rstb.2007.2167

Wilkinson, J., and S. Herrera. 2010. Biofuels in Brazil: Debates and impacts. *The Journal of Peasant Studies* 37, no. 4: 749–768. doi:10.1080/03066150.2010.512457

World Bank. 1986. *Poverty and hunger: Issues and options for food security in developing countries*. Washington, DC: World Bank. http://www-wds.worldbank.org/external/default/WDSContentServer/WDSP/IB/1999/09/17/000178830_98101901455676/Rendered/PDF/multi_page.pdf

Zoomers, A. 2010. Globalisation and the foreignisation of space: Seven processes driving the current global land grab. *The Journal of Peasant Studies* 37, no. 2: 429–447. doi:10.1080/03066151003595325

On the margins of soy farms: traditional populations and selective environmental policies in the Brazilian Cerrado

Ludivine Eloy, Catherine Aubertin, Fabiano Toni, Silvia Laine Borges Lúcio and Marion Bosgiraud

In the Cerrado, the expansion of soybean cultivation since the 1990s has coincided with the strengthening of environmental regulations. We analyze how the two main environmental policies – Protected Areas and the Forest Code – have played out at the ground level in western Bahia state. These policies in Cerrado have not been designed to curb the expansion of this agricultural frontier. These norms have, on the contrary, accommodated this expansion because the way environmental managers selectively choose environmental problems and publicize them through specific information systems depreciates traditional fire-dependent production systems. These 'politics of selection' are likely to increase competition for resources in the margins of soybean agriculture, which is where traditional populations have now become confined.

1. Introduction

Confrontations and interactions between conservation and agricultural development have intensified during recent decades in Latin America. Several countries have emerged as global centers of environmental conservation while, at the same time, encouraging land-use modernization under neoliberal policies, such as the widespread expansion of soybean agriculture (Zimmerer 2011). In the mid- and late 1990s, Brazil witnessed an acceleration in the spread of large-scale soybean monoculture. At the same time, the country has also seen a significant expansion of protected areas since 1985 (Rylands and Brandon 2005), accompanied by successful efforts to enforce environmental legislation to curb deforestation (Nepstad et al. 2002).

The most dynamic agricultural frontier in Brazil is the central savanna, called the 'Cerrado'. This biome was the source of more than 60 percent of the soybean harvest of 2012–2013 (Terra-AFP 2013). In 2013, Brazil was the largest soybean and sugarcane producer in the world. Since the 1990s, most of the country's beef and grain has been produced in this region. This trend has been recognized domestically and internationally, and the Cerrado has been praised as Brazil's breadbasket. During the 2000s, agriculture and ranching made a remarkable contribution to the country's economic growth, and exports of primary goods helped balance its current account. This economic success, however, came at a price. In the mid-2000s, the Cerrado overtook Amazonia as the leader in deforestation in Brazil (MMA 2011a).[1]

[1]Legal Amazonia covers nearby 4 million km^2, of which 19 percent of native forest has already been deforested. The Cerrado has half the size of Legal Amazonia (2 million km^2) and, in 2012, 49 percent of its natural vegetation had been cleared (IBGE 2012).

The Cerrado is also home to traditional populations[2] whose livelihoods involve subsistence farming, extensive cattle ranching and multiple forms of extractivism. Although these populations have been living in the Cerrado since long before the advent of soybean agriculture, most of them do not have titles to the lands they live on. In a context in which land is being rapidly converted to agro-industrial production or targeted by conservation policies, what are the impacts on traditional populations? In this contribution, we analyze how these policies play out at the ground level in the Cerrado in western Bahia state, the main area of rapid expansion of high-input soybean agriculture in the Cerrado.

Our main argument in this paper is that the recent strengthening of environmental policies in the Cerrado relies on 'politics of selection' that converge on fire fighting and set-aside forest reserves, while glossing over other major environmental issues, which do not affect soybean producers and traditional populations in equal measure.

The paper is organized as follows: In section 2, we present our theoretical and methodological framework. In section 3, we briefly describe existing land use in the Cerrado biome, with an emphasis on the role of public policies and the direct and indirect impacts of soybean agriculture in this process. We also discuss how environmental concerns have been raised about the Cerrado, in particular because of increased deforestation. We show that despite the strengthening of environmental regulations, the soybean frontier is still advancing. We briefly discuss two important aspects of Brazilian environmental policy: the Forest Code and the National System of Protected Areas. Using the data collected during our field studies, in section 4 we show how these policies converge on fire fighting and their various implications for traditional populations and soybean producers. Drawing from these results, we discuss the causes and consequences of these politics of selection for the formation and the future of the territories that have been forced into 'the soybean margins'.

2. Theoretical framework and methods

2.1. *Environmental norms and politics of selection*

As part of a global phenomenon (Aubertin and Rodary 2011), the Cerrado biome basin has experienced the diversification and expansion of protected areas since the early 1990s, a process which has allowed some communities to reassert their collective rights over territories and resources. As a result, there has been a notable shift in the governance of protected areas from a model of strict protection to one of participatory management, which involves peasants, indigenous people and resource-user groups in decision-making.

At the same time, however, environmental policies also respond to the evolution of representation of stakeholders and their power relationships. As a result, environmental norms can be seen as indicators of policy choices, what we call the 'politics of selection', which can be divided into three interrelated processes: spatial selection, selection of environmental problems to address and selection of information.

[2]We define traditional populations (or traditional farmers) in a broad sense, independently from their political mobilization and their cultural affirmation: communities that are historical heirs of different forms of use, management, knowledge and symbolic representations of the Cerrado, characterized by their historical relationship with the territory they claim, and by low-input production systems. This expression clubs together different social groups, such as indigenous and quilombola (descendents of Maroon slaves) communities, farmers (in agricultural settlements or members of rural communities), artisans, fishermen, extractivists, etc.

First, conservation development initiatives increasingly rely on spatial representations that emerge from 'hidden' power networks (Zimmerer 2011). Methods of 'prioritization' serve to define the spatial targeting of protected areas, but they take little account of social issues and the diversity of land uses (Milian and Rodary 2010). After being applied at a global scale, these methods were used at national and smaller scales (biomes), like in Amazonia or the Cerrado (Scaramuzza, Machado, and Rodrigues 2005; MMA 2011b). Furthermore, participatory management programs place great emphasis on planning and zoning within conservation territories. In practice, this spatial fixing is usually designed to allow certain activities and restrict numerous others. Moreover, protection is clearly concentrated in tropical forest environments (Zimmerer 2000).

Second, environmental norms may derive from the selection of environmental problems to address and the formulation of simplified and official discourses to explain these problems (Bassett and Zueli 2000; Mathews 2005). Conservation booms in Latin America have generally focused on biodiversity conservation and fighting deforestation. Forest protection has taken the form of an undiscriminating fight against fires, irrespective of their objectives and conditions. Other environmental concerns, such as soil conservation and water management, have become less important in the mainstream environmental-conservation agenda (Zimmerer 2000).

Third, selection also takes place through the choice and design of environmental information systems – that is, the material inputs and outcomes of knowledge that are produced and disseminated. Indeed, advances in geographical information systems (GIS) and remote sensing now power the formulation and implementation of environmental norms. But a focus on remotely sensed data is not strictly a methodological choice; it is also a political one, since pixels can be interpreted in different ways, depending on stakeholders (Harwell 2000; Robbins 2003). Furthermore, the recent emphasis on creating transparency through mapping can also obscure some management practices, frauds and abuses. Foucault's theory postulates that the power of government is based on the standardization and control of knowledge, which dictate good practices (Foucault 2001). Certain standards, such as the general prohibition of fire in savanna ecosystems, are also based on ignorance, since practices of silencing and concealment are inherent to the process of knowledge production (Mathew 2005).

2.2. *Methods*

In order to understand the spatial expansion of protected areas in conjunction with recent trends of deforestation and soybean expansion in the Cerrado, we superimposed data originating from different public databases and processed them through a GIS software application (ArcMap 10).

In order to verify how institutional changes are playing out at the ground level, we undertook fieldwork between 2012 and 2013 at three research sites, around and in three Protected Areas (PAs) which border the agricultural frontier of western Bahia state (Figure 1).[3]

The first PA is the Veredas do Acari Sustainable Development Reserve (RDS), which covers 609 km^2, in the municipality of Serra das Araras, northern Minas Gerais state. The RDS is a Sustainable Use Conservation Unit, managed by the environmental agency of the state of Minas Gerais (IEF). The second one is the Nascentes do Rio Vermelho

[3]Fieldwork was undertaken in collaboration with Brazilian institutions (see Acknowledgements).

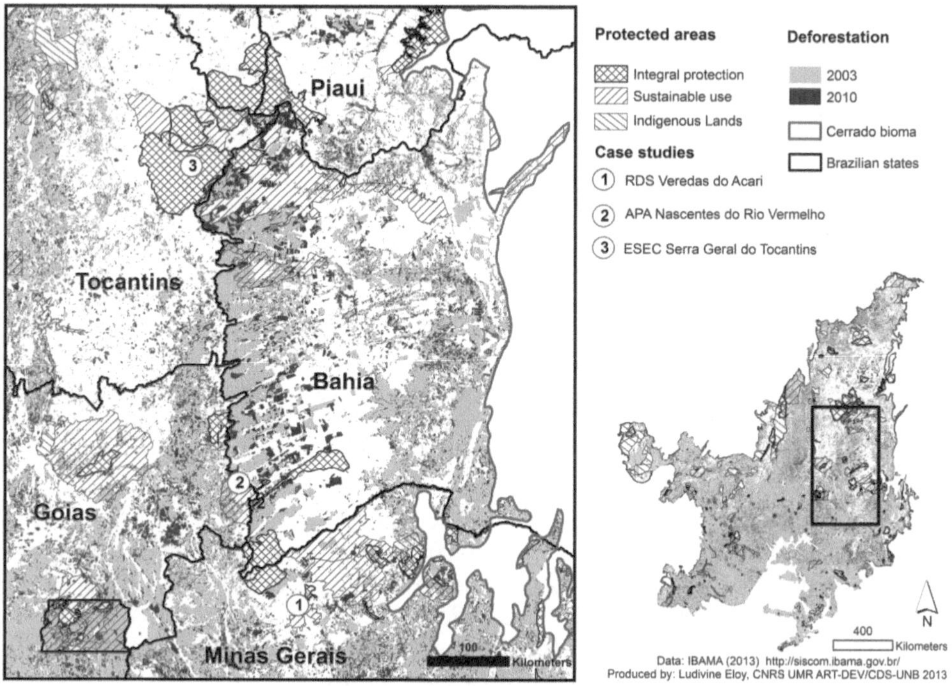

Figure 1. Protected areas and deforestation in the eastern Cerrado and location of study sites. Source: IBAMA (2013).

Environmental Protection Area (APA). Its 1761 km^2 are spread over four municipalities, but the majority of this PA is in the municipality of Mambaí (Goiás state). This APA is a Federal Sustainable Use Protected Area, managed by the Chico Mendes Institute of Biodiversity Conservation (ICMBio). The last research site is the Serra Geral do Tocantins Ecological Station (EESGT). It covers 7070 km^2, spanning two municipalities, mainly Mateiros, in the Jalapão region of the state of Tocantins. This ESEC is also federal and, therefore, managed by ICMBio.

Most of the farmers who live within or depend on the use of these PAs are descendants of migrants from the northeast (Bahia state) or from southern Goiás state who settled in this remote region in the late nineteenth or early twentieth centuries to escape slavery and drought. Once they arrived in the region, they started working for ranchers and clearing some land for subsistence agriculture. Soybean producers originate, for the most part, from southern Brazil (Rio Grande do Sul, Paraná, São Paulo) and arrived in the 1980s and 1990s (Haesbaert 1997).

We combined socio-economic data at the household scale and land-cover data from satellite images to investigate local farming systems. Surveys were used to estimate the sources of income of each domestic unit. In order to understand the economic strategies of extended families, we also assessed the location of houses and sources of income of all the children of the head of the household. Participant observation, landscape interpretation and semi-structured interviews were used to understand the perceptions of traditional farmers about the environmental impact of their farming practices and about environmental policies that influence them. We interviewed a total of 57 heads of households (25, 20 and 12, respectively, in the three study sites; Table 1).

Table 1. Sample characteristics.

	Number of household heads	Minimum age	Maximum age	Mean age
Serra das Araras, MG	25	27	79	51
Mambaí, GO	20	37	71	52
Mateiros, TO	12	45	71	56

Source: Field data.

The sampling criterion was to choose farmers who depend on access to and use of natural resources located within each PA. In Serra das Araras, we interviewed randomly selected members of the local association of herders of the RDS (nearly 50 members). In Mambaí, we randomly selected farmers who lived in four agricultural settlements with a gradient of distance from the town. In Mateiros, we interviewed all heads of households who belonged to the association of users of the EESGT – Serra Geral do Tocantins Ecological Station (Ascolombas-Rios); they represented nearly 90 percent of all the families that use resources in this protected area (Eloy and Lúcio 2013).

To study the formulation and implementation of environmental norms at the local scale, we analyzed official texts, carried out participant observation during information meetings, and conducted six semi-directive interviews with the managers of these PAs. We also conducted five interviews with soybean producers in western Bahia (municipality of Jaborandi) and northern Minas Gerais (municipality of Chapada Gaúcha).

3. Land use, soybean expansion and the environment in the Cerrado: an historical overview

3.1. *Soybeans in the Cerrado*

The Portuguese treated the Cerrado, as they did most of the Brazilian backlands, as commercial rather than colonial territory. From these vast lands they extracted valuable resources, such as gold, silver and gemstones. They also had the intention of hunting and enslaving the indigenous populations. Over time, extensive cattle ranches occupied parts of the Cerrado and the Caatinga backlands.

The occupation of the Cerrado gained momentum during the second Vargas government, known as the Estado Novo, when the *Marcha para o Oeste* (March to the West) was launched in the 1940s. The *Marcha* was a bundle of polices, programs, projects and institutional innovations inspired by a modernizing ideology (Maia 2013). During this period, the government promoted land survey expeditions and established colonization sites in Dourados (Mato Grosso do Sul state) and Ceres (Goiás state) (Shiki, da Silva, and Ortega 1997). The most significant efforts made by the government to advance the frontier took the form of infrastructure building. This trend had been initiated a couple of decades earlier, when a railway connecting São Paulo to Anápolis spurred the occupation of southern Goiás and western Minas Gerais states. The integration of the Cerrado into Brazil's major urban markets was completed after the new capital city, Brasília, was constructed in 1960 (Silva 2000).

By the 1970s, the central–western portion of Brazil, which comprises most of the Cerrado, was the region with the largest area of pasturelands in the country (Rezende 2002). Cattle ranching had become remarkably extensive. The intensification of ranching, as well as the introduction of high-output agriculture, required public investments that went far beyond the building of road networks. In the mid-1960s, the Brazilian government launched a series of programs for agricultural modernization and the occupation of Amazonia and the Cerrado.

Two federal programs specifically targeted the Cerrado: the Program for Development of the Cerrado (POLOCENTRO) and the Japanese Cooperation Program for Development of the Cerrado (PRODECER). POLOCENTRO channeled agricultural and ranching credit to rural areas where infrastructure was already present. Between 1975 and 1983, it reached some 3.7 million ha (aggregating credit of USD 577 million). POLOCENTRO mainly benefited large and medium producers: properties larger than 200 ha (81 percent of the projects) received 88 percent of credit resources, with properties over 1000 ha (39 percent of projects) absorbing 60 percent of the total resources (Coelho 2001). PRODECER, on the other hand, targeted smallholders and promoted colonization through cooperatives. Even though the program was partially aimed at curbing deforestation – a serious problem in POLOCENTRO – its goal was also to increase the production of commodities. These programs promoted the cultivation of soybeans, eucalyptus, grains, coffee, sugarcane and other crops.

The long-term policies that had the most impact in promoting the occupation of the Cerrado were those pertaining to science, technology and innovation, particularly as regards soybean production. Colonists from the South were already involved in the soybean commodity network, and were determinant in the introduction of the crop in the Cerrado (Sousa and Busch 1998). However, adapting soybeans to the Cerrado required a massive research effort, undertaken by an extensive network of private and public institutions, under the leadership of the Brazilian Agricultural Research Corporation (Embrapa). This research had two main components. The first one was the selective breeding of new cultivars adapted to low latitudes (Wilkinson and Sorj 1992). Between 1993 and 2007, the public research system introduced some 195 cultivars (FAO, 2007). Embrapa was instrumental in testing and adapting soybean and other grains to the Cerrado. In particular, it played a key role in developing new technologies for soil management to deal with the acidity and high aluminum content of Cerrado soils (Pessôa and Bonelli 1997). The adaptation of soybeans to the Cerrado was so successful that in the early 1980s, the average yield was comparable to that in the South. By 2001, the cultivated area in the Centro Oeste region surpassed that in the South (CONAB 2013). In addition, there has been a very strong expansion of sugarcane cultivation for ethanol on grazing lands.

Since the 2000s, soybean production grew in the region known as MAPITOBA (the southern parts of Maranhão, Tocantins and Piaui, and western Bahia), where the flat uplands (*chapadão*) of the Serra Geral offer prime farmlands (Figure 2). According to the Brazilian Institute of Geography and Statistics (IBGE), Mato Grosso state remains the national leader in the production of soybean with 20,800,544 tonnes produced in 2011, against 3,512,568 tonnes in Bahia state. However, within the boundaries of the Cerrado biome, the increase in production between 2005 and 2011 is higher in the MAPITOBA region (3,876,609 tonnes) than in the eastern part (Mato Grosso state with 3,372,569 tonnes). In this same period, harvested soybean areas expanded three times more in the eastern part than in the western part, which indicates a much more extensive production in the MAPITOBA region. Western Bahia is considered an example of the 'Miracle of the Cerrado', with farms there sustaining Brazil's agroexports (Brannstrom and Brandão 2012).

Soybean expansion is similar to the eucalyptus expansion that took place in northern Minas Gerais during the 1970s and 1980s. Agro-industrial farmers preferred flat uplands, situated at altitudes between 700 and 1200 m, where the topography and conditions

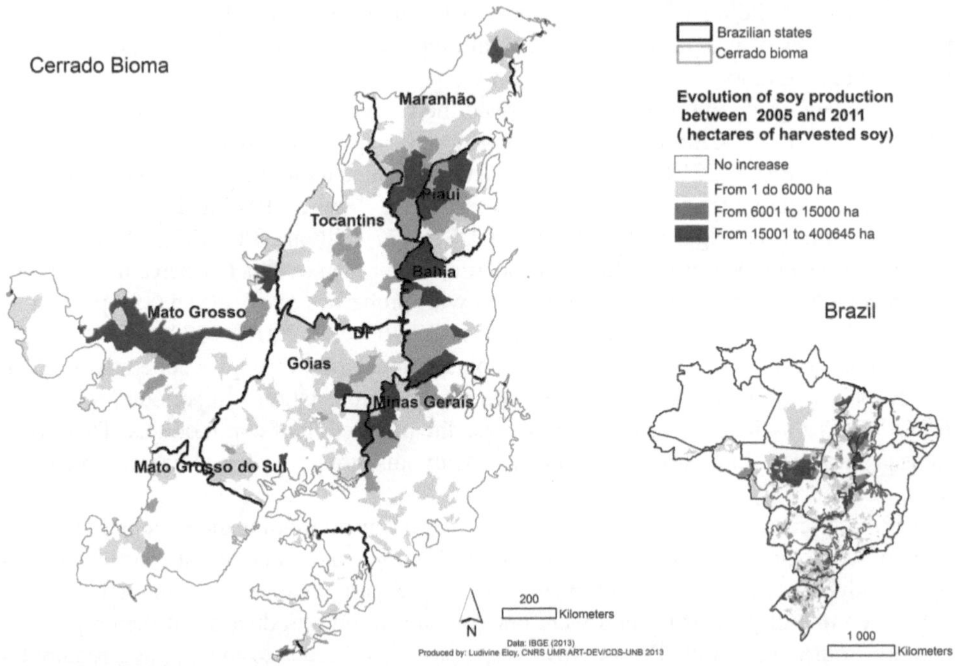

Figure 2. Soybean expansion in the Cerrado biome between 2005 and 2011.
Source: IBGE (2013).

– such as regular precipitation, mild temperatures and well-weathered and drained soils – are suitable for mechanized farming. They tended to ignore the wet lowlands (*Veredas*), which were occupied by traditional populations displaced by agricultural expansion (Ribeiro 2008, 2010). In the mid-2000s, 94 percent of agricultural areas were on slopes up to 5 percent, with the rest on lands sloping between 5 percent and 10 percent. In contrast, cultivated and native pastures, less dependent on soil fertility and mechanization, were located over a wider variety of topographic gradients (Ferreira, Ferreira, and Latrubesse 2008).

Until the arrival of large-scale agriculture, the uplands were common areas, occasionally used by traditional populations as natural pastures and for extractivism. These populations were pushed out in the early and mid-1980s, creating marginalized groups of resource users, whose practices came into conflict with conservation policies adopted in the late 1990s. Traditional communities lost common-use access to significant portions of the Cerrado in the flat uplands and thus increased their use of forest environments in the lowlands, often already divided by their distribution among heirs, or by the expansion of large estates. Environmental legislation, meanwhile, imposed a number of restrictions on the use of resources by these populations, affecting hunting, fishing, slash-and-burn practices, removal of timber and fiber, extraction of wild honey, picking of flowers, collection of medicinal plants, production of coal and gold mining (Ribeiro 2008). Furthermore, the encouragement the federal and state governments provided to industrial eucalyptus plantations, and later to those of soybean, led to major environmental impacts such as deforestation, loss of biodiversity, soil erosion and water depletion (Ribeiro 2008). The movement of traditional populations of the Cerrado took place in the late 1990s in this context of land conflicts and environmental degradation (Nogueira 2009).

3.2. *The emergence of the Cerrado as an environmental concern*

The Cerrado covers about 22 percent of Brazil's landmass. It spans a large area of central Brazil and borders other major biomes: Amazonia to the north, Caatinga to the northeast, Pantanal to the west and the Atlantic Forest to the east and southeast,[4] which gives it a very rich biodiversity. The headwaters of three major South American basins are located in the Cerrado: the Amazonas/Tocantins, the São Francisco and the La Plata. Despite its size and importance, the Cerrado biome is one of the most threatened environments in the world. Even though it was designated a biodiversity hotspot in the early 2000s (Myers et al. 2000), deforestation has reached an alarming average rate of three million ha/year during the 2000s, in particular due to the expansion of eucalyptus, soybean and sugarcane plantations.

Social movements and environmentalists took up the defense of the Cerrado following the 1992 Earth Summit. Those groups came together in a network of socio-environmentalist non-governmental organizations (NGOs), known as *Rede Cerrado*. A first national program of conservation and sustainable use of the Cerrado (*Programa Cerrado Sustentavel* – PCS), funded by the Global Environmental Fund and the World Bank, was launched in 2005 (Sawyer and Souza Lobo 2008). However, it was only four years later, in 2009, that the Brazilian government included the fight against deforestation in Cerrado as part of its strategy to cope with climate change.

The Cerrado clearly emerged as a political and environmental issue in 2009, at the Copenhagen Conference on Climate Change (CoP15). Brazil presented there an ambitious program which included: (1) the reduction by 36.1 percent to 38.9 percent of CO_2-equivalent emissions by 2020; (2) an 80 percent reduction in the rate of deforestation in the Amazon based on the 1996–2005 period; (3) a reduction by 40 percent of deforestation in the Cerrado biome based on the 1999–2005 period; and (4) the restoration of 15 million ha of degraded pastures. These commitments shaped the national policy on climate change, which established action plans, including a plan for the prevention and control of deforestation and fires in the Cerrado (*PP Cerrado*).

At this conference, it became clear that the problem of deforestation in Brazil was far worse in the Cerrado than in Amazonia (Sawyer 2009). Between 2002 and 2008, the average annual rate of deforestation in the Cerrado was twice as high as in Amazonia. By 2008, 986,247 km^2, or 48.37 percent, of the Cerrado's original cover was already cleared, against only 19 percent for Amazonia (IBGE 2010).

Three main factors explain this situation. First, land in the Cerrado is more suited to agricultural modernization than that in Amazonia. Investors have thus applied pressure on the federal government not to limit the expansion of the Cerrado agricultural frontier. Second, unlike the lush Amazon rainforest, the savannas have never attracted public and media attention. Even more dramatic is the 'invisibility' of the local populations in the region. Traditional populations in the Cerrado are very diverse in terms of ethnicity, culture, land use and livelihoods. They are not organized politically and they have only recently started using their collective identities to fight for their territorial rights (Nogueira 2009). They are not linked to global environmental networks and funds, as has been the case of traditional and indigenous populations of Amazonia since the 1990s (Hecht 2012). However, there are very few, if any, spots in the Cerrado that have not been occupied by local communities (Ribeiro 2008).

[4]The Cerrado biome spreads across the states of Goiás, Tocantins, Mato Grosso, Mato Grosso do Sul, Minas Gerais, Bahia, Maranhão, Piauí, Paraná and São Paulo, and the Federal District.

Third, controlling deforestation in the Cerrado, where private property prevails, is highly problematic: restricted-use Conservation Units can only be created through costly expropriations. The major political force of the *fazendeiros*, who have come together under the 'ruralist' banner, also has its roots in the resistance to all environmental policies of the federal government (Aubertin and Pinton 2013). Brazil has focused its conservation efforts on Amazonia, as shown by the soybean moratorium initiative. Under this pact, signed in 2006, the Brazilian Association of the Vegetable Oil Industry (Abiove) and the Brazilian Association of Cereal Exporters (ANEC) pledged not to market soybean from deforested areas in Amazonia Legal (encompassing all nine states in the Amazon basin). Large companies like Cargill, Bunge and ADM were therefore forced to concentrate their investments in the Cerrado. This moratorium has been a success and, according to the industry representatives, soybeans are responsible for only 0.39 percent of the total area cleared in the Amazonian states of Mato Grosso, Pará and Rondônia after the moratorium was adopted (ABIOVE 2013).

In 2009, the federal government issued a decree approving the agro-ecological zoning for sugarcane. Most of the expansion area – 64 million ha – is in the Cerrado. According to the zoning, sugarcane cultivation is banned from the Amazon, the Pantanal and the upper Paraguay basin. As a consequence, the expansion of agro-industry in the Cerrado and the promotion of green technologies for land sparing allowed the protection of Amazonia, with deforestation rates there decreasing significantly since 2004 (Aubertin and Pinton 2013).

At the same time, the policies of the federal and state governments for the protection of the Cerrado have relied on two main strategies: the stricter enforcement of the Forest Code and the expansion and diversification of PAs. We argue, however, that these strategies do not affect soybean producers and traditional populations equally.

3.3. *The implementation of environmental legislation in the Cerrado*

Brazilian environmental legislation has two important – and controversial – components aimed at conserving biodiversity: the Forest Code and the National System of Conservation Units (SNUC). The Forest Code regulates land use on private properties. The SNUC, on the other hand, mainly concerns public lands.

3.3.1. *The Forest Code*

The Brazilian Forest Code, passed in 1965 and revised many times thereafter, requires that landowners protect forest cover along all watercourses and steep hill slopes (Permanent Protection Areas – APP) and maintain a part of their lands under natural vegetation, in the so-called 'Legal Reserve'. In the Cerrado, the Legal Reserve must cover 20 percent of each property (against 80 percent in most parts of the Legal Amazon). Since the 2000s, remote sensing and GIS have helped reveal which farmers were engaged in illegal deforestation, as well as facilitating the issue of Environmental Licenses for Rural Properties.

Environmental licensing has been spreading in all Brazilian rural areas since 2008, when the government prohibited public and private banks from issuing credit to landowners who are not in compliance[1] with land-tenure legal requirements and environmental rules. Partnerships between state and non-state actors at the municipal level have provided incentives to farmers to bring their lands into compliance with environmental laws in Amazonia, like in the municipalities of Lucas do Rio Verde, in Mato Grosso State and Paragominas in Pará State. These experiences have been upscaled to the state level in Mato Grosso, and also replicated in the state of Bahia (Brannstrom and Brandão 2012). While these partnerships decrease the costs of environmental compliance, due to economies of scale they have so far mainly benefitted large landowners (Ferreira 2010).

3.3.2. *Protected areas*

Law No. 9985, of 18 July 2000, created the SNUC, which regulates all public Conservation Units (PAs), be they federal, state or municipal. Integral Protection Units, in which the exploitation of natural resources is completely prohibited, differ from Sustainable Use Units, which can be exploited in accordance with a management plan. In theory, PAs, which include Conservation Units defined by the SNUC, as well as indigenous lands and *quilombola* lands, can help curb deforestation.[5] Currently, 8.2 percent of the Cerrado land is classified as PAs, against 26.1 percent in Amazonia (CNUC/MMA 2013). Indigenous lands represent 4.4 percent of the area of the Cerrado biome, while covering 21.7 percent of Amazonia (CNUC/MMA 2013).

Nevertheless, there has been a remarkable expansion and diversification of PAs in the Cerrado since the late 1990s. This period corresponds to the creation of large PAs for sustainable use, as opposed to the previous period, when integral protection areas were most common (Figure 3). The majority of these new PAs are Environmental Protection Areas (APAs), which, in 2013, covered 108,498 km^2. That figure represents 62.23 percent of all conservation units in the Cerrado and 97.25 percent of the sustainable use ones (CNUC/MMA 2013)

The expansion of APAs can be explained by the fact that these areas are the less strictly protected areas in Brazil. They aim at protecting biodiversity, controlling land use and ensuring the sustainable use of natural resources, considered as category IV in the International Union for Conservation of Nature (IUCN) Protected Areas Categories System. Yet they allow individual properties as well as urban and rural settlements, and, in practice, authorities have imposed relatively low obstacles or regulations relating to resource use. If an APA does not have a management plan (which is often the case), landowners just need to abide by the general environmental legislation – the Forest Code in the case of rural areas. Therefore, APAs do not prevent the expansion of soybean agriculture since landowners can deforest up to 80 percent of their properties for planting crops, and use irrigation water if they have a license to do so. On this latter point, Brannstrom (2005) showed how the agribusiness sector in western Bahia (represented by its main cooperative) controls the legal process for obtaining these licenses.

The large-scale creation of APAs brings political dividends to state and municipal governments. As APAs do not pose restrictions on land use, administrators meet little resistance from farmers and ranchers. Municipalities and states are able to increase their areas under protection and thus qualify to receive additional funding. This is particularly significant in states where there is a mechanism known as ICMS-ecológico (Imposto sobre Circulação de Mercadorias e Serviços), where the state government returns part of the sales taxes it collects to municipalities according to the proportion of PAs in their territories.

[5]The federal government is responsible for demarcating indigenous lands and is the formal owner of those lands. However, indigenous people have full rights to the natural resources on their lands and can use the land according to the Forest Code, as can any private landowner. Even so, indigenous lands in the Amazon are the least deforested areas in the region, particularly on the agricultural frontier (Nepstad et al. 2006).

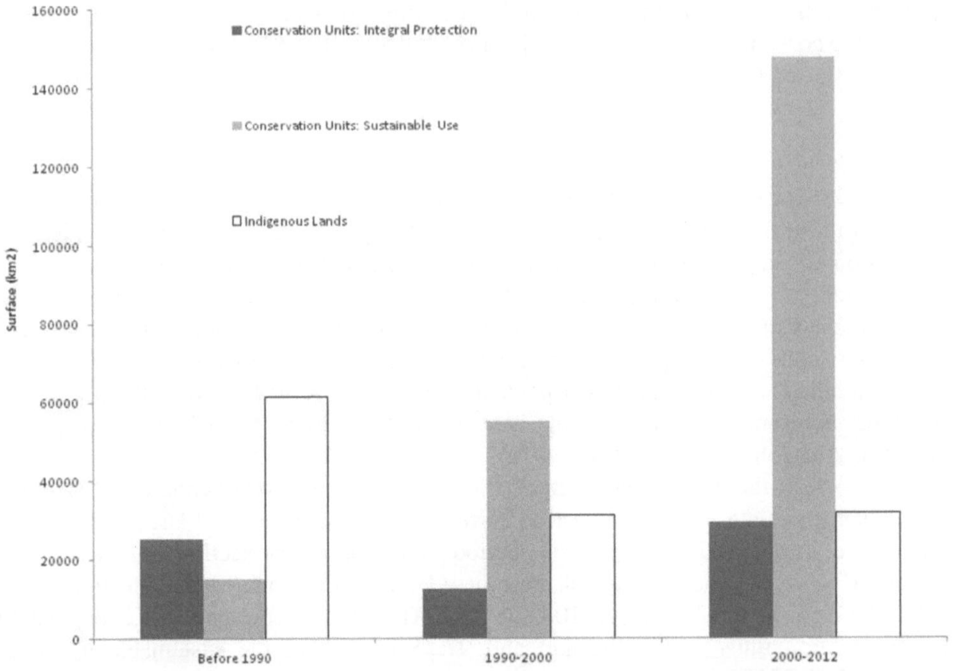

Figure 3. Evolution of the surface of protected areas in the Cerrado, within different categories. Source: IBAMA (2013).

4. Results: environmental policies around the soybean frontier – politics of selection and their implications for traditional farming systems

Large-scale farming and traditional farming systems have historically occupied different areas in the Cerrado, so conflicts between them arose only occasionally. In this section, we show how the selection of areas for conservation and recent changes in the Forest Code, as well as the focus on fire suppression, are likely to increase struggles for resources in the margins of the areas of expansion of soybean cultivation. To this end, we present our results in four parts, the first three according to our theoretical framework on the 'politics of selection' and the last on the implications of these policies for traditional populations and soybean producers.

4.1. *Politics of selection*

4.1.1. *The spatial targeting of protected areas legitimizes the expansion of soybean agriculture*

Since 1998, the Ministry of Environment and NGOs have undertaken studies for the prioritization of protected areas and environmental monitoring within Brazilian biomes (Probio project). On the basis of three fundamental principles – representativeness, persistence and vulnerability – these studies identified soybean expansion areas (western Bahia state, southern Maranhão and Piauí, and eastern Tocantins) as priority areas (MMA 2011b). However, PAs in the Cerrado are mainly delimitated on lowlands, where soils and topography are not suitable for mechanization, and do not intrude on soybean expansion areas.

Figure 1 shows that around the soybean frontiers of the eastern Cerrado, in the states of Goiás, Minas Gerais and Tocantins, the majority of the PAs have been demarcated on the west of the plateau formed by the Serra Geral (in Bahia, Piauí and Maranhão states), where deforestation and soybean expansion have been prominent since the 2000s (cf. Figure 3). So PAs are mostly delimitated on lowlands, in rough or hilly areas, where soils and topography are not suitable for the mechanization of agriculture. The only large PA that overlaps with the soybean frontier is an APA, which hardly affords any protection to the native vegetation. Parts of restricted-use PAs do encompass riverbanks, but such areas are, in any case, protected by the Forest Code.

Table 2 shows that the municipalities with the largest percentage of their territories under soybean cultivation have an average of 2.5 percent of their territories under PAs, while the average for all municipalities in the Cerrado is 4.6 percent.

As in other regions, the majority of PAs are created without regard to prioritization guidelines, mainly because of historical and political factors (Milian and Rodary 2010).

Moreover, the creation of PAs in areas of undulating topography offers new opportunities for soybean expansion, because such areas, even if located in another state, became target areas for Legal Reserve (LR) compensation. The new Forest Code (2013) also allows forest recovery (APP and LR) through payments for environmental services (article 41). It established a trading system for Legal Reserve compensation based on environmental assets (*cota de reserva ambiental*) which can operate at the scale of the entire biome (articles 44, 48, 66). In other words, farmers who do not want to preserve or reforest their Legal Reserve on prime farmland can purchase land elsewhere and set it aside as Legal Reserve. If this land is located inside a PA, farmers can avoid the responsibility of managing it (preventing fire, logging, etc.) by making a donation to the environmental institution that is in charge of the PA (*desoneração de reserva legal*). Areas of 'natural' vegetation acquire a new value, greater than the opportunity cost that would represent their clearing. This change leads to a new type of competition between large-scale farming and traditional agriculture.

4.1.2. *The focus on fire suppression undermines participative management of PAs*

The APA is a less strict conservation unit and is meant to be managed in collaboration with farmers. The RDS category was also created to promote participative management. In addition, the RDS and the EESGT belong to wider territorial-management units, called Mosaics of Protected Areas. Similar to ecological corridors, the mosaics are sets of adjacent or overlapping PAs and indigenous lands placed under integrated and participatory management. Mosaics are supposed to be not only tools for conservation but also land planning mechanisms which encourage stakeholder participation, strengthen local identities and help add value to local products and knowledge. Participatory local management used to be mediated by community interactions with non-local actors, such as NGOs, universities and international organizations. But in the Cerrado, this kind of alliance is mediated by the selection of productive activities that are perceived to be compatible with conservation. They include fire-free production techniques, like fruit and fiber gathering, handicrafts and ecotourism. Traditional agricultural systems, on the other hand, are usually ignored or regarded as harmful due to their use of fire.

Farming systems within PAs rely mainly on shifting cultivation, extensive grazing on native grasses and, in the case of Jalapão, extractivism of ornamental flowers. These activities depend on access to and use of riverside vegetation (*matas de galeria*), especially in

Table 2. Coverage of protected areas in areas of soybean expansion in the Cerrado.

		Restrict use PAs			Sustainable use PAs			Indigenous lands			
	Total area (km^2)	Number	Total area	% of the area	Number	Total area	% of the area	Number	Total area	% of the area	Average % of the area
Within the most rising soy production regions*	253553.1	10.0	3830.2	1.5	20.0	12453.9	4.9	10.0	2698.7	1.1	2.5
Within all the Cerrado biome	2189180.1	151.0	63658.0	2.9	314.0	146372.4	6.7	95.0	96789.3	4.4	4.6

Note: *Municipalities where the area of harvested soy expanded by more than 15,000 ha between 2004 and 2011 (33 in the Cerrado biome).
Sources: IBGE (2013) and IBAMA (2013).

upstream wetlands, called 'veredas', where soils are more fertile and humid, even during the long dry season (June to October). Fire is a key element in the management of these activities: it stimulates the regrowth of native pastures and flowers during the dry season and clears new agricultural fields.

However, environmental managers are focused on fighting fire, because fire is commonly regarded as a crime. They are particularly sensitive about the *veredas*, and their flagship palm species *Mauritia flexuosa*, which must be preserved, especially for its role in water regulation and biodiversity. In fact, *vereda* wetlands are vulnerable ecosystems. They host water springs, are rich in biodiversity and are fire-sensitive. Depending on their frequency and intensity, fires can affect the original vegetation, biodiversity and water regimes (Bahia et al. 2009; Maillard, Pereira, and Souza 2009). But the total prohibition of the use of agro-pastoral fire in these areas does not take into account the studies that suggest that regular fire practices can encourage the conservation of biodiversity and assist forest regeneration in savannas (Jeltsch et al. 2000; Laris 2002; Pivello 2006; Eloy and Lúcio 2013, Welch et al. 2013).

As a result, the ban on agro-pastoral fire and the prohibition of forest clearing along the riverbanks (APP), which are strictly enforced in PAs, turn traditional productive activities illegal. Since most smallholders have lost access to lands in downstream valleys (now used mainly by cattle ranchers) and on the plateau (now used by industrial plantations), these upstream riparian forests (*veredas*) are the only areas that can provide access through the managed use of fire to, at the same time, fertile soils for small-scale, low-input cultivation, access to water and natural grassland during the dry season. Only a minority of farmers can afford to modernize their farming systems to cultivate the less fertile and sandy soils of the open savannas.

Herders can be banned from the RDS if they use unauthorized fire. In the APA, the Forest Code is applied strictly: fire and clearing, as well as irrigation require a license from SEMARH, the State Secretariat of Environment, located in the distant state capital, Goiania. Obtaining a license is a complex administrative process, out of reach for the majority of farmers due to the lack of money, education or land title. In the EESGT, an agreement (*termo de compromisso*) was signed in 2011 between ICMBio and users to solve conflicts. According to this agreement, burning is tolerated under specific conditions for shifting cultivation, cattle grazing and extractivism. But this institutional innovation only the second case in protected areas of the Cerrado.

In practice, the ban on fire in PAs is hard to enforce because fires are common in the savannas. Also, the extension of PAs makes law enforcement expensive. In any case, the official anti-fire discourse is explicitly publicized, so some farmers continue to make use of fire discreetly while avoiding official fire control regulations. They do so with guilty feelings and under fear of denunciations, as was observed in our case studies (Figure 4). In particular, in APAs without management plans, the uncertainty arising from the complex and variable environmental rules creates a feeling of permanent insecurity among smallholders (see Matthews 2005 for similar cases in Mexico).

As a result, conflicts arise through varying identifications and interpretations of environmental problems. Environmental managers point at ranching and slash-and-burn agriculture as the sources of the region's main environmental problem: uncontrolled wildfires. Smallholders, on the other hand, blame changes in land use, mainly the introduction of eucalyptus and soybean plantations, for the depletion of water resources, especially due to the use of central-pivot irrigation and soil erosion (Eloy and Lúcio 2013; Lúcio, Eloy, and Ludewigs 2014).

Figure 4. The criminalization of fire in the sustainable development reserve (RDS) Veredas do Acari.
Source: Photo by Ludivine Eloy, 2012.

At the same time, or maybe as compensation, environmental managers employ the sons of farmers in the fire brigades and involve them in the management of the PAs. They also promote fire-free activities, such as fruit and fiber gathering, handicrafts and ecotourism. Environmental managers perceive the membership of farmers and their sons in fire brigades as a way of implementing participatory management while providing new income for these families. But farmers remain subordinate to park managers. In general, farmers' needs and knowledge are not incorporated into decision making about fire management. Following Mistry and Bizerril's (2011) analysis, we observed that collective systems of fire control were undermined. The criminalization of fire, along with the prohibition to undertake primary productive activities, explains the increase in criminal wildfires, set intentionally during the peak of the dry season.

Three reasons can explain why traditional agro-pastoral practices are excluded or, at best, ignored in most participatory management policies in Cerrado. First, according to a federal environmental official, 'the problem is that fire and cattle are the main taboo of environmentalism in Brazil'. In fact, the use of fire for cattle grazing is historically linked to deforestation in most of the country, especially in tropical forests (Amazonia and Atlantic Forest) and southern Cerrado, where cultivated pastures replaced natural vegetation. But this is not the case for many places in northern Cerrado, usually incorporated into PAs, where shifting cultivation and pastoralism on native grasses are not associated with the suppression of the savanna and is central to production systems (Lúcio, Eloy, and Ludewigs 2014). Here, again, the Amazonian environmental model has played against the interests of the populations of the Cerrado. National and international conservation interests consistently dominate fire-management policy (Sorrensen 2009). The importance accorded to the topic of climate change and carbon markets has led to a total rejection of fire in development policies, as clearly illustrated by the PP-Cerrado. As a result, for environmental managers at the local level, preventing fire seems to be the easiest environmental strategy to apply, and the best way to obtain economic resources.

Second, the relative weakness of civil society (traditional populations and family farmers) in the Cerrado vis-à-vis other regions may explain why such environmental norms are not challenged (see section 3.2).

Third, the fire suppression and set-aside discourse is supported by soybean producers since it legitimizes their activity. In other words, the focus on fire management and set-aside does not hamper the expansion of soybean plantations, as soybean farmers do not rely on the use of fire. The enforcement of the APP rule does not seem to be a big challenge to soybean farmers, for they can afford to buy marginal lands outside the plateau to compensate for the deforestation above the 80 percent limit. As mentioned in section 3.3, since the 2000s, large farmers in western Bahia have been benefitting from new public–private partnership programs that help them comply with the Forest Code. Moreover, even where PAs overlap with the soybean frontier, such PAs are mainly APAs, which permit agriculture. Large-scale farmers do not find it difficult to obtain deforestation and irrigation licenses,[6] especially for central-pivot irrigation.

4.1.3. *Information systems help conceal other major environmental issues*

Agro-industrial production is causing major environmental damage in western Bahia, especially to water resources and soils (Gaspar 2006). Environmental agencies do not monitor and sanction these problems to the same extent that they do fire. This situation creates a feeling of injustice among the smallholders who suffer from strict environmental norms, and who have to bear the impact of plantations on water regimes, water quality and soils within PAs.

Recent technological and institutional changes have further spurred fire suppression. First, the introduction of Modis (Moderate-Resolution Imaging Spectroradiometer) made fire detection more effective and less expensive. Second, the registry of rural properties became mandatory after the 2012 reform of the Forest Code. Thus, it became easier for environmental managers to identify fire spots and penalize landowners.

The growing sophistication of fire detection tools has been accompanied by an increased access to data by ordinary citizens. The National Institute for Space Research (INPE) offers free access to a public system of fire monitoring. It provides precise geographical data, updated every three hours, on hot spots detected by more than a dozen satellites whose resolutions can go down to 30 m × 1 m. A particular system is proposed for protected areas. INPE calls users of the system to assist in the improving information to fight against fires through denunciations. It considers that fires are the result of 'bad faith associated with laziness and ignorance' (INPE, 2012).

In comparison, basic hydrological data are extremely poor, especially in western Bahia, where the state water agency (*Superintendência de Recursos Hídricos*, SRH) faces considerable problems in licensing stream-water use (Brannstrom 2005). Due to this lack of information, SRH over-licensed irrigation water to hundreds of central pivots, which has considerably reduced water flow. Indeed, on a few occasions, a downstream hydroelectricity plant had to shut down its turbines during the dry season. Soybean farmers, in turn, have begun drilling for groundwater, taking advantage of vague licensing procedures and poor regulation. In the area above the Urucuia aquifer, there was an increase of 125 percent in the number of licenses for the use of groundwater, and of 76 percent for surface waters between 2003 and 2006 (Gaspar 2006). Most wells do not have water meters so it is impossible to control the water uptake. Furthermore, the main organization of soybean producers in western Bahia (AIBA) seeks to control the production of information on water resources, working with state agencies to fund and implement research and monitoring of climate, land cover, stream flow and soil erosion (Brannstrom and Brandão 2012).

[6]In the 1990s, irrigation in Western Bahia began without proper licensing from the state (Brannstrom 2005).

4.2. *Implications for traditional production systems*

Over the past 20 years, the focus on fire suppression, in combination with the targeting of PAs, has discouraged traditional production systems. Shifting cultivation in gallery forests and cultivation through drainage in *veredas* wetlands have been, for the most part, abandoned, except in the remotest areas. As small-scale ranchers are not allowed to clear the savanna to plant pastures without licenses, nor to use fires to stimulate the regrowth of native grasses near *veredas*, they are left with few options to avoid overgrazing: (1) reduce the size of the herd, which means a decrease in assets and income; (2) rent pastures outside their lots, which incurs extra costs; or (3) plant grasses under the canopy of wooded areas to avoid monitoring. With this last option, the carrying capacity of the pastureland is severely reduced due to the shade. As a result, two trends were observed: while communities near cities decreased their production and increased purchases of food from nearby markets, families living in remoter areas, whose incomes depend mainly on agriculture, continued their fire-dependent practices. They continue to use fire at night, illegally.

Those changes have resulted in a reduction of agricultural production and may have accelerated a process of rural exodus that started in the 1970s. This situation has intensified in recent years because of new job opportunities in urban areas, as a result of public investments in infrastructure and the various social transfers implemented or intensified since 2003, particularly the *Bolsa familia*. The main source of income has now become employment in urban areas, often in the public sector. Land rental or land sale, often in land reform settlements, has also become a source of income. Table 3 shows that in the three study sites, families who depend primarily on agricultural income (class 3: 67–100 percent) are a minority ($n = 11$ in a total of 57). Agricultural income usually represents less than 50 percent of total income in most domestic groups. In this changing context, social inclusion and citizenship are obtained through social transfers rather than by agriculture (Sawyer 2011).

Table 4 shows that most women and more than 50 percent of children are engaged in urban services (domestic maid, construction, transportation). Youngsters from Serra das Araras work mainly in distant large urban centers, like Brasilia, whereas youngsters from Mambaí and Mateiros have migrated mainly to Goiânia or Palmas. Seasonal employment of farmers' sons as fire wardens (classified as 'local urban service') is an important source of income for some families, especially in Mateiros (TO), where the ICMBio funds a fire brigade with nearly 40 firefighters. Many family members rely also on social welfare. Only a few family members are employed regularly in the agricultural sector (10–15 percent). The highest proportion is found in Mambaí, a municipality close (30–50 km) to the large-scale soybean farms in Bahia. In sum, most families rely on off-farm activities, mainly in regional or national urban centers, while soybean plantations offer seasonal jobs.

Table 3. Share of farm income within domestic groups.

Share of farm income in total income (%)	Serra das Araras ($n = 25$)	Mambai ($n = 20$)	Mateiros ($n = 12$)
Class 1: 0–33	13	7	6
Class 2: 34–66	9	8	3
Class 3: 67–100	3	5	3

Source: Authors' field data.

Table 4. Main source of income of traditional farmer families (household head, spouse and all children).

		Serra das Araras, MG			Mambaí, GO			Mateiros, TO		
Main source of income	Localization	Household heads ($n = 25$)	Spouse	Children	Household heads ($n = 11$)	Spouse	Children	Household heads ($n = 12$)	Spouse	Children
Farm activities (agriculture, ranching and handicraft)	Local*	16	2	6	10	2	4	9	4	5
Retail/rental properties		1	1	2						
Urban service (construction,	Local	5	1	5		1	8			9
carpentry, fire fighting,	Regional**			8			13			6
domestic maid)	National***			39			8			12
Agricultural services	Local*	2		3						6
	Regional**			15			8			3
Civil servant	Local*	1		3	1	2	4		2	3
Social subsidies (*Bolsa Familia*, retirement, etc.)	Local*		15	9		5	4	3	5	6
Without source of income			6	16			8			11
Total		25	25	106	11	10	57	12	11	61

Source: Authors' field data.
*Same municipality **Same state, other municipality, ***Other state.

5. Discussion and conclusions

The enforcement of the Forest Code and the expansion of protected areas in the Cerrado coincided with the emergence of the soybean frontier in the late 1990s. Our results show how environmental policies in Cerrado rely on the combination of three processes of selection: spatial targeting of PAs in the lowlands, focus on fire as the main environmental problem and construction of related information systems that conceal major other environmental problems in flat uplands, where soybean agriculture is expanding, such as water depletion and soil erosion.

The model of fire fighting and forest set-aside applied in savanna ecosystems is influenced by the conservation models imported from forest ecosystems (in particular, from Amazonia) and by the negotiations on climate change. These selective policies put a disproportionate strain on smallholders and traditional populations. Although these groups are not responsible for most of the deforestation and the associated loss of biodiversity in the Cerrado, they have been penalized due to their traditional use of fire. As a result, these 'politics of selection' undermine participative management of PAs, while legitimizing the expansion of soybean agriculture around them.

These results permit us to discuss the socio-spatial effects of the concurrent processes of 'conservation booms' and the modernization of land use in Latin American countries (Zimmerer 2011). Some authors have described the political and social mechanisms that sustain the rapid expansion of monocultures in the Brazilian Cerrado since the 1980s, and the containment of traditional populations to the economic and geographic margins of soybean and eucalyptus farms (Brannstrom 2005; Ribeiro 2010). Hecht (2005) showed how agricultural lobbies influence the Brazilian states to allow unfettered use of resources in environmental 'sacrifice zones', which are created as territories for the development of agribusinesses. But few studies address the struggles and competitions for resources that take place in the margins of these sacrifice zones, where low-input production systems are located and where conservation policies focus their actions.

In fact, politics of selection are not new to the Cerrado: large-scale farming and traditional farming systems have historically occupied different areas, so conflicts between them arose only occasionally. The reduction of areas available to traditional farming systems and the rural exodus do not differ from what has happened elsewhere in Brazil, although the process in the Cerrado is more dramatic due to the rapid expansion of the soybean frontier. But the recent political, environmental and scientific changes, such as the reform of the Forest Code (flexibilization of compensation systems), the progressive rehabilitation of fire as a management tool, water depletion and new claims for land from traditional populations (especially *quilombolas*), may increase struggles at the margins of the soybean expansion zones.

These changes may have led to new relationships with PAs. As they are typically located in areas less suitable for agricultural mechanization, these PAs become targets for legal-reserve compensation, helping farmers who want to expand their plantations in the '*chapadão*'. This system of forest compensation may reduce even further the space available for agro-pastoral activities, usually carried out without land title.

According to Brannstrom and Brandão (2012), nature is central in the discourse of soy farmers of western Bahia when they emphasize their commitment to complying with existing state and federal environmental legislation. This discourse encourages environmental compensation that enables future soybean expansion through environmental compliance of the new Forest Code. States like Goiás are drafting their own legislation to organize a stock exchange system of environmental assets. A large amount of speculative activity takes place in this new market (Aubertin and Pinton 2013).

The production of knowledge, particularly the cartography of land use in the Cerrado, helps exacerbate these tensions. Simplified and official land-use maps classify land use in binary terms – deforested area or natural vegetation – thus obscuring traditional land uses, which are now located in areas eligible for environmental compensation through set-asides. As observed by Sorrensen (2009) in the Amazon, the predominance of geospatial technologies in environmental monitoring activities limits the problematization of fire hazard (and deforestation) to the behavior of individuals, and does not consider the regional development arrangements under which individuals operate. Hybrid technologies that incorporate remotely sensed data at the macro scale with local ethnographic inquiry may be more effective in understanding human-landscape relationships.

As a result, for traditional populations, the challenge is not only to tap niche markets, such as of extractivist and organic products, and to obtain or secure their land rights, but also to play new roles and occupy new spaces in scientific and political networks that help determine what constitutes an environmental harm and what is an acceptable farming system.

Acknowledgements

We thank the communities that welcomed us and took part in the research, our academic research partners and institutions that have supported this work. One part of the Environmental Governance (ENGOV) project was conducted by the Centro de Desenvolvimento Substentável (CDS) of the National University of Brasilia (UnB), in collaboration with the Chico Mendes Institute of Biodiversity Conservation (ICMBio) and the Instituto Estadual das Florestas (IEF), with funding from the European Union (FP7-2010, Agreement no. SSH-CT-2010-266.710). The project "*Visões contemporâneas do cerrado e intersecção de politicas sociais e ambientais*" (RESBIO) was conducted by the Instituto de Estudos Socioambientais (IESA) of the Federal University of Goiás (UFG), by the CDS and by the Institute of Research for Development (IRD-France), in collaboration with ICMBio, with funding by the Conselho Nacional de Desenvolvimento Científico e Tecnológico (Edital UNIVERSAL CNPq No. 14/2011). Finally, the research in Jalapão (Tocantins) was conducted by ICMBio, with funding from the German Agency for International Cooperation (GIZ). This research also received the support of the Institute for Humanities and Social Sciences (INSHS) at the the Centre National de la Recherche Scientifique (CNRS) through the 'Support for International Mobility – 2014' program.

References

ABIOVE - Associação Brasileira das Industrias de Oleos Vegetais, 2013. T*he Soy Moratorium will be renewed for another year*. http://www.abiove.com.br/english/sustent/abiove_release_moratoria_13out11_us.pdf (accessed January 10, 2014).

Aubertin, C., and F. Pinton. 2013. L'invention du biome Cerrado. *Confins [Online]*. http://confins.revues.org/8218

Aubertin, C., and E. Rodary. 2011. *Protected areas. Sustainable land?*. Marseille: IRD/Farnham (GBR): Ashgate.

Bahia, T.O., G.R. Luz, L.L. Braga, G.C.O. Menino, Y.R.F. Nunes, M.D.M. Veloso, and W.V. Neves. 2009. R. M. Florística e fitossociologia de veredas em diferentes estágios de conservação na APA do Rio Pandeiros, norte de Minas Gerais. *MGBiota* 2: 14–21.

Bassett, T.J., and K.B. Zueli. 2000. Environmental discourses and the Ivorian Savanna. *Annals of the Association of American Geographers* 90, no. 1: 67–95.

Brannstrom, C. 2005. Environmental policy reform on north-eastern Brazil's agricultural frontier. *Geoforum* 36, no. 2: 257–71.

Brannstrom, C., and P.R.B. Brandão. 2012. Two hundred hectares of good business: Brazilian agriculture in a themed space. *Geographical Review* 102, no. 4: 465–85.

Coelho, C.N. 2001. 70 Anos de política agrícola no Brasil (1931–2001). *Revista de Política Agrícola, Brasília*, 10, no. 3: 3–58.

CONAB. 2013. Séries Históricas Relativas às Safras de 1976/77 a 2013/14. http://www.conab.gov.br/conteudos.php?a=1252&t=2&Pagina_objcmsconteudos=3#A_objcmsconteudos (accessed January 3, 2014).

CNUC/MMA. 2013. Cadastro Nacional de Unidades de Conservação. http://www.mma.gov.br/areas-protegidas/cadastro-nacional-de-ucs (accessed January 3, 2014).

Eloy, L., and S.L.B. Lúcio. 2013. *Caracterização agronômica e socioeconômica das roças de toco e de esgoto na Estação Ecológica Serra Geral do Tocantins.* Brasilia: GIZ/ICMBio.

FAO. 2007. *Future expansion of soybean 2005 -2014: Implications for food security, sustainable rural development and agricultural policies in the countries of Mercosur and Bolivia. Synthesis document.* FAO Policy Assistance Series no. 3. Santiago: Food and Agriculture Organization of the United Nations.

Ferreira, I.N.R. 2010. *Parcerias para a Gestão Ambiental em Propriedade Rurais: o Caso de Lucas do Rio Verde - MT.* Master's dissertation, Centro de Desenvolvimento Sustentável, Universidade de Brasília.

Ferreira, M.E., L.G. Ferreira, and E.M. Latrubesse. 2008. Current Land Use and Conversion Trends in the Savanna Environments of Central Brazil: a Preliminary Assessment from a Landscape perspectives. http://www.lapig.iesa.ufg.br/lapig/index.php/produtos/material-didatico/viewdownload/8-periodicos/487-current-land-use-and-conversion-trends-in-the-savanna-environments-of-central-brazil-a-preliminary-assessment-from-a-landscape-perspective (accessed December 5, 2013).

Foucault, M. 2001. La 'gouvernementalité', *Dits et écrits II, 1976–1988.* Paris: Gallimard.

Gaspar, M.T.P. 2006. *Sistema Aqüífero Urucuia : caracterização regional e propostas de gestão.* PhD thesis in Geology, Universidade de Brasília, Brasilia.

Haesbaert, R. 1997. *Des-territorialização e identidade: A rede 'gaúcha' no nordeste.* Niterói, RJ: Editora da Universidade Federal Fluminense.

Harwell, E. 2000. Remote Sensibilities: Discourses of Technology and the Making of Indonesia's Natural Disaster. *Development and Change* 31, no. 1: 307–40.

Hecht, S.B. 2005. Soybeans, development and conservation on the Amazon frontier. *Dev and change* 36: 375–404.

Hecht, S.B. 2012. From eco-catastrophe to zero deforestation?. Interdisciplinarities, politics, environmentalisms and reduced clearing in Amazonia. *Environmental Conservation* 39, no. 1: 4–19.

Instituto Brasileiro de Goegrafia e Estatistica – IBGE. 2010. Indicadores de Desenvolvimento Sustentavel. Estudos & Pesquisas, Brasilia, n. 7.

Instituto Brasileiro de Goegrafia e Estatistica – IBGE. 2012. Indicadores de desenvolvimento sustentável do Instituto Brasileiro de Geografia e Estatística. http://www.ibge.gov.br/home/geociencias/recursosnaturais/ids/default_2012.shtm. (accessed December 23, 2014).

Instituto Brasileiro de Goegrafia e Estatistica – IBGE. 2013. Produção agrícola Municipal 2005–2011. http://www.ibge.gov.br/home/estatistica/pesquisas/pesquisa_resultados.php?id_pesquisa=44 (accessed January 5, 2014).

Instituto Brasileiro do Meio Ambiente e dos Recursos Naturais Renováveis- IBAMA. 2013. Sistema compartilhado de informações ambientais. http://siscom.ibama.gov.br (accessed December 4, 2013).

Instituto Nacional de Pesquisas Espaciais – INPE. 2012. Portal do Monitoramento de Queimadas e Incêndios. Perguntas Frequentes, n.33. http://www.inpe.br/queimadas/faq.php (accessed February 14, 2015).

Jeltsch, F., G. Weber, and V. Grimm. 2000. Ecological buffering mechanisms in savannas: A unifying theory of long-term tree-grass coexistence. *Plant Ecology* 150, no. 1–2: 161–71.

Laris, P. 2002. Burning the seasonal mosaic: Preventive burning strategies in the wooded savanna of southern Mali. *Human Ecology* 30, no. 2: 155–86.

Lúcio, S.L.B., L. Eloy, and T. Ludewigs. 2014. O gado que circulava: desafios da gestão participativa de Unidades de Conservação no gerais de norte de Minas. *BioBrasil* 1: 130–56.

Maia, J.M.E. 2013. A Fundação Brasil Central e as relações entre Estado e território no Brasil. In *Fronteira Cerrado: Sociedade e Natureza no Oeste do Brasil,* ed. S.D. Silva, J.P. Pietrafesa, J.L.A. Franco, J.A.L. Drummond, and G.G. Tavares, 75–86. Goiânia: PUC Goiás.

Maillard, P., D.B. Pereira, and C.G. Souza. 2009. Incêndios florestais em veredas: conceitos e estudo de caso no Peruaçu. *Revista Brasileira de Cartografia* 61: 321–30.

Mathews, A.S. 2005. Power/Knowledge, Power/Ignorance: Forest Fires and the State in Mexico. *Human Ecology* 33, no. 6: 795–820.

Milian, J., and E. Rodary. 2010. La conservation de la biodiversité par les outils de priorisation : entre souci d'efficacité écologique et marchandisation. *Revue Tiers Monde* 202: 33–56.

Mistry, J., and M. Bizerril. 2011. Por Que é Importante Entender as Inter-Relações entre Pessoas, Fogo e Áreas Protegidas?. *Revista Biodiversidade Brasileira* 2: 40–49.

MMA. 2011a. *Monitoramento dos biomas brasileiros por satélite. Acordo de cooperação técnica MMA/IBAMA. Monitoramento do Bioma Cerrado, 2009–2010.* Brasilia: MMA. http://www.mma.gov.br/estruturas/sbf_chm_rbbio/_arquivos/relatoriofinal_cerrado_2010_final_72_1.pdf. (accessed December 23, 2014).

MMA. 2011b. *Plano de Ação para Prevenção e Controle do Desmatamento e das Queimadas no Cerrado (PPCerrado).* Brasilia: MMA.

Myers, N., R.A. Mittermeier, C.G. Mittermeier, G.A.B. da Fonseca, et al. 2000. Biodiversity hotspots for conservation priorities. *Nature* 403, no. 6772: 853–58.

Nepstad, D., D. McGrath, A. Alencar, A.C. Barros, et al. 2002. Frontier Governance in Amazonia. *Science* 295, no. 5555: 629–31.

Nepstad, D., S. Schwartzman, B. Bamberger, M. Santilli, D. Ray, P. Schlesinger, P. Lefebvre, A. Alencar, E. Prinz, G. Fiske and A. Rolla 2006. Inhibition of Amazon Deforestation and Fire by Parks and Indigenous Lands. *Conservation biology*, 20: 65–73.

Nogueira, M.C.R. 2009. *Gerais a dentro e a fora : identidade e territorialidade entre Geraizeiros do Norte de Minas Gerais.* PhD Thesis in anthropology, Universidade de Brasília, Brasilia.

Pessôa, E., and R. Bonelli. 1997. O papel do estado na pesquisa agrícola no Brasil. *Cadernos de Ciência & Tecnologia* 14, no. 1: 9–56.

Pivello, V.R. 2006. Fire management for biological conservation in the Brazilian cerrado. In *Savannas and dry forests: linking people with nature*, ed. J. Mistry and A. Berardi, 129–54. Aldershot: Ashgate.

Rezende, G.C. 2002. *Ocupação Agrícola e Estrutura Agrária no Cerrado: o Papel do Preço da Terra, dos Recursos Naturais e da Tecnologia.* Rio de Janeiro: IPEA. Texto para Discussão no 913.

Ribeiro, R.F. 2008. Da Amazônia para o Cerrado: as reservas extrativistas como estratégias socioambientais de conservação. *Sinapse Ambiental* 5, no. 1: 12–32.

Ribeiro, E.M. 2010. *História dos gerais.* Editora UFMG: Belo Horizonte.

Robbins, P. 2003. Beyond Ground Truth: GIS and the Environmental Knowledge of Herders, Professional Foresters, and Other Traditional Communities. *Human Ecology* 31, no. 2: 233–53.

Rylands, A.B., and K. Brandon. 2005. Brazilian Protected Areas. *Conservation biology* 19, no. 3: 612–18.

Sawyer, D. 2009. Fluxos de carbono na Amazônia e no Cerrado: um olhar socioecossistêmico. *Sociedade e Estado* 24, no. 1: 149–71.

Sawyer, D. 2011. O Cerrado numa perspectiva Eco-social. *Ateliê Geografico* 5, no. 2: 339–46.

Sawyer, D.R., and A.d. Souza Lobo. 2008. O Papel da Sociedade no Estabelecimento de Políticas Públicas para as Savanas. In *Savanas: desafios e estratégias para o equilíbrio entre sociedade, agronegócio e recursos naturais*, ed. F.G. Faleiro and A.L. Fraias Neto, 1153–80. Planaltina: Embrapa – Cerrados.

Scaramuzza, C.A.d.M., R.B. Machado, and S.T. Rodrigues. 2005. Áreas prioritárias para conservação da biodiversidade em Goiás. In *Conservação da biodiversidade e sustentabilidade ambiental em Goiás: Prioridades, estratégias e perspectivas*, ed. L.G. Ferreira, 88–99. Goiânia: Universidade Federal de Goiás/Agência Ambiental/Banco Mundial.

Shiki, S., J.G. da Silva, and A.C. Ortega. 1997. *Agricultura, meio ambiente e sustentabilidade do cerrado brasileiro.* Uberlândia: Centro de Documentação e Pesquisa Sócio-econômica, departamento de Economia e Departamento de Geografia da Universidade Federal de Uberlândia.

Silva, L. 2000. O Papel do Estado do Processo de Ocupação das Áreas de Cerrado entre as Décadas de 60 e 8. *Caminhos de Geografia* 1, no. 2: 24–36.

Sorrensen, C. 2009. Potential hazards of land policy: Conservation, rural development and fire use in the Brazilian Amazon. *Land Use policy* 26, no. 3: 782–91.

Sousa, I.S., and L. Busch. 1998. Networks and agricultural development: the case of soybean production and consumption in Brazil. *Rural Sociology* 63, no. 3: 349–71.

Terra-AFP. 2013. Soja é responsável indireta pelo desmatamento no Brasil. http://noticias.terra.com.br/ciencia/sustentabilidade/soja-e-responsavel-indireta-pelo-desmatamento-no-brasil,3febd717c9c21410VgnCLD2000000ec6eb0aRCRD.html (accessed September 17, 2013).

Welch, J.R., E. Brondizio, S.S. Hetrick, and C.E.A. CoimbraJr. 2013. Indigenous Burning as Conservation Practice: Neotropical Savanna Recovery amid Agribusiness Deforestation in Central Brazil. *PLOS ONE* 8, no. 12: e81226.

Wilkinson, J., and B. Sorj. 1992. *Structural adjustment and the institutional dimensions of agricultural research and development in Brazil: Soybeans, wheat, and sugar cane*. Paris: OECD.

Zimmerer, K.S. 2000. The Reworking of Conservation Geographies: Nonequilibrium Landscapes and Nature-Society Hybrids. *Annals of the Association of American Geographers* 90, no. 2: 356–69.

Zimmerer, K.S. 2011. 'Conservation booms' with agricultural growth?. Sustainability and Shifting Environmental Governance in Latin America, 1985–2008 (Mexico, Costa Rica, Brazil, Peru, Bolivia). *Latin American Research Review* Special Issue: 82–113.

Genetically modified soybeans, agrochemical exposure, and everyday forms of peasant collaboration in Argentina

Pablo Lapegna

Since the 2000s, both the production of genetically modified (GM) soybeans and the cases of agrochemical exposure have grown exponentially in Argentina. Drawing on ethnographic research, I analyze how peasant social movements understand the socio-environmental problems caused by the expansion of GM soybeans. I argue that at national, provincial, and local scales, the institutional recognition of peasant social movements and the performative actions of authorities discourage contentious collective action through subtle yet powerful mechanisms. The article contributes to social movement research and to the literature on peasant resistance by analyzing the cultural dynamics that constrain contention and shape processes of peasant collaboration, which are arguably as important as peasant resistance, although much less studied.

Introduction

Based on 12 months of fieldwork carried out between 2003 and 2013, this paper examines the growth of agrochemical use in Argentina due to the expansion of genetically modified (GM) soybeans. I zoom in on the province of Formosa, in northern Argentina, to better understand how rural populations and peasant social movements negotiate and accommodate these socio-environmental and agrarian processes. Specifically, I analyze the ways in which they interpret the discourses and actions of state authorities and interact with them at different political scales. I propose that two mechanisms – performative governance and institutional recognition – help explain processes of accommodation that create obstacles to contentious collective actions.

GM soybeans were initially planted in Argentina in 1996, and by 2011, they covered close to 19 million hectares of farmland (more than 46 million acres), over half of Argentina's agricultural land (MAGyP 2014). Between 1996 and the early 2000s, GM soybean production rapidly expanded in the Pampas, a region in central Argentina historically characterized by its commercial and export-oriented agriculture. Beginning in 2003, the transgenic agriculture that started in the Pampas began to expand north (Pengue 2005; Goldfarb and Zoomers 2013; Leguizamón 2014). Agribusiness companies and large-scale farmers sought to reach 'economies of scale'; that is, expand their operations by leveraging their machinery, know-how, and access to seeds and agrochemicals while profiting from cheaper land in the subtropical climate of the northern provinces.

With the expansion to northern Argentina, GM soybean production entered areas where peasants and indigenous communities comprise most of the rural population. The expansion of GM soybeans has come hand in hand with a steep increase in agrochemical use in Argentina. At the end of the 1990s, 42 percent of the agrochemicals used in the country were utilized for the production of GM soybeans (Casabé et al. 2007). In 1997/1998, 28 million liters of glyphosate were sold in Argentina and in 1998/1999, this quantity doubled to 56 million liters (Pengue 2004b). By 2003/2004, conservative estimates indicate that annual glyphosate use reached 100 million liters (Pengue 2004a). What's more, by 2012, this quantity doubled again, with almost 200 million liters of glyphosate sprayed across Argentina.[1]

The agronomical features of GM soybeans and the practices of agribusiness in Argentina have resulted in a higher use of herbicides. Unlike traditional crops that require closer monitoring to control pests or stimulate growth, GM soybeans follow a simple, pre-established, step-by-step process. Contrary to popular belief, GM soybeans do not necessarily increase yields but rather simplify agricultural production: they can be sown without plowing the land because glyphosate eliminates weeds but not GM plants, which are engineered to resist this herbicide. In northern Argentina, agribusinesses usually rent fields for two or three consecutive seasons and grow GM soybeans without rotating crops. The GM soybeans that fall to the ground during the harvest are engineered to resist glyphosate, which means that other more toxic herbicides (often times, 2,4D) must be used to 'clear' the field before replanting new soybean GM seeds.[2]

The increasing use of agrochemicals has created a series of environmental and public health problems across Argentina. Since the early 2000s, researchers have reported cases in which herbicides drifted from GM soybean plots to people's houses and/or to farms, affecting their health and contaminating water sources (Joensen, Semino, and Paul 2005; Domínguez and Sabatino 2010). In 2006, environmental activists launched a campaign and organized meetings in several provinces to raise awareness about agrochemical exposure.[3] Researchers have documented the effects of glyphosate and other herbicides (for instance, 2,4D) associated with soybean production (Vitta, Tuesca, and Puricelli 2004; Zabaloy, Garland, and Gómez 2008; Binimelis, Pengue, and Monterroso 2009), and studies have linked the use of agrochemicals with a series of public health problems (Oliva et al. 2008).[4]

Peasant populations have been particularly affected by the expansion of transgenic agriculture and the exposure to agrochemicals. In the early 2000s, peasant movements organized contentious collective actions to address these problems. The Peasant Movement of Formosa (MoCaFor) is a case in point. MoCaFor activists mobilized contentiously in 2003 to complain about the effects of agrochemical drifts in rural communities, organizing

[1]Based on the report of the Argentine chamber of agrochemical companies (Kleffman Group 2013). I assume that glyphosate sales equal glyphosate use.

[2]The World Health Organization considers glyphosate as 'Unlikely to be Hazardous' and it is listed as 'Unlikely' in the US EPA Carcinogens list. The 2,4D, in contrast, is assessed as 'Moderately Hazardous' and presents 'Unclassifiable, ambiguous data' in EPA list; see the Pesticide Action Network (PAN) Pesticide Database. Glyphosate is the *active ingredient* of Roundup, and its combination with adjuvants and surfactants may increase toxicity (Bonn 2005). On the heated debate about Roundup's toxicity see Séralini et al. (2013).

[3]See the websites of Grupo de Reflexión Rural (GRR) and of the Centro de Protección de la Naturaleza (CeProNat).

[4]See also Aranda (2008a, 2010a), Ramírez (2012), and Fernández (2014).

several roadblocks and demonstrations. Even though Formosan peasants still suffer the consequences of agrochemical exposure, they have no longer been able to organize contentious collective actions to address this problem (Lapegna 2013, 2014).

The case of Formosa illustrates similar processes suffered by other indigenous peasant populations in northern Argentina. MoCaFor is one of the several organizations representing peasants and indigenous peoples that were formed in the northern provinces in the midst of the process of neoliberalization of the 1990s.[5] Since the mid-2000s, however, peasant organizations have played a limited role in protesting transgenic agriculture and agrochemical exposure. How do peasant organizations and their members understand agrochemical exposure? How do they adapt to transgenic agriculture in a socio-political context different from neoliberalism?

To answer these questions, I first situate them within broader discussions about peasant resistance and social movements. After discussing this literature, I provide an overview of Argentine agriculture and how it shifted from neoliberalization in the 1990s to a 'national-popular' orientation since the mid-2000s. In the last two sections of the paper, I draw on ethnographic data to understand how peasants and their organizations accommodate transgenic agriculture and agrochemical exposure. I propose that the mechanisms of performative governance and institutional recognition encapsulate processes of peasant collaboration and create obstacles for contention (in Formosa specifically, but also among Argentine peasants more broadly). In seeking to connect my localized ethnography with larger processes, I pay attention to the articulation of different political scales, showing how these mechanisms work at the national, provincial, and local level.

Peasant resistance, peasant collaboration, and the constraints to contention

Since the mid-1980s, the work of James Scott (1985, 1990) has inspired a stream of research on everyday forms of resistance (for discussions of this extensive literature see Hollander and Einwohner 2004; Sivaramakrishnan 2005; Seymour 2006; Ho 2011). At the core of the literature on everyday forms of resistance is the idea that the lack of open, direct confrontation between subordinate groups and more powerful social actors does not necessarily indicate acquiescence on the part of the former. Through ethnography, Scott and others have argued that peasants and other subordinated groups may engage in foot dragging, sabotage, character assassination, and other hidden, less visible forms of resistance. The influence of this approach is mirrored in the criticisms that it inspired. Scholars took issue with the alleged 'neo-populist' and post-modern view of the literature on everyday forms of resistance (e.g., Brass 2002; Petras and Veltmeyer 2001; cf. Beverley 2004). Others criticized what they understood as a binary representation of resistance/domination (arguably missing how power relationships shape both), and questioned the autonomous character of resistance (e.g., Abu-Lughod 1990; Mitchell 1990; Fletcher 2001).

In this article, I build on these debates to better understand why and how Argentine peasants negotiate the negative effects of transgenic agriculture and agrochemical exposure. In doing so, I engage three sub-themes within the extensive literature on everyday forms of peasant resistance. First, the ethnographic data that I present below explores the dilemmas

[5]A non-exhaustive list includes: MOCASE (Peasant Movement of Santiago del Estero); UNPEPROCH (Union of Small Farmers of Chaco); MAM (Agrarian Movement of Misiones); MOCAFOR (Peasant Movement of Formosa); UST (Landless Union of Mendoza); APENOC (Small Farmers of Northern Córdoba); and Red Puna (Jujuy province). See Domínguez (2009).

created by the 'internal politics of the dominated' (Ortner 2006), showing how different peasants at different moments have different positions vis-à-vis agricultural change and their relationships with authorities. Previous research has scrutinized the question of why some peasants engage in resistance while others do not, providing answers that focus, for instance, on regional variation (Brosius 1997) and on the diversity of contacts with external actors, access to information, and self-understandings of identity (Fletcher 2001, 48–55). Here, I take heed of this research by focusing on the variegated actions of peasants in terms of temporal transformations, as agrarian policies in Argentina changed from neoliberal to 'national-popular'.

Second, I tackle issues of everyday forms of peasant collaboration. As Christine Pelzer White (1986, 56) argued when analyzing rural development in Vietnam: 'we must add an inventory of "everyday forms of peasant collaboration" to balance our list of "everyday forms of peasant resistance": both exist, both are important.' Furthermore, this insight is even more relevant to my study given that 'in "normal" periods there is more everyday collaboration than everyday resistance. Indeed, this is why new regimes seek rapid "normalisation"' (White 1986, 56). This resonates with the Argentine context, where the new political regime that began to take shape in 2003 sought to normalize the country after a deep crisis in 2001–2002, while also attempting to break away from neoliberal policies and work with social movements to implement policies and programs.

I suggest that peasants deploy their agency to *negotiate* with rather than confront authorities or powerful opponents, and that they *accommodate* the negative environmental impacts associated with the expansion of GM soybeans. To be clear, I am not claiming that peasants are duped into eschewing confrontation. Rather I argue that peasants face constraints to contention as an outcome of the alliances, frictions, and oppositions between social movement leaders, constituents, and authorities at national, provincial, and local scales. This focus resembles the forms of 'rightful resistance' identified by Kevin O'Brien (1996, 2013) when analyzing the ways in which subordinate actors employ institutional channels to address their grievances. My study thus speaks to previous research on the role of powerful actors producing consensus about agricultural biotechnology at the global level (Andrée 2011) and about GM soybeans in Argentina (Newell 2009). I complement this work by concentrating on how this articulation takes place at the subnational scale.

Third, I seek to bridge the literature on peasant studies and social movement scholarship. Previous research has taken notice of the lack of dialogue between these two literatures, criticizing that the work on everyday forms of resistance does not pay enough attention to how they can evolve into open confrontations and/or social movements (Escobar 1992; McAdam, Tarrow, and Tilly 1997). The meager exchanges between these bodies of research also has to do with an academic division of labor in which anthropologists concentrated on 'peasants, the urban (especially Third World) poor, ethnic minorities, and millenarian or syncretic religious sects and allocated other types of mobilization (and national-level phenomena) to sociologists, political scientists or historians' (Edelman 2001, 286). Although some researchers have addressed the interplay between everyday resistance and open contention (Adas 1981; Korovkin 2000) this article does so by focusing on everyday forms of *collaboration* and the *obstacles* posed to contentious collective action.

I connect studies on peasant resistance and recent social movement scholarship by engaging their mutual interest in the role of culturally informed practices. Agrarian and peasant studies have a long history of drawing inspiration from an understanding of culture related to social class formation and conflict (Thompson 1971; Williams 1977), and some social movement scholars have built on this tradition (e.g., Steinberg 1999). I follow a similar path, retaining James Scott's ethnographic interest in 'putting the experience of human agents at the center of the analysis' (Scott 1985, 43; see also Scott 2005; Sivaramakrishnan 2005, 348–49). This interest in culturally informed practices is increasingly gaining traction with social movement scholars, who have criticized canonic views of seeing culture 'as a residual category that is invoked to explain what structure does not explain' (Polletta 2008, 78–79). I build on the scholarship interested in social movements and meaning-making processes (Goodwin and Jasper 2004; Kurzman 2008), particularly on the understanding that culture not only enables contention but can also *constrain* collective action (Polletta 2008). Similarly, I take heed of scholars arguing that social movement scholarship has overwhelmingly focused on how organizations, political opportunities, and frames *facilitate* collective actions, which overlooks the many instances where collective action is *constrained* by organizations (McAdam, Tarrow, and Tilly 2008, 325; McAdam and Boudet 2012), political opportunities (Jasper and Goodwin 2011), and framing (Benford 1997, 412). Not unexpectedly, researchers have offered insights about the resistance against GM crops (Bridge, McManus, and Marsden 2003; Otero 2008; Scoones 2008; Eaton 2009; Schurman and Munro 2010; Fitting 2011; Kinchy 2012; Klepek 2012; Heller 2013; Grandia 2014), but less so in the way of understanding the constraints to protesting them.

I dissect issues of constraints to contention and subordination through a close analysis of language and discourse (Roseberry 1994; Steinberg 1999; Bröer and Duyvendak 2009), proposing the concepts of institutional recognition and performative governance while paying special attention to how they are structured at different scales. The term 'institutional recognition' borrows from two sources: Nancy Fraser's (2003) elaboration of recognition in connection to redistribution, and social movement research arguing that culture as expressed in practices is connected to institutional settings (Armstrong and Bernstein 2008; Polletta 2008).[6] My argument is that institutional recognition and performative governance capture processes of accommodation and expose obstacles to collective action in protest of agrochemical exposure. Ethnographic research, building on Erving Goffman (1959), proposed the concept of performative governance for situations 'in which impressions of committed governance are staged and maintained by officials, yet effective inclusion of citizenry in the decision-making process is negligible' (Futrell 1999, 495). I understand performative governance as a composite of these ideas and J.L. Austin's (1962) 'performative speech acts', referring to those discourses that not only describe the world but also remake it. Performative speech acts may 'bring about or achieve *by* saying something, such as convincing, persuading, deterring, and even, say, surprising or misleading' (Austin 1962, 108). I draw inspiration from this concept to indicate that authorities can 'do things with words' and to show how they deploy performative actions that create an effect for their audience.

Below, I examine the effects of institutional recognition and performative governance at the national, provincial, and local scales, exploring how the experience of agrochemical exposure is socially and politically produced. By zooming in on how the discourses and actions of authorities are read and acted upon by activists, I show that the availability of information about toxic exposure may not necessarily translate into collective action

[6]In Fraser's (2003, 27) understanding, recognition is not a form of 'identity politics', but rather part of 'a politics aimed at overcoming subordination by establishing the misrecognized party as a full member of society'.

(Auyero and Swistun 2009). I thus build on the research unveiling the numerous but less studied instances where contention *does not* emerge in the face of environmental risks (McAdam and Boudet 2012), particularly in connection to experiences of subordination and the micro-politics of these processes.

Argentine agriculture: from neoliberal to national-popular

During the 1990s, Argentina underwent a profound process of neoliberalization during the administrations of President Carlos Menem (1989–1995 and 1995–1999). The agricultural sector was at the forefront of this process as production was re-oriented towards export markets, the institutions and mechanisms that regulated production for the internal market were eliminated, and global agribusiness corporations expanded their operations. In this context, in 1996, the national government approved the commercialization of herbicide-resistant soybeans.

In the 2000s, Argentina experienced, yet again, profound social, political, and economic changes. In the aftermath of the political and economic crisis that exploded at the end of 2001, Néstor Kirchner was elected president as one of the candidates of the Peronist party in 2003. His administration and those of his wife and successor, Cristina Fernández de Kirchner (2007–2011, re-elected until 2015), reoriented the macroeconomic policies of the country towards a Keynesian model, replacing the overarching policies of structural adjustment that dominated the country in the 1990s. The state re-intervened in the economy, particularly in agricultural export markets.

With the devaluation of the Argentine peso after the 2001 crisis, the national government established taxes on agricultural exports to appropriate part of the extraordinary difference between national prices (in pesos) and export revenues (in dollars).[7] By 2009, the exports of the soybean agro-industrial complex (including soybeans, oil, and soybean meal) accrued to nearly one-fourth of total Argentine exports, worth close to 14 billion US dollars (INDEC 2012). From 2003 and onwards, the administrations of Néstor and Cristina Kirchner tapped into the income generated by export taxes on GM soybeans, creating what some call an 'export-oriented populism' (Richardson 2009).

In March 2008, the national government raised export taxes on soybeans, triggering opposition from medium to large landowners, farmer associations, and agribusinesses. Supported by media conglomerates, these actors blocked roads throughout Argentina and demanded the complete elimination of *all* export taxes, deploying a 'free trade' discourse (Giarracca and Teubal 2010). Cristina Kirchner reacted by sending a bill to congress, 'Resolution 125', which modified the export tax on soybeans and adjusted its taxable percentage to the ups and downs of international prices. The bill was approved in the lower house (*Diputados*), but the opposition parties and agribusiness lobby fiercely opposed the bill in the Senate, where it was overturned.

The defeat of Resolution 125 had broad political implications. During the protests, soybean growers and the mass media exhibited a keen capacity to deploy their hegemonic position, installing the idea that 'what is good for agribusiness is good for the country' by successfully presenting themselves as *el campo*, the countryside (thus sidelining peasant

[7]Export taxes (*retenciones*) were reinstated in 2002 at a rate of 10 percent. During Néstor Kirchner's government (2003–2007), export taxes for soybeans reached 24 percent and then were raised to 35 percent: Richardson 2009.

and indigenous smallholders).[8] The events exemplified what Peter Newell (2009, 38) calls (bio)hegemony: 'the alignment of material, institutional, and discursive power in a way which sustains a coalition of forces which benefit from the prevailing model of agricultural development.' While the conflict revealed the dispute between the national government and agribusiness, the environmental and social consequences of transgenic agriculture were never part of the debate.

Following this confrontation with large agribusiness, the national government launched a series of initiatives aimed at garnering support from small farmers and peasants. I argue that these initiatives are expressions of institutional recognition and performative governance that, as they create territorial effects at different scales, contribute to stifling popular mobilization against agrochemical exposure. Table 1 schematically summarizes this argument, which I flesh out in the remainder of the paper.

Peasant movements and institutional recognition

Néstor Kirchner won the 2003 presidential election by a narrow margin, and in the initial years of his administration, he built his political powerbase by enlisting the support of several social movements and incorporating their leaders into the government. One of these alliances was constructed with the MNCI (*Movimiento Nacional Campesino Indígena*), a network of peasant and indigenous organizations at the provincial and local scale that were created in the early 2000s. In 2005, the national government offered the MNCI the management of the PSA (*Programa Social Agropecuario*, Social Agricultural Program), a national program targeting the rural poor administered by the national Secretary of Agriculture. After an internal debate, the MNCI leaders accepted the offer and a leader of a peasant organization became the head of the PSA.

During the 1990s, the policies of neoliberalization combined 'roll back' deregulation and 'roll out' interventions through rural welfare (Peck and Tickell 2002). Several national programs targeted the rural poor, adopting the language and views of the international funding agencies (chiefly, the World Bank). The underlying aims of these welfare programs (the PSA being the most prominent) were to 'soften' the impact of neoliberalization and stimulate in smallholders a disposition towards 'efficiency, competitiveness, and profitability' (DDA-PROINDER 2003, 9). The PSA funded projects directed at groups, which would pay for supplies and equipment with government subsidies and then donate eventual surpluses to community organizations (schools, hospitals, etc.).

Towards the end of the 1990s, peasant movements were increasingly dissatisfied with the PSA, arguing that the program atomized their constituents by creating 'tiny groups', discouraging processes of collective organization and instilling a sense of acquiescence. Upon taking control of the PSA in 2005, peasant leaders in their new role as national public officials decided to develop a new profile. Instead of distributing money among small groups of *individuals*, PSA funds were distributed directly to *organizations*. The goal of this change was to strengthen grassroots, peasant, and indigenous organizations vis-à-vis provincial governments that were opposed to local social movements.

A new instance of institutional recognition materialized in the aftermath of the 2008 conflict between the government and agribusiness when President Cristina Kirchner decided to transform the PSA, which was renewed annually, into a permanent office. The PSA was 'upgraded' and turned into the *Sub-Secretaría de Desarrollo Rural y*

[8] See Aranda 2008b and Dandan 2008.

Table 1. Institutional recognition and performative governance at three scales.

	Institutional Recognition	Performative Governance
National Agencies	SSDRAF	CNIA
Provincial Actors	Provincial office of SSDRAF Provincial Roundtables	Provincial Ministry of Human Development
Local Effects	Tensions between and within organizations; redirection of time and energy	CNIA report on Monte Azul; showed peasants that 'something was being done'

Agricultura Familiar (SSDRAF, Under-Secretary of Rural Development and Family Farming). The inauguration of the SSDRAF in October of 2008 was staged in a union-run sports center located in a working-class suburb of greater Buenos Aires. The place is called *17 de Octubre* (October 17th), in reference to the day in 1945 when massive mobilizations reclaimed the freedom of then-Secretary of Labor, Juan Perón. The streets were filled with buses that brought people from all over Argentina, paid for by the SSDRAF and the FoNAF (*Foro Nacional de Agricultura Familiar*, National Forum of Family Agriculture), a network of rural organizations created to support the SSDRAF (more on this below). Before the ceremony, I bumped into members of several peasant movements. An activist from a peasant organization in the Chaco province told me: 'the government sometimes gives money, but instead of uniting us to reach agreements about land, water, and other important issues, no, they give us money to divide the organizations. For the public officials, those issues are not a priority.'

In the packed stadium, the head of the SSDRAF delivered a speech repeatedly mentioning the 'small producers and indigenous peoples' (*pequeños productores y pueblos originarios*) and eliciting spontaneous applause after offering Peronist slogans ('our aim is to achieve social justice, political independence, and economic sovereignty'). At the conclusion of the speech, the audience sang the national anthem, and many raised their fingers with a 'V', a typical Peronist gesture. President Cristina Kirchner closed the event, mentioning the 'rural workers, peasants, and the rural family', which prompted enthusiastic applause and ovations from the crowd. 'I feel it's a *great* honor being here today, with those who work the land from sunrise to sunset, *from sunrise to sunset!*' said the President, emphasizing the latter phrase in the midst of the applause. 'And with the men and women that, throughout the country, contribute every day to building food sovereignty for the Argentine people.' She mentioned the importance of creating the SSDRAF as providing 'an institutional area within the state that takes into account the small producers of our country'. The President closed her speech with a remembrance of her grandfather, 'who had a dairy farm, and worked from sunrise to sunset'.

For anyone reading between the lines, the celebratory reference to people 'working the land from sunrise to sunset' was a not-so-subtle critique of soybean growers, who during the conflict with the government, were depicted by public officials as absentee proprietors who did not want to contribute (via taxes) to the rest of the nation. Furthermore, the concept of 'food sovereignty' that she mentioned several times during her speech was a cue to the Argentine peasant organizations associated to *Via Campesina*.

My description of the ceremony launching the SSDRAF underscores two points. First, the ceremony reveals the authorities' engagement in 'impression management' as part of the institutional recognition of peasant movements. Second, it shows the cultural context in which discourses take place, which is critical to understanding their relevance. The ideas of Austin on language and action have been criticized on the grounds that they did not pay enough attention to 'the cultural context out of which particular signs are drawn forth by a speaker' (Alexander and Mast 2006, 3). The ritual performed in the launching of the SSDRAF, constantly punctuated by Peronist symbols, suggest the cultural (and political) dimensions of this institutional recognition.

The following excerpt illustrates the dilemmas of institutional recognition. I took these field notes at an internal meeting of a peasant social movement while Nélida and Silvio, two vocal members, debated participation in the institutional spaces opened by the national government.

Nélida:	We went to countless forums in Buenos Aires and it was useless. We only went to eat and travel. [...]
Silvio:	The organizations are useless, because they [the government] have the power.
Nélida:	Because Kirchner used it. We were like animals they brought to Buenos Aires; I've lost track of how many forums I have gone to in Buenos Aires.
Silvio:	Okay, fine, but you knew that when you went there. You're the first culprit.
Nélida:	No, sir. I went to learn and to see what was going on.
Silvio:	On one hand, that's okay, but at the same time, it serves them [the authorities].

These bittersweet tones also emerged when members discussed the role of organizations and their participation in and influence of government policies.

Silvio:	You know, the movement says, 'The nation [the national government] says that we must organize to access things, we have to organize in order to have a voice and vote.' But the truth is that you neither have a voice, nor a vote, not anything.
Nélida:	I saw that the nation [the national government] at one time gave lots of attention to organizations in many places. They allowed them to participate, to talk. And afterwards, they faded when they did what they wanted, they solved one thing, they fixed another thing, and it is a strategy they have to win you over. In Buenos Aires we met with the people from the Secretary of Agriculture and we took part in the Ministry for Social Development. We are still there today, but they conditioned you.

As these dialogues indicate, members of the peasant movement felt sidelined and saw their participation in public events as a self-serving strategy of politicians. Yet they also believed that it was hard to disengage from their connections to the government, lest they be isolated, lose their voice, or be left without resources. The counterpoint between Nélida and Silvio shows that peasants are keenly aware of the pitfalls of institutional recognition while they also find it hard to turn their backs on an influential ally. Next, I further elaborate on the *effects* of the government's institutional recognition at the provincial and local scales, showing how it redirects time, energy, and resources from contention to management.

Institutional recognition at provincial and local scales

Some weeks after the launching of the SSDRAF, in November 2008, I traveled to Formosa, a province in the north of Argentina bordering Paraguay. In Curuzú, a town in the hinterland of the province, I had dinner with Julio, a local leader of MoCaFor, the *Movimiento Campesino de Formosa* (Peasant Movement of Formosa). While catching up, he told me that the PSA 'was a patch, but it really didn't solve anything, it never worked', and that MoCaFor was participating in a provincial roundtable organized by the FoNAF (*Foro Nacional de Agricultura Familiar*, National Forum of Family Agriculture). Still, all the resources coming from the national SSDRAF (the former PSA) were channeled through provincial and regional roundtables organized by the FoNAF. 'Even though we aren't Kirchneristas [supporters of the president], it's better to change things from within', Julio said, justifying their participation in the roundtables. He complained that the Formosan representative of the PSA (still using the old denomination) had invited some organizations to the provincial roundtable that were like 'a rubber stamp ... like the MAF, totally coopted by the government', in reference to the *Movimiento Agrario de Formosa* (Formosan Agrarian Movement), an organization that acts like the rural branch of the political machine controlled by the provincial governor. MoCaFor and the MAF compete for the resources coming from the national government; thus, MoCaFor leaders fear that they may be neglected since a portion of these resources is first allocated to the provincial government.

A month later, I traveled from Curuzú to the city of Formosa with two MoCaFor members to attend the meeting of the provincial roundtable organized by the FoNAF. We met other members of MoCaFor at the house of Elisa, who lives in the city. While we sipped mate on Elisa's sunbathed patio, Oscar, a MoCaFor representative, complained that the national government 'didn't give us even one peso yet, but we're already busy thinking what [agricultural] machinery we're going to buy, how are we going to spend the money coming from the under-secretary [the SSDRAF], and that distracts us from the struggle'. The next day, we gathered to discuss MoCaFor's strategies for the FoNAF roundtable. Most of the conversation revolved around concerns over resources and debates about which organization would have the right to represent 'family farmers' at the roundtable. A back-and-forth dialogue between Arturo (a leader of MoCaFor near Moreno), Isidoro, and Julio (both leaders of MoCaFor in Curuzú) encapsulated these concerns.

Arturo: We don't want the guys from the PAIPPA [Program of Integral Assistance for Small Farmers, a provincial welfare program] at the roundtable, nor the MAF [Agrarian Formosan Movement, allied to the provincial government]. The roundtable has to work well [representatives of the provincial roundtables gathered at regional roundtables]. And we have to aim for that at the regional roundtable, because if the regional [roundtable] is strong, we'll be stronger. If we slacken [*si aflojamos*], they [PAIPPA or MAF representatives] will get in because they're skillful.

Isidoro: And they have money.

Arturo: They have economic power.

Julio: They have mobility; they have economic power.

Arturo: And they can talk, they can really talk [*mucha lengua tienen*]. We have to be aware that the roundtable is a political decision. And the national government doesn't want to lose face with the provincial government; it will be a big struggle.

Julio: We have to have a representative at the regional [roundtable].

Arturo: The provincial roundtable is weak, let's be realistic. Because there wasn't any money. And the same thing happened with the regional. If we have a strong position within the provincial roundtable, we'll also be strong at the regional [roundtable].

The next day, I joined MoCaFor representatives at the FoNAF provincial roundtable. We gathered with representatives of several provincial organizations on the outskirts of the city of Formosa from morning until late in the afternoon. Most of the hours were spent discussing the bylaws of the roundtable, who had the right to represent family farmers, and the steps to ensure that resources would 'come down' (*que bajen*) from the national government. The meeting extended to the next day, and I witnessed how representatives of organizations spent endless hours mired in discussions about bureaucratic procedures and fictional resources (that is, resources that were promised by the government but, as of 2013, never materialized).

While doing fieldwork in the rural community of Monte Azul, I was able to witness the ways in which the tensions in the air at the meetings in Formosa translated into disagreements and internal divisions. Furthermore, the participation of MoCaFor in those spaces created a dilemma that became a centerpiece for social movement leaders, who were increasingly torn between confrontation and negotiation. The following excerpts culled from my 2009 field notes reflect these dynamics, involving the relationships between MoCaFor and other organizations of the province, and between MoCaFor and public officials.

> We arrive with Julio to the farm of his mother, Nélida, and while we are sipping mate, he briefs her about the meeting he attended in Formosa. The provincial Minister of the Community, Zorrilla, praised the provincial welfare program for the rural poor, called PAIPPA, in front of Braulio, the president of the provincial roundtable of peasant organizations. 'Zorrilla said that the issues of the small farmers were solved by PAIPPA … I don't know how much bullshit he said; that we should join the PAIPPA, and that the PAIPPA was being discriminated against because it wasn't allowed to join the roundtable [resources coming from the national government were going to be channeled through the roundtable composed by community and peasant organizations; since PAIPPA is a government institution, peasants argue that it should not participate in the roundtable].
>
> 'If I was there, I would have told him to his face that he was lying,' Nélida replies, upset.
>
> 'And Braulio was sitting there, without saying anything' Julio continues, angered, 'I understand him, he has a salary of 2,500 pesos for being the chair of the roundtable, I wouldn't criticize him because everybody needs a salary, but how come he didn't say anything!' (June 18, 2009).

The next day, Julio recounted the meeting in Formosa to Isidoro, nurturing the tensions between MoCaFor and other organizations:

> While Julio describes the situation, Isidoro snorts and shows his disapproval with a tsk-tsk. 'If Braulio is not up to the task, he should resign,' Isidoro says.
>
> Julio adds that Braulio told him that they shouldn't fight, that if they barked, they wouldn't get anything. 'But it isn't fighting, it's telling it as it is!' says Julio. (19 June 2009)

Some days later, I was at the warehouse that doubles as the MoCaFor offices in Curuzú. Nélida, Isidoro, and Horacio were preparing for an upcoming meeting with the representative of the national SSDRAF, and they invited me to join their discussion on the best strategy to implement.

> Horacio says that they have to be more diplomatic; that the way they talk may sound aggressive to other people.
>
> 'I'm uncouth [*bruta*], I don't know how to say things', says Nélida apologetically.
>
> I add that it is a matter of saying the same things but in a different way.
>
> 'We won't stop saying the things we want to say', Isidoro chimes in, 'but, as Pablo says, we should think about the way we say them.'

[They talk briefly about the roundtable] 'And they see us as the troublemakers' (*como los que hacemos quilombo*), continues Isidoro. 'At the roundtable they think discussing is fighting' (25 June 2009).

The following week, this discussion was amplified at the local to the provincial level when representatives of MoCaFor from different regions of Formosa gathered to discuss the best approach to meeting with the representative of the SSDRAF.

'We need to be more diplomatic in the way we talk to the PSA' Horacio says to the group. 'We have to make clear that we want advice, but we don't want them to tell us what to do. And if they tell us what to do and we don't want to do it, we are not discrediting or attacking them; they just have to accept that we don't want to do it. [...] We need to propose to the PSA that we want to be sincere, they should tell us what they don't like about us, but they also should respect our decisions. And this while being diplomatic, without fighting.'

'So we need to put a gag on Isidoro', intervenes Nélida, to which we all laugh. 'I'm going to shut up', Nélida continues in an apologetic tone, 'sometimes I say things badly and then I regret it.'

When the representative from the SSDRAF arrives with two aides and joins the meeting, Horacio takes the lead. He begins, 'Sometimes we are not diplomatic, we don't talk in a nice way, but that's the way we talk, is not that we want to clash with the PSA.' (29 June 2009)

What these interactions amount to is the perception among MoCaFor leaders that a discursive field was closing in on them: some of the grassroots members of their movement, as well as the leaders of other organizations and public officials, were suggesting that they subdue their discourse and be less contentious (the sociologist, i.e., myself, included). Put another way, institutional recognition molded the discourse of social movements, leading them to privileging negotiation over confrontation.

In summary, the creation of the SSDRAF was a powerful sign from the national government that recognized peasant families as a valid actor in food production and rural life. Nevertheless, as the up-close analysis of the effects of this recognition at the local level demonstrates, this recognition also has its perils. The necessary funds to strengthen local organizations through the FoNAF roundtables were, as of 2013, arriving very slowly or never materialized. Furthermore, as hinted by Oscar's comment in Elisa's house, the endless meetings, constant bickering, and tensions created between organizations distracted peasants 'from the struggle'; in other words, constraining their ability to mobilize contentiously.

Agrochemical exposure and performative governance

Another prong of the initiatives launched by the national government in the aftermath of the 2008 conflict with agribusiness was the creation of the *Comisión Nacional de Investigación de Agroquímicos* (National Commission of Research on Agrochemicals, or CNIA). President Cristina Kirchner created the CNIA in 2009 on the heels of significant events such as the mobilization of the *Madres de Ituzaingó*, a group of women organized in Córdoba province against agrochemical drifts and their health effects.[9] In addition, researcher Andrés Carrasco publicized his work on the impact of glyphosate on embryos in 2009 (Paganelli

[9]See Aranda 2009, Piqué and Aranda 2009, and the decree 21/2009, available at the website of the *Comisión Nacional de Investigación de Agroquimicos* (CNIA).

et al. 2010), which bolstered the claims of activists and prompted the presentation of a bill in Congress (Aranda 2010b). By the end of 2009, however, the national Minister of Health and the official in charge of the CNIA were replaced. The newly appointed official continued the commission's meetings, but the language about agrochemicals was more 'industry-friendly' and no concrete actions were taken to curb the indiscriminate use of agrochemicals nationwide. In 2012, the *Auditoría General de la Nación*, the office auditing the national state, issued a report concluding that the CNIA has effectively been stalled since 2010.[10]

I argue that the creation of government agencies and commissions is an example of performative governance, which refers to a response to demands yet without necessarily ameliorating the underlying causes of problems. My claim, in other words, is that the state's use of spectacles and performance can result in people's fragile alignments with incumbents, resulting in their compliance with the state (Wedeen 1999). This also points to the relational nature of performative governance, which involves both authorities' discourses and actions (in a display of impression management: Goffman 1959) and their audience, as these performances influence activists' subsequent actions (or lack thereof), revealing the force of performative speech acts (Austin 1962).

Performative governance at the provincial and local scales

In February 2009, an agrochemical drift affected peasant families of Monte Azul. Nélida, a local leader of MoCaFor, had a burning sensation in her neck, blisters on her skin, and dry, sore lips; a neighbor got blisters all over her back. Hundreds of chickens, which peasants let roam freely, died suddenly. To cure the pimples and skin rashes, the people of Monte Azul went to the hospital, where they were told that their ailments were caused by scabies. Nélida and her brother Isidoro went to the hospital and told an authority from the hospital that the tanks with 'poisons' (agrochemicals) were being cleaned near people's houses and the community health center. 'The hospital is not in charge of agricultural matters', was the response they received. This was not the first time that this community had been victim to an environmental onslaught. In 2003, MoCaFor organized a series of roadblocks to protest other instances of agrochemical drifts. In 2009, in contrast, MoCaFor launched a communiqué denouncing the situation, but no contentious collective action resulted (Lapegna 2013, 2014).

Shortly after these events, provincial authorities arrived unexpectedly in Monte Azul. Dozens of doctors, employees, and authorities from the provincial Ministry of Human Development gathered Monte Azul's inhabitants in the local community health center. Authorities told the people there was no reason to be alarmed, that they would take blood samples to determine if they had traces of agrochemicals in their bodies, and then they proceeded to distribute anti-parasite drugs to everyone. Authorities did not allow people to ask questions, but Nélida and Horacio, peasant leaders of MoCaFor, spoke up to say that the reason for their ailments was not parasites but the agrochemicals used in the nearby soybean fields. A heated debate ensued, with authorities saying that soybeans were grown around the world and that the government authorized the use of agrochemicals. Nélida and Horacio replied that soybeans have also created problems around the world and that the government was not sufficiently controlling agrochemical use. The employees from the ministry administered surveys, took blood from 140 people, and sent the samples to be

[10]See RENACE (2013).

analyzed in Buenos Aires. The people in Monte Azul never received the results of the analyses.

After the contamination, people in Monte Azul 'were growing *very* anxious', according to Mario Hispano. He had worked for the national PSA for many years, but his sympathies towards peasant organizations forced him to walk a thin line, considering that the provincial government was constantly clashing with peasant organizations. By 2009, he had three years of experience as head of the PSA in Formosa, and was reappointed as provincial representative when this federal program was promoted to the SSDRAF. The newly created SSDRAF received a note from Monte Azul peasant leaders, listing how many chickens each person had lost (700 in total). Hispano organized meetings with the affected peasants and told them the SSDRAF was going to compensate the peasants for their losses. The SSDRAF also brought peasants to the city of Formosa to be tested for the presence of specific agrochemicals, although Hispano was skeptical about the results: 'I knew the tests would turn out negative ... and they did.' The tests that can be done in Formosa can detect insecticides, but not the herbicides that are used in GM soybean production. 'And back then was not the season when insecticides are used. But even if I explain that, they [the people of Monte Azul] won't hear me. They are very anxious.' Throughout our conversation, he repeatedly mentioned people's anxiety: 'And we needed to do something ... Peasants told me, "we are tired of seeing people coming, taking blood samples, and then nothing happens".' The SSDRAF also organized meetings to discuss provincial environmental laws because, according to Hispano, 'a lawyer needs to train them on what to do in cases of contamination because peasants always take the wrong path: they go to the police, or to the mayor, or they want to burn the fumigating machine.'

To address their grievances, the peasants appealed to the institutional channels recently created by the national government. Horacio, a leader of MoCaFor, sent a note to the National Ministry of Health asking for the intervention of the CNIA. In March, Monte Azul families received a visit from a representative of the CNIA. He wrote a report that clearly backed peasants' complaints about the contamination, mentioning that provincial authorities had not studied the environmental impact and pinpointed a series of environmental hazards affecting local families. The conclusions and recommendations of the report closely followed the demands repeatedly made by peasants, specifying that there were 'enough elements to assert the existence of problems of environmental contamination due to the use of pesticides, resulting in a risk for the local population' and that 'local inhabitants should not demonstrate the intoxications but instead the State should create proper and impartial institutional procedures to assess health damages, and to demonstrate that the local habitat does not pose a health risk'. The report also recommended the suspension of pesticide use, closer oversight of the soybean growers, and the creation of 'instances of participation for the control of pesticide management'. The report even recognized the clash between the peasants' mode of production and that fostered by soybean growers, saying that 'the coexistence of a productive model for self-consumption and an extensive industrial agriculture is very hard' and the introduction of soybean production had disrupted the lifestyle of peasant families.[11]

The reactions of the provincial authorities, the support of the SSDRAF, and the report of the CNIA all recognized the health problems of the local population and the environmental hazards created by soybean production and the use of agrochemicals. Combined, the actions of provincial and national authorities showed peasants that 'something was being

[11]Excerpts from CNIA, *Comisión Nacional de Investigación de Agroquímicos* (2009).

done' about the health and environmental problems affecting them. Unlike in 2003 when peasants blocked the provincial roads because they 'were not being heard', the government was now paying attention to their complaints. In terms of popular contention, the actions of public offices effectively curtailed local leaders' ability to frame the situation as a denial of contamination.

I use the term performative governance to refer to the actions of authorities (in Formosa and elsewhere) who respond to people's demands without addressing the underlying causes motivating those claims. I use the idea of performance in the sense outlined by Goffman (1959). That is, not as cynical or manipulative action, but as a means of managing the ways others interpret interactions in public life. In the specific events presented above, the responses of the authorities are an illustration of performative governance for two reasons. First, the results of the analyses of blood samples taken by the provincial Ministry of Human Development were never shared with local inhabitants. Provincial authorities never attempted to regulate soybean production or create a more systematic or efficient control of the use of agrochemicals. Yet, as Hispano underlined, the SSDRAF administered blood tests mostly due to the pressure of peasants, which suggests the interplay between actions and audience (Alexander and Mast 2006). The official reports can also be seen as performative in the sense of speech acts that intervene in a situation rather than simply describing it (Austin 1962). The report of the CNIA never translated into concrete actions to stop fumigations or enforce regulations about agrochemical use. It was, however, a way of 'doing things with words', by way of 'convincing, persuading, [and] deterring' (Austin 1962, 108).

In short, authorities' discourses and public actions conveyed the sense that the government was dealing with the demands regarding agrochemical exposure. The authorities' actions curtailed movements' discursive opportunities (Bröer and Duyvendak 2009) while sending the message that there was no need for a protest in order to address the issues at hand.

Conclusion

Since the mid-2000s, peasants and rural populations have been increasingly exposed to the herbicides that GM soybeans are engineered to endure. I proposed two concepts – institutional recognition and performative governance – to better understand how Argentine peasants accommodate the sweeping expansion of GM soybeans and agrochemical exposure, which ultimately create obstacles to organizing contentious collective actions. I argue that institutional recognition and performative governance are symbolic mechanisms that nonetheless have material consequences in the lives of populations affected by agrochemical exposures and in social movement trajectories. These mechanisms are articulated at multiple scales (ranging from national, to provincial, to local) and result in peasant social movements directing their time and resources towards negotiation rather than confrontation, while reducing the discursive field available for social movement organizations.

By focusing mostly on how the discourses and actions of political actors influence the action of social movements, I am aware that I paid less attention to important processes like mobilization *within* movements (Wolford 2010, 1–34), the organizational relationships between social movement organizations and external allies (but see Lapegna 2013, 2014), and the relationships between the state and social movements. In spite of these limitations, the concepts of institutional recognition and performative governance may shed light on the barriers to protests against socio-environmental problems beyond Argentina, and suggest lines for comparative research. For example, in Canada and Germany, governments also open institutional spaces to address concerns about GM crops, but NGOs reject invitations (Levidow 1998; Prudham and Morris 2006; Kinchy 2012, 105). For peasant organizations in northern Argentina, however, the government's offer was harder to reject since leaders had to respond to the pressures and needs of their constituents (Lapegna 2013).

The concepts proposed here may potentially illuminate research on contemporary South America where left-of-center governments have become allies for many social movements. These governments have relied economically on the recent commodity boom of extractive activities (i.e., agriculture, mining, and hydrocarbons) but without addressing their social and environmental consequences. Institutional recognition and performative governance may thus provide clues to understanding the interfaces between governments and social movements beyond Argentina.

Acknowledgements

I would like to express my gratitude to the Formosans who generously volunteered their time and opened their houses for my fieldwork. I am also grateful to the anonymous reviewers of *JPS* for their comments and to the Willson Center for Humanities and Arts (University of Georgia), which supported part of this research. An earlier version of this article was presented at the meeting of the Latin American Studies Association in 2014, where I benefitted from the comments of Javiera Barandiarán, Maiah Jaskoski, and Kathryn Hochstetler. Katie Sobering also provided extremely insightful suggestions. Any omissions or errors are my own.

Disclosure statement

No potential conflict of interest was reported by the authors.

References

Abu-Lughod, L. 1990. The romance of resistance: tracing transformations of power through Bedouin women. *American Ethnologist* 17, no. 1: 41–55. doi:10.1525/ae.1990.17.1.02a00030.

Adas, M. 1981. From avoidance to confrontation: Peasant protest in precolonial and colonial southeast Asia. *Comparative Studies in Society and History* 23, no. 2: 217–47. doi:10.2307/178734.

Alexander, J.C., and J.L. Mast. 2006. Introduction. Symbolic action in theory and practice: the cultural pragmatics of symbolic action. In *Social performance: symbolic action, cultural pragmatics, and ritual*, ed. J.C. Alexander, B. Giesen and J.L. Mast, 1–28. New York: Cambridge University Press.

Andrée, P. 2011. Civil society and the political economy of GMO failures in Canada: a neo-Gramscian analysis. *Environmental Politics* 20, no. 2: 173–91. doi:10.1080/09644016.2011.551023.

Aranda, D. 2008a. Varias generaciones están comprometidas. *Página 12*, May 14.

Aranda, D. 2008b. Otras entidades, otras demandas. *Página 12*, July 27.

Aranda, D. 2009. El veneno que asoló el barrio de Ituzaingó. *Página 12*, January 12.

Aranda, D. 2010a. La salud no es lo primero en el modelo agroindustrial. *Página 12*, June 14.

Aranda, D. 2010b. Para frenar los agrotóxicos. *Página 12*, September 22.

Armstrong, E.A., and M. Bernstein. 2008. Culture, power, and institutions: A multi-institutional politics approach to social movements. *Sociological Theory* 26, no. 1: 74–99. doi:10.1111/j.1467-9558.2008.00319.x.

Austin, J.L. 1962. *How to do things with words*. London: Oxford University Press.

Auyero, J., and D.A. Swistun. 2009. *Flammable: environmental suffering in an Argentine shantytown*. New York: Oxford University Press.

Benford, R.D. 1997. An insider's critique of the social movement framing perspective. *Sociological Inquiry* 67, no. 4: 409–30.

Beverley, J. 2004. Subaltern Resistance in Latin America: A Reply to Tom Brass. *Journal of Peasant Studies* 31, no. 2: 261–75.

Binimelis, R., W. Pengue, and I. Monterroso. 2009. 'Transgenic treadmill': Responses to the emergence and spread of glyphosate-resistant johnsongrass in Argentina. *Geoforum* 40, no. 4: 623–33. doi:http://dx.doi.org/10.1016/j.geoforum.2009.03.009.

Bonn, D. 2005. Roundup Revelation. *Environmental Health Perspectives* 113, no. 6: A403–A404.

Brass, T. 2002. Latin American peasants-new paradigms for old? *Journal of Peasant Studies* 29, no. 3–4: 1–40.

Bridge, G., P. McManus, and T. Marsden. 2003. The next new thing? Biotechnology and its discontents. *Geoforum* 34, no. 2: 165–74.

Bröer, C., and J.W. Duyvendak. 2009. Discursive opportunities, feeling rules, and the rise of protests against aircraft noise. *Mobilization* 14, no. 3: 337–56.

Brosius, J.P. 1997. Prior transcripts, divergent paths: Resistance and acquiescence to logging in Sarawak, East Malaysia. *Comparative Studies in Society and History* 39, no. 3: 468–510. doi:10.2307/179156.

Casabé, N., L. Piola, J. Fuchs, M.L. Oneto, L. Pamparato, S. Basack, R. Gimenez, R. Massaro, J.C. Papa, and E. Kesten. 2007. Ecotoxicological assessment of the effects of glyphosate and chlorpyrifos in an Argentine soya field. *Journal of Soils and Sediments* 7, no. 4: 232–239.

Centro de Protección de la Naturaleza (CeProNat). Campaña paren de fumigarnos. http://www.cepronat-santafe.com.ar/editorial.html

Comisión Nacional de Investigación de Agroquímicos (CNIA). Decreto 21/2009, Ministerio de Salud. http://www.msal.gov.ar/agroquimicos/decreto-21-2009.php

Comisión Nacional de Investigación de Agroquímicos (CNIA). 2009. *Informe de reconocimiento y observación*. Buenos Aires: Ministerio de Salud.

Dandan, A. 2008. Demandas del otro campo. *Página 12*, August 31.

DDA-PROINDER. 2003. *Los programas de desarrollo rural ejecutados en el ámbito de la SAGPyA*. Buenos Aires: Dirección de Desarrollo Agropecuario.

Domínguez, D. 2009. La lucha por la tierra en Argentina en los albores del siglo XXI. La recreación del campesinado y de los pueblos originarios. PhD Dissertation. Buenos Aires: Universidad de Buenos Aires.

Domínguez, D., and P. Sabatino. 2010. La muerte que viene en el viento. La problemática de la contaminación por efecto de la agricultura transgénica en Argentina y Paraguay. In *Los señores de la soja. La agricultura transgénica en América Latina*, ed. Ana Lucía Bravo, 31–121. Buenos Aires: CLACSO.

Eaton, E. 2009. Getting behind the grain: the politics of genetic modification on the Canadian prairies. *Antipode* 41, no. 2: 256–81. doi:10.1111/j.1467-8330.2009.00672.x.

Edelman, M. 2001. Social movements: Changing paradigms and forms of politics. *Annual Reviews in Anthropology* 30, no. 1: 285–317.

Escobar, A. 1992. Culture, practice and politics: Anthropology and the study of social movements. *Critique of Anthropology* 12, no. 4: 395–432.

Fernández, M. 2014. Cáncer en Córdoba: en el este provincial, la mortalidad más alta. *La Voz del Interior*, May 29.

Fitting, E. 2011. *The struggle for maize: Campesinos, workers, and transgenic corn in the Mexican countryside*. Durham and London: Duke University Press.

Fletcher, R. 2001. What are we fighting for? Rethinking resistance in a Pewenche community in Chile. *Journal of Peasant Studies* 28, no. 3: 37–66. doi:10.1080/03066150108438774.

Fraser, N. 2003. Rethinking recognition: overcoming displacement and reification in cultural politics. In *Recognition struggles and social movements: contested identities, agency and power*, ed. Barbara Hobson, 21–32. New York: Cambridge University Press.

Futrell, R. 1999. Performative governance: Impression management, teamwork, and conflict containment in city commission proceedings. *Journal of Contemporary Ethnography* 27, no. 4: 494–529. doi:10.1177/089124199129023316.

Giarracca, N., and M. Teubal, eds. 2010. *Del paro agrario a las elecciones de 2009*. Buenos Aires: Antropofagia.

Goffman, E. 1959. *The presentation of self in everyday life*. Garden City, NY: Doubleday.

Goldfarb, L., and A. Zoomers. 2013. The drivers behind the rapid expansion of genetically modified Soya production into the Chaco region of Argentina. In *Biofuels: Economy, environment and sustainability*, ed. Fang Zhen, 73–95. InTech. Available from http://www.intechopen.com/books/

biofuels-economy-environment-and-sustainability/the-drivers-behind-the-rapid-expansion-of-genetically-modified-soya-production-into-the-chaco-region

Goodwin, J., and J.M. Jasper, eds. 2004. *Rethinking social movements: Structure, meaning, and emotion*. Lanham: Rowman & Littlefield.

Grandia, L. 2014. Modified landscapes: vulnerabilities to genetically modified corn in northern Guatemala. *Journal of Peasant Studies* 41, no. 1: 79–105. doi:10.1080/03066150.2013.872631.

Grupo de Reflexión Rural (GRR). Paren de Fumigar: Campaña de concientización sobre los impactos de los agrotóxicos. http://www.grupodereflexionrural.com/campanapdf/index.php

Heller, C. 2013. *Food, farms, and solidarity: French farmers challenge industrial agriculture and genetically modified crops*. Durham and London: Duke University Press.

Ho, W.-C. 2011. James Scott's resistance/hegemony paradigm reconsidered. *Acta Politica* 46, no. 1: 43–59.

Hollander, J.A., and R.L. Einwohner. 2004. Conceptualizing resistance. *Sociological Forum* 19, no. 4: 533–54.

INDEC. 2012. *Exportaciones del complejo soja, por zonas económicas, según componentes. Años 2006-2010*. Buenos Aires: Instituto Nacional de Estadísticas y Censos (INDEC).

Jasper, J., and J. Goodwin, eds. 2011. *Contention in context: political opportunities and the emergence of protest*. Stanford, CA: Stanford University Press.

Joensen, L., S. Semino, and H. Paul. 2005. *Argentina: A case study on the impact of genetically engineered soya*. London: Gaia Foundation.

Kinchy, A. 2012. *Seeds, science, and struggle: The global politics of transgenic crops*. Cambridge, MA: MIT Press.

Kleffman Group. 2013. Mercado Argentino de Productos Fitosanitarios 2012. http://www.casafe.org

Klepek, J. 2012. Against the grain: knowledge alliances and resistance to agricultural biotechnology in Guatemala. *Canadian Journal of Development Studies/Revue canadienne d'études du développement* 33, no. 3: 310–25.

Korovkin, T. 2000. Weak weapons, strong weapons? Hidden resistance and political protest in rural Ecuador. *Journal of Peasant Studies* 27, no. 3: 1–29. doi:10.1080/03066150008438738.

Kurzman, C. 2008. Introduction: meaning-making in social movements. *Anthropological Quarterly* 81, no. 1: 5–15.

Lapegna, P. 2013. Social movements and patronage politics: Processes of demobilization and dual pressure. *Sociological Forum* 28, no. 4: 842–63. doi:10.1111/socf.12059.

Lapegna, P. 2014. Global ethnography and genetically modified crops in Argentina: On adoptions, resistances, and adaptations. *Journal of Contemporary Ethnography* 43, no. 2: 202–27. doi:10.1177/0891241613516629.

Leguizamón, A. 2014. Modifying Argentina: GM soy and socio-environmental change. *Geoforum* 53: 149–60. doi:http://dx.doi.org/10.1016/j.geoforum.2013.04.001.

Levidow, L. 1998. Democratizing technology -or technologizing democracy? Regulating agricultural biotechnology in Europe. *Technology in Society* 20, no. 2: 211–26. doi:http://dx.doi.org/10.1016/S0160-791X(98)00003-7.

MAGyP. 2014. Sistema Integrado de Información Agropecuaria. http://www.siia.gov.ar

McAdam, D., and H. Boudet. 2012. *Putting social movements in their place. Explaining opposition to energy projects in the United States, 2000–2005*. New York: Cambridge University Press.

McAdam, D., S. Tarrow, and C. Tilly. 1997. Toward an integrated perspective on social movements and revolutions. In *Comparative politics: Rationality, culture, and structure*, ed. Mark I. Lichbach and Alan S. Zuckerman, 142–71. New York: Cambridge University Press.

McAdam, D., S. Tarrow, and C. Tilly. 2008. Methods for Measuring Mechanisms of Contention. *Qualitative Sociology* 31, no. 4: 307–31.

Mitchell, T. 1990. Everyday metaphors of power. *Theory and Society* 19, no. 5: 545–77.

Newell, P. 2009. Bio-Hegemony: The political economy of agricultural biotechnology in Argentina. *Journal of Latin American Studies* 41, no. 1: 27–57.

O'Brien, K.J. 1996. Rightful resistance. *World Politics* 49, no. 1: 31–55.

O'Brien, K.J. 2013. Rightful resistance revisited. *Journal of Peasant Studies* 40, no. 6: 1051–62. doi:10.1080/03066150.2013.821466.

Oliva, A., R. Biasatti, S. Cloquell, C. González, S. Olego, and A. Gelin. 2008. ¿Existen relaciones entre los factores ambientales rurales y la salud reproductiva en la Pampa Húmeda Argentina? *Cadernos de Saúde Pública* 24, no. 4: 785–92.

Ortner, S.B. 2006. Resistance and the problem of ethnographic refusal. In *Anthropology and social theory. Culture, power and the acting subject*, 42–62. Durham and London: Duke University Press.

Otero, G., ed. 2008. *Food for the few: neoliberal globalism and biotechnology in Latin America*. Austin, TX: University of Texas Press.

Paganelli, A., V. Gnazzo, H. Acosta, S.L. López, and A.E. Carrasco. 2010. Glyphosate-based herbicides produce teratogenic effects on vertebrates by impairing retinoic acid signaling. *Chemical research in toxicology* 23, no. 10: 1586–95.

Peck, J., and A. Tickell. 2002. Neoliberalizing space. *Antipode* 34, no. 3: 380–404.

Pengue, W.A. 2004a. El modelo de agricultura industrial intensivo. *Revista Saber Cómo* no. 16.

Pengue, W.A. 2004b. La ingeniería genética y la intensificación de la agricultura argentina. In *Los transgénicos en América Latina y el Caribe: un debate abierto*, ed. Alicia Bárcena, Jorge Katz, César Morales and Marianne Schaper, 167–190. Santiago de Chile: CEPAL.

Pengue, W.A. 2005. Transgenic crops in Argentina: The ecological and social debt. *Bulletin of Science Technology & Society* 25, no. 4: 314–22. doi:10.1177/0270467605277290.

Pesticide Action Network (PAN), Pesticide Database. http://www.pesticideinfo.org/

Petras, J., and H. Veltmeyer. 2001. Are Latin American peasant movements still a force for change? Some new paradigms revisited. *Journal of Peasant Studies* 28, no. 2: 83–118.

Piqué, M., and D. Aranda. 2009. Para que no se extienda la mancha venenosa. *Página 12*, January 15.

Polletta, F. 2008. Culture and movements. *The Annals of the American Academy of Political and Social Science* 619, no. 1: 78–96.

Prudham, S., and A. Morris. 2006. Making the market 'safe' for GM foods: The case of the Canadian biotechnology advisory committee. *Studies in Political Economy* 78: 145–75.

Ramírez, M. 2012. Relación entre el uso de agroquímicos y el estado sanitario de la población en localidades de los Departamentos Bermejo, Independencia y Tapenagá de la Provincia del Chaco. Ministerio de Salud de la Nación. http://redaf.org.ar/wp-content/uploads/2014/05/agroquimicos_salud_informechaco_minsalud.pdf

Red Nacional de Acción Ecologista (RENACE). 2013. Auditoría General de la Nación y su impactante informe sobre SENASA y los agrotóxicos. http://renace.net/?p=3105

Richardson, N. 2009. Export-oriented populism: commodities and coalitions in Argentina. *Studies in Comparative International Development (SCID)* 44, no. 3: 228–55.

Roseberry, W. 1994. Hegemony and the language of contention. In *Everyday forms of state formation: Revolution and the negotiation of rule in modern Mexico*, ed. Joseph Gilbert and Daniel Nugent, 355–66. Durham, NC: Duke University Press.

Schurman, R., and W.A. Munro. 2010. *Fighting for the future of food: Activists versus agribusiness in the struggle over biotechnology*. Minneapolis: University of Minnesota Press.

Scoones, I. 2008. Mobilizing Against GM Crops in India, South Africa and Brazil. *Journal of Agrarian Change* 8, no. 2–3: 315–44.

Scott, J.C. 1985. *Weapons of the weak: Everyday forms of peasant resistance*. New Haven: Yale University Press.

Scott, J.C. 1990. *Domination and the arts of resistance: Hidden transcripts*. New Haven: Yale University Press.

Scott, J.C. 2005. Afterword to "Moral economies, state spaces, and categorical violence". *American Anthropologist* 107, no. 3: 395–402.

Séralini, G.E., R. Mesnage, N. Defarge, S. Gress, D. Hennequin, E. Clair, M. Malatesta, and J. Spiroux de Vendômois. 2013. Answers to critics: Why there is a long-term toxicity due to a Roundup-tolerant genetically modified maize and to a Roundup herbicide. *Food and Chemical Toxicology* 53: 461–468.

Seymour, S. 2006. Resistance. *Anthropological Theory* 6, no. 3: 303–21. doi:10.1177/1463499606066890.

Sivaramakrishnan, K. 2005. Some intellectual genealogies for the concept of everyday resistance. *American Anthropologist* 107, no. 3: 346–55. doi:10.1525/aa.2005.107.3.346.

Steinberg, M.W. 1999. *Fighting words: Working-class formation, collective action, and discourse in early nineteenth-century England*. Ithaca: Cornell University Press.

Thompson, E.P. 1971. The moral economy of the English crowd in the eighteenth century. *Past and Present* 50: 76–136.

Vitta, J., D. Tuesca, and E. Puricelli. 2004. Widespread use of glyphosate tolerant soybean and weed community richness in Argentina. *Agriculture, Ecosystems & Environment* 103, no. 3: 621–4.

Wedeen, L. 1999. *Ambiguities of domination: politics, rhetoric, and symbols in contemporary Syria.* Chicago: University of Chicago Press.

White, C.P. 1986. Everyday resistance, socialist revolution and rural development: the Vietnamese case. *Journal of Peasant Studies* 13, no. 2: 49–63.

Williams, R. 1977. *Marxism and literature.* Oxford: Oxford University Press.

Wolford, W. 2010. *This land is ours now: Social mobilization and the meanings of land in Brazil.* Durham and London: Duke University Press.

Zabaloy, M.C., J.L. Garland, and M.A. Gómez. 2008. An integrated approach to evaluate the impacts of the herbicides glyphosate, 2,4-D and metsulfuron-methyl on soil microbial communities in the Pampas region, Argentina. *Applied Soil Ecology* 40, no. 1: 1–12. doi:http://dx.doi.org/10.1016/j.apsoil.2008.02.004.

‘More soy on fewer farms’ in Paraguay: challenging neoliberal agriculture’s claims to sustainability

Laureen Elgert

The knowledge claims of sustainable commodity discourses are often presented as ‘fact’ in policy debates. Such claims, however, are better understood as coproduced with neoliberal values and power, in the context of ‘more soy on fewer farms’ – the concentration of land and productive resources – in Paraguay. I analyze five claims that link ‘responsible soy’ to reduced deforestation, good agricultural practices, national economic growth, increased food security and public participation in soy governance. I examine ways in which each of these claims is contingent and contested. With an explicit commitment to equity, I argue that alternatives to ‘responsible soy’, that include rather than exclude small-scale producers in Paraguay’s agricultural development trajectory, are likely to culminate in stronger claims to sustainability by redressing the equity issues that have been marginalized by neoliberal agriculture.

Neoliberal sustainable commodity discourses and the marginalization of alternatives

The production of commodities such as soy, timber, palm oil and biofuel has been linked to large-scale agricultural investment and agribusiness in developing countries. In many parts of the developing world, these investments have increased the concentration of productive capacity and wealth (Borras et al. 2012; Galeano 2012), caused environmental problems (Spieldoch and Murphy 2009), exacerbated conditions of insecurity and violence (Garcia-Lopez and Arizpe 2010; Grajales 2011), provoked or aggravated food insecurity (Brigham 2004) and, through consolidating wealth and power, have undermined inclusive decision-making in societies (Moog et al. 2014). Ironically, however, large-scale commodity producers, buyers, governments and investors increasingly bill these investments as sustainable through prominent discourses of sustainable commodity production, discourses that are often fomented and buttressed by roundtable organizations and certification programs such as the Roundtable on Sustainable Palm Oil (RSPO), and the Roundtable on Sustainable Biomaterials (RSB). Discourses of sustainable commodities are hallmarks of neoliberal governance: they propose non-state, voluntary production standards in response to social and environmental critiques of large-scale production, distancing such production from problematic ramifications (Guthman 2007; Klooster 2010; Bailis and Baka 2011; Bloomfield 2012; Johnson 2012). The terms on which neoliberal sustainable commodity discourses engage with sustainability, however, fail to examine, question or disrupt patterns of ownership and investment that entrench and intensify dramatic inequality.

Since its onset more than three decades ago, Paraguay's soy boom has resulted in *increasing* soy production on *fewer*, *larger* farms, increasingly excluding peasant farmers from the rural landscape whilst providing them few livelihood alternatives. Meanwhile, large-scale producers, commercial buyers, environmentalists and policy-makers tout a myriad of benefits of soy production for society, including reduced deforestation, sustainable agricultural practices, national economic growth, improved global food security and public participation in soy governance. Their optimism about the promise of soy for Paraguay is in part fueled by the Roundtable on Responsible Soy (RTRS), an organization that has created a certification system for 'responsible soy'. The discourse of responsible soy, however, has extended well beyond the certification system, and although only a tiny minority of soy producers has been certified, responsible soy is often invoked in the contexts of Paraguayan media, business, policy and society.

Like other neoliberal discourses, sustainable commodity production eschews what proponents suggest are inefficient and ineffective top-down regulatory frameworks, and instead promotes reliance on non-state, voluntary and market-based mechanisms to incentivize modes of production (Robertson 2007; Besky 2008; Higgins et al. 2008; Klooster 2010). Neoliberal approaches to agriculture governance, therefore, demand farm-level reforms, as opposed to a more radical restructuring of the agricultural sector, leaving issues such as the concentration of land and productive resources, the terms of foreign investment and historical-structural injustices unexamined, and 'off the table' in mainstream debates. The increased currency of 'responsible soy' has thus marginalized (though not silenced) elements of a more radical stance against Paraguay's soy industry by legitimizing 'soy expansion while failing to bring to the front more pressing issues of unequal land distribution and the democratic representation of peasants' interests in decision-making, [and] the excessive power of agri-business' (Garcia-Lopez and Arizpe 2010, 204). Currents of dissent and protest against the soy industry exist, highlighted by the general strike of 2014 that drew attention to violence against peasants and the lack of progress towards agrarian reform. The RTRS, however, is often understood by the domestic and international communities as part of the redress of soy-induced controversy, despite that such governance mechanisms 'have questionable environmental benefits at best and at worst work to reenforce the hegemony of ENGOs [environmental non-governmental organizations], to legitimize agribusiness multinationals, and to destabilize strategies of resistance' (Baletti 2014, 7). Furthermore, sustainability discourses have fomented the shift of allegiances of some stakeholders away from the political grievances of peasants, towards large soy producers. For example, ENGOs, once allies in the peasant struggle, have come to see new windows of opportunity to achieve goals like reduced deforestation and sustainable soil management through developing partnerships with agri-business, rather than resisting their expansion (Elgert 2012).

Researchers have found similar processes of marginalization in the context of other neoliberal sustainable commodity discourses that have curtailed the power of alternatives. Klooster, for example, has noted that as sustainable forestry discourses become mainstream, they effectively 'restrain political struggle' and 'limit the scope of action' outside the neoliberal governance framework (Klooster 2010). Brosius has observed that as sustainability becomes professionalized and quantified through indicators and criteria, protest and dissent become 'not merely irrelevant but disruptive' (1999, 49), as sustainable commodity discourses position the task at hand as troubleshooting and problem solving in immediate and seemingly uncomplicated (i.e., nonstructural and market-friendly) ways. Understanding claims to sustainability requires an understanding of how facts, values and power shape each other and combine in discourses that define and limit legitimate and credible policy and action.

Understanding policy discourse: facts, values and power

The dominance of neoliberal sustainable commodity production and the marginalization of dissent create an urgent need for greater critical reflexivity about how such discourses emerge, become influential and persist. While some critics fear that a focus on discourse diverts attention away from material issues (Bryant and Bailey 1997), others point out that that discourses are vital elements of policy debates that *do matter* in policy outcomes (Litfin 1994; Hajer 1995; Feindt and Oels 2005). As the material and political means through which stakeholders establish shared ways of understanding the world, discourses reflect *particular* assumptions, judgments, values and interests (Meadowcroft 2007; Walker and Shove 2007; Elgert 2009) and, in turn, shape environmental policies and their empirical and material outcomes (Adger et al. 2001; Hajer and Versteeg 2005). Such outcomes incite struggles for political recognition and equity.

A focus on policy discourses means looking beyond knowledge claims, often presented as 'facts' in arguments, to the broader discursive context which includes the values and power structures that undergird such claims, but that often remain less visible. While 'facts' are a powerful means of gaining ground in 'credibility contests' (Gieryn 1999), they do not speak for themselves or present in a detached, objective manner. The production of facts is, rather, a socially situated and contingent activity (Lövbrand 2007) supported by values and power that are, in turn, supported by facts: 'society cannot function without knowledge any more than knowledge can exist without appropriate social supports' (Jasanoff 2004, 2–3). This process of *coproduction* means that facts, values and power iteratively condition, constitute and validate each other (Jasanoff and Wynne 1998; Jasanoff 2004); nature, society, facts and values must thus be considered concurrently (Latour 1993; Goldman 2000; Castree 2014). Such links between knowledge and power explain why certain ways of reading and relating environmental problems (and solutions) gain authority and credibility, while others are marginalized or even silenced altogether (Jasanoff and Martello 2004).

Attention to the coproduction of facts, values and power in policy discourse therefore requires empirical scrutiny of knowledge claims, and their mutually supportive relationship with values and power structures. Political ecologists, for example, have studied how colonial interventions were often based on simplistic, erroneous, yet privileged assumptions about the relationship between environmental change and population, migration, and 'unscientific' and 'non-modern' land-use management (Fairhead and Leach 1996; Neumann 2003; Davis 2007). Such policies often harmed local environments and livelihoods, but were nonetheless maintained for decades and, in some cases, persist. Today, 'to a degree that seemed unimaginable only two decades ago', it is neoliberalism that defines the dominant values and sources of power in environmental policy (Arsel and Büscher 2012, 53–54). Free-market values bolster the promotion and implementation of payments for ecosystem services, carbon trading, private conservation and the certification and marketization of sustainable production practices. Despite numerous case studies that indicate the political origins and implications of such interventions (Heynen et al. 2007), they are often advanced as mere instruments of sustainability. Within neoliberal discourses, therefore, technical and 'constructive' expertise often eclipses political critique and deconstruction, leaving debates about what constitutes, and what does not constitute sustainability to be simplistically argued on pragmatic grounds (Fischer 2009; Silva-Castañeda 2012).

Despite their resilience, powerful discourses are not immune to contestation, and remain vulnerable to challenge and deconstruction by those they exclude (Howarth and Griggs 2012; Ingram et al. 2014). Indeed, 'what a narrative leaves out is often as important as what it contains' (Ingram et al. 2014, 990). To invigorate deliberation about sustainability in commodity production, and to ensure that resistance to sustainable commodity discourses is taken seriously, policy analysts must expose such discourses to interrogation by excluded actors and perspectives (Hajer and Versteeg 2005; Bäckstrand et al. 2010; Forsyth 2011). Such critical discourse analysis has the potential to erode the authority and credibility of dominant policy idioms over time by exposing 'the political' beneath 'the factual' (Adger et al. 2001; Hajer and Versteeg 2005; Forsyth 2011) and presenting viable alternatives.

This paper argues that the knowledge claims of the responsible soy discourse must be considered in the context of neoliberal values of privatized governance, deregulation and liberalized trade, and the power of the soy lobby. It begins by establishing the context of more soy on fewer farms – the concentration of land and wealth – in Paraguay. The paper then moves to its central concern by analyzing five claims about soy production, now commonplace in the media, government and business, that are used to laud the growth of the soy industry in Paraguay: (1) soy production has been historically linked to deforestation, but has been recently 'decoupled' from deforestation; (2) soy production is associated with 'good agricultural practices' that maintain soil quality and avoid land degradation; (3) soy production is the engine of economic growth; (4) soy production promotes global food security by 'feeding the world'; and (5) processes of establishing standards for responsible soy production have promoted inclusion by inviting participation from multiple and diverse stakeholders. I examine ways in which each of these claims is contingent and contested. With an explicit political commitment to equity, I argue that alternatives to 'more soy on fewer farms', that include rather than exclude small-scale producers in Paraguay's agricultural development trajectory, are likely to culminate in stronger claims to sustainability by redressing the equity issues that have been marginalized within the responsible soy discourse.

This research is based on two years of fieldwork, between 2004 and 2008, in Canindeyú, a department (the name used for administrative and political subdivisions of Paraguay) in the North-East of Paraguay, where small-scale subsistence and cash-crop farming on parcels of around 10 hectares sit uncomfortably beside soy farms 10 to > 100 times their size. During this period, I met and interviewed dozens of small-scale farmers and several soy producers and employees of the soy industry. I also attended the 2006 RTRS conference in Buenos Aires. The work also draws from a wide array of existing research, news reports and analysis of data from Paraguay's 2008 National Agricultural Census (DCEA 2009) and the Ministry of Agriculture's 2010 report on soy production in Paraguay (MAG 2010). My analysis focuses mainly on the issues raised by soy production in Paraguay. The soy industry, however, has a distinctly regional character, owing to such cross-border dynamics as the migration of Brazilian soy farmers to Paraguay, and the export of raw soybeans from Paraguay to Brazil and Argentina for crushing and shipping. Much of this discussion, therefore, has relevance for soy-producing regions of South America more generally.

More soy on fewer farms and the emergence of 'responsible soy'

Paraguay's 'soy boom' (Carter et al. 1996; Dros 2004), continues to grow in the context of increasingly concentrated land holdings in a country already widely recognized as having one of the most unequal land distributions in the world. Data from the Ministry of Agriculture's report on Paraguay's soy harvest (MAG 2010) and the 2010 Paraguayan census (DCEA 2009) show that the amount of land dedicated to soy production in Paraguay

Table 1. Land surface dedicated to soy production in Paraguay (Hectares), 2008-2011

	2008–9	2009–10	2010–11
Land surface dedicated to soy production (in Hectares)	2.570.000	2.671.059	2.805.467

Source: Ministerio de Agricultura y Ganaderia (MAG), Paraguay, 2012.

Table 2. Land area dedicated to soy cultivation- by farm size (Hectares), 1991–2008

	Total land area	Farm size (Hectares)				
		<20	20–<50	50–<100	100–<1000	1000+
Total 2008	2,463,510.4	98,442.2	97,014.5	133,906.7	1,408,693.4	1,085,453.7
Total 1991	552,656.9	110,740.0	91,597.6	86,904.6	203,050.2	60,364.4
Change 1991-2008 (%)	345.8	−11.1	5.9	54.1	416.5	1,698.2

Source: Dirección de Censo y Estadística Agropecuaria (DCEA) 2009.

grows every year, increasing in 2011 to more than 400 percent over 1991 levels (see Tables 1 and 2). Soy, furthermore, is increasingly a large-scale crop (Table 2). While the total land area in Paraguay dedicated to soy production grew by nearly 350 percent between 1991 and 2008, the amount of land dedicated to soy production on farms smaller than 20 hectares contracted by 10 percent. The larger the farm size, the faster the growth in size of soy cropland: from just under 6 percent growth in cropland on farms between 20 and < 50 hectares, to a staggering near-1700 percent rate of growth in cropland on large farms over 1000 hectares.

In the early 2000s, increasing publicity linked soy production in South America to deforestation in the Amazon and the Interior Atlantic Forests, and to horrific 'soy stories' that turned up on activist blogs and in the European and North American media. These stories featured animals and people poisoned by pesticides (Balch 2007; Hetherington 2013), brutal and violent land evictions (Barrionuevo 2008b), and bulldozers and fires bringing down massive areas of forest (Rohter 2002, 2003). Soy producers in South America – dubbed by some 'agro-bandits' (Rocha 2005) – became a collective pariah for activists and concerned citizens worldwide. Boycotts and sanctions, particularly from Europe, threatened some of the soy industry's most important markets.

In response to this threat, the World Wildlife Fund for Nature (WWF) invited soy producers and other stakeholders to consider developing a certification scheme for soy – a non-state, voluntary governance mechanism – that would provide third-party assurance to buyers that the crop was produced 'sustainably'. The process that ultimately led to a standard for 'responsible' soy production originated in 2006 under the name the Roundtable on Sustainable Soy (RTSS). Later, at the behest of several non-governmental organizations (NGOs), concerned about how vague and ubiquitous the term 'sustainable' had become, the initiative was renamed the Roundtable on Responsible Soy (RTRS). By 2013, RTRS membership stood at 162, with representation from industry, trade and finance (49 percent) such as the International Finance Corporation, Shell, Cargill and Monsanto; soy producers (18 percent), including farmer associations and cooperatives; civil society (11 percent), including Conservation International and the Nature Conservancy; and observers (22 percent).[1]

[1] A complete list can be found on the RTRS webpage (www.responsiblesoy.org) under 'membership'.

In 2010, after four years of discussion and debate both within and outside of the official proceedings, the RTRS standard for 'responsible' soy production, version 1.0, was released. The RTRS standard includes 21 criteria that reflect five main principles: legal compliance and good business practice, responsible labor conditions, responsible community relations, environmental responsibility and good agricultural practice. Proponents say that the standard will improve the environmental and social performance of soy production in Paraguay and other countries where the soy industry has become controversial, including Brazil, Bolivia and Argentina (WWF 2010). Opponents of the RTRS say that the creation of such a standard only provides a 'greenwash' for the industry and detracts from the fundamental ways in which the soy industry confounds environmental, social and developmental goals (ASEED 2006). In 2013, the RTRS won the 'greenwash' award, organized by the Belgian climate campaign group Climaxi, who claimed that 'soy producers hardly have to make any changes at all to obtain a "responsible" label' (Corporate Europe Observatory 2012). Some opponents draw attention to what they call the 'development model for soy' in Paraguay as 'irresponsible and unsustainable', based on industrial monocultures and export commodities – a model that physically, socially and economically excludes Paraguayans from rural society (ASEED 2006). In preparation for the first RTRS gathering in Paraguay in 2006, dozens of NGOs signed a declaration denouncing the soy model and its inherent propagation by the RTRS (ASEED 2006).

By 2012, only one soy producer in Paraguay had become an RTRS-certified grower. Another was certified in 2013, and together in that year they produced and certified 'responsible soy' on around 22,000 hectares (RTRS 2014). Uptake of certification has been limited in Paraguay, yet the RTRS has influenced debate and policy by catalyzing and institutionalizing a discourse of responsible soy, a discourse that prior to this would not have been imaginable. Broadly speaking, the RTRS has facilitated the emergence of a discourse of responsible soy that is deployed both inside *and outside* the formal roundtable proceedings and certification processes.

The next section is committed to critically examining five claims featured in this discourse about how 'more (responsible) soy on fewer farms' can contribute to national environmental and socio-economic developmental goals. The point here is not necessarily to disprove the claims, but to show that they are not unequivocal, and that alternative approaches – approaches that embody a commitment to equity – have potential to achieve sustainability in a broader sense.

Challenging five claims about responsible soy

Decoupling soy from deforestation

Paraguay's Interior Atlantic Forest is considered among the world's most bio-diverse and threatened eco-regions, containing a minimum of 1500 species of vascular plants (comprising more than 0.5 percent of the world's species) and having lost at least 70 percent of its original habitat (Myers 2000; Conservation International 2006). Landsat data showed that the clearing of the country's forested area accelerated between 1990 and 2000, reducing forest cover to 176,741 km² from 202,202 km² over the decade (Huang et al. 2009). Deforestation in Paraguay accelerated further between 2001 and 2010, a period over which satellite data showed a net loss of woody vegetation (including forests and shrubs) of 42,778 km² (Aide et al. 2013). Agriculture has been associated closely with deforestation in Paraguay's Interior Atlantic Forests, as in the Brazilian Amazon, with analytical attention shifting from the role of small-scale agriculture between 1960 and 1980 to large-scale agriculture in the 1980s (Richards 2011). Along with cattle production, soy production has been singled out as a particularly aggressive driver of deforestation (Fearnside 2001; Altieri and Pengue 2005). Forested area in Canindeyú, one of Paraguay's largest soy-producing departments, reportedly shrank from 8262.97 km² in the 1990s to 4903.05 km² in the 2000s – a change in forest cover of 40.66 percent in roughly 10 years (Huang et al. 2009). By 2001, the Atlantic Forest had been declared one of the earth's 25 official 'biodiversity hotspots' (to which nine were added in February 2005, announced by Nature: '9 More Crisis Areas for Biodiversity'; Ebert 2005).

In response to agriculturally driven, rapid deforestation in Paraguay, the WWF drafted a proposal in 2004 for a forest conversion moratorium, and rallied support for greater enforcement of the existing '25 percent law'. The 'zero deforestation law', as the forest conversion moratorium came to be known, was passed by the Paraguayan congress in 2004 and 'prohibited the transformation or conversion of forested areas in the Eastern region … ' (WWF 2011, 18); it was brought into effect for a period of two years, and, due to its reported success, was extended twice. The '25 percent law', legislated in 1973, requires landowners with parcels over 50 hectares to maintain 25 percent of the land as forest. The law, however, is often ignored and rarely and sporadically enforced (Hutchinson 2008). In conjunction with their efforts to articulate and implement the zero deforestation law, the WWF has been working with landowners to increase voluntary compliance with the 25 percent law, even developing a tradable rights mechanism where farmers with forest 'credits' (more than 25 percent forest on their land) can sell these credits to farmers with forest 'debt' (Hutchinson 2008). Through the zero deforestation and 25 percent laws policy-makers aimed to conserve forests and biodiversity through 'land sparing', an approach that promotes the physical separation of intensified agriculture production from conservation areas (Phalan et al. 2011).

In the years following their work to implement the zero deforestation and 25 percent laws, the WWF claimed a 90 percent decline in deforestation, against a 2002 baseline (WWF 2011). This corresponds with findings from Brazil where the 'soy moratorium' – a consumer boycott against soy grown in the Amazon, on areas deforested after 2006 – was credited with massive reductions in deforestation. These successes, widely understood as the outcomes of effective communication and increased compliance with policies, were understood to signal the 'decoupling' of soy production from deforestation (Rudorff et al. 2011; WWF 2011; Macedo et al. 2012; Nepstad et al. 2014). Contributing to an emergent optimism about the new role of large-scale producers as forest *protectors* was that these dramatic declines in deforestation coincided with *increases* in soy production (MAG 2010).

One assumption of the 'land sparing' approach embodied by the 25 percent law, and the forest conversion and soy moratoriums, is that increasing productivity on existing agricultural land through intensification can help farmers to avoid extending production by clearing forested lands (Macedo et al. 2012). Critics have rallied against this logic of land sparing, arguing that research has furnished little evidence that intensification can 'spare' substantial quantities of land for conservation over the longer term (Ewers et al. 2009; Byerlee et al. 2014). Such increases in agricultural yields are more likely to attract greater attention and investment as profitability increases (Angelsen and Kaimowitz 1999; Ewers et al. 2009; Angelsen 2010). The impact of other factors on deforestation is also overlooked (Richards et al. 2012), and may be erroneously attributed to land-sparing policies, inflating their perceived effectiveness. For example, market prices and exchange rates have a role in slowing rates of soy expansion. High market prices for soy stimulate production, and because 'most agricultural commodities are traded in US dollars, a favorable price depends not only on the market price of the commodity, but also on the exchange rate of the domestic currency against the dollar' (Richards et al. 2012, 456). Comparing outcomes to counterfactual simulations, Richards et al. estimate that if the Paraguayan Guarani had been pegged to 1993 exchange rates, soybean production would have held at half of 2012 levels (Richards et al. 2012).

Others join the cautiously optimistic, regarding the potential of land sparing for decoupling soy production from deforestation over the longer term. The Union of Concerned Scientists conceded that at least part of the continued growth in soy production after the moratorium was due to increased productivity rather than expansion (Boucher 2011). Production gains, they argued, however, would be limited, and would slow once intensification had brought levels of soy production in line with the best performers. This closing of the 'yield gap' can be seen by comparing the productivity of US farms – where the world's highest productivity levels are found – against that of Paraguayan farms between 1991 and 2008 (see Table 3). During this period, Paraguayan soy productivity reached US levels, with productivity gains of 0.8 tons/hectare, nearly twice the productivity gain of US soy. This supports the idea that productivity gains are more forthcoming as countries are 'catching up', and that gains may be more difficult to achieve and more slowly realized after the yield gap has been closed (Boucher 2011). The dramatic benefits of intensification are thus likely to have a time-limited effect. Once the yield gap is closed, expansion becomes, once again, the primary means of raising productivity (Licker et al. 2010).

The veracity of the decoupling phenomenon has been shaded by research that goes beyond the immediate and direct relationship between soy and deforestation to look at indirect land-use change. Leakage, for example, can shift deforestation around a country or around the globe, as agricultural expansion moves away from countries with relatively strict regulation towards more vulnerable countries without (Meyfroidt et al. 2010; Lambin and Meyfroidt 2011). Detecting indirect land-use change on a wide geographical scale can be difficult due to the 'conceptual difficulties in linking distal land cover drivers to the point of impact' (Arima et al. 2011, 1). Research suggests, however, that agricultural behaviors in one place can drive significant deforestation in places hundreds of miles away (Arima et al. 2011). Leakage may also occur between different land uses, such as when land is cleared for cattle grazing, after soy production expands into pasture (Barona et al. 2010; Arima et al. 2011). Land-use changes may also be attributable to agricultural expansion away from protected areas into ecosystems that are not protected by existing policy, such as the displacement of soy production from tropical forests to savannahs (Merry Baker and Small 2005). Under all of these scenarios of indirect land-use change, soy production is distanced from deforestation, but is still part of the broad causal relationship.

Another assumption of the land-sparing approach is that conservation activities and production activities need to be spatially separated. In contrast, proponents of 'land sharing' suggest that combining small-scale agriculture with efforts to maintain biodiversity can

Table 3. Average soy production (tons/hectare).

	1991	2008
USA	2.3	2.7
Paraguay	1.9	2.7

Source: Soystats (http://www.soystats.com/2012/page_08.htm; and, author calculations using are of soy cropland and production data (MAG, 2010).

be effective in terms of farm productivity, environmental outcomes *and* social justice (Perfecto and Vandermeer 2010). The paradigm of 'land sharing' suggests that food production and nature conservation can converge on-farm and that productivity does not have to be compromised (Perfecto and Vandermeer 2010). National and regional policy actors, however, including within the RTRS, have shown little interest in promoting combined productivity and conservation goals, particularly on small farms. In fact, national-level investment in small-scale farming was slashed by more than half between 2005 and 2009 (Itriago 2012). This hit Paraguayans hard, considering that nearly half of the population is counted as rural, that well over half of these rural inhabitants count agriculture among their livelihoods, and that rural poverty is estimated to exceed 50 percent (DGEEC 2012). These demographics suggest that integrating environmental and productive concerns on small farms would be more appropriate, and would more effectively address equity in rural Paraguay, than a sole focus on land sparing.

Policy actors such as environmental NGOs widely view increased compliance with the zero deforestation and 25 percent laws as an effective means of decoupling large-scale soy production (and even growth in the soy industry) from deforestation. Increased production through intensification and mechanization 'spare' land for conservation, turning the tables on environmental critiques of industrial agriculture, and creating environmental stewards out of large-scale soy farmers. But researchers have suggested various ways in which the impact of land sparing on deforestation is likely overestimated. Not surprisingly, the responsible soy discourse entirely overlooks the potential for achieving both productive and environmental goals on smaller landholdings.

Good agricultural practices

The second claim made within the responsible soy discourse is that the soy industry promotes 'good agricultural practices' that protect resources (water, riparian zones and soil quality), promote integrated pest management, control and manage agro-chemical use and enable the continued existence of multiple production systems. Practices that protect soil quality are of paramount importance because they ensure that land can be cultivated indefinitely, eliminating the need for land abandonment (due to degraded soils and diminishing returns) and resultant expansion into forested areas. Research has suggested that improved soil management practices may even increase agricultural output per hectare of land. Under regimes of soil maintenance, then, expansion occurs only when additional fields are dedicated to soy production; new land is not required to replace depleted land.

One such practice, used widely in Paraguay, is no-till cultivation, which reduces or eliminates the need for tillage or plowing and avoids the release of soil nutrients. A crop is planted as 'green cover', meaning it is grown, cut and left on the soil surface, and the soybean seed is then planted directly into this cover, with little or no soil disturbance. The increased uptake of no-till cultivation in North and South America has been shown to maintain and even enhance soil quality (Derpsch 1999; Huggins and Reganold 2008). No-till is called a 'revolution' by proponents (Huggins and Reganold 2008), because of its promise for making agriculture sustainable.

No-till cultivation, however, is not without controversy, particularly in South America, and specifically in Paraguay, because it requires increased herbicide use to control weeds in the absence of tillage (Elgert 2013). The Monsanto-engineered 'Round-up Ready' soy is genetically designed to resist the application of glyphosate (Round-up is its commercial name in English; in Spanish it is called 'Mata-todo', or 'Kill All'), which controls weeds by killing 'everything green' that is not genetically conditioned to resist. The combination of Round-up Ready soy and no-till agriculture increases the need for herbicides. Given the large scale of soy production, these herbicides are often applied using efficient but imprecise aerial crop-dusters. Herbicide leakage from soy farms to surrounding small-scale farms has resulted in tragedies ranging from loss of subsistence crops to illness and death of livestock and people. I interviewed three small-scale producers, living and farming along the perimeters of soy fields, who reported their entire subsistence crop eradicated by a neighbor's glyphosate application. Their stories referenced the case of 11-year-old Silvino Talavera, who died in January 2004 from exposure to agrochemicals which were routinely applied to the soy crops around the periphery of which Silvino lived with his family (Palau et al. 2010). With the support of NGOs and alliances, the 2004 acquittal of the two agribusiness owners responsible was appealed. In November 2006 they were found guilty of 'creating public risks' and committing homicide, each receiving a sentence of two years in prison.

Enthusiasm for agricultural practices that yield particular environmental benefits can overshadow potential trade-offs. No-tillage cultivation, used widely in large-scale soy production in Paraguay, is only 'sustainable' from a soil management perspective, as it can result in maintained and even improved soil quality. More contextualized research, however, can expose 'cracks' or contingencies in supposed 'win–win' scenarios, suggesting reasons why promoting sustainable agriculture should be conditional on alleviating unintended consequences.

The engine of national economic growth

The third claim of the responsible soy discourse is that large-scale soy is the engine of national economic growth, and the country's best chance for economic development. Indeed, soy has been high among Paraguay's most important exports over the past few decades, and alone accounted for more than three quarters of the nation's impressive 15 percent gross domestic product (GDP) growth in 2010 (MAG 2010). Despite being a flash point for conflict between peasants and large landholders in Paraguay, the soy industry is often portrayed by producer organizations, the media and government agencies as the engine of national economic growth and development, and even a source of national identity and pride, propelling the country into a prosperous and modern future. Alberto Yanosky, Director of Paraguayan environmental NGO Guyra Paraguay and supporter of the RTRS, commented in the WWF newsletter:

> Soy production is a leading activity in Paraguay's economy and contributes to national development. The RTRS and its members could make us all very proud of producing soy and remaining a soybean producing country with clear principles and criteria to make this production happen in a responsible way. (WWF 2006, 1)

In Paraguay (as elsewhere), however, GDP and national-level economic growth indicators say little about distribution. This is particularly significant in Paraguay, with one of the most inequitable distributions of wealth of South American countries. In 1999, 'the wealthiest 20% of the population accounts for 62.4% of incomes, and the 10% poorest just 0.7%' (Benegas 1999, 278–279). In 2008, the top 10 percent of the population earned 42.3 percent of Paraguay's total income, compared to the bottom 10 percent that earned 1.1 percent (UNDP 2009). The richest 10 percent earned 38.8 times more than the poorest (UNDP 2009). Compare this with Sweden (the world's most equal country) where the same proportion earned 6.2 times more, and neighboring Brazil where the difference was

11 times (UNDP 2009). The Gini coefficient for income inequality in Paraguay, where 0 represents perfect equality and 1 represents perfect inequality, was 0.52 (UNDP 2009).

Paraguay's population is highly rural and dependent on agriculture for subsistence and income (DGEEC 2012). Thus, land distribution and control over land and other productive resources is an important dimension of inequality. Indeed, research has shown that 'poverty among farm households in Paraguay is closely related to lack of access to land by many farmers' (López and Thomas 2000, 257; see also López and Valdés 2000). Data suggest that land is becoming more concentrated in Paraguay. For example, between 1991 and 2008, the total number of farms decreased, yet the number of large farms increased substantially by nearly 57 percent (Table 4). Similarly, the total land area accounted for by farms smaller than 100 hectares shrunk, while area covered by farms over 100 hectares increased by over 43 percent (Table 5).

The trend in Paraguay towards fewer, larger farms raises the question of the extent to which such farms generate labor opportunities. The agricultural sector in Paraguay provides the greatest promise for job creation because of the high rural population, the importance of agriculture for the Paraguayan economy, and the high levels of rural poverty (Berry 2010). Large-scale soy production, however, is capital-intensive and land-intensive, characterized by highly mechanized farms that offer little employment except to a handful of skilled farm managers and equipment operators (Masi and Ruiz Diaz 2010; Palau et al. 2010). Growth in the soy industry, therefore, 'is likely to have a net negative impact on labor demand ... the larger these sectors ... the worse are labor market indicators likely to be' (Masi and Ruiz Diaz 2010, 219). A case in point is a company called Desarrollo Agricola del Paraguay (DAP), a soy producer in the Paraguayan department of San Pedro. As of 2012, DAP cultivated over 13,700 hectares of soy on four separate properties, employing 'over 50 people ... in the administrative offices and agricultural units of production' (DAP 2012). This amounts to about one job per 275 hectares of land. In comparison, the 'family farm', averaging between 10 and 20 hectares (in Canindeyú, for example), can employ more than four or five people. Although the income for a few skilled individuals is likely to be much higher on mechanized soy farms than on the family farm, the outlook for job creation and absorption of Paraguay's rural labor force in the soy industry is dismal. Furthermore, the development of linkages and agro-industrial clusters in L

Another means by which soy production could contribute to national development goals and some level of wealth redistribution is through export taxes. Export taxes enable a government to raise general revenue, but in Paraguay, taxes on agricultural exports have remained extremely low (Itriago 2012; Guerena 2013). Accounting for 36 percent of the Paraguayan GDP, agricultural exports only account for 2.5 percent of the government's tax-take. The tax rate on soy exports remains a paltry 2.5 percent compared to the 35 percent export tax levied on soy exports in neighboring Argentina, where a proposed hike to 44 percent was contemplated by President Kirchner, but scrapped after industry-wide strikes and protests (Barrionuevo 2008a). In 2012, then-President Lugo's attempts to increase the export tax marginally to 6 percent were 'balked at' by soy producers, and met not with debate, but with flat-out refusal (Desantis and Cristaldo 2011a). One industry representative called the proposal to increase export taxes on soy, 'an absurd form of intervention', saying that 'Paraguay has great potential and could be a big food producer for the world, but this could slow the sector's development' (Desantis and Cristaldo 2011b).

The focus on the contribution of large-scale soy production to Paraguay's economic growth (but not to sufficient employment creation or Paraguay's tax base) supersedes discussion of the potential of small-scale productivity to support national growth to a comparable degree. This discussion came out of the shadows, albeit temporarily, in 2008, when

Table 4. Quantity of farms by size (Paraguay).

		Farm size (Hectares)						
	Total number of farms	<5	5–<10	10–<20	20–<50	50–<100	100–<500	500+
Total 2008	289, 666	118,003	66,218	57,735	22,866	6,879	10, 487	7478
Total 1991	307,221	122,750	66,605	66,223	31,519	7,577	7 782	4765
Change 1991–2008 (%)	–5.7	–3.9	–0.6	–12.8	–27.5	–9.2	34.8	56.9

Source: Censo Agropecuaria Nacional, 2008 (DCEA 2009).

Table 5. Land area of farms by size (Paraguay)(Hectares).

		Farm size						
	Total land area of farms	<5	5–<10	10–<20	20–<50	50–<100	100–<500	500+
Total 2008	32,527,075	238,013	416,702	685,381	620,016	459,555	230,0193	27,807,215
Total 1991	23,817,737	231,305	430,658	806,802	857,909	502,648	1,61,9203	19,369,213
Change 1991–2008 (%)	36.6	2.9	–3.2	–15.0	–27.7	–8.6	42.1	43.6

Source: Censo Agropecuaria Nacional, 2008 (DCEA 2009).

Joseph Stiglitz traveled to Paraguay to participate in then-newly elected Fernando Lugo's presidential inauguration. Stiglitz, a Nobel Laureate in economics and critic of neo-liberal economic policy, made a speech to a crowd including cattle ranchers and soy growers, advocating land reform. Drawing on the East Asian economic successes, in part credited to aggressive land reform and national programs for improved productivity (Kay 2002), Stiglitz said that improving access to land and other resources of the poor would lay the foundation for increased growth and equity (Setrini 2008). Others agree that small-scale agriculture must play a prominent role in any plan for equitable growth and poverty reduction in Paraguay (Berry 2010; Luis Galeano 2010).

Stiglitz's bold statement invokes one of the most powerful economic arguments for equitable landholdings – that farm size is inversely related to farm productivity (Sen 1975). Large agricultural enterprises have been defended in terms of their higher levels of productivity owing to capital investments, mechanization, rational farm management and access to economies of scale. But critics point out that low productivity on small farms is less a matter of *inherent* productive capacity than a policy environment that privileges large landholders, with little monitoring of compliance with environmental regulation, low taxes and cheap land. Galeano indicates, for example,

> it is important to emphasize that the relative success of each group depends not only on its own characteristics but also on factors partially or totally beyond its control, like access to land, the level and quality of support provided by external agents, the way markets function, etc. (2010, 109)

Significant contributions to the productivity and competitiveness of the small farm can be made by 'supportive social, economic and political contexts' (Luis Galeano 2010, 105).

Past performance suggests that small farmers can indeed realize productivity gains similar to those of large farmers, and become the collective basis for national growth. The efficiency of soy production, in Paraguay and globally, has increased since the early 1990s. The 1991 Agricultural Census showed that the average level of soy production for Paraguay was around 1.9 tons per hectare. This increased to around 2.7 tons per hectare in 2008 (Table 6). Under the assumption that large farms are more efficient, there is a temptation to attribute some of this increased yield to the increased domination of large producers in Paraguayan soy production. Based on the desegregated data from the 1991 and 2008 censuses, however, small farms appear to have similar levels of soy production per hectare as large farms in both periods. Indeed, other studies of agricultural production in Paraguay show similarly favorable results. Toledo analyzed data from the 2000–2001 National Household Survey, and found that they confirmed the inverse relationship between farm size and productivity (2010). He concluded: 'Agricultural policies that maximize the utilization of labor accompanied by the provision of technological support and access to capital *and, of course, land* improve the likelihood of growth with equity in the sector' (Toledo 2010, 93, emphasis added).

The ways in which policy environments facilitate growth in different agricultural sectors is not an accident but a deliberate political choice. Despite the feasibility of

Table 6. Soy production (Tons) per Hectare of cultivation by size of soy crop, 1991–2008.

	<1–<20	20–<50	50–<100	100–<1000	1000–<10000	10000+
2008	2.6	2.7	2.7	2.7	2.8	2.7
1991	1.9	1.9	1.8	1.9	1.9	1.8

Source: Calculated by author using area of soy cropland and production data from (MAG 2010).

small-scale agriculture as an engine of productivity and growth in Paraguay, there has been a stubbornly low level of public support for the small farmer. According to an analysis undertaken by Oxfam, support to family-based agriculture has accounted for a declining percentage of public expenditure, from 12 percent of GDP in 2005, to 5 percent in 2009 (Itriago 2012).

Coinciding with shrinking investment in small-scale agriculture is the remarkable ineffectiveness of attempts at agrarian reform in Paraguay. Such attempts have, historically and recently, been stalled or made impotent by a powerful landowning elite that wields significant control over the state (Nickson and Lambert 2002): 'the lack of reform reflects the absence of representatives of peasant interests in the political system' (Lambert 2000, 388). Even the so-called populist policies of land redistribution in the 1950s and 1960s were only superficially devised in the interests of the landless and land-poor. The national institution charged with land redistribution, the Instituto de Bienestar Rural (IBR), was established at the demand of, and in the interests of, large landholders in the central and eastern parts of the country in order to protect their land holdings, and in the interests of expanding state presence and authority into the northern and western regions (Kleinpenning and Zoomers 1988; Nagel 1999). Large landholders were looking to mitigate social tension and maintain respect for their own interests in light of the threat posed by increasingly dissatisfied peasant groups (Kleinpenning and Zoomers 1988), groups that were becoming increasingly organized through the Christian Agrarian Leagues (Ligas Agrarias) against inequality and the systemic patronage of political and economic elites (Nagel 1999).

Ultimately, the IBR processes for implementing the land reform policies were incongruent with redistributive aims, being largely ineffective, cumbersome or corrupted. Much of the land slated for reform was gifted to associates and supporters of the government (Hetherington 2011). For example, Blas Riquelme, one of the richest men in Paraguay until his death in 2012, and political ally to Dictator Stroessner in the 1960s and 1970s, is reported by the Paraguayan media to have illegally acquired, in 1969, 50,000 hectares of land that was destined for distribution to peasants as part of land reform (E'a: Periódico de Interpretación y Análisis 2012). Consistent with this and other accounts of 'tierra malhabida' ('ill-gotten land'; Hetherington 2012), Nagel argues that the reform policies were largely a façade – rhetorical political tools – with few economic and redistributive effects (1999).

Land reform returned to prominence in public debate in 2008. Fernando Lugo won the presidential election on a platform of pro-poor policies and land reform, raising peasants' expectations that historic inequalities and injustices in land distribution would be redressed. But once again, promises of reform fell far short of expectations as President Lugo faced a number of challenges in implementation. The notion of land reform was opposed, predictably, by the landowning elite. Opposition also emanated from the Paraguayan congress, where the opposition party maintained a majority, and effectively managed to shut down any move towards redistribution. The Brazilian government issued strong warnings and threatened sanctions against any appropriation of Brazilian-owned land in Paraguay. Lugo's attempts to raise government revenues to fund land acquisitions were also thwarted. The institutionalization of the country's first ever income tax was delayed until 2013, and efforts to renegotiate the terms of Brazil's purchase of energy from Paraguay's share of the Itaipu dam, jointly owned but highly favorable towards Brazil, were also unsuccessful.

Most of Paraguay's 6.8 million people receive few benefits from the country's soy boom. While the soy industry generates tremendous wealth, and is commonly celebrated as Paraguay's engine of growth, this wealth remains in very few hands, carrying on a Paraguayan legacy of inequality and concentration of wealth and resources. Highly mechanized agriculture creates little employment. Extremely low rates of taxation limit benefits and development potential more broadly beyond the soy industry. Land ownership is increasingly concentrated, as stubborn and powerful resistance to land reform persists. The responsible soy discourse distinctly lacks attention to mechanisms that would increase more generalized benefits of large-scale soy to society such as taxation, or to rethinking the need for such large-scale production and supporting broader participation in soy production through agrarian reform.

Feeding the world's growing population

The fourth claim of the responsible soy discourse draws on a well-worn assertion about the complementary relationship between industrial agriculture, intensified production and global food security: that industrial agriculture equals greater food production, and more food equals more food security. Indeed, proponents claim that the large-scale soy farming strengthens global food security by satisfying the basic needs of a growing global population. For example, showing his support for South American soy producers during his keynote presentation at the 2005 RTRS meeting in Buenos Aires, Roberto Peiretti (President-elect of the American Confederation of Farmers Organization for a Sustainable Agriculture, 2002–2006) exclaimed: 'We should keep moving forward and we should not rest until the complete elimination of hunger is achieved in the entire world!!!!' (Peiretti 2005). The link between soy production and global food security is based on the sheer production of food, as Roger Beachy (Director of the US National Institute for Food and Agriculture, 2009–2011) says: 'Agriculture faces a serious challenge as it strives to produce food for a global population expected to reach 9 billion by 2050' (USDA 2011). But, Beachy adds, soy occupies a privileged place in this link: 'Today soybeans are the largest source of protein and the second largest source of vegetable oil in the world, so improving soybean production has important implications for food security' (USDA 2011).

The 'food availability thesis' – the notion that food security is dependent mainly on food production – is still widely deployed in support of large-scale agriculture, but has been undermined in the development economics literature for decades (Sen 1981). Great variation exists in the degrees to which food security – and insecurity – is experienced across populations, outdating policy discussions of food security that mainly revolve around increased production and productivity (Sen 1981; Jenkins and Scanlan 2001). Today, for example, food insecurity exists despite production levels that are more than sufficient to meet global needs (FAO 2011). Food security, rather, is determined by a host of contextually specific social, economic and political factors that determine both patterns of demand (Cohen 1995) and access (Sen 1981). Access variables include employment opportunities, education and skills, income, income support programs, food subsidies, access to land and agricultural inputs, and access to markets. Hunger, in short, 'is not so much linked to the quantity of food that is globally produced but to poverty' (Tscharntke et al. 2012).

The soy industry's role in producing more food for a growing global population is likely little comfort to Paraguay's small-scale farmers who have a small and decreasing share in the country's soy boom, and as food security is effectively shifted from producer countries to consumer countries. The soy industry is overwhelmingly concentrated on the export market, with negligible production for domestic consumption. The majority of exports go from land-locked Paraguay, where there is little industrial capacity for processing, to Brazil and Argentina for processing and/or shipping. Final destinations for Paraguayan soy include China and Europe, where much of the import is used for animal feed and thus contributes to 'meat security' in those countries. The FAO has argued that such a dominant focus on export crops can undermine local food security by replacing subsistence and food crops for domestic consumption:

> 90% of the value of exports in 1995 relate to the agricultural and ranching sector. The agro-export model has favored monocultures, principally of cotton on reduced holdings of small producers and of soy on farms at the industrial level. However, this agro-export model has been responsible for reducing agricultural diversification and reduction of subsistence crops of small producers. (FAO 2001)[2]

The displacement of subsistence crops with export crops can also mean that a greater proportion of basic food staples need to be imported from outside the country, potentially leading to higher food prices and lower food security (Altieri 2009). Export crops do not necessarily have negative impacts on food security. Indeed, greater engagement with cash crops for export is associated with greater food security for poor farmers through income generation, when marketing support and infrastructure allow access to markets for both buying food and selling crops. In Paraguay, however, we see that small producers are being increasingly excluded from soy production. In 1991, producers planting fewer than 20 hectares of soy accounted for 20 percent of all soy production; in 2008 they accounted for less than 4 percent (MAG 2010). This is consistent with a picture of Paraguay as increasingly a producer of soy on fewer, larger land holdings, with declining participation of small-scale producers in agriculture.

Food security has little to do with sheer food production, having much more to do with poverty and more generalized access to food via purchase or access to land and inputs to grow subsistence crops. Just as more wealth does not necessarily create a wealthier population, it is equally true that more food production does not necessarily translate to greater food security. Food security, like wealth, is distributed unequally and, thus, it is important to raise the question of whose food security is being strengthened by soy production and soy exports from Paraguay and other South American producers. Little of this production is consumed in producer countries, and, as discussed in the previous section, few benefits of the soy industry are converted to food entitlements for Paraguayans in the form of income-earning opportunities or tax-funded social programs. Soy, on the other hand, does strengthen food security and, more specifically, access to meat, for importer countries. Meanwhile, the association of the soy industry with low levels of job creation and the concentration of productive resources into fewer hands does not bode well for domestic food security.

Inclusive politics

The fifth and final claim analyzed in this paper is that non-state governance initiatives such as certification standards development and the accompanying discourse of corporate social responsibility have increasingly emphasized the role of stakeholder (as opposed to shareholder) in industry decision-making. Participatory local engagement is seen as a means of ensuring that, beyond corporate rhetoric, certification standards incorporate local

[2]Translated by the author from the following: 'El 90 percent del valor de las exportaciones de 1995 corresponden al sector agrícola y ganadero. El modelo agro exportador ha favorecido el monocultivo, principalmente de algodón en parcelas reducidas de pequeños agricultores, y de soja, en fincas de nivel empresarial. Sin embargo, este modelo agro exportador ha sido el responsable del desaliento de la diversificación agrícola y de la disminución de la producción de cultivos para autoconsumo de los pequeños agricultores'.

perspectives, interests and priorities. In addition to outcomes, thus, the processes of stakeholder engagement and participation that are pursued in the context of non-state governance initiatives are an important part of the overall assessments of such initiatives. This emphasis is aimed at increasing the democratic legitimacy of and public support for such initiatives in the absence of state regulation.

The RTRS was based on what was claimed to be a participatory process (Garcia-Lopez and Arizpe 2010; Elgert 2012). Discussions and debates on how the certification standards should take shape were open to not only those directly involved in the soy industry, but diverse stakeholders, including those representing 'local' interests. Indeed, RTRS support for stakeholder participation in standards development took many forms. Funds were set aside for developing country NGOs to attend the initial RTRS meetings. Participation has been built into the codes and norms for diverse representation in the institutional structures of the RTRS. Finally, throughout the standards development process, mechanisms were implemented to facilitate feedback from not only RTRS members, but the public at large.

The RTRS did not manage to achieve broad-based participation in the standards development process, not unlike other sustainable commodity production roundtables (Jones 2003; Bailis and Baka 2011; Silva-Castañeda 2012). In fact, closer examination of the RTRS case brings the potential for authentic stakeholder engagement and participation into question (Elgert 2011). Rather than joining in the RTRS debate, groups that opposed the RTRS formulated a coherent campaign against it. They held demonstrations outside the RTRS meetings, and various groups within this opposition have published reports featuring a refusal of even the conceptual possibility of producing soy 'sustainably' or 'responsibly' in the current socio-political environment that characterizes Paraguay and Brazil (ASEED 2006).

The RTRS proceedings gave rise to strong and often polemic protests in the wider public discourse about how the certification standards developed, but inclusive politics did not become a defining feature of the standards development process. Extreme socio-economic and power differences between stakeholders in the soy industry impaired the possibility of equal access to deliberation and decision-making. At one extreme, participants included representatives from some of the world's largest international corporations that direct and control soy production in South America. These included the Archer Daniels Midland Company, one of the largest agricultural processors in the world; the Andre Maggi Group, the world's largest private producer of soybeans; and Cargill, an international food producer and marketer with over USD 116 billion in sales in 2009. In February 2009, the biotechnology giants Monsanto and Syngenta were admitted as members to the RTRS. At the other extreme, the typical Paraguayan peasant has around 10 hectares of land (to which there may or may not be formal title) and earns less than USD 3000 per year. It is not difficult to see how such disparities, even if all stakeholders had been 'around the table', likely would have led to situations of intimidation and thus impacted on the quality of 'inclusive' debate.

The dominance of soy producers and large environmental organizations in decision-making processes was particularly evident in discussions around what knowledge was regarded as 'relevant' to standards development. Indeed, within the RTRS, problems associated with soy were framed as largely technical concerns, to be addressed through sustainable agriculture, environmental stewardship and compliance with labor laws. These technical criteria appear to be apolitical, pragmatic and uncontroversial (Elgert 2012). This emphasis distinctly marginalized more political concerns such as land distribution and inequality.

Recent events in the formal political sphere have cast an even a greater shadow on the relationship between Paraguayan soy interests and democratic politics more generally. On

22 June 2012, President Fernando Lugo was impeached on the grounds of 'poor performance of his duties', after the opposition party in congress issued a vague and troubling challenge to Lugo's ability to exercise or lead 'effective' governance. Jeremy Hobbs of Oxfam was not alone when he associated the impeachment with the soy boom, claiming Lugo to be 'the latest victim of Paraguay's "soy war"' (Hobbs 2012).

Lugo's democratic victory in 2008 'was hailed as a crucial step forward for democracy [in Paraguay]' (*The Economist* 2012). The victory ended the 60-year rule of the Colorado party, a period plagued by oppressive dictatorship, corruption, censorship and violence; a period which led to the second highest rates of inequality in South America, despite economic growth. Lugo won the election on a platform of agrarian reform and pro-poor policies – not universal political sympathies in Paraguay. Tensions between agro-industry, large farmers (particularly soy producers) and left-leaning Lugo have not been secret. One outspoken critic has been Blas N. Riquelme, previously mentioned as one of the richest landowners in Paraguay, and an ex-Senator in the Colorado government during the Stroessner dictatorship. Many Paraguayans closely associate Riquelme with the 'old political order'.

It was on land to which Riquelme claimed ownership that 17 people were massacred on 15 June 2012. The 2000-hectare property, known as 'Marina Cue', is located in the northeastern department of Canindeyú, near the town of Curuguaty. The deaths resulted from gunfire in a clash between civilians and police after Riquelme called in police to evict the peasant group accused of 'invading' the land. Such occupations are not an uncommon form of protest for the landless poor in Paraguay and Brazil. Though not the first clash of its kind in Paraguay or in this region, this tragic end was the most fatal in decades.

The Curuguaty massacre catalyzed the Presidential impeachment, as the largely Colorado opposition in congress claimed that it was proof that Lugo was unable to maintain order in the country. Instantly, the governments of Argentina and Brazil invoked provisions contained in the treaties of Mercosur, the South American trade bloc, that suspend any member assessed to have broken the rules of democratic governance. The impeachment was labeled a 'coup' by much of the news media. The Organization of the American States (OAS) – acting in accordance with its mandate to promote and consolidate representative democracy – sent a delegation to Paraguay on a fact-finding mission, and concluded that the impeachment bore formal legality, through provisions of the Paraguayan constitution. Observers, however, remained uncomfortable with the basis on which, and speed with which, the impeachment took place. The whole process lasted a mere 31 hours from beginning to end, scarcely offering Lugo a chance to muster a defense. In the weeks following the ousting of Lugo, Wikileaks released cables suggesting that even the US embassy knew about the opposition's plan for Lugo's fall, as early as 2009.

The extent to which Lugo's impeachment has embodied a 'rupture' of democracy in Paraguay is a matter of debate and perspective. The extent to which it is directly linked with the soy industry is unclear, and it is apparent that whatever the extent, there are other interests and power relations, involving other large businesses such as cattle ranching, involved. The soy industry, however, certainly represents power in Paraguay and its intimate link – both contemporary and historical – with land ownership. This casts serious doubt on the extent to which a Paraguayan democracy can be taken seriously, without some redress of land inequality.

Conclusion: challenging knowledge claims, considering alternatives

Sustainable commodity discourses are increasingly used to disassociate commodity production from environmental, economic and social problems in many developing countries where national and global authorities promote large-scale agriculture as a pathway towards development. This study has highlighted one such discourse, responsible soy, which has become a prominent, even dominant, lens through which many powerful policy actors understand the role of soy in Paraguay's environment, economy and society. But responsible soy is contested. Resistance comes from stakeholders who see the soy boom as facilitating the physical, economic and political exclusion of small-scale producers from the rural landscape. Indeed, given recent data on Paraguay's growing soy production, on the one hand, and the increasing size but shrinking number of soy farms, responsible soy must be considered in the context of concentrated landholdings and declining participation in the production of the country's top export. Such resistance, however, and alternatives to responsible soy that prioritize greater equity in Paraguay's rural development trajectory, has been marginalized in mainstream policy debates about sustainability and sustainable agriculture.

The resilience of the responsible soy discourse is made possible by the credibility and authority that is amassed as advocates promote knowledge claims – five of which I have analyzed in this paper – as facts. The veracity of these 'facts', however, is dependent on their association with the neoliberal values that are at the core of sustainable commodity discourses, including voluntary, private actions incentivized by a growing market for 'sustainable commodities'. This paper has presented empirical scrutiny of the knowledge claims along with the consideration of alternatives that are explicitly concerned with greater equity in Paraguayan society, that together provide a much stronger basis for sustainability policy than is offered by the responsible soy discourse. The decoupling of soy production from deforestation may be time-limited as the yield gap closes, or may hide the environmentally degrading effects of leakage to other ecosystems, other countries or other sectors. Recent research on 'land sharing' challenges the wisdom of 'land sparing' by showing how environmental and economic (production) goals can be achieved on the same land. The promotion of 'good agricultural practices' can involve trade-offs between positive environmental outcomes and negative social impacts. Pesticide use, for example, is an integral part of no-till farming, which promotes soil conservation at the expense of neighboring areas where water, soil, crops, livestock and people can become contaminated. The creation of wealth within the soy industry has been poorly distributed, remaining in the hands of an elite few because of inadequate job creation and the sector's meager contributions to the nation's tax base. Analyzing production data suggests that, given adequate institutional attention and resources, there is no reason why small-scale production could not make comparable contributions to economic growth. Claims about the social contributions of large-scale responsible soy, such as food security and political inclusion, are also contested. Many of Paraguay's soy exports find their way to Europe where they are used in animal feed. Rather than creating greater food security, soy production redistributes food security away from rural Paraguayans to northern consumers. Strategies to enhance political inclusion in the decision-making processes around responsible soy standards have had only limited success. Many poor, rural Paraguayans have remained outside of the RTRS discussions, rejecting even the possibility of 'responsible soy'. Such groups have questioned the possibility for participation of the poor and powerless when sitting alongside some of the world's most powerful corporate stakeholders. Likewise, when technical arguments about sustainable practices are privileged over political-historical discourses of inequality, there may be little chance for meaningful inclusion.

The coproduction of facts, values and power must be accounted for in policy that does not privilege powerful discourses to the exclusion of others. Support for knowledge claims often means support for certain values. The nature of these values must therefore be made explicit. As concerns about equity become featured in debates about sustainability, the responsible soy discourse (and others like it, that pertain to other commodities such as palm oil and biofuels) becomes less viable; more soy on fewer farms cannot be sustainable. Ultimately, any hope for sustainable soy in Paraguay will depend on a radical shift that highlights the distribution of land and resources as a key feature of environmental, economic and social goals.

Disclosure statement

No potential conflict of interest was reported by the author.

References

Adger, W.N., T.A. Benjaminsen, K. Brown, and H. Svarstad. 2001. Advancing a political ecology of global environmental discourses. *Development and Change* 32: 681–715.

Aide, T.M., M.L. Clark, H.R. Grau, D. López-Carr, M.A. Levy, D. Redo, and M. Muñiz. 2013. Deforestation and reforestation of Latin America and the Caribbean (2001–2010). *Biotropica* 45, no. 2: 262–271.

Altieri, M. 2009. The ecological impacts of large-scale agrofuel monoculture production systems in the americas. *Bulletin of Science, Technology and Society* 29, no. 3: 236–244.

Altieri, M., and M. Pengue. 2005. Roundup ready soybean in Latin America: A machine of hunger, deforestation and socio-ecological devastation. http://www.biosafety-info.net/article.php?aid=284 (accessed November 1, 2010).

Angelsen, A. 2010. Policies for reduced deforestation and their impact on agricultural production. *Proceedings of the National Academy of Science* 107: 19639–44.

Angelsen, A., and D. Kaimowitz. 1999. Rethinking the causes of deforestation: Lessons from economic models. *World Bank Research Obs.* 14: 73–98.

Arima, E.Y., P.D. Richards, R.T. Walker, and M.M. Caldas. 2011. Statistical confirmation of indirect land use change in the Brazilian Amazon. *Environmental Research Letters.* doi:10.1099/1748-9326/6/2/024010.

Arsel, M., and B. Büscher. 2012. Nature™ inc.: Changes and continuities in neoliberal conservation and market-based environmental policy. *Development and Change* 43, no. 1: 53–78.

ASEED. 2006. Declaration: The development model for soy in Paraguay: Irresponsible, unsustainable and anti-democratic. http://www.aseed.net/index.php?option=com_content&task=view&id=285&Itemid=107 (accessed September 6, 2007).

Bäckstrand, K., J. Khan, A. Kronsell, and E. Lovbrand. 2010. The promise of new modes of environmental governance. In *Environmental politics and deliberative democracy: Examining the promise of new modes of governance*, eds. K. Backstrand, J. Khan, A. Kronsell, and E. Lovbrand, 3–27. Cheltenham: Edward Elgar.

Bailis, R., and J. Baka. 2011. Constructing sustainable biofuels: Governance of the emerging biofuel economy. *Annals of the Association of American Geographers* 101, no. 4: 827–38.

Balch, O. 2007. Paraguay's soya barons. *The Guardian*,. www.theguardian.com/world/2007/nov/2026/paraguay-farming (accessed July 2008, 2013).

Baletti, B. 2014. Saving the Amazon? Sustainable soy and the new extractivism. *Environment and Planning A* 46, no. 1: 5–25.

Barona, E., N. Ramankutty, G. Hyman, and O.T. Coomes. 2010. The role of pasture and soybean in deforestation of the Brazilian Amazon. *Environmental Research Letters* 5, no. 1: 1–9.

Barrionuevo, A. 2008a. Argentina blocks farm export tax. *New York Times*, July 18. http://www.nytimes.com/2008/07/18/world/americas/18argentina.html?_r=0.

Barrionuevo, A. 2008b. Left-leaning president's election gives hope to landless Paraguayans. *The New York Times.* http://www.nytimes.com/2008/2010/2014/world/americas/2014paraguay.html?pagewanted=print&module=Search&mabReward=relbias%2003Ar%2002C%2007B%2222%2022%2003A%2022RI%2003A2014%2022%2007D.

Benegas, G. 1999. Análisis de la coyuntura económica. In *Derechos humanos en Paraguay 1999*, 273. Asunción: CODEHUPY.

Berry, A. 2010. The role of agriculture. In *Losing ground in the employment challenge: The case of Paraguay*, ed. A. Berry, 61–84. New Brunswick and London: Transaction.

Besky, S. 2008. Can a plantation be fair? Paradoxes and possibilities in fair trade darjeeling tea certification. *Anthropology of Work Review* 29, no. 1: 1–9.
Bloomfield, M.J. 2012. Is forest certification a hegemonic force? The FSC and its challengers. *The Journal of Environment and Development* 21, no. 4: 391–413.
Borras, S.M. Jr, C. Kay, S. Gómez, and J. Wilkinson. 2012. Land grabbing and global capitalist accumulation: Key features in Latin America. *Canadian Journal of Development Studies/ Revue Canadienne d'Études du Développement* 33, no. 4: 402–16.
Boucher, D. 2011. *Soybeans the root of the problem: What's driving tropical deforestation today?* Cambridge: Union of Concerned Scientists.
Brigham, A.M. 2004. *Just hungry? Analyzing the relationship between land concentration and food insecurity.* phd thesis., Trondheim, Norway: Norwegian University of Science and Technology, Department of sociology and political science.
Brosius, J.P. 1999. Green dot, pink hearts: Displacing politics from the Malaysian rain forest. *American Anthropologist* 101, no. 1: 36–57.
Bryant, R.L., and S. Bailey. 1997. *Third world political ecology*. London: Routledge.
Byerlee, D., J. Stevenson, and N. Villoria. 2014. Does intensification slow crop land expansion or encourage deforestation. *Global food security*. http://dx.doi.org/10.1016/j.gfs.2014.04.001.
Carter, M.R., B.L. Barham, and D. Mesbah. 1996. Agricultural export booms and the rural poor in Chile, Guatemala, and Paraguay. *Latin American Research Review* 31, no. 1: 33–65.
Castree, N. 2014. *Making sense of nature*. London and New York: Routledge.
Cohen, J. 1995. Population growth and earth's human carrying capacity. *Science* 269, no. 5222: 341–46.
Conservation International. 2006. Biodiversity hotspots. http://www.biodiversityhotspots.org/xp/ Hotspots/hotspotsScience/hotspots_defined.xml.
Corporate Europe Observatory. 2012. 'Responsible soy' nominated for Belgian greenwash award.
DAP. 2012. http://www.dap.com.py/index.php (accessed August 3, 2012).
Davis, D.K. 2007. *Resurrecting the granary of Rome: Envrionmental history and French colonial expansion in North Africa*. Athens, OH: Ohio University Press.
DCEA. 2009. *Censo agropecuaria nacional 2008*. San Lorenzo, Paraguay: Dirección de Censo y Estadística Agropecuaria.
DCEA/MAG. 2012. *Análisis del comportamiento de rubros agrícolas*. San Lorenzo, Paraguay: MAG.
Derpsch, R. 1999. *Frontiers in conservation tillage and advances in conservation practice*. Paper presented at the 10th ISCO Conference, West Lafayette In.
Desantis, D., and M. Cristaldo. 2011a. Paraguay wants 6 pct soy export tax, farmers balk. *Reuters*, August 25. http://uk.reuters.com/article/2011/08/25/paraguay-soy-tax-idUKN1E77O0HS20110825.
Desantis, D., and M. Cristaldo. 2011b. Paraguay wants 6 pct soy export tax; farmers balk. http://uk.reuters.com/article/2011/08/25/paraguay-soy-tax-idUKN1E77O0HS20110825 (accessed August 1, 2012).
DGEEC. 2012. *Compendio estadístico 2010*. Fernando de la Mora, Paraguay: Dirección General de Estadística Encuestas y Censos.
Dros, J.M. 2004. *Managing the soy boom: Two scenarios of soy production*. Amsterdam: AIDEnvironment.
E'a: Periódico de Interpretación y Análisis. 2012. Blas n. Riquelme y las tierras malhabidas de curuguaty. http://ea.com.py/blas-n-riquelme-y-las-tierras-malhabidas-de-curuguaty/ (accessed June 25, 2012).
Ebert, J. 2005. Nine more crisis areas for biodiversity. *Nature*. doi:10.1038/news050131-10.
Elgert, L. 2009. Politicizing sustainable development; the co-production of globalized evidence-based policy. *Critical Policy Studies* 3, no. 3–4: 375–90.
Elgert, L. 2011. Certified community development? Local participation in corporate governance. In *Deliberations in community development: Balancing on the edge*, eds. J.P.R. Rothe, L.J.C. Carroll, and D. Ozegovic, 277–88. Hauppauge: Nova Science Publishers.
Elgert, L. 2012. Certified discourse? The politics of developing soy certification standards. *Geoforum* 43, no. 2: 295–304.
Elgert, L. 2013. Hard facts and software: The coproduction of indicators in a land-use planning model. *Environmental Values* 22, no. 6: 765–86.
Ewers, RM., J.P.W. Scharlemann, A. Balmford, and R.E. Green. 2009. Do increases in agricultural yield spare land for nature?. *Global Change* 15: 1716–26.

Fairhead, J., and M. Leach. 1996. *Misreading the African landscape: Society and ecology in a forest-savanna mosaic.* Cambridge: Cambridge University Press.

FAO. 2001. *Perfiles nutricionales por paises: Paraguay.* Rome: FAO.

FAO. 2011. *The state of food insecurity of the world: How does international price volatility affect domestic economies and food security?* Rome: FAO.

Fearnside, P.M. 2001. *Soybean cultivation as a threat to the environment in Brazil. Environmental Conservation* 28, no. 1: 23–38.

Feindt, P.H., and A. Oels. 2005. Does discourse matter? Discourse analysis in environmental policy making. *Journal of Environmental Policy and Planning* 7, no. 3: 161–73.

Fischer, F. 2009. *Democracy and expertise: Reorienting policy inquiry.* New York: Oxford University Press.

Forsyth, T. 2011. Expertise needs tranparency not blind trust: A deliberative approach to integrating science and social participation. *Critical Policy Studies* 5, no. 3: 317–22.

Galeano, L. 2010. Potential and future of Paraguayan family farming. In *Losing ground in the employment challenge: The case of Paraguay*, ed. A. Berry, 101–22. New Brunswick and London: Transaction.

Galeano, L. 2012. Paraguay and the expansion of Brazilian and Argentinian agribusiness frontiers. *Canadian Journal of Development Studies* 33, no. 4: 458–70.

Garcia-Lopez, G.A., and N. Arizpe. 2010. Participatory processes in the soy conflicts in Paraguay and Argentina. *Ecological Economics* 70, no. 2: 196–206.

Gieryn, T. 1999. *Cultural boundaries of science: Credibility on the line.* Chicago: University of Chicago Press.

Goldman, M., and Rachel A. Schurman. 2000. Closing the "great divide": New social theory on society and nature. *Annual Review of Sociology* 26: 563–84.

Grajales, J. 2011. The rifle and the title: Paramilitary violence, land grab and land control in Colombia. *Journal of Peasant Studies* 38, no. 4: 771–92.

Guerena, A. 2013. *The soy mirage: The limits of corporate social responsibility: The case of the company Desarrollo Agrícola del Paraguay.* Oxfam.

Guthman, J. 2007. The polanyian way? Voluntary food labels as neoliberal governance. *Antipode* 39, no. 3: 456–78.

Hajer, M. 1995. *The politics of environmental discourse.* Oxford: Clarendon Press.

Hajer, M., and W. Versteeg. 2005. A decade of discourse analysis of environmental politics: Achievements, challenges, perspectives. *Journal of Environmental Policy and Planning* 7, no. 3: 175–84.

Hetherington, K. 2011. *Guerrilla auditors: The politics of transparency in neoliberal Paraguay.* Duke University Press.

Hetherington, K. 2012. 'Tierra malhabida' y el engaño de la institucionalidad. *Revista Novapolis*, 5 (Abril-Octubre), 31–54.

Hetherington, K. 2013. Beans before the law: Knowledge practices, responsibility, and the Paraguayan soy boom. *Cultural Anthropology* 28, no. 1: 65–85.

Heynen, N., J. McCarthy, S. Prudham, and P. Robbins, eds. 2007. *Neoliberal environments: False promises and unnatural consequences.* Routledge.

Higgins, V., J. Dibden, and C. Cocklin. 2008. Neoliberalism and natural resource management: Agri-environmental standards and the governing of farming practices. *Geoforum* 39, no. 5: 1776–85.

Hobbs, J. 2012. Paraguay's destructive soy boom. *New York Times.* http://www.nytimes.com/2012/07/03/opinion/paraguays-destructive-soy-boom.html.

Howarth, D.R., and S. Griggs. 2012. Poststructuralist policy analysis: Discourse hegemony and critical explanation. In *The arugmentative turn in policy analysis*, eds. F. Fischer and J. Forester, 305–42. Durham and London: Duke University Press.

Huang, C., S. Kim, K. Song, J.R.G. Townshend, P. Davis, A. Altstatt, and J. Musinsky. 2009. Assessment of Paraguay's forest cover change using landsat observations. *Global and Planetary Change* 67: 1–12.

Huggins, D.R., and J.P. Reganold. 2008. No till: The quiet revolution. *Scientific American* 299, no 1: 70–77.

Hutchinson, S. 2008. Innovation in conservation: The case of Paraguay's atlantic forests. www.wwf.org.uk (accessed July 6, 2014).

Ingram, M., H. Ingram, and R. Lejano. 2014. What's the story? Creating and sustaining environmental networks. *Environmental Politics* 23, no. 6: 984–1002.

Itriago, D. 2012. *Taxation in Paraguay: Marginalization of small-scale farming*. Oxfam. http://www.oxfam.org/en/policy/taxation-paraguay.

Jasanoff, S. 2004. The idiom of co-production. In *States of knowledge: The co-production of science and social order*, ed. Jasanoff, 1–12. London: Routledge.

Jasanoff, S., and M.L. Martello. 2004. Earthly politics: Local and global in environmental governance. In *Politics, science, and the environment*, vol. viii, 356. Cambridge, Mass: MIT Press.

Jasanoff, S., and B. Wynne. 1998. Science and decisionmaking. In *Human choice and climate change, vol.1: The societal framework*, eds. S. Rayner and E. Malone, 1–87. Batelle.

Jenkins, J.C., and S.J. Scanlan. 2001. Food secruity in less developed countries, 1970 to 1990. *American Sociological Review* 66, no. 5: 718–44.

Johnson, A. 2012. Green governance or green grab? The biopolitics of the roundtable on sustainable palm oil (RSPO) and its processes in Ecuador. International Conference on Global Land Grabbing II.

Jones, H.C. 2003. Participation in FSC certified community forest management projects in the Brazilian Amazon. Paper presented at the International Conference on Rural Livelihoods, Forests and Biodiversity, 19–23 May, in Bonn, Germany.

Kay, C. 2002. Why East Asia overtook Latin America: Agrarian reform, industrialisation and development. *Third World Quarterly* 23, no. 6: 1073–02.

Kleinpenning, J.M.G., and E.B. Zoomers. 1988. Internal colonisation as policy instrument for changing a country's rural system: The example of Paraguay. *Journal of Economic and Social Geography* 79, no. 4: 257–65.

Klooster, D. 2010. Standardizing sustainable development? The Forest Stewardship Council's plantation policy review process as neoliberal environmental governance. *Geoforum* 41, no. 1: 117–29.

Lambert, P. 2000. A decade of electoral democracy: Continuity, change and crisis in Paraguay. *Bulletin of Latin American Research* 19: 379–96.

Lambin, E.F., and P. Meyfroidt. 2011. Global land use change, economic globalization, and the looming land scarcity. *Proceedings of the National Academy of Science* 108: 3465–72.

Latour, B. 1993. *We have never been modern*. Cambridge: Harvard University Press.

Licker, R., M. Johnston, J.A. Foley, C.J. Barford, C. Kucharik, C. Monfreda, and N. Ramankutty. 2010. Mind the gap: How do climate and agricultural management explain the 'yield gap' of croplands around the world?. *Global Ecology and Biogeography* 9: 769–82.

Litfin, K. 1994. *Ozone discourses: Science and politics in global environmental cooperation*. New York: Columbia University Press.

López, R., and T. Thomas. 2000. Rural poverty in Paraguay: The determinants of farm household income. In *Rural poverty in Latin America*, eds. R. López and A. Valdés, 244–57. Houndmills, Basingstoke, Hampshire and London: Macmillan Press Ltd.

López, R., and A. Valdés. 2000. Fighting rural poverty in Latin America: New evidence of the effects of education, demographics, and access to land. *Economic Development and Cultural Change* 49, no. 1: 197–211.

Lövbrand, E. 2007. Pure science or policy involvement? Ambiguous boundary-work for Swedish carbon cycle science. *Environmental Science and Policy* 10, no. 1: 39–47.

Macedo, M.N., R.S. DeFries, D.C. Morton, C.M. Stickler, G.L. Galford, and Y.E. Shimabukuro. 2012. Decoupling of deforestation and soy production in the southern Amazon during the late 2000s. *Proceedings of the National Academy of Sciences* 109, no. 4: 1341–46.

MAG. 2010. *Produccion de soja en el Paraguay: Zafra 2008–2010*. Asuncion: Ministerio de Agricultura y Ganaderia.

Masi, F., and F. Ruiz Diaz. 2010. The role of international trade in growth and employment generation in Paraguay. In *Losing ground in the employment challenge: The case of Paraguay*, ed. A. Berry, 215–39. New Brunswick and London: Transaction.

Meadowcroft, J. 2007. Who is in charge here? Governance for sustainable development in a complex world. *Journal of Environmental Policy and Planning* 9, no. 3–4: 299–314.

Merry Baker, M., and D. Small. 2005. Brazil's agricultural 'sucess' in the cerrado is a disaster. *Executive Intelligence Review* 32, no. 8: 20–23.

Meyfroidt, P., T.K. Rudel, and E.F. Lambin. 2010. Forest transitions, trade, and the global displacement of land use. *Proceedings of the National Academy of Science* 107: 20917–22.

Moog, S., A. Spicer, and S. Böhm. 2014. The politics of multi-stakeholder initiatives: The crisis of the Forest Stewardship Council. *Journal of Business Ethics*: 1–25.

Myers, N. 2000. Biodiversity hotspots for conservation priorities. *Nature* 403, no. 24: 853–58.

Nagel, B.Y. 1999. "Unleashing the fury": The cultural discourse of rural violence and land rights in Paraguay. *Comparative Studies in Society and History* 41, no. 1: 148–81.

Nepstad, D., D. McGrath, C. Stickler, A. Alencar, A. Azevedo, B. Swette, et al. 2014. Slowing Amazon deforestation through public policy and interventions in beef and soy supply chains. *Science* 344, no. 6188: 1118–23.

Neumann, R.P. 2003. The production of nature: Colonial recasting of the African landscape in Serengeti national park. In *Political ecology: An integrative approach to geography and environment-development studies*, eds. K.S. Zimmerer and T.J. Basset, 240–55. New York: The Guilford Press.

Nickson, A., and P. Lambert. 2002. State reform and the 'privatized state' in Paraguay. *Public Administration and Development* 22: 163–74.

Palau, T., G. Ortega, L. Rojas Villagra, M. Pereira, and R. Doughman. 2010. *Socioenvironmental impacts of soybean in Paraguay - 2010*. São Paulo and Asuncion: Repórter Brasil and Base Investigaciones Sociales.

Peiretti, R.A. 2005. The necessity of a sustainable agriculture: The global focus and the Mercosur-la Confederación de Asociaciones Americanas para la Producción de la Agricultura Sustentable (CAAPAS) view. Paper presented at the First Roundtable on Sustainable Soy Conference, Foz de Iguacu.

Perfecto, I., and J. Vandermeer. 2010. The agricultural matrix as an alternative to the land-sparing/agricultural intensification model: Facing the food and biodiversity crises. *Proceedings of the National Academy of Science* 107: 5786–91.

Phalan, B., M. Onial, A. Balmford, and R.E. Green. 2011. Reconciling food production and biodiversity conservation: Land sharing and land sparing compared. *Science* 333, no. 6047: 1289–91.

Richards, P.D. 2011. Soy, cotton, and the final Atlantic forest frontier. *The Professional Geographer* 63, no. 3: 343–63.

Richards, P.D., R.J. Myers, S.M. Swinton, and R.T. Walker. 2012. Exchange rates, soybean supply response, and deforestation in South America. *Global Environmental Change* 22: 454–62.

Robertson, M. 2007. Discovering price in all the wrong places: The work of commodity definition and price under neoliberal environmental policy. *Antipode* 39, no. 3: 500–26.

Rocha, J. 2005. Brazil declares forest havens after nun's killing. *The Guardian*, February 19. http://www.guardian.co.uk/environment/2005/feb/19/endangeredhabitats.activists.

Rohter, L. 2002. Amazon forest still burning despite the good intentions. *The New York Times*,. http://www.nytimes.com/2002/2008/2023/international/americas/2023BRAZ.html?pagewanted=print.

Rohter, L. 2003. Relentless foe of the Amazon jungle: Soybeans. *New York Times*, September 17.

RTRS. 2014. Volumenes y productores certificados. http://www.responsiblesoy.org/mercado/volumenes-y-productores-certificados/?lang=en (accessed July 8, 2014).

Rudorff, B.F.T., M. Adami, D.A. Aguiar, M.A. Moreira, M.P. Mello, L. Fabiani, and B.M. Pires. 2011. The soy moratorium in the Amazon biome monitored by remote sensing images. *Remote Sensing* 3: 185–202.

Sen, A. 1975. *Employment, technology and development*. Oxford: Clarendon Press.

Sen, A. 1981. *Poverty and famines: An essay on entitlement and deprivation*. Oxford: Oxford University Press.

Setrini, G. 2008. Stiglitz goes to Paraguay: Move over Chicago, a Cambridge boy's in town. http://upsidedownworld.org/main/paraguay-archives-44/1439-stiglitz-goes-to-paraguay-move-over-chicago-a-cambridge-boys-in-town (accessed July 5, 2009).

Silva-Castañeda, L. 2012. A forest of evidence: Third-party certification and multiple forms of proof – a case study of oil palm plantations in Indonesia. *Agriculture and Human Values* 29, no. 3: 361–70.

Spieldoch, A., and Murphy, S. 2009. Agricutlural land acquisitions: Implications for food security and poverty alleviation. In *Land grab? The race for the world's farmland*, eds. M. Kugelman and S. L. Levenstein, 39–54. Washington, DC: Woodrow Wilson International Center for Scholars.

Toledo, R. 2010. Farm size-productivity relationship in Paraguay's agricultural sector. In *Losing ground in the employment challenge: The case of Paraguay*, ed. A. Berry, 85–100. New Brunswick and London: Transaction.

The Economist. 2012. Out Lugo. http://www.economist.com/blogs/americasview/2012/06/paraguayan-politics (accessed August 3, 2012).

Tscharntke, T., Y. Clough, T.C. Wanger, L. Jackson, I. Motzke, I. Perfecto, and A. Whitbread. 2012. Global food security, biodiversity conservation and the future of agricultural intensification. *Biological Conservation* 151, no. 1: 53–59.

UNDP. 2009. *Human development report 2009. Overcoming barriers: Human mobility and development.* New York: United Nations Development Program.

USDA. 2011. USDA Awards Grant to Virginia Tech to Improve Soybean Production and Sustainability. http://www.csrees.usda.gov/newsroom/news/2011news/03281_vt_soybean.html (accessed August 8, 2012).

Walker, G., and E. Shove. 2007. Ambivalence, sustainability and the governance of socio-technical transitions. *Journal of Environmental Policy and Planning* 9, no. 3–4: 213–25.

WWF. 2006. Consensus reached on impacts of soy production at workshop. *Forest Conversion News*, 11June.

WWF. 2010. RTRS finalizes standards for responsible soy at annual meeting. *Forest Conversion News*, (26), 1.

WWF. 2011. Making a pact to tackle deforestation in Paraguay. *WWF report*, 1–18 March.

The moving frontiers of genetically modified soy production: shifts in land control in the Argentinian Chaco

Lucía Goldfarb and Gemma van der Haar

This article attempts to understand how control over land (power in practice) is built, achieved and contested in the context of land transfers involving pressures over possession rights in Santiago del Estero in northern Argentina. Here new forms of land control – due to expansion of the speculative, soy and cattle frontiers – are changing and involve new relationships while using novel mechanisms to gain and maintain control. The article adopts the notion of shifting 'frontiers of land control' as an analytical lens, following Lund and Peluso (2011). We can say that the process of soy and cattle expansion into the new frontiers happens through a group of different mechanisms which range from voluntary purchase to violent evictions. As shifts in land control in the frontier involve pressures on possession rights, we observe different mechanisms of control, mainly in the direction of dispossession and enclosure. The paper adds to the debate on 'land grabbing' by (a) showing how domestic investors operate to advance industrial agriculture and (b) showing how this frontier advances in a context of (rather unsecured) possession rights where rights are being shifted through transfers (sales, leases, evictions) and compensation mechanisms as well as conflict and judicial procedures.

Introduction

One of the most striking transformations of the Argentinean countryside over the last two decades is the rapid expansion of soy plantations. Around 20 years ago it would have been hard to imagine that soy would invade the margins of the roads and reach the green limits of rural towns and poor suburbs of larger Argentine cities, replacing millions of hectares of food crops and cattle. The type of landscape typical of the Pampas has now extended to the native forest and monte[1] of the Chaco region, long deemed unfit for soy production.

The expansion of soy plantations in Argentina can be seen as one of the most impressive cases of the current global rush for agricultural land, in terms of the number of hectares and resources involved but also in terms of its economic, environmental, and social consequences. In particular, the expansion into the Chaco region shows a process that is more directly transformative of social and environmental relationships than in any other region in the southern cone (Goldfarb and Zoomers 2013). The 'soy boom' in Argentina since the mid-1990s is favoured by the rise in international prices as an effect of a growing

[1]'Monte' popularly designates non-cultivated land with native vegetation – which may include a range of types and density of trees, bushes, and plants, but also designates a full eco-system – and is constitutive of peasant livelihoods in northern Argentina.

global demand for food, feed and energy crops (so-called flex crops), in combination with key technological innovations,[2] and in the context of flexible access to land through a private property regime.

Our intent in this article is to analyse how pressures deriving from the expansion of soy production have affected the Chaco region, in particular Santiago del Estero province. In the Chaco region, increasing commercial interest in land for soy production (and, in its wake, other agricultural and livestock production) has put enormous pressures especially on land under so-called 'possession'[3] status, causing displacement and dispossession of rural dwellers,[4] as well as social tensions and open conflict. Santiago del Estero is one of the provinces most affected by the expansion of soy production, in terms of land use change as well as in terms of number of investors. Most investors are successful soy producers from the so-called 'core region' who want to diversify, or expand, or to relocate cattle to spare more land for soy in the Pampas. Others are capitalized individuals as well as national and foreign companies seeking investment opportunities and fast-track/short-term benefits, through direct purchase or rent (arrendamiento) of land or through participation in investment funds or sowing pools (pools de siembra, see below).

Building on the notion of 'shifting frontiers of land control', as introduced by Peluso and Lund (2011), we analyse how the frontier of soy expansion advances in the Chaco region, through what mechanisms, and what dynamics of contestation this engenders. For this region, it is now evident that soy expansion mostly affects forest areas, indigenous communities, and small-scale farmers (campesinos) with vulnerable land rights. In Santiago del Estero province, approximately 75 percent of peasants do not hold formal titles to the land they live and work on, which they hold under possession status.[5] These small-scale producers are subject to a range of pressures and mechanisms of enclosure, which include (inequitable) legal disputes, violent evictions, intimidation, and killing of animals, as well as pollution of water and soil, and the clearance of forests. Peasants resist the advance of the agricultural frontier in this province by means of the struggle for the legal recognition of possession rights, as well as direct physical resistance to dispossession and enclosure.

The problematic and contested nature of the case of soy expansion in Argentina is often downplayed, and, internationally, it is generally portrayed by the World Bank (2010) and the media amongst others, as a rather 'successful' case of integration to the global commodities market, resulting in a win-win-win model for the private sector, the government and society at large. This portrayal suits the justification of new frontiers of capital expansion in terms of the discourse of 'agriculture for development' and 'Responsible Agricultural Investment' (World Bank 2010; De Schutter 2012; McMichael 2012). Earlier discourses on making 'empty lands' productive, have given way to more subtle arguments about

---

[2]The most significant of these innovations is genetically modified (GM) soy and associated technology packages, but also its unique combination with direct tillage techniques, storage bags and organizational innovation through sowing pools (pools de siembra).

[3]The Argentine Civil Code acknowledges possession rights when a person has occupied a [the right to property of a] land parcel [to whom has occupied it] for at least 20 years and made improvements to it.

[4]To the present MCNI (Movimiento Nacional Campesino Indígena) estimates that approximately 200,000 rural families have been forced away from their lands owing to the advance of soy in Argentina.

[5]Argentina has a special measure in the 2001 National Agricultural Census for farms without defined limits, which indirectly captures producers who have 'ownership' behaviour, but not property titles. These holdings include 'communal fields', occupants with and without permission, and indigenous communities, and therefore present a methodological challenge to estimate the number of hectares.

the inclusion of poor peasants with little hope for progress. We argue for the case of the Chaco that this operates to disguise forms of enclosure and dispossession.

Even the more critical perspectives about the global wave of investments in land seem to miss the dispossession involved in soy expansion in Argentina. Critical scholars have been hesitant to treat land control transformations in the case of Argentina – and South America, in general – as 'land grabbing'. The processes of land concentration and foreignization (FAO 2013) and trans-regionalization of investments (Borras et al. 2012) are not labelled as 'land grabbing' given that these transfers are mostly market-based, and involve relatively limited numbers of people displaced and dispossessed if compared to other continents. As we argue in this paper, this is only part of the picture. We will show that for Santiago del Estero province the market-based transactions and legal procedures conceal more problematic and violent dynamics of enclosure and dispossession.

With this article we intend to correct the picture that dispossession, taking place at the frontiers of soy expansion in Argentina, is only marginal or incidental. In our analysis we want to recover the notion of 'grab' – in the sense of land grab or control grab – for what is happening in the Chaco region. We do this by working with the notion of 'shifting frontiers of land control' that was introduced by Peluso and Lund (2011), showing how the advance of the frontier operates through specific mechanisms that are more contested than those known for soy expansion in Argentina's core region,[6] and include – next to (semi) voluntary transactions – violent evictions. This article attempts to understand how control over land (power in practice) is built, achieved and contested in the context of land transfers involving pressures on possession rights in Santiago del Estero in northern Argentina. As we will show, in this province the expansion of the speculative, agricultural and cattle frontier implies that forms of land control are changing, and we analyse the mechanisms involved to gain and maintain control.

Our argument is built on field-based research in Santiago del Estero. We have documented cases of land transfers between new claimants and existing occupants holding possession rights as well as cases of competing claims and conflict. We draw on 20 semi-structured interviews with qualified informants – privileged actors in and/or observers of the processes under study – from state institutions, civil society, peasant community members and leaders and private actors, as well as participant observation of community meetings, and recordings of negotiations between investors and community members. The interviewees were selected through a combination of purposive[7] and snowball sampling. The research also involves the analysis of documents, local press, secondary sources of information and review of previous research on the subject.[8]

This paper is organized as follows. First, we elaborate on the notion of 'shifting frontiers of land control' (Peluso and Lund 2011) and how we use it to analyse the mechanisms through which land control changes due to the pressures of soy expansion. Connecting the idea of 'control grab' introduced by Borras to the notion of 'new enclosures' (White et al. 2012) we hope to contribute to the understanding of how the frontiers of land control shift (Peluso and Lund 2011). We then outline the dynamics of the soy expansion in the Chaco region, linking structural factors of change to local mechanisms of land

[6]Expansion in the core region been more purely market-based – willing buyer/willing seller.
[7]Purposive sampling is described as a random selection of sampling units within the segment of the population with the most information on the characteristic of interest. The inherent bias of the method contributes to its efficiency.
[8]The field work was conducted by Lucía Goldfarb in 2011–2012.

control. Next, we take a closer look at land acquisition logics in Santiago del Estero province showing how land control is established through a combination of market-based transactions, legal procedures, semi-voluntary negotiations, and (threats of) eviction. Finally, we present conclusions recapitulating the main mechanisms through which land control is being established at the soy frontiers in the Chaco region, highlighting that dispossession and enclosure and take place through judicial, legal and 'market-based' negotiations in combination with (threats of) violence.

What happens at the soy frontier? Grabs, enclosures and shifts in land control

Our analysis of soy expansion in the Chaco builds on the notion of 'shifting frontiers of control' proposed by Peluso and Lund to advance theorizing on land grabbing. Peluso and Lund have defined 'frontiers of land control' as 'sites where authorities, sovereignties, and hegemonies of the recent past have been or are currently being challenged by new enclosures, territorializations, and property regimes' (Peluso and Lund 2011, 668). From this perspective, we want to understand the pressures exerted through the expansion of soy production and associated changes, notably the introduction of livestock, and we want to unravel how these pressures operate in practice, challenging existing access regimes and forging shifts in land control. In line with Peluso and Lund (2011, 669), we conceptualize the shifting frontier not as 'one' singular land grab but as a 'series of forces, contestations, that produce new conditions and facilitate shifts in both the facto and de jure land control'. In the case of soy expansion, we find that it triggers interlinked processes of further expansion, bringing in new expectations and pressures and multiplying its expansion power.

Frontiers imply expansion, pressure and continuous change. We are interested in the nature and the sense of processes of 'frontiering' associated with soy production. Physical metaphors of the frontier allow us to make a representation of how frontier zones produce and reproduce social realities. The raw material of frontiers is friction (Tsing 2004). Shifting frontiers are the result of opposing forces like waters pushing both sides of a dam; the forces in play in general are competing and do not develop in linear fashion. In this sense, it is relevant to understand frontier shifts in terms of drivers but also in terms of resistance, in terms of 'push' and 'push back'.

We explore what types of control shifts the soy frontier implies and how these are achieved. Ribot and Peluso (2003) have related access to the ability to gain, control and maintain access to a certain resource, beyond the issue of who holds the legal right to that resource, the formal 'property'. The idea of 'land control' provides a lens to observe the mechanisms through which access is achieved and challenged:

> By land control we mean practices that fix or consolidate forms of access, claiming, and exclusion for some time. Enclosure, territorialization, and legalization processes, as well as force and violence (or the threat of them), all serve to control land. The mechanisms of land control need not always align, nor proceed in a singular linear direction. They may be wielded in concert or competition with one another. (Peluso and Lund 2011, 668)

This framework suggests a range of mechanisms through which frontier shifts work. From the mechanisms proposed by Peluso and Lund, we pick up especially on the role of 'dispossession', 'new enclosures' and 'legalization' to characterize how soy expansion affects control over land, as it progresses especially at the expense of land under 'possession right' status. As we will show, shifts in the soy frontier depend on particular mechanisms that allow control over land to be gained (through material and perceived pressures and dispossession) and maintained (through enclosures). Playing into these processes is the process of re-valuation of land as the soy frontier is also a commoditization frontier.

The mechanisms by which the frontier advances and land is being transferred to different actors alternate between governance and violence (Blomley 2003). We distinguish dispossession, enclosure and legalization, but these are interlinked in practice. In the case of Santiago del Estero in the Chaco region, we find that market-based transfers based on legally sanctioned property rights are accompanied by semi-voluntary negotiations and threats of violence and eviction. We find instances of 'land grab' situations of material displacement of entire communities, through violence or persuasion. In addition, rural communities suffer other forms of dispossession which rather than implying land being physically grabbed from them, imply 'grabbing control' instead (Borras et al. 2012), affecting their capacity to control and maintain access (Ribot and Peluso 2003; Peluso and Lund 2011). In this context, the main effect of the different types of 'enclosures' is the naturalization of violence (Blomley 2003). Regulation is then a double sided coin which can both serve to stop and re-direct commoditization of land, but also it provides frontiers with particular shapes, when it protects the rights of nature and locals, or when it sets up policies that attract investors.

Our exploration of the frontier requires engaging with discursive constructions about development. The advance of private property over possession rights (which we discuss as 'property frontier' below) is also discursively constructed in terms of progress advancing over 'wilderness'. As noted by Tsing (2004, 29), the frontier is commonly constructed as

> an edge of space and time: a zone of not yet – not yet mapped, not yet regulated. They are projects in making geographical and temporal experience (its wildness is both material and imaginative [...] frontiers are not places, not even processes but projects capable of moulding places and processes).

The idea of frontier has historically referred to the conquest of new geographical areas. Within the literature about the advance of monoculture and industrial agriculture over native forests through deforestation, and on marginal lands with low productivity, the notion of frontier as a material border has been taken almost literally, separating progress from non-developed areas. As noted by Peluso and Lund (2011, 668) frontiers are not simply geographic places where 'progress and development meet wildness' but rather areas where particular material and symbolic notions of development are being socially constructed. In this sense, we are looking into social, economic and political forces shaping the limits, giving a particular pace and direction to shifting control over land. This constructionist view about how globalization shapes local geographies allows us to understand the contingency and, therefore, political character of frontier construction.

In the case of the Chaco region, the 'frontier' metaphor has practical implications and works effectively to justify and legitimize the incorporation of so-called 'empty' lands into a global agro-extractive production model with the justification of maximizing opportunities or achieving win-win outcomes. This discourse – in our case operating within the judicial discourse, as we will see later – tends to present global capital expansion into almost every (empty, idle) territory as an inevitable process.

The 'soy frontier' advancing into the Chaco region

By the end of the 1990s, high international soybean prices and the 'saturation' of the core soy production area (Región Pampeana) led to shifts in the investment patterns: soy expansion moved from the Pampas region to the northwest of the country and into the Chaco forest. Land in the Chaco was acquired for both soy and cattle production, through land use change (displacing to some extent cotton, tobacco and beans), through clearing of native forests (Slutzky 2008) and through encroaching on land used by campesino and indigenous communities. The potential incorporation of the Chaco region into the global land market attracted new actors making claims on the land, including real estate speculators, cattle growers, and soy producers of local, extra- regional, and international origin. The formal authorization of GM soy production was a key moment, allowing soy production to grow and become a permanent crop in the core region, pressing for an increase of the scale of production and driving up land prices, and, as a consequence, a growing interest in finding lands outside of the core region.

As in the core Pampean region, soy became a permanent monoculture; it replaced rotation with other crops and pushed cattle from the core to other areas, particularly the Chaco region. It is important to note that the expansion of agriculture, and particularly livestock and soy, has not necessarily been linked to large properties in the cadastral sense, but rather to large production units, incorporating the need of flexible forms of achieving scale. This flexibility operated in terms of taking advantage of existing mixed forms of access to land, the dissemination of new technologies and innovative business models, as well as through the commoditization and financialization of soy in the national and international markets. As a consequence, between 1988 and 2002,[9] the agricultural area increased by 15 percent in the whole country and 50.3 percent in the regions outside the Pampas. Genetically modified soy area expanded by two million hectares on average per year between 1997 and 1999 (Goldfarb and Zoomers 2014) and IISD (2012) calculates that around three million hectares were incorporated between 1996 and 2010 with a close link to availability of new GM soy varieties. Chaco region became a target for the expansion of both soy and livestock production to free land for soy cultivation in the Pampa region, partly because of the quality of its lands with seasonal but sufficient rains. Moreover, land prices had increased so much in the core region, that investors saw the potential of incorporating the Chaco to both national and global land markets, which is how the soy frontier also became a financial and land commoditization frontier. The Chaco region presented to these investors a less regulated terrain to access land through indiscriminate deforestation and increasing pressures on holders of possession rights, as we will later see. Soy production also 'subsidized' cattle expansion into the more marginal areas by using soy exports benefits to acquire land for cattle ranging. In this sense, the cultivated soy area in Santiago del Estero increased from 250,000 hectares to 650,000, reaching around 1 million hectares (Ministry of Agriculture 2012) in the 1988–2002 period, while cattle heads increased by 40 percent between 1988 and 2002 (González and Román 2009).

During the peak of a first wave of investments (around 2004), the Chaco region was filled with 'newcomers' from the core region provinces, pools and investment funds, and some foreign investors. Land prices in the core region rose from an average U$S 5000 per hectare in 2004 to USD 17,000 in 2012 (Compañía Argentina de Tierras 2014). Most of the investors were already successful soy producers, but amongst them were also cattle producers from the Pampas moving their stock to the Chaco, sparing land for soy production or rent in the provinces of Buenos Aires, Santa Fe and Córdoba. During this first wave of soy expansion to the Chaco, many small – but rich – producers (200/300 hectares) sold land in core provinces such as Santa Fe and Córdoba, where prices up

[9]2002 is the latest reliable National Agricultural Census available.

to USD 20,000 per hectare were paid, and invested in Santiago del Estero for USD 200 to 300 per hectare in lands with 'development' potential, becoming large landowners. Moreover, a range of non-local and foreign investors was also gradually attracted (Goldfarb and Zoomers Forthcoming). In this sense, not necessarily foreign but extra-regional investors have dominated the soya-ization of the Chaco region (Goldfarb and Zoomers 2014)

This expansion to the north was only possible, however, because of the combination of innovations in genetic engineering, further flexibilization of land management systems, and the introduction of new varieties of crops and pastures, producing low yields, but with low costs. The 'soy frontier' in the Chaco moved along three dimensions, which we discuss below, related to three main structural forces. We distinguish (1) a 'financial frontier', driven by the growing demand for soy and soy-derived products and involving the commoditization of land in the Chaco, (2) a 'technological frontier', driven by technological innovations related to GM soy and zero tillage cultivation, and (3) a 'property frontier', advancing private property over possession lands, facilitated by the particular make-up of the property regime in Argentina (Goldfarb and Zoomers 2014).

Capital expansion needs: soy and the 'financial frontier'

Some international organizations, developing countries' governments and policymakers perceive soy as part of the answer to the global food and energy crisis. Increasing soy demand also has to do with changes in the consumption patterns of the middle and upper classes mainly in Europe and in Asia. In addition soy received a boost through the growth of an international biofuels market.[10] Therefore soy has expanded through its frontiers in the direction of becoming a global commodity. Recently the national government launched the Federal and Participative Agri-food and Agribusiness Strategic Plan, 2010/2016, through which Argentina seeks to strike a privileged position as a global provider of food. Moreover the National Cattle Plan aims to stimulate production up to 77 million heads aiming at exporting 12 percent of total production. For the province of Santiago del Estero this means thousands of hectares of peasants' and forest lands put into production for export commodities (Goldfarb and Zoomers 2014).

Large-scale land acquisition and commercial pressures on land in general, and soy expansion in particular, must be situated in a broader restructuring of the corporate food regime (McMichael 2012). McMichael understands land grabbing as a reflex of changing conditions of capital accumulation. We find this notion relevant for the case of soy in Argentina, where soy expansion has responded to a need of international capital to find new ways of reproduction through the commodification and financialization of natural resources. The Argentine soy expansion model is an example of how land is being controlled in order to make these capital flows more flexible, while these flows need to find ad hoc mechanisms of land control when entering into friction with local dynamics (Tsing 2004). This has been the case of the pool model that was revolutionary in terms of the organization of production, increasing production scale, flexibility of access to land and financialization of investment sources and benefits. These pools were successful in the core areas, but once expansion of soy headed to the Chaco region, new mechanisms of controlling land emerged from the frictions between external pressures and existing land-based relations. The combination of property and other forms of access to land, such as

[10]Mainly created by the setting of EU targets for progressively reducing greenhouse emissions until 2050.

lease or 'accidental contracts', increased at the national level by 25 percent in the period 1988–2002, and these mixed forms, which were typical of the Pampean area or core region, started to spread to the Chaco (pampeanization of the Chaco as an access-to-land practice).

Therefore, financial speculation plays an increasingly important role: there is a gamble for soya prices in the international market, while land and crops become increasingly a function of commodity prices, expanding the gap between conceptions of economic rationality on the one hand, and social and ecological sustainability on the other (McMichael 2012). The extension and concentration of the production in the hands of former latifundia owners as well as sowing pools was made possible through the combination of different tenure figures within a land market characterized by flexibility – mainly through land leasing, with the appearance of the figure of 'small rentier'. This type of relation shows that access and control to land do not necessarily mean property. In this sense, control is more likely to be determined by the injection of financial capital, technology and a particular way of organizing production based on flexible lease contracts and outsourcing production tasks to service providers.

GM soy, zero tillage: soy and the 'technological frontier'

The use of the GM soy technological package expanded very quickly. In Argentina, the total soy bean area has expanded more than that of any other crop since 1996, and especially since the introduction of direct tilling techniques and Roundup (glyphosate) Resistant soy, combined with the intensive use of fertilizers and agrochemicals. This technological innovation made soy cultivation economically feasible, even in agro-ecological zones where soy would not grow before, like the Chaco.

This turning point was closely related to the legal approval, by the Argentinean government, of GM soy production[11] while this was still forbidden in all the neighbouring countries. With this move, Argentina became an outlet for massive and rapid expansion of 'illegal' transgenic soybeans across the region, which today is also one of the main export crops in Brazil and Paraguay, while increasing in importance in Bolivia and Uruguay. In combination with the dissemination of no-tillage technologies among producers, the new GM seeds facilitated pest control over large areas of land and made expansion in scale much more viable.

The combination in one package of direct tillage techniques and machinery with Roundup Ready soy has been fundamental for the expansion of soy production into other ecological regions out of the core areas. Large-scale enterprises have better possibilities to 'correct' environmental and social costs, which they consider externalities of the business model. In conclusion, technological innovation in Argentina has changed the way in and scale at which land can be devoted to a single crop, cultivated in large production units, and managed by few hands and with almost no human labour (Goldfarb and Zoomers 2013).

The introduction of GM seeds facilitated pest control over large areas of land and made expansion in scale much more viable. Meanwhile the development of new GM seeds allowed soy cultivation on lands with very different soils and fertility types, as well as in a diversity of climates, increasing the security of investments. With the massive

[11]Today Argentina is the second producer of GM crops in the world after the US, veering a total area of around 22 million hectares (soy, corn, and cotton, among others).

Table 1. Argentinean land tenure regime.

Type of property	
Private Property [propiedad] (owner)	Direct/rent purchase in a free market or through inheritance
Tenure [tenencia] (user)	Recognition of ownership in another person
Possession [posesión] (possessor)	Owner's behaviour: 20 years occupation and demonstrated improvements to plot

introduction of the GM soy technological package, this production model started to expand towards areas that had not been considered suitable for agriculture. Examples are Argentina's northeast and the Chaco region, which are vulnerable in both environmental and social terms.

Soy and the 'land property frontier'

To expand soy cultivation or cattle production into the Chaco, investors developed claims on land considered, or constructed as, 'unclaimed'. This concerned land that had apparently been abandoned by its former legal owners and titleholders, sometimes decades before, and was now occupied by either 'monte' (bush vegetation) or used by indigenous (Salta) or campesino communities (Santiago del Estero). In Santiago del Estero rural dwellers mostly had only 'possession rights', a historical occupation without formal titles, but recognized in Argentinean law (see Table 1). The arrival of investors, mostly with formal titles in hand, put possession rights under immediate pressure. The new claimants obtained property titles in a variety of ways: inheritance or purchase from the legal titleholders, but also non-transparent manoeuvres in connivance with notaries, cadastre and registry of property offices.

One of the most visible expressions of the shifting frontier of land control has to do with tensions and open conflict that occur as these two types of land rights – property rights and possession rights – encounter each other, or more precisely, as formal land is being claimed over which possession rights have developed. Argentinean law backs both forms of land rights.

Beyond the legality of the property title, holders of possession rights can also appeal the law to claim. In Argentina, land can be accessed through three different legal categories: property, tenure and possession (Table 1).

These three categories have cohabited in the Chaco. With the increasing commercial pressures on land, next to other developments such as rural labour decline after the cotton crisis in the 1990s, the tensions between 'property' and possession' have been growing. Legal titles – however acquired – are confronted with existing claims, which may be backed by the '20 years law' included in the Argentinean civil code, which formally protects those who have lived and worked the land for more than 20 years by giving them possession rights.

Holders of possession rights can appeal to the law to claim the lands they occupy. In response to the claims of investors, this has been one central avenue by which peasant families and communities have sought to defend their control over land. In this sense shifts in control over land are being contested individually but also collectively in an increasingly institutionalized strategy. For many peasant communities and families in Santiago del Estero the capacities to exercise land control appear to be linked to the category of possession, which involves a complex process of legal and political construction and

Table 2. Legal definition of possession rights.

Property	Possession
Implies possession Only requires a title to be proved	Does not necessarily imply property Property can be claimed and obtained if proof of: • Possession actions • Ownership behaviour/intention

defence. Next to legal avenues (involving complex and costly court cases), in some cases peasant communities have engaged in collective mobilizations, confronting investors. In many other cases, however, they have come to a settlement with the investor, ceding some of the lands – and commons – they occupy to safeguard the rest (Table 2).

The shifting frontier in Santiago del Estero

Santiago del Estero province is paradigmatic in terms of how the agricultural frontier has expanded in recent years, and provides an excellent place to examine the mechanisms involved in pushing as well as resisting the frontier. But it is also unique for being the only province in Argentina where peasant resistance to the soy model – and the agroindustry in general – became explicitly political. In this province the land claims engendered by the expansion of soy production and livestock heighten the land tenure vulnerability experienced by most of the rural peasant population. The current land occupation pattern was established in the wake of logging activities, paramount during the nineteenth and much of the twentieth centuries, extracting quebracho and carob for railway sleepers, as well as charcoal and tannins. With the removal of the railway, logging activity decreased and part of these 'degraded' lands were abandoned by their concession holders, both individuals and companies, and started to be occupied by the labourers who worked for logging companies. Now jobless, former forest industry workers in some cases paid a small rent to an administrator, while others simply stayed on the land that was dedicated primarily to extensive cattle and goat rearing. In some cases, the families that stayed were former workers in timber companies who got paid with land rather than receiving a salary, but without, however, receiving titles. As investors started to claim such lands, as of 2004, conflicts about lands under possession right have exponentially increased.

As the value of lands in Santiago del Estero increased, so-called 'absentee owners' who had abandoned the region started to return, attracted by rising land prices and the interest in investing in agricultural or cattle activities with good international prices. As holders of property titles, these absentee owners, and their heirs, use their legal title to reclaim land that is now, in many cases, occupied by possession right holders. Where absentee owners face situations where they find that holders of possession rights occupy the land, different types of conflicts evolve. Moreover, foreign capital has been expanding in Chaco region provinces, attracted by comparatively cheap land. Our research showed a range of alliances or joint ventures between local and extra-regional entrepreneurs: locals providing land and knowledge about local land control needs, and outsiders providing know-how about technology use, field preparation, flex-crops management and markets.

We found that in Santiago de Estero holders of possession rights have become especially vulnerable to commercial pressure on the lands they occupy. Mechanisms to acquire land from this group ranged from purchase of possession rights (which presents the advantage to the buyer that the value to be paid is considerably under market value,

Figure 1. Argentina. Santiago del Estero province.

since no property titles are involved) to negotiation and legal disputes, to threats of evictions, and enclosures, for example through acquiring surrounding lands. We have synthesized the range of deals/transfers through which land is acquired, accessed and controlled in Table 3. It is striking that most of the deals – or the 'not yet deals' – for land transfer involve private parties, whether they take place in a voluntary 'buyer-seller' relation or between holders of possession rights *vis à vis* potential land investors. In contrast with other provinces in Argentina, in Santiago del Estero today the provincial state does not play an important part in the deals, though it did in previous decades.

Transfer repertoire and mechanisms of land control on the ground

Beyond the transfer itself, there are certain mechanisms of control that determine the terms of the deals, which are the result of relative power positions among actors. These mechanisms condition the encounters between parties, the terms of negotiation and the outcomes of the land transfer. The typology represented in Table 2 shows how access does not necessarily mean control in Santiago del Estero. It also illustrates the conditions in which land control is transferred (pressure of different natures and inequalities in terms of knowledge, access to justice, social organization and economic resources) and the results in terms of one of the main mechanisms through which control over land is maintained: enclosure. In this sense enclosure is not only an outcome but also a mechanism for maintaining land control and frontier shifts.

The soy frontier in Santiago del Estero progresses through a series of interconnected mechanisms. Table 3 provides an overview of the mechanisms of transfer. Below we discuss in detail the way these transfers work on the ground and how they imply shifts in land control. Our discussion illustrates that shifts in land control include different types of pressures and inequalities in terms of knowledge, access to justice, and economic resources, arguing that all of the types of transfer combine legal procedures with the threat or use of violence, though to different degrees. The transfers are shaped by the relative power positions among actors, which condition the encounters between parties, the terms of negotiation and the outcomes of the land transfer. We discuss the mechanisms through which control over land is shifted or maintained in terms of enclosure. Enclosure

Table 3. Land transfers and shifts in land control.

Modality of transfer	Type of rights transferred	Mechanism of land control	Shifts in land control
Purchase/sale/ inheritance	Individual property rights	Myth of voluntariness	Re-concentration/ Commoditization/ Homogenization and segmentation of space/ Indirect enclosures (increasing land prices, pollution, access to commons blocked)/ Appearance of new brokers
Lease/temporary rent	Use rights	As above	As above
Ceding/donation of possession rights	Possessor becomes tenant	Persuasion/pressure/ unequal access to justice/direct violence	Direct and indirect enclosures (access to neighbouring parcels is facilitated)
Mediation/ '*comodato*' agreements (give a bit, receive a bit)	Possession rights ceded/property rights ceded	Persuasion/Unequal access to justice/ Negotiation	Direct and indirect enclosures (access to neighbouring parcels is facilitated)/land as a commodity
Conflict through judicial process (acquisitive) prescription trials (claim) re-vindication trial (defensive)	Property/possession rights in legal contradiction	Legalization/unequal access to justice/ development worldviews of judicial elites/ Access and use of specialized knowledge	Construction of possession actions/ Permanent pressure/ land as commodity/land already in production
Territorial conflict (the no-deals)	Possession rights in contradiction with property rights or in conflict with illegal grabbers	Direct violence/ Resistance/In situ knowledge and violence brokers	More conflict/Political organization, alternatives and resistances strategies/ land as a commodity/ land in production

is in this sense not only an *outcome* but also a *mechanism* for maintaining land control and producing frontier shifts. This section mostly relies on primary data collection through interviews.

The typology represented in the table below shows how access does not necessarily mean control in Santiago del Estero.

Private owners were interested in selling land to extra-regional investors in the context of what was at the time a rather paralyzed land market. The dynamization of the market consequently attracted new investors speculating on rising land prices.[12] Many of these

[12]According to Compañfía Argentina de Tierras-CAT (a real estate company and base of the sectoral association) in core soy producing areas land increased from USD 4,700/hectare on average in 2003 to USD 15,000/hectare in 2013. Nowadays land can range between USD 250 and USD 3500 per hectare in the best area in the Chaco region, while in the best area in the core region land price ranges between USD 4000 and 20,000/hectare.

direct purchase situations took place between medium/large land owners and new incoming soy companies, investment funds and sowing pools aiming at gaining scale and diversifying production zones. They also included middle-sized producers from the Pampas region, seeking new lands in the Chaco to invest their surplus, and who then became larger producers. More recently land has also been purchased to invest in productive mixes, combining agriculture, intensive cattle grazing and payment of environmental services. This has concerned especially the so-called marginal lands, affected by the recent changes in the Forest Law that now defines them as conservation value zones, putting limits to the productive use of the resource, although without explicitly protecting the population that lives on this land.

While in the core Pampean region land market forces mostly resulted in production concentration, in the Chaco they are also creating new enclosures, bringing market pressures to lands previously considered of no economic value. Increasingly, peasant communities find themselves surrounded, fenced in, polluted, cleared and threatened by usurpation [encroachment] and dispossession. The seeming voluntariness of market-based transactions becomes especially problematic when it comes to lands occupied by peasants holding possession rights. The transfer of control over land here assumes a different character, since it involves evident power inequalities of the two parties, and a very high risk for possession-right-holders of losing their land.

We will see that whether there is purchase or lease (market-based transaction), negotiation, or (pressure for) eviction, mechanisms of land control taking place are enclosures, legalization and violence.

Market-based transfers: the myth of voluntariness

In the context of existence of a liberalized land market like that of Argentina, there is an obvious assumption about a field – the market – where rational, equal actors encounter each other ruled only by demand and supply forces. While property transfers in the Chaco region seem to concern voluntary transactions within free market relations, the deals conceal the imperfections of the market and the pressures that exist both to sell (rising land prices) and to buy (intrinsic need of scale for profit-making in the soy business). The process of commoditization of land that emerged as the land market was dynamized, created pressures and produced specific forms of enclosure. While in the core Pampean region land market forces mostly resulted in the concentration of production in fewer hands, in the Chaco it brought market pressures to lands previously considered of no economic value. As a result, peasant communities increasingly find themselves surrounded and fenced in by large-scale producers, suffer from pollution of their water sources, are accused of usurpation and encroachment and are threatened with eviction.

Presently, the vulnerability of possessors is still linked first and foremost to the lack of property titles in combination with a lack of legal knowledge and political awareness about their land rights. New land claimants have used possessors' vulnerability to their own advantage, combining persuasion and deception, intimidation and violence to get people with possession rights to give up their claims.

Transfer of possession rights to a private investor may take place through various mechanisms, which imply pressures based on unequal bundles of powers (i.e., access to justice, need of monetary resources, knowledge imbalances). Therefore, rather than being voluntary, land transfers occur under different kinds of pressures, more specifically:

Economic extortion, by offering an unequal encounter of supposedly equal parts. Although it is possible to find a number of cases where peasants do not regret having got a few 'pesos' for a good number of hectares, loss of control over land frequently happens when the rights of possession are sold for bargain prices or exchanged for use (tenant) rights, ceded or donated. This constitutes a relatively common way in which communities and peasant families lose their control over land and in many cases also their access, with very limited compensation. Peasants confronting land claimants often lack knowledge of and access to the law and are not aware of the legal possibilities to claim possession of the land. This leads to highly unfair negotiation capacities.

Legal vulnerability of possession rights (threat of high costs of legal procedures, of holding a property title vis-à-vis possession rights), as well as interference with legal procedures takes place: refusal of juridical institutions to process peasants' claims, such as the refusal to accept complaints of usurpation by the police, false reports to criminalize farmers and arbitrary detention. Land claimants bringing title deeds confront peasants with possession rights. Some actually are descendants of former owners, having been absent from the region for decades. But there are also many 'fake absentee owners' who have obtained property rights through non-transparent processes, e.g., by 'purchasing' titles through employees of the Land Property Register. Increasing land prices attract real estate companies, a range of intermediaries and plain land speculators – often involving attorney bureaus and notaries – whose business it is to buy cheap land and sell it at a higher price. Some of them are dedicated to obtaining land rights from possession holders through deceptive arrangements, e.g., by providing stolen satellite information to interested potential buyers.

Enclosures ranging from persuasion to sell or cede possession rights, or indirect forms of enclosure and isolation by purchasing or obtaining rights from neighbours, surrounding communities and individual families,[13] to physical enclosure by fencing, clearance of common pasture lands, blocking access to water sources, killing and theft of animals and intimidation with firearms by the police and para-police employed by businessmen.[14]

Leasing deals in the Chaco region follow roughly the same dynamics of activation of the land market due to the soy boom. Recently, however, rising prices have made sale more attractive to landowners and leasing has declined, pushing renters out to produce in their own fields. While property transfers in the Chaco region seem to concern voluntary transactions within free market relations, the deals conceal the imperfections of the market and the pressures that exist both to sell (rising land prices) and to buy (intrinsic need of scale for profit-making in the soy business). In addition, we found intimidation and threats were common mechanisms to persuade people to cede the land. Roodhuizen (2010) also shows this for the case of islanders in the Argentinean Delta region, where lies, deceptive contracts and procedures, the use of intimidation and violence are

[13]According to some community leaders, in some of the departments of Santiago del Estero, such as Jimenez, purchase of possession rights from families or communities may work as a pressure mechanism for remaining peasant families who become surrounded by private owners, showing how one enclosure may lead to further enclosures.

[14]Direct violence and intimidation mechanisms towards campesino families and communities led to the death of two peasant activists from Movimiento Campesino de Santiago del Estero (MOCASE) – Cristian Ferreyra in November 2011 and Miguel Galván in October 2012.

common practice. Outright harassment also takes place. As described in Barbetta's (2009) work and as also emerges in the interviews with peasants' leaders, NGOs and social pastoral in December 2011, these harassment mechanisms include destruction of native forests (source of grazing, fuel and food), as well as destruction of improvements done in the fields (poles, fences), which in many cases are part of the proof needed to make claims of possession in the judicial field. These actions are mostly taken in order to force signing of agreements to cede or sell possession rights (see also next section). Harassment may lead to the abandonment of the land or, in the best cases, to political resistance or judicial strategies to defend possession rights.

In response to the vulnerability of peasants without formal titles, social and political organizations in Santiago del Estero have invested in peasants' awareness and access to justice. However, the GM soy boom found the peasant sector largely unprepared to defend their rights, although land conflicts had been common since the early 1980s and despite the presence of important peasant organizations such as MOCASE (Movimiento Campesino de Santiago del Estero). But the level of organization and the capacity to cope with conflict varies throughout the region, with some communities remaining very vulnerable to pressures.

Possession rights under pressure: negotiation and legal(ization) procedures

Conflicts between claimants holding private titles and possession rights holders are formally solved through (re)vindication and prescription trials against the latter holders and, in a few cases, through pre-judicial arrangements. Information obtained from 43 trial verdicts from the Archive of the provincial Palace of Justice and collected by Barbetta (2009) shows that in Santiago del Estero the outcome of trials initiated before 2002 (around half of the cases) were decided in favour of possession rights holders, mostly because the person registered as legal owner never showed up. This judicial dynamic changes considerably with the 'soy boom'. Trials become much more complex when alleged owners started to appear in person or are represented by intermediaries, and a whole new range of 'brokers' – e.g., gatekeepers, dummies, lawyers, administrators (Goldfarb, forthcoming) – becomes necessary in order to identify parcels or to obtain titles.

Not all conflicts are solved through trials. In many cases, negotiations between alleged owners and possession rights holders take place to seek an agreement. The outcome of these deals is usually that the holders of possession rights receive less land than they originally claimed, and lose access to resources such as pastoral lands, water, routes and improvements made. On the other hand, the 'legal owner' of land accepts rearranging the title, reducing his area under property. Negotiation does not necessarily mean consensus or absence of pressure. In general, outcomes are related to differences in power positions and access to different types of capitals and networks of support. The vulnerable situation of possession rights holders and lack of knowledge about their rights and how to defend them makes them surrender when facing this type of 'second best dilemma':

> What will happen if you don't make an agreement with me? I will sell my part to another one who will come with the bulldozers ... (Absentee owner in negotiation meeting with community for the control of 1400 hectares in Santiago del Estero, 2011)[15]

[15]Recordings of this sort of meeting were collected by Nienke Busscher in Santiago del Estero in 2011.

> You are faced with a personal choice, to say hey! is an arrangement or mediation useful or not? And it obviously implies a reduction in the possession of families. But being in the situation of assessing this question, how is the community, what's the level of organization; and like always to make clear to the families, look, the defence of the land involves not sleeping, notifications there, notifications here, courts, shooting at night, visits of vans, threats, but also the other side means you continue to live as you live, and there, with the assurance that your children can continue to live there. (Interview with peasants' organization lawyer, 2011)

Advocates of this type of argument claim that ceding from both sides could lead to a better situation for all, instead of the zero-sum situations that characterize judicial processes over recognition of possession rights. As Barbetta (2009, 24) clearly shows, the equation 'give a bit in order to receive a bit' is rarely solved. The agreements work as strategies to break peasant resistance: once accepted, they undermine the arguments about the legitimacy of possession rights, de-politicizing peasants' claims to land. Ceding a part of what corresponds to them implies a hidden message of the illegality of peasants' occupation (Barbetta, 2009), creates an illusion of equal parts and at the same time puts them in a place of receiving a favour from the investor. Negotiations are highly unequal due to the historically-grown power relations between different social groups in the province. The patronage type of relationships[16] that have governed the relations between political and social classes has much to do with the outcome of these supposedly successful deals.

The cases in Santiago del Estero show that when an entrepreneur negotiates access to land with possession rights holders, this is often followed by attempts to acquire more land beyond the initial agreement, as illustrated by this quote:

> if the agricultural frontier advances it is because there is always somebody willing to negotiate; I don't know one case of mediation where the entrepreneur remains [after a certain time] with what has been agreed. (Interview with peasant activist and local government official, 2011)

Another generalized practice to gain access to land from holders of possession rights is the signature of a '*comodato*'. This is a particular type of contract – with a huge range of variations – through which the possessor formally cedes their land while he is given the right to live on and use part of the land for a specified period of time. It implies relinquishing possession rights and accepting a new status as a user (tenant). In some cases, former possession rights holders end up paying the new land owner for, e.g., grazing rights on the land they used and lived on for decades.

When competing claims ensue in legal conflicts the contradictory and unequal character of property and possession rights comes out strongly. Legal procedures can take two forms:

- *Re-vindication trials* are those initiated by an entrepreneur who has the titles to land that is occupied by holders of possession rights, who are accused of being 'usurpers'. The legal process then centres around proving that these peasants do not fulfil the requirement of *ánimo de dueño* – the intention to own or behave as an owner – a prerequisite for the legal recognition of possession rights. In many cases these trials are accompanied by destruction of material improvements, another key element for recognition. The next quote illustrates how sometimes the initiator of the trial holds titles that changed hands in non-transparent processes.

[16]Patronage relationships have historically characterized rural and political life in most of Argentinean provinces. In northern ones, the political leadership of 'caudillos' has its origin on land 'ownership' and rural labour in sui generis plantations, mostly based in and on fear, subservience, authoritarian attitudes and protection.

> In cases that they [the newcomers] report the name of the holder of the title, then we know we have to sue Juan Perez. Nobody showed up, and then they [possessors] end up winning. What happens after 20 years? The son says: Look, my oldman has not registered verdict, I need it registered... What happens? The Property Register says 'no', look, you have done a mistake in your demand, because you sued Juan Perez but the registered owner is Carlos Gomez. Then you say, that has happened? I have a report from 1982 which tells me the owner was Juan Perez, then you realize that there has been a fraudulent enrolment of year 68 in which Juan Perez allegedly sold it to Carlos Gomez and this is sold after the Gutierrez who are the people from Córdoba who want to get the land. That case is clear, the manoeuvres of property registration office. (Interview with peasants' organization lawyer, 2011)

• *Acquisitive Prescription* are the tools for holders of possession rights to prove their possession over the right given to someone holding a property title.[17] There are considerable obstacles to initiating and pursuing such trials, related to high costs as well as lack of knowledge of legislation. Peasants need to succeed in what Barbetta (2009, 149) calls the 'presentation of the facts in a judiciable way', in which they assume the judicial figure of 'poseedor' and construct the truth of their possession actions and their intention to own ('ánimo de dueño') within the field civil justice. This involves a complex juridical as well as political process. Peasants need to link to intermediaries, to develop organizational capacity and collective identification as a peasant community. High costs of – and lack of access to – justice are among reasons why the strategy of acquisitive prescription trials generally has a reactive rather than a preventative character, and in general these processes are a response to a vindictive judicial action by the alleged owner of the land (Barbetta 2009). Chances of an outcome in favour of the possessor depends a lot on the access to legal support and organizational capacity in order to cope with the costs (money, time, physical defence) of carrying on with the judicial process.

What our discussion shows is that dispossession and enclosure are not extra-judicial processes but in part take place through judicial struggle and pre-judicial negotiations. We also can see how the 'judicial' is necessarily political. As in the previous section, in apparently consensual negotiations that result in pre-judicial agreements, as well as in judicial processes for legalizing the situation of possessors as owners or as mere tenants, we also observe how mechanisms of land control such as enclosures, legalization and violence in the terms of Blomley (2003) are present. Finally, we turn to processes in which claims on land are established through the use of direct violence.

Violent dispossession and conflict shaping the frontier

In Santiago del Estero we also found evidence of enclosures being established through direct violence, pointing to a process of accumulation by dispossession. Examples of tactics are intimidation and/or eviction of holders of possession rights, encroachment, and illegal occupation. The use of violence is interwoven with legal strategies, as illustrated by this peasants' attorney:

> There are other cases in which newcomers do not even have titles, just start fencing with armed guards, they bring the wire, as a case that I have in the department Banda. They go from wire to wire to wire and make people sign papers by threatening them, sign the transfer of possession

[17] Peasant organizations are involved in around 500 such cases in the judicial system, and with increased political organization almost none of the cases are lost, though this may include the loss of the initial number of hectares claimed.

> rights. They buy these rights from several families and them they become themselves holders of possession rights over 2000 hectares. Afterwards, they initiate a prescription trial and so on … Why? How? Why in one case this will require so much to prove and this other case almost nothing? So there are clearly manoeuvres as well as the ideological position of judges. Corruption is installed; from the Property Register Office, cadastre, and the same justice officials. (ONG and peasants' lawyer, 2011)

Increasingly, these types of actions result in open conflict as organized peasants resist eviction and destruction.[18] According to the Human Rights Sub-secretariat in Santiago del Estero, by 2005 these conflict situations affected around 5000 families and 360,000 hectares of land – and possibly their number may increase considering that most of the affected families and communities did not yet have the possibility to make a formal demand to this institution (Barbetta 2009).

Recent intensification of land disputes in Santiago del Estero is clearly an expression of a dynamic of advance of the agricultural frontier related to the soy boom. The pattern of land disputes is quite different in character from earlier periods (1980s and early 1990s), in terms of speed, scale, intensity of violence, types of land control strategies, the origins of capital invested and the diversity of actors involved in 'grabbing the land'. The conflict involves violent displacement, threats, wounds and deaths and more than 500 peasants have been criminalized due to their protests. The database provided by REDAF (2013) on land conflicts registers 123 conflicts[19] only in Santiago del Estero province, most of them related to violent evictions, total or partial misappropriation of land and enclosure through fencing, deforestation, killing of animals, armed groups organized by the landowners, etc.

> … we can theorize … the judge may disagree about the concepts, the extent of possession and everything else, the fact is that in the land, in the 'monte', there are armed groups [of provokers] and people resist. (Peasants' organization lawyer)

Pressures on land and the ensuing land conflicts open a new chapter in peasant organizing in Santiago del Estero. The defence of possession rights has become part of a broader political strategy of re-vindication of the peasant economy, way of life and identity. It is also becoming a central tool in the creation of a semi-institutionalized network (Mesas de tierras, Emergency Committee, Possessors Register and grassroots peasant organizations across the province) that brings together an array of social and state actors at the provincial level, picking up on the land governance challenges that the conflictive situation presents plays an important role in the governance of land.

Conclusions

We have attempted to show how the GM soy boom in Argentina has involved a far-reaching and problematic shift in land control, negatively affecting peasants in vulnerable tenure situations in the northern provinces, which operated through transformations in the capital, technology and land property relations frontier. The expansion of GM soy production beyond the core region of the Pampas has attracted investors and speculators

[18]Article 2470 of the Argentinean Civil code also states the legitimate physical defence right if the state fails in protecting the '20 years right'.
[19]In this province the number of conflicts increased in 58 in 2011, affecting 3528 families and 400,000 hectares (REDAF 2013).

who seek to establish claims on land under possession status, with dwellers lacking solid legal recognition.

We have found the notion of 'shifting frontiers of land control' as coined by Peluso and Lund (2011) particularly useful to analyse the encounters between possession and property rights generated by soy expansion. Though we may explain why and how GM soy became hegemonic and pushed the frontier forward, with reference to abstracted drivers, related to developments in capital needs, technological innovations and the make-up of a property regime, a look into the particular mechanisms at work helps understand why frontiers went in a particular direction and adopted a certain shape, and which forces are actually in tension when defining transfers from one hand to the other. Rising international commodity prices, land market liberalization and financial capital crisis matter, but on the ground the construction of frontier dynamics, their pace and shape, are the result of a more complex, contingent and contested field of social and political interactions. This is why we looked deeper into mechanisms of land control: enclosure, legalization and violence.

Building on land transfers and lawsuits in the province of Santiago del Estero we have explored through what mechanisms and strategies private property advances over possession rights. Reactivating or forging title deeds, land claimants have sought to dislocate or disown possession right holders, or to severely limit their access to and control over land. Through a combination of interlinked and progressive mechanisms, which include legal litigation, negotiated settlement, various forms of non-physical enclosures, intimidation and overt violence, possession holders have suffered various forms of dispossession.

We found enclosure to take place in the following ways:

- When possession holders win a prescription trial but manage to prove only part of the 'ownership' through 'possession actions', losing the rest;
- When possession holders sign a contract of 'use' in which lands are transferred to the formal owner, acquire renter-status and are charged for the use of the land, e.g., for permission to graze or use certain resources;
- When possession rights holders need to accept a reduction in the area in possession as part of a negotiated settlement to safeguard at least part of their land; and
- In the case of *comodato* contracts, enclosure does not only have to do with the loss of concrete physical resources such as land, water and forests, but also with the loss of the legal entitlement to claim their rights as possessors.

With our analysis we hope to correct the image that the expansion of commercial agriculture, in this case soy in Argentina, proceeds through legal mechanisms based on voluntary transactions and arrangements, placing it outside of the concern of 'land grabs'. Indeed, we argue that the subtle and brutal forms of enclosure and dispossession registered in Santiago del Estero qualify as 'grabs'. The transfers and arrangements embody different forms of violence and power inequality that systematically work to the disadvantage of the peasants with possession rights. These inequalities include the types of property enshrined in the law and what is needed to claim or defend these, unequal access to justice and the agrarian bureaucracy, but also historically-grown social relations. The frontier advances by making use of and exacerbating the vulnerabilities of possession rights holders. We have shown how dispossession and enclosure are not only isolated, illegal and technical, but also processes that take place through judicial, legal (sometimes violent) struggle and 'market-based' negotiations, while dispossession and enclosure take place thanks to the politicization of the judicial.

Pressures on possession rights have, as we have suggested, translated into more conflictive social relations, open conflict, and a higher degree of political organizing. Some observers have tried to un-link the occurrence of conflict from the larger trends of shifting land control, blaming it unilaterally on activist peasant unions in an attempt to de-politicize the understanding of what is happening. We hope to show that the conflicts are a response to the pressures of the frontier and one of the few avenues for possession rights holders to defend their access to land.

The debate on the current land rush is starting to connect to a broader and longer-term political ecology discussion about the globalization and commoditization of natural resources. The case of Argentina in particular is a paradigmatic case in understanding the historical commoditization of nature. This has advanced on the basis of a production model that moves towards new geographical, productive and cultural spaces, relying on technological innovation sustained on a modernization narrative, which also depends on a specific constellation of agrarian relations.

Argentine soy expansion is an example of how land control follows the pressure of capital flows and the kind of friction this causes on the ground, where existing access is affected. Only recently has awareness grown in Argentina that these processes call for new forms of regulation, moving away from the earlier liberal trend to rely on market forces. Soy expansion seems an irreversible process as strong forces are driving GM soy frontier expansion: expectations of increasing international prices, reliance on eco-efficiency and modernization discourses, new certification standards, official support, and a hegemonic rural industrialization development model. What are the emergent governance dynamics and institutional arrangements at the frontiers? We have shown current arrangements ranging from the appeal to justice to protect property or possession rights, Territorial Planning processes, human rights protection, economic nationalism, judicial decrees, land regularization and territorialization, political resistance and organization. In dealing with social and economic impacts of frontier expansion, there is no policy oriented to the defence of peasant production systems and the particular relation of them with land. In this context, current efforts to resolve tensions around possession rights, and defend these rights through social organization in combination with judicial and political strategies, have become themselves drivers of an emerging field of land governance.

Acknowledgements

We would have not been able to write this paper without the help in of Nienke Busscher, Ana Karol from Fundapaz and the feedback from Annelies Zoomers.

Disclosure statement

No potential conflict of interest was reported by the authors.

Funding

This research was conducted in the framework of the IS Academy Land governance for development.

References

Barbetta, P. 2005. El movimiento campesino de Santiago del Estero: luchas y sentidos en torno a la problemática de la tierra. In Giarraca, N. and Teubal, M. (coords.) *El Campo Argentino en la encrucijada. Estrategias y resistencias sociales, ecos en la ciudad*, Alianza Editoria, Buenos Aires.

Barbetta, P. 2009. *En los bordes de lo jurídico. Conflcitos por la tenencia legal de la tierra en Santiago del Estero*. Buenos Aires: UBA.

Blomley, N. 2003. Law, property and the geography of violence: The frontier, the survey, and the grid. *Annals of the Association of American Geographers* 93, no. 1: 121–141.

Borras, S.M. Jr., J.C. Franco, S. Gómez, C. Kay and M. Spoor. 2012. Land grabbing in Latin America and the Caribbean. *The Journal of Peasant Studies* 39, no. 3–4: 845–872.

Compañía Argentina de Tierras. 2014. www.cadetierras.com.ar

De Schutter, O. 2010. Responsible destroying the world's peasantry, 4 June. http://farmlandgrab.org/13528 (accessed September 30, 2013).

FAO. 2013. *Reflexiones sobre la concentración y extranjerización de la tierra en America Latina el Caribe.*

Goldfarb, L. and E.B. Zoomers. 2013. The drivers behind the rapid expansion of genetically modified soya production into the Chaco region of Argentina. In *Biofuels - Economy, environment and sustainability*, ed. Z. Fang, 73–95. InTech.

Goldfarb, L. and E.B. Zoomers. Forthcoming, February 2014. The rapid expansion of genetically modified soy production into the chaco region of Argentina. In *The global land grab beyond the hype*, ed. M. Kaag and E.B. Zoomers, 71–85. Zed Books.

González, M.C. and M. Román. 2009. Expansión agrícola en áreas extrapampeanas de la Argentina. Una mirada desde los actores sociales. *Cuadernos de Desarrollo Rural* 6, no. 62: 99–120. Bogotá.

International Institute for Sustainable Development (IISD). 2012. Inversion externa en agricultura en los paisen miembros del MERCOSUR.

Manciana, E., M. Trucco, and M. Pineiro. 2009. *Large-scale acquisition of land rights for agricultural or natural resource-based use: Argentina.* Buenos Aires: World Bank.

McMichael, P. 2012. The land grab and corporate food regime restructuring. *The Journal of Peasant Studies* 39, no. 3–4: 681–701.

Ministry of Agriculture. 2012. La superficie de soja en la República Argentina.

Peluso, N.L. and C. Lund. 2011. New frontiers of land control: Introduction. *The Journal of Peasant Studies* 38, no. 4: 667–81.

Red Agroforestal Chaco Argentina (REDAF). 2013. *3° Informe Conflictos sobre la Tenencia de la Tierra y Ambientales en la Región Chaqueña Argentina.*

Ribot, J.C. and N. Peluso. 2003. A theory of access. *Rural Sociology* 86, no. 153–81, June 2003.

Roodhuizen, D. 2010. *Welcoming the agrocultural frontier? A case studiy into the social impacts of growing land pressures in the Paraná River Delta, Argentina.* Master study Thesis, WUR.

Sili, M. and L. Soumoulou. 2011. *La problemática e la tierra en Argentina – Conflictos y dinámicas de uso, tenencia y concentración*, FIDA/Cooperación Italiana/Ministerio de Agricultura, Ganadería y Pesca de Argentina.

Slutzky, D. 2008. *Situaciones problemáticas de tenencia de la tierra en Argentina.* PROINDER. Secretaría de Agricultura, Ganadería, Pesca y Alimentos. Ministerio de Economía y Producción.

Tsing, A.L. 2004. *Friction: An ethnography of global connection.* Princeton: Princeton University Press.

White, B., S. M. Borras, R. Hall, I. Scoones and W. Wolford. 2012. The new enclosures: Critical perspectives on corporate land deals. *The Journal of Peasant Studies* 39, nos. 3–4: 619–47.

World Bank. 2010. *Rising global interest in farmland. Can it yield sustainable and equitable benefits?* Washington, DC: World Bank.

Bolivia's soy complex: the development of 'productive exclusion'

Ben McKay and Gonzalo Colque

A significant transition is underway in Bolivia where both domestic and foreign capital are monopolizing commercial agriculture and leading a highly mechanized, capital-intensive production model which has considerably diminished the need for labour. This paper explores mechanisms and processes of 'productive exclusion' in the soy-producing zones of Santa Cruz in relation to the expansion, concentration and mechanization of the 'soy complex'. We provide an analysis of how the agrarian structure has developed since soy was adopted – from 'putting land into production' to 'expanding the agricultural frontier' and 'controlling the agro-industrial chain'. We explore how and the extent to which the penetration of new capital is leading to new processes of agrarian change which exclude the rural majority from accessing the means of production. While a process of 'foreignization' of land began to take shape in the early 1990s, new processes of capital accumulation are eroding the ability of small farmers to engage in productive activity, potentially leading to 'surplus' populations no longer needed for capital accumulation.

Introduction

Expanding soybean plantations in Bolivia's eastern lowlands have led to dramatic changes in the agrarian structure over the past 25 years. Initial expansion of the agricultural frontier was driven by state policies influenced and funded by a United States-led economic development plan in the 1940s which 'recommended that the population be shifted from the poor lands of the Altiplano to the fertile lands of the east'[1] (Malloy and Thorn 1971, 165). This large-scale migration, known as '*la marcha hacia el oriente*' (march to the east) was driven by an intentional, unequal distribution of land by the state. Andean colonists were given 20–50 hectares to produce traditional crops for domestic food supply, while capitalist entrepreneurs and local elites were given between 500 and 50,000 hectares to produce crops for export (Malloy and Thorn 1971; Valdivia 2010). This planned two-track agricultural development strategy shaped Bolivia's agrarian structure for years to come, as highlands colonists produced traditional crops for domestic consumption while also providing large-scale landowners with an abundant supply of cheap labour necessary for initial frontier expansion and land preparation. But it was not until the 1980s and 1990s that soybean plantations really started to expand, with an influx of foreign producers and capital and the introduction of new technologies. Favourable soybean prices, mechanization and increased

[1]This is known as the Bohan Plan after State Department official Merwin L. Bohan.

capital investment began to rapidly alter the region's productive relations as labour power became much less necessary, and industrial crops for export widespread.

More recently, new and increasing global demands for soybeans and their derivatives, particularly from China and the EU, have exacerbated this process as international market prices send positive signals for increased investment and further expansion. This investment and interest in soybean production goes well beyond the farm (land), to a much larger 'soy complex' which includes genetically modified seeds, chemical inputs, agricultural machinery, storage facilities, processing, transportation and the financialization of the agro-food system (Isakson 2014). There is also a growing multitude of 'flex' uses and end products traded on the global market, as soybeans are increasingly being used (or promoted) in flexible ways – for food, animal feed, edible oils and industrial products such as biodiesel (see Borras et al. 2015; Oliveira and Schneider 2015). The substantial increases in agro-capital,[2] in terms of both quantity and costs necessary for production, have rendered accessing such factors of production extremely difficult for small-scale farmers.[3] As a result, many small landholders are becoming separated from farming activity and unable to take the risk of putting land into production. Instead, processes of semi-proletarianization (see Kay 2000) and petty bourgeois rentierism are unraveling, whereby small-scale landowners engage in wage labour activities or other rural non-farm employment (such as small-scale informal enterprises, taxi/bus drivers) while renting their parcels to medium- and large-scale farmers who have access to the necessary agro-capital to put land into production. For some, access to sufficient amounts of land and title still remain a problem; for many, however, the problem is the lack of access to agro-capital necessary to work the land.

Technological advancement, mechanization and a concentration of control of the soy complex are also putting a squeeze on labour. This squeeze on labour, combined with the inability to access agro-capital and land, is threatening future farming prospects for the rural majority, and more so the youth. Whether or not this trajectory of agrarian change is leading to a 'truncated trajectory of agrarian transition' whereby land rents are no longer a viable livelihood and employment opportunities in rural areas diminish with no pathway to the city or industrial activity, remains to be seen (Li 2011, 296). Rural youth make up the majority in Bolivia's expansion zone – the greater part of whom have no access to land or live in households where their land is rented to other producers. In such a context, how, where and whether people find other employment opportunities are important concerns.

This paper traces the development of Bolivia's 'soy complex' since the 1980s when public policies started looking outwards for foreign investment, the economy (and the agricultural sector) became deregulated and frontier expansion via soybean production began to increase. We frame this development through three somewhat distinct but overlapping phases – from 'putting land into production' to 'expanding the agricultural frontier' and 'controlling the agro-industrial chain'. The paper then reveals the more contemporary dynamics of land/resource access and control relations whereby land ownership has

[2]This includes inputs such as seeds, fertilizers, herbicides, pesticides; heavy machinery such as tractors, sowers, fumigators, harvesters, transport trucks and storage/processing facilities such as silos and processors.

[3]While we acknowledge that small-scale farmers/landowners and peasants can be categorized in different socio-economic groups, we will use the terms interchangeably to refer to people who own parcels of land of less than 50 hectares since many are '*colonizadores*' who self-identify as 'peasants' due to their histories but are now a mix of small-scale capitalist producers and small-scale rentiers who produce, on average, no more than 10 percent of their crops for their own consumption.

become a less significant aspect for agribusiness as the capital-intensive agricultural model bypasses the need for labour and leads to processes of 'productive exclusion'. While land ownership remains very important for small-scale farmers as well as for large-scale landowners, value-chain relations have enabled agribusiness to maintain access to land without necessarily having tenure rights (ownership). This latter section is primarily based on insights gained through in-depth semi-structured interviews with key informants in two communities located in Santa Cruz's soy expansion zone: Cuatro Cañadas and San Julián. In the final sections, we point to some implications of such agrarian changes and raise some serious concerns which we hope will contribute to the debate concerning this development model and the fate of small-scale family farmers.

Mechanisms of access and productive exclusion

To understand productive relations in the soy expansion zone, it is useful to apply Ribot and Peluso's 'theory of access' (2003), whereby they challenge rights-based approaches to property (and resources) by differentiating between one's *right* to benefit from things and one's *ability* to benefit from things. While people may hold the right to certain property, they may not necessarily have the ability to use the property/resource in a productive way in order to reap benefits from it. Having the ability to benefit entails several access mechanisms, not just legal rules which, as Houtzager and Franco (2005) point out, 'are neither self-enforcing nor self-interpreting' (2005, 37). Lacking such access inevitably entails exclusion, but as Hall, Hirsch, and Li emphasize, 'all land use and access requires exclusion of some kind' (2011, 4). Yet, while exclusion may form a necessary component of landed relations, it often exacerbates inequalities and marginalizes the poor, therefore not always (or commonly) excluding everyone *equally*. We draw from Ribot and Peluso's 'mechanisms of access' to explore how and the extent to which the rural majority have become excluded from productive activity, despite an increasing agricultural frontier and the ability to retain their formal access to land. This type of 'access analysis' not only focuses on property relations but also delves into relations of production and entitlements, illicit actions and 'the histories of all of these' (Ribot and Peluso 2003, 157).

Exclusion in Bolivia's eastern lowlands is not entirely the result of a lack of access to land since many (but certainly not all) small-scale family farmers have maintained their property rights and/or access to land. This enables them to retain some benefits from their property relations through rent collection, but their inability to access agro-capital (and credit), technology and labour markets outside of commercial farming is increasingly leading to their exclusion from working their land. This is in part due to the mechanization and industrialization of agricultural production, but also due to historical land relations and a 'landlord bias' in the eastern lowlands, which have led to structural inequalities in rural areas (see Kay 2009; Kay and Urioste 2007). Labour has become ever more subordinate to the needs of capital in such a highly mechanized and capital-dependent agricultural system and inequality has been further exacerbated by what Ribot and Peluso call 'access to authority' (2003, 170). This mechanism of access 'shapes an individual's ability to benefit from resources' by privileging certain individuals or institutions due to their economic, political, and/or social status (Ribot and Peluso 2003, 170).

Lacking these mechanisms of access has resulted in exclusionary dynamics in Bolivia's most productive agricultural zone. But while the majority of rural households are faced with these processes of productive exclusion, there remain no major overt, organized social 'countermovements' by which people are struggling with and rejecting the terms of their exclusion/inclusion. This is due, in part, to the type of exclusionary dynamics by which small family farmers are faced. Their conversion into semi-proletarians and petty bourgeois rentiers has seemingly subsided forms of resistance as they continue to reap benefits, albeit in a much smaller way, relative to those who are working their land. This is mainly due to the fact that the large majority of their income is derived from land rent, meaning that their economic interests are aligned with the agro-capitalist class as better harvests result in higher rent payments. For a better understanding of how such mechanisms of access developed in the context of Bolivia's agrarian structure, we provide a brief overview of land and productive relations followed by three stages of development of the soy complex.

A brief overview of Bolivia's agrarian context

Bolivia's insertion into what we might call a corporatized agro-commodity and food regime is a relatively new development. It was only in 2005 that genetically modified (GM) seeds became legalized, and it was not long before that when many of the small farmers who make up the vast majority of farm units in the soy-producing region became integrated into the 'soy complex'. With this integration brought an increased dependence on external inputs – that is, chemical-based pesticides and fertilizers, GM seeds, heavy machinery – as well as an increasingly monopolized control over the storage (silos), processing and access to external markets necessary for the export-oriented crop. This coincided with the election of leftist President Evo Morales and the Movimiento al Socialismo (MAS) who had strong support from many of Bolivia's rural and indigenous social movements, and promised an 'agrarian revolution'.[4] Nine years later, dynamic state–capital relations in conjunction with a changing food regime and the convergence of multiple crises around food, fuel, climate and finance are leading to profound and perhaps unexpected (considering the initial political context) agrarian change. These changes are driven by both domestic and foreign (particularly Brazilian) capital, altering the social relations of production, property and power in the country's eastern lowlands (see Urioste 2012).

Mennonite and Japanese settlers were the pioneer producers who introduced soybean crops after testing them for a few years in smaller areas (Medeiros 2008, 183). Commercial opportunities began in 1985, when Bolivia officially adopted the free-market model as part of new public policies in order to overcome the crisis of hyperinflation that was occurring at the time. During the second half of the 1980s, liberalization policies were increasingly adopted, and given that the mining sector had collapsed (due to a decline in prices), subsequent governments considered that agricultural commodities and export products were important economic alternatives. The state helped in the process both by implementing infrastructure projects and by providing economic backing to the private financial sector in order to facilitate access to credit for agro-exporters (Perez 2007).

During the 1990s, the World Bank (WB)'s 'Eastern Lowlands Project' played an important role in expanding soybean production and determining the main characteristics that this type of agriculture would come to take. A 'landlord bias' resulted in landlord favouritism and the funnelling of public resources of up to 1000 hectares of land for industrial oilseed production, mainly soy (Kay and Urioste 2007). New technical systems for classification of land use (arable land, grazing, mixed) and types of producers (small, medium, large) were introduced and implemented with the assistance of international agencies. Andean peasants began to produce commodities for export in settlement zones on the frontier. Moreover, the 1990s was a decade with visible incursions of foreign capital, going not only towards soy

[4]See Urioste (2007) for a brief analysis of Bolivia's 2006 'Agrarian Revolution'.

production, but also towards acquiring cattle ranches in remote areas of the frontier. At the end of this decade, the total area of soy cultivation reached around a half million hectares, increasing more than eight times in just 14 years (1986–2000; Urioste 2011; Perez 2007; Killeen et al. 2008; Thiele 1995).

The rapid growth of the agricultural frontier over a relatively short time period was carried out with little regulatory oversight in terms of legal land tenure. The land market was 'open for business', as one Brazilian farmer working in Bolivia explained (Klaus, personal communication, April 2014). In the 1990s, the Bolivian government was so eager to attract investment that land was almost given away for free. Farmers were buying land in what is now called the 'expansion zone' for USD 20–USD 30/ha. Today, these same lands sell for USD 2000 to USD 5000/ha. Such expansion and lack of regulation rendered the agrarian law inoperable. In 1992, a government commission was set up to intervene in the functions of the *Consejo Nacional de Reforma Agraria* (CNRA) or National Council on Agrarian Reform and to implement a 'moratorium' on new land grants and the redistribution of public land. After four years of intense discussions and negotiation, a new land law was approved in 1996: the *Ley del Instituto Nacional de Reforma Agraria* (INRA) or National Institute of Agrarian Reform Law. Its main objective was to give technical, legal and political power to the state in order to recover authority over the frontier land through a process called *saneamiento de tierras* (land regularization). This was essentially a new land titling process that focused on the revision of legal precedents of properties granted by the CNRA (Urioste 2007).

From 1990 to 2010, soybean production in Santa Cruz increased more than eight times, from 232,743 to 1,917,150 tonnes (FAO 2013). Further, from 1990 to 2007, cultivated land in Santa Cruz increased almost five-fold from 413,320 to 1,821,631 hectares, including up to nearly one million hectares directed for soybeans and other oilseed crops (Urioste 2012, 30). This expansion coincided with an increase in foreign capital, particularly from Brazil, Argentina, and Colombia. However, as many studies conclude (Urioste 2012; Zoomers 2001; Mackey 2011), the increasing importance of soy and other commodities on the frontier is linked to groups of fragmented private investors rather than to the type of state-led/facilitated large-scale land grabbers supported by states and with legal protections (Borras et al. 2012).

A recent study by Urioste, for example, reveals the high concentration of land and influx of foreign, particularly Brazilian, capital since the 1990s (Urioste 2012). He suggests that in 2006/2007, Brazilians controlled 40.3 percent of total soybean area in Bolivia, while controlling another 700,000 hectares of land for cattle ranching (2012). Other studies suggest similar findings, though the lack of availability and reliability of more recent data presents difficulties in updating these findings (see Pérez 2007; Marques Gimenez 2010; Mackey 2011). Environmentally, a myriad of studies point to the expansion of soybean plantations as a driving force behind deforestation, soil degradation and water contamination (see Hecht 2005; Redo, Millington, and Hindery 2011; Muller et al. 2011). Castañón (2014) also provides an insightful study concerning the contradictions of Bolivia's soy production. On the one hand, the economic changes due to soy have provided a higher level of food security for many families, at least in the short term. On the other, processes of social differentiation are increasing as small farmers are becoming more and more subject to the terms of their unequal relations of production, and dependent on external inputs and international markets. In this sense, they have lost control over the terms of both their production and their consumption, becoming absorbed into the soy complex and opting to buy rather than produce their own food. This process of substituting traditional for industrial crops via a more mechanized and monocrop agricultural production model is also unravelling throughout other parts of Bolivia (see Colque 2014b; Castañón 2014; Urioste 2014;

Eyzaquirre 2014; Jaldín 2014; Pérez 2014). This is leading to a rapid transition away from traditional and peasant-based crop production for domestic consumption to an export-oriented industrial model increasingly reliant on food imports. While Bolivia is no stranger to a model of development based on the extraction of raw materials for export (mining, hydrocarbons; see Veltmeyer 2014), this agro-industrial model is increasingly becoming characteristic of a type of 'agrarian extractivism' (see Alonso-Fradejas 2015; Giarracca and Teubal 2014; Veltmeyer and Petras 2014).

The many contradictions that exist regarding Bolivia's agricultural development model in the context of state rhetoric for an 'agrarian revolution', a law of mother nature and commitment to food sovereignty (see McKay, Nehring, and Walsh-Dilley 2014) reflect the broader process of what some describe as 'reconstituted neoliberalism' taking place throughout the country (Webber 2011). This process is characterized by social and economic policy changes at the margins without major structural changes of the political economy (Brabazon and Webber 2014). State rhetoric has become more of a 'legitimating discourse' (Kerssen 2015) than a structural transformation, while key social movements have been co-opted by the state, resulting in a loss of autonomy and lack of empowerment among those in the movements (McKay, Nehring, and Walsh-Dilley 2014). As Brabazon and Webber point out, the trajectory of agrarian change in Bolivia is 'reinforcing rather than dismantling the concentration of quality productive land amongst medium- and large-scale agrarian capitalist' (2014, 461–62). Instead of breaking with the past, current state policies have actually 'reproduced dependency relationships with agro-industrial capital' (Cordoba and Jansen 2013, 18), thereby not providing any alternative pathways for small farmers or peasants through 'neocollectivist agrarian development' but rather by reinforcing a model attuned to the WB's proposed pathways out of poverty: (1) advance as a capitalist farmer within the agro-industrial system; (2) become rural wage labourers working on or off the farm; (3) migrate to the city (World Bank 2007).

However, many of Bolivia's small-scale landowners are not engaging in any of the three pathways. Instead, many are transitioning to semi-proletarian and petty bourgeois rentiers, unable to advance as agro-captalists and faced with too much uncertainty to migrate to the city. Further, labour opportunities are sparse since the agro-industrial model has become highly mechanized and no longer requires much labour power. This process has developed over time through several 'stages of development' which have gradually excluded small-scale farmers and peasants from engaging in productive activity.

Bolivia's 'soy complex': stages of development

In this section we present three stages of the development of the 'soy complex'. While it is possible to distinguish these three stages, they often overlap in certain areas and with other processes. For example, the gradual expansion of the frontier implies that, while some areas are already incorporated into production, others farther away from settlement areas are still in the early stages. Also, changes are neither homogenous along the frontier nor take place at the same time. In addition, it is important to note that there are no unidirectional changes or rigid cause–effect relationships. However, what gives meaning and direction is that there are concrete economic motivations to expand and consolidate domains of profitable land.

Putting land into production

This first stage, which we term 'putting land into production', can be characterized by increasing economic activity on the frontier, which mainly took place from 1985 to

1992. It was a clearly differentiated process from the previous vigorous, but limited, large farming operations that produced sugarcane, cotton and livestock (Thiele 1995). During the 1970s cotton producers benefited from high market prices, direct and indirect subsidies and easy access to credits funded by the state (Gill 1987). Even though external markets were unstable and there were structural limitations (lack of roads, high transport costs), large-scale farmers, as well as small-scale farmers in 'colonization zones,' achieved better economic returns by adopting labour-intensive agriculture. For the latter, rice was a strategic or 'subsistence crop' because it could easily be marketed and used for self-consumption (Fifer 1982). Moreover, this stage was preceded by Hugo Banzer's military dictatorship from 1971 to 1978 where 'millions of dollars of cheap credit subsidised agro-capitalists in Santa Cruz' (Webber 2011) and 'hundreds of thousands of hectares of land were fraudulently distributed to political cronies for free (some up to 50,000 hectares)' (Urioste, 2010, 2). Privileged individuals with ties to the military government benefitted immensely during this period, as their *access to authority* enabled them to benefit from Bolivia's untouched arable resources in the eastern lowlands (Ribot and Peluso 2003).

From 1986 to 1992, the illegal expansion of cultivated areas by means of what is now known to be the result of 'massive corruption in the distribution and titling of lands' led to rapid rates of deforestation and a rise in agricultural production. According to Steininger et al. (2001), annual deforestation due to agriculture in Santa Cruz's expansion zone went from 68,196 hectares in 1986 to 225,018 hectares in 1992 – a 330 percent increase – with agro-industry (103,623 hectares) and Mennonite farmers (89,954 hectares) accounting for 86 percent of deforestation in 1992. Further, in the expansion zone, cotton increased by 135 percent, soybeans by 194 percent, sorghum by 108 percent and wheat by 539 percent (see Table 1). By 1992, soy crops covered 200,000 of the 354,000 hectares in the expansion zone.[5]

Data on annual forest clearings (Table 1, part A) show that both Andean and Mennonite colonists and settlements were noticeably involved in deforestation, many of them using slash-and-burn methods. During this period, colonists increased their role as major food suppliers, producing rice, corn, wheat and other 'subsistence crops'. Production of sunflowers, sorghum and maize also increased as part of the evolution of the oilseed complex and the agro-industry in general. However, other commodities such as sugarcane remained a slow-growing sector, mainly because there was no sufficient labour force. According to Gill (1987), labour-intensive crops could not develop quickly because potential workers (frontier colonists and small-scale peasants) preferred to cultivate crops on their own available land, instead of becoming rural wage labourers. Peasants and small-scale farmers therefore chose to exploit their own labour by working and maintaining control of their own landholdings. Access to labour and labour opportunities were therefore available for the rural majority as both their own land and large-scale landholders demanded a labour supply.

During this same period, soybean cultivation area expanded from 63,000 to 217,000 hectares, while its export value increased from USD 19 to 57 million (Perez 2007). In other words, soy production was no longer a marginal activity. This successful beginning of soy production relied on increased deforestation and the gradual mechanization of large farms on the frontier. Killeen et al. (2008, 6) show that over time not only soybean

[5]Other crops such as wheat and cotton were also growing quickly, but towards the end of this period both declined recurrently, returning nearly to 1980s levels.

Table 1. Annual clearing of the forest by type of actors and expansion of cultivated areas[a].

	1986	1988	1990	1992	Total, 1986–1992	Percentage of total deforestation from 1986–1992
A. Annual clearing (in ha)	**68,196**	**83,539**	**149,152**	**225,018**	**525,905**	**100**
Peasant agriculture	9282	11,095	16,184	17,772	54,333	10
Andean colonist	6956	11,573	14,424	13,669	46,622	9
Mennonite colonist	22,501	24,649	52,060	89,954	189,164	36
Agroindustrialist	29,457	36,222	66,484	103,623	235,786	45
B. Crop area (1000 ha) in expansion zone	**132.3**	**149.8**	**291.8**	**389.3**	**963.2**	**100**
Cotton	11.2	10.0	3.9	26.3	51.4	5
Rice	13.7	16.2	18.2	18.2	66.3	7
Corn/maize	17.1	14.2	19.7	35.4	86.4	9
Soy	68.2	85.4	179.3	200.2	533.1	55
Sorghum	12.1	20.0	30.0	25.2	87.3	9
Wheat	10.0	4.0	30.0	63.9	107.9	11
Sunflower	–	–	10.7	20.1	30.8	3

Source: Steininger et al. (2001) and Hecht (2005).

[a]The categorization of agrarian actors as Andean, Mennonite Colonists and others can be questioned because it denotes their cultural and religious adherence rather than their role as economic actors. This is often overlooked in data classification. That is why these categorizations require further revision, not only to overcome potential misleading interpretations but also to identify different processes within and among groups from class-based analysis. Noting this important issue, in this paper we use available data (and categorization) in terms of frontier actors according to their economic role as small-, medium- or large-scale producers. For instance, 'Andean colonists' are mainly recognized as small farmers in colonization areas. 'Mennonite colonists' first came to Bolivia in the 1960s; they are Bolivians established as small family farms, today, producing oilseed crops. 'Japanese colonists' are mostly 'medium-scale farmers' (51–1000 ha) associated together in cooperatives. 'Agro-industrialists' refers to national and foreign investors closely connected to large-scale farming (Killeen et al. 2008).

producers, but also other producers, such as Andean colonists and cattle ranchers, have cleared forests intensively.

One key factor for putting land into production was the WB's 'Eastern Lowlands Project,' initiated in 1991. The main aims of this project were to expand the production of profitable agricultural commodities by increasing soy for export by about 200,000 tons/year and by substituting imported wheat by about 30,000 tons/year (World Bank 1998). This project was explicitly oriented to consolidate large-scale soy production under the rationale that it would accelerate economic growth and lead to 'sustainable agricultural development'. Seven years later, the WB reported the results of agricultural production, as follows:

> Bolivia's real annual agricultural growth since 1987 of 1.5 percent has been strongly influenced by the expanded production in the Eastern Lowlands, the most salient features of which are as follows: between 1990 and 1996, agricultural exports from Santa Cruz increased 400 percent; the gross value of the Department's agricultural output rose from USD350 million to USD685 million during the period 1990–96. It has been estimated that 37 percent of the increased output could be credited to the project [...]. (World Bank 1997, iii)

However, these outcomes were overshadowed by the failure to control deforestation. Perez (2007) concludes that it is most likely the WB programme that caused the deforestation of the primary forests rather than encouraging production in existing cultivated areas.[6] The WB (1997, 4) reported that

> [i]n the process unfortunately, deforestation increased considerably, e.g., almost one million ha between 1989 and 1996. These actions far exceeded expectations, e.g., the project plan forecast only 25,000 ha of new land clearance in the expansion zone over five years.

This is 40 times over their original plan.

This first stage entailed a rapid incorporation of frontier land into areas of soybean farming and also clearly demonstrates the shaping of certain patterns and defining characteristics of this process. Deforestation was mainly caused by increased agro-industrial production of soybeans and similar oilseeds due to the mechanization of production (Barber et al. 1996; Muller et al. 2013). Other crops, such as sugarcane and cotton, remained slow-growing commodities. Andean and Mennonite settlers also began to expand cultivated areas to increase rice, maize and wheat production for the internal market. While large properties began to adopt capital-intensive farming, settlers still had the advantage of controlling labour-intensive farming, maintaining their productive capacity and their access to land and labour. Peasant agriculture in the Altiplano, however, suffered immensely as the state's structural adjustment programme's 'landlord bias' focused on developing an agro-industrial export sector in the east. As Kay and Urioste point out,

> up until the policies of structural adjustment of 1985, the internal (food) supply was able to meet the demand of the Bolivian population but, with the ability to freely import foodstuffs from 1985, the peasant economy declined as it could not compete with the better quality and cheaper imports. (2007, 53)

The subsequent stagnation of agricultural production in the western Altiplano coincided with rapid increases in the eastern lowlands – in terms of cultivated land, productivity and migration, as Andean peasants went east in search for land or labour opportunities, providing large-scale farmers with a stable supply of rural wage labourers.

While this stage of 'putting land into production' contributed to the structural inequalities of access and control relations, it was the expansion of the agricultural frontier from the early 1990s onwards that really shaped rural social relations and led to exclusionary dynamics that exist today. We examine these dynamics in the following section.

Expanding the agricultural frontier

To move forward it is worth distinguishing Santa Cruz's agricultural frontier according to different zones of expansion and settlements. From 1993 to 2004,[7] commercial crops began to spread around the traditional agricultural area (Integrated Zone). The map below (Figure 1) identifies five zones on the frontier[8]: The Integrated Zone (A), around the city

[6]As Perez (2007) and Kreidler et al. (2004) noted, one facilitator factor for expansion of soya was the trade and tariff agreements of the *Comunidad Andina de Naciones* (CAN) or Andean Community of Nations.

[7]The time period of this stage is not meant to indicate that the expansion of the frontier stopped in 2004. Expansion continues today, but this period marks a rapid increase in cultivation area and deforestation before GM seeds were legalized and before mechanization became ubiquitous.

[8]The five zones have been defined by adapting expansion zones identified by Fifer (1982) and Pacheco (2006) to current municipalities (INE 2001). The classification of actors and their relation

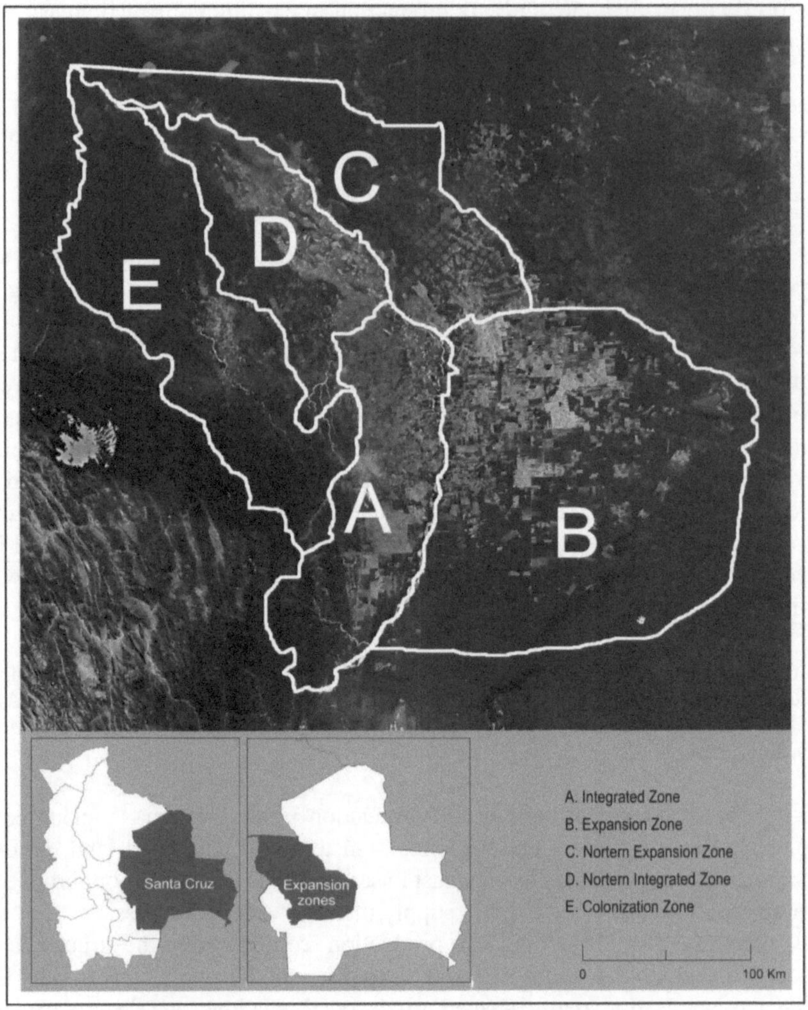

Figure 1. Map of Santa Cruz locating five zones of expansion (A–E).
Source: Adapted from GAPSC (2013), Killeen et al. (2008) and Google Maps.

of Santa Cruz, which pioneers and early settlers put into production; the Expansion Zone (B), mentioned above, which by the middle of the 1980s became the most representative case of soy planting expansion (Pacheco 2006; Killeen et al. 2008); the Northern Expansion Zone (C), with more extensive farming and also historical settlement areas such as San Julian; the Northern Integrated Zone (D) which is another vigorous regional economy; and the Colonization Zone (E) which mostly represents those settlement areas of Yapacaní created by the Instituto Nacional de Colonización (INC) (see Table 2 and Figure 1).

By 2004, all five zones had experienced massive deforestation, were occupied and were put into production. 'Cruceño farmers' (traditional landowners from Santa Cruz) had clear and dominant control over zone A. They also controlled zone D and had a

with deforestation are adapted from Killeen et al. (2008). The information about land use has been disaggregated according to these five zones and by types of actors.

Table 2. Land occupation by actors and five zones (documented until 2004).

Zones		Cruceño farmers (ha)	Agro Industrialists (ha)	Andean colonists (ha)	Mennonite and Japanese colonists (ha)	Cattle ranchers (ha)	Forestry (ha)	Restricted areas/ others (ha)	Total by zones (ha)	Total by zones %
A	Integrated Zone	584,905	45,639	23	192,592	151,101	140,801	22,601	**1,137,662**	**17**
B	Expansion Zone	29,941	530,731	42,648	259,847	964,310	0	257	**1,827,736**	**27**
C	Northern Expansion Zone	7716	191,821	433,133	13,634	186,282	425,574	525	**1,258,684**	**18**
D	Northern Integrated Zone	374,175	348,711	141,990	4872	5228	92,432	208	**967,617**	**14**
E	Colonization Zone	317,824	0	351,725	67,966	69,421	624,311	203,382	**1,634,630**	**24**
	Total by actors	**1,314,562**	**1,116,902**	**969,519**	**538,912**	**1,376,343**	**1,283,118**	**226,973**	**6,826,330**	**100**
	Percentage by actor	**19%**	**16%**	**14%**	**8%**	**20%**	**19%**	**3%**	**100%**	

Source: Adapted from Killeen et al. (2008), Pacheco (2006) and INE (2001).

Table 3. Agribusiness established in Bolivia in the value chain of the oilseed economy.

Agro-business	Date founded	Main characteristics	Relation with foreign capital
Gravetal Bolivia	2003	The largest soy processors in Bolivian agro-industry. Produces crude oil and soybean meal. 100% export oriented. Controls 31% of Bolivia's soy and soy derivative exports. Represents approximately 10% of foreign currency income derived from Bolivian exports in 2011. Direct employment generated in Bolivia reached 4500 positions per year.	Since 2008, 99% owned by Capital Inversoja, a transnational company based in Venezuela.
Industria de Aceites (FINO)	1944	Considered the second major soy and sunflower exporter and also produces cooking oil, butter, margarine, soap and other cosmetic products for the internal market. Controls 22% of Bolivia's soy and soy derivative exports.	74% controlled by Urigeler International, a transnational company that is part of Grupo Romero from Perú.
ADM	1923 (USA)	ADM is one of the world's largest transnational agro-industrial companies operating in more than 75 countries with sourcing, transportation, storage and processing assets. In Bolivia, ADM sells and exports vegetable oils and protein meals from soybeans and sunflower seeds. It started operating in Bolivia in 1998, buying 50% of the Bolivian SAO company. Controls 13% of Bolivia's soy and soy derivative exports.	Multinational based in the United States
Industrias Oleaginosa	1967	According to their website (http://www.iol-sa.com), Industrias Oleaginosas is an agribusiness that is 100% owned by Bolivians. It is an oilseed processor that handles grain purchases, storage, processing facilities and marketing. Controls 9% of Bolivia's soy and soy derivative exports. Main external markets are in the 'Andean Community', North America and European countries.	Owned by the notorious Marinkovic family (Croatian immigrants). Branko Marinkovic is a Bolivian politician and businessman who fled Bolivia after being accused of planning an armed rebellion to overthrow the current government.

(*Continued*)

Table 3. Continued.

Agro-business	Date founded	Main characteristics	Relation with foreign capital
Cargill Bolivia	1865 (USA)	Beginning in 1998, this company has operated for 15 years in Bolivia. Sells industrial food, exports agricultural commodities and also offers financial services. In Bolivia, Cargill has silos and warehouses where it can store up to 27,000 tons of grain. It also has partnerships with other silo owners in 12 locations. Controls 11% of Bolivia's soy and soy derivative exports.	Multinational based in the United States. Cargill is an international producer and marketer of food, agricultural, financial and industrial products and services. This company employs 140,000 people in 65 countries. In 2012, their income reached USD 116,000 million.
GRANOS	1991	Controls 9% of Bolivia's soy and soy derivative exports.	Established in Bolivia. International investors/capital unknown. Exports to Peru.

Source: Adapted from Pacific Credit Rating (PCR 2012), Nueva Economia (2011), AEMP (2012), Jubileo (2013) and the companies' respective websites.

significant presence in zone E. Agro-industrialists mainly controlled zone B, but also occupied a large portion of the land in zone D. Cattle ranchers, economically part of the group of elites, controlled over half of zone B, but since 2004 they have had a greater tendency to change land use for agriculture and move toward new areas on the frontier.

Furthermore, zone E and to some extent zone C were territories controlled by Andean immigrants ('*colonizadores*'), but not exclusively. This is partly because there were overlapping areas disputed by many actors since the 1960s. At this stage, Andean settlers were already consolidated as producers for the internal food market because the agro-industry began to focus more on commodities for export. However, as we see later, this was not the final situation for settlers.

A review of aggregated actors reveals important differences between settlements and large farms. Three main actors together, Cruceño farmers, agro-industrialists and cattle ranchers, had control of over 55 percent of the total area taken into account and, excluding forests and restricted areas, this number increases to up to 70 percent of land dedicated to agriculture and livestock. Forests tended to disappear not only because there were pressures to expand arable lands but also because logging was a profitable activity, even though it was done illegally (Pacheco 2006). Mennonite and Japanese colonists were also connected to market production, but they are generally not large-scale farms (over 1000 hectares). They had control over 8 percent of the land, mainly in zone A and zone B. Another group is made up of Andean colonists who occupied over 14 percent of the frontier land near two main settlement zones (San Julian in zone C and Yapacaní in zone E). The slash-and-burn methods that this group used to deforest land have been criticized for a long time in Bolivia; however, as Killeen et al. (2008, 13) concluded, colonists tended to reduce their impact by investing in intensive cropping systems to produce rice, maize, citrus and other traditional crops.

Through the clearing of forests and increased investment, by the end of this second stage, the classification of this land moved from being 'unproductive *latifundios*' with questionable property rights, to becoming effective and productive land for agricultural purposes. The expansion of the agricultural frontier had occurred in all five zones by the early 2000s, and agriculture became highly mechanized.[9]

During this second period (1994–2004), the Bolivian state implemented a new land law in an attempt to control the indiscriminate expansion of the agricultural frontier. In 1996, the state approved Law 1715 with the aim of recovering illegal landholdings to redistribute to peasant and indigenous communities. On the one hand, this initiative pursued legitimate principles of social justice for the peasant and indigenous majority. On the other hand, economically and politically powerful groups launched an open confrontation with the intention of neutralizing the state-led initiative to expand its control over the lowlands of Santa Cruz. The legal struggle became a political struggle (Valdivia 2010).

The implementation of the new Land Law implicated that all fallow lands should be reverted to the state. However, large-scale landowners represented by the National Agricultural Confederation (CONFEAGRO) reinforced their demands for the protection of private property and fought against any redistribution of pre-existing property (Urioste 2007). This sectorial political resistance, with heavy economic and political clout in Santa Cruz, was extremely effective as the land titling process produced marginal results below 10 percent by the end of 2004. As Valdivia (2010) explains, the Santa Cruz elite formed a regional hegemony representing themselves as 'successful producers' built on legitimation narratives proclaiming that small-scale producers and peasants, too, could become successful capitalist agricultural entrepreneurs. In addition to the political resistance of the agri-business, another factor which led to the Land Law's ineffectiveness in redistribution was that the unproductive lands of the early 1990s became controlled by Brazil, Argentina and Bolivians who bought huge tracts of land and expanded their landholdings. This hampered the reversal of large- and medium-sized properties since previously idle land was now being put into production, meeting all requirements to protect their private property. The boom in oilseed production contributed to the increased commodification of land. As a result, land redistribution failed to be realized, since the growing economic and productive interests resulted in and expanded concentration of control and a new mechanized, industrial oilseed production model came to dominate. As production, productivity and land expansion increased, smallholders and peasants became excluded from these processes, unable to access more land but able to capture a marginal share of the productive surplus via their position as small-scale rentiers – what we call 'productive exclusion'.

Several structural and relational mechanisms of access developed fully during this period, which began to exclude the majority. With access to capital, markets and technology, the Santa Cruz elite maintained and controlled much of the access to 'knowledge', forming cross-sectoral alliances and increasing their political and economic influence through what Ribot and Peluso call 'bundles and webs of power'. The National Statistics Institute (INE), for example, receives much of their 'official' data on agriculture, land and production from groups such as the Cámara Agropecuaria del Oriente (CAO), Cámara de Industria, Comercio, Servicios y Turismo de Santa Cruz (CAINCO) and Asociación de Productores de Oleaginosas y Trigo (ANAPO) which represent medium-large scale farmers and agribusiness. These groups represent the interests

---

[9] Up to 624,000 hectares of uncultivated land remain in zone E, which is the Protected National Park of Amboró.

of agribusiness and have now become very politically influential – successfully lobbying for the legalizing of GM soybean seeds and more recently for other GM crops. They also helped write the most recent agricultural policy (Ley de la Revolución Productiva Comunitaria Agropecuaria, LRPCA) with a tri-council of ministries, successfully shaping the law to benefit agro-industrial expansion (see Francescone 2012; Villegas 2011; Araujo 2011). Most recently, representatives of the CAO and ANAPO, together with Vice President Garcia Linera, launched the 'expansion of the agricultural frontier' (ampliación de la frontera agrícola) in which the government has pledged to help facilitate the expansion of 1 million hectares of agricultural land every year until 2025 (Heredia Garcia 2014).

The political influence of the Santa Cruz elite is also evident in the 2009 Political Constitution of the State (*Constitución Política del Estado, CPE*). In an attempt to control the expansion of large-scale landholders (*latifundium*), the Bolivian population voted in a referendum (*referéndum dirimitorio*) to set a maximum land-size ceiling of 5000 hectares (Article 38, CPE). Although this popular mandate was constitutionalized, the government also negotiated with powerful agro-industrial groups to incorporate an additional provision permitting an unlimited number of business associates to hold up to 5000 hectares, essentially rendering the land ceiling futile (Article 315.II, CPE).

Relating back to access analysis, the Santa Cruz elite were able to use their economic influence through their control and access to productive resources, markets and knowledge in the agricultural sector to extend their influence ('webs of powers') to the political sphere. This exemplifies how certain access mechanisms – particularly the structural and relational mechanisms – directly influence one's ability to gain, maintain and control other access mechanisms. As Ribot and Peluso put it, 'some actors pool their powers, forming bundles of owners, workers, or beneficiaries acting in concert to assert greater control or to maintain their resource access' (2003, 173). The situation in Santa Cruz exemplifies how economic bundles of power are used to leverage political power and thus solidify economic power by influencing regulatory mechanisms in a dialectical manner.

Controlling the agro-industrial chain

In terms of the trajectory and dynamics of frontier production, this third stage features the legalization of GM soybean seeds and a greater dependence on external, chemical-based inputs (see Catacora-Vargas et al. 2012).

The question of large-scale capitalist farming and its linkages with transnational investments has been recently discussed as 'foreignization', a phenomenon led primarily by Brazilians and Argentinians (Mackey 2011; Urioste 2011, 2012; Zoomers 2003). With the initial influx beginning in the 1990s, a large foreign presence remains and continues to extend its control over the soy complex, including land. Globally, demand for soybean continues to rise due to its 'flexible' utility in response to various crises (food, fuel, climate, finance) and especially its necessity to feed a growing animal complex in 'newer hubs of global capital' and rapidly growing economies such as Brazil, Russia, India, China, South Africa (BRICS) and some middle income countries (MICs) (Borras et al. 2012, 850–51). But while Brazil is by far the largest foreign 'entity' – in terms of foreign capital and land ownership – there are no Bolivia–Brazilian state partnerships involved in land investments or control of the soy complex. Rather, there are clear capital alliances between Bolivian and Brazilian agribusinesses and investors which have successfully created an alliance with the Bolivian government by transferring the experience of EMBRAPA,[10] and will likely continue to expand their control if the new 'expansion of the agricultural frontier' programme stays its course (Heredia Garcia 2014).

The difficulty of estimating the extent to which frontier land is owned by foreign or transnational capital arises not just from incomplete data or the lack of a reliable database but, above all, because these disputed lands are located in areas where illegal and informal land deals prevail. Moreover, the '*saneamiento*', or land registration, is still ongoing in this region. But while land is certainly an important factor, the larger components of the 'soy complex' are indicative of the changing agrarian dynamics and processes of control. Those who control storage, processing, distribution and exports have much more influence over the soy industry than landowners do. The following shows the main actors controlling Bolivia's soy complex – six companies control the export of 95 percent of Bolivia's soy (Figure 2).

Excluding Industrias Oleaginosas and Granos, the rest of the six listed companies are owned by transnational agribusinesses, which include US-based multinationals Archer Daniels Midland (ADM) and Cargill (Table 3). Many began to operate at the end of 1990 in Bolivia through the acquisition of local companies in Santa Cruz using their previous Brazilian and Argentinean subsidiaries to enter the country. Their connections with direct primary production, land ownership, and the leasing of land as well as their relationships with soy producers such as Grupo Monica Norte, El Tejar and others involved directly in land control are not clear (Urioste 2011). The companies are mainly characterized by activities such as grain purchases, storage, processing facilities, marketing and export. According to evaluations by Pacific Credit Rating PCR (2012), these transnational companies often operate by contract farming, where they provide seeds and credit to producers who in turn agree to sell them their harvest.

The case of Industrias Oleaginosas needs a brief additional consideration. This is the only important Bolivian agribusiness in oilseed production, processing and trade. The owning family, Marinkovic, particularly Branko Marinkovic, was an active political opponent of Morales' government. Marinkovic was accused of instigating an armed uprising against the state and consequently his family abandoned Bolivia in 2012.

The companies ADM South America (S.A.) and Industria de Aceites had their origin in large scale Cruceños farms during the cotton boom era, but when their economic importance increased, transnational companies became their major shareholders.

In this stage, Andean colonizers increasingly became soy producers as well. Many of them have substituted 'subsistence crops' (rice, maize, roots and tubers) with soy due to the better market conditions of the oilseed complex (AEMP 2012; Catacora 2007; Amigos de la Tierra 2007; Medeiros 2008). Towards the end of the 2000s, small-scale producers continued to be involved in oilseed farming for sales mediated by a few agribusinesses installed along the agro-industrial chain. Many structural elements of the 'soy complex' such as dependency on mechanization, imported seeds, chemical fertilizers and credits have exposed this sector to cyclical risks and put them at a disadvantaged position vis-à-vis large-scale farming (Castañon 2012; Catacora 2007; Urioste 2011). Their inability to access the capital and technology necessary to participate and compete as soy producers has marginalized their ability to fully benefit from their land. Access to markets and other

[10]Empresa Brasileira de Pesquisa Agropecuária (EMBRAPA; Brazilian Corporation of Agricultural Research) is a state-owned company to develop applied research on agriculture. It conducts agricultural research on many topics including animals, agriculture and crops.

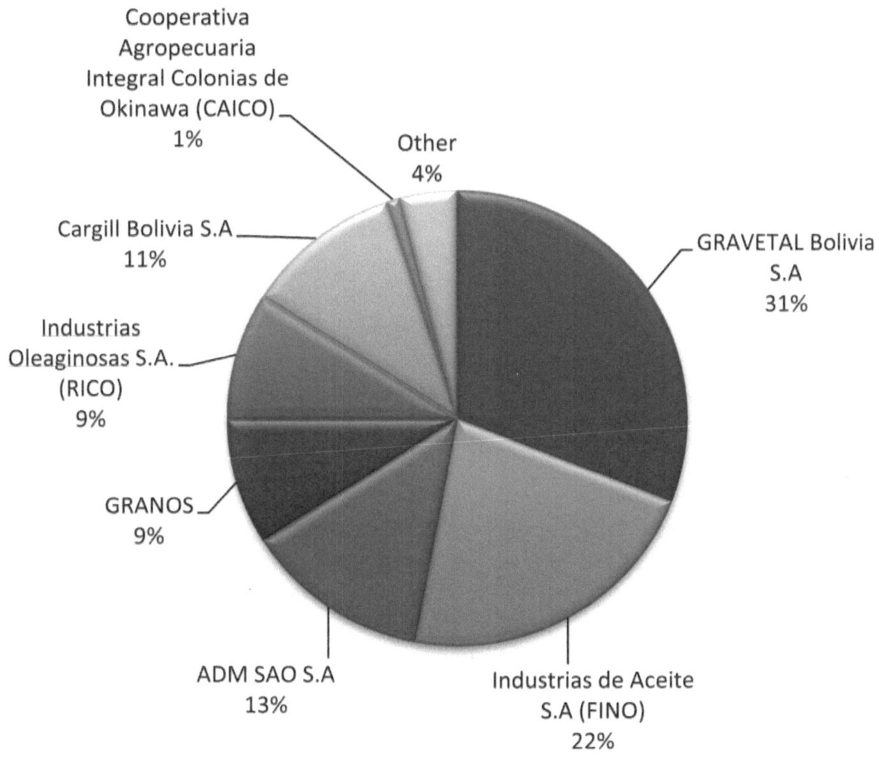

Figure 2. Market share of Bolivia's soy (+ derivatives) export market, 2012.
Source: Adapted from AEMP (2013).

exchange relations are also monopolized by multinationals controlling many facets of the soy complex – from GM seeds to agro-chemical inputs, machinery, land, storage facilities and export markets. The industrialization of agricultural production has also eliminated labour opportunities. The adoption of Monsanto's glyphosate herbicide, for example, has replaced the need to hire workers. As one farmer explained, 'we used to employ 60–70 people to clean the fields after harvest; now the glyphosate kills everything so we don't need to hire anybody' (Freddy, personal communication, October 2014). This is commonplace across the entire soy expansion zone. The development of highly mechanized agro-industrial production continues to exclude smallholders and peasants in a double sense: their inability to access capital, technology and therefore machinery to put land into production; and their inability to access viable labour opportunities in a highly productive rural area.

Mechanisms of social and economic exclusion

The development of the 'soy complex' and the growing concentration of control are substantially changing the social and economic relations of production compared with the previous stage. Andean colonists already settled in Santa Cruz and highland peasants seeking land access in the expansion zone are confronted with a new situation in which the mechanisms of access to land and agro-capital are more complex and inherently

Table 4. Population growth in the Santa Cruz (SC) region from 1950–2012.

			% Increase		Average yearly rate (%)	
	Inhabitants/ km^2	Increase in pop.	Santa Cruz	Bolivia	Santa Cruz	Bolivia
1950–1976	3.5	466,066	190.5	70.6	7.3	2.7
Pop. 1950 (24,658)						
1976–1992	10.1	653,665	92.0	39.2	5.7	2.4
Pop. 1976 (710,724)						
1992–2001	19.3	665,082	48.7	28.9	5.4	3.2
Pop. 1992 (1,364,389)						
2001–2012	28.7	625,613	30.8	21.2	2.8	1.9
Pop. 2001 (2,029,471)						

Source: Adapted from INE (2001, 2012).

exclusive to those with certain *access mechanisms*,[11] while excluding the capital-less workforce. While in the previous stages mentioned, migration occurred in which Andeans relocated to Santa Cruz to become small landholders and/or rural wage labourers for large-scale producers, the current situation does not present such opportunities. The majority of those who were fortunate enough to gain a small parcel during the previous 'marcha hacia el oriente', despite being small-scale landowners, lack other access mechanisms to advance as small-scale capitalist producers due to their dependence on agribusiness in terms of access to technology, capital, markets, etc. Andean or other peasants seeking land access are not only excluded from land (as it now sells for USD2000–USD5000/ha) but are no longer required as a supply of cheap labour since the model of production has changed from labour intensive to capital intensive.

Migration flows and workforce dynamics over time exemplify this pattern, as data suggest that migration to the lowlands has decreased substantially as has the labour force employed in the agricultural sector.

First, the economic rise of the Santa Cruz region starkly contrasts with the deceleration of internal migration flows from the highlands to lowlands. Table 4 shows that, historically, Santa Cruz's population grew faster than other regions of Bolivia. During 1950–1976, and in line with internal colonization, the average rate was 7.3 percent annually while the rest of Bolivia was growing at an annual rate of 2.7. During the next intercensus period, the situation was quite similar. This was also the period when settlement programmes were officially closed and when the new commercial era in the frontier was emerging.

The last two intercensus periods present not only a declining trend, but a significant reduction in recent years with the annual rate of Santa Cruz's population growth dropping from 5.4 to 2.8 percent. While the national population growth declined substantially, what is most notable here is that the gap between Santa Cruz and Bolivia as a whole declined drastically, from 2.2 to 0.9 percentage points. The initial population growth in the lowland corresponds mainly to rural–rural migration, where highland peasants became settlers or colonizers in rural Santa Cruz. However, reports from the 1992 census showed that Bolivia was more populated in urban areas and that rural migration flows were increasingly

[11]We refer here to the structural and relational access mechanisms, such as technology, capital, markets, labour, knowledge, authority, identities and social relations (Ribot and Peluso 2003, 162).

towards urban areas (INE 1992, 2001). In other words, assuming that the gap is attributed to internal migration, people were migrating more often to urban areas of Santa Cruz than they were acquiring new land on the frontier.

Second, the labour force employed in the agricultural sector has decreased over time. According to census data, the decrease in the labour force employed in agricultural activity during the intercensal period (1991–2001) is not just in relative (percentage) terms but also in absolute terms, mainly due to the 104,260 people who fled agricultural activity in the highlands.[12] In the five zones of expansion the changes in population employed in agriculture are positive, but their relative importance compared with other economic non-agricultural sectors has declined. Comparing the two intercensal periods, we can observe that agricultural activity is absorbing less of the workforce over time (from 6.3 to 2.8 percent). These data show that the agricultural sector in Santa Cruz is based on an economic model that continues to grow economically without the need for additional labour supply (Colque 2014a). One important qualitative feature that these quantitative data do not reveal is the semi-proletarianization of small farmers and peasants. Official data consider those who lease their lands as 'small farmers' – but they are not engaging in productive activity, as the capital-intensive model has rendered them subject to processes of productive exclusion. This is a situation where the fundamental problem is not a direct dispossession of land, but the denial of access to agro-capital for small-scale landholders, and the separation of the workforce from the accumulation dynamics of agrarian capitalism.

Comparing and contrasting the frontier and the rest of Bolivia from a labour perspective exposes that a large number of highland peasants have abandoned their farms, not to get land on the frontier or to become part of the rural labour force, but to migrate to the city. In fact, many peasants left their highland farms largely due to impoverishment and a lack of state support, and have been excluded from access to frontier land. It has also been made evident that the severe reduction in the internal migration flow to the lowlands is a structural change and is consistent with our findings that show how land is systematically controlled by agrarian elites. Next, we turn to the current agrarian changes taking place in Santa Cruz in which the original Colonizadores are also undergoing another transition in their insertion into the soy complex.

The 'partida' arrangement: a new form of mechanism of exclusion[13]

The development of the soy complex in Bolivia has certainly brought new investments to rural areas. New technologies, actors and agro-inputs have penetrated into the lowlands of Santa Cruz as soy production continues to expand. With the official adoption of GM seeds in 2005, the mechanization of the soybean production process has intensified and what was once a labour-intensive agrarian production region has become dominated by high agro-capital. Soybean production today requires very little labour power, eliminating

[12]Much of this migration was related to the worsening social and economic conditions due to neoliberal policies implemented in the 1990s which rolled back public services for peasant farmers. As urban economies expanded, rural–urban migration flows followed to fill in increasing demand for wage labour. For example, the city of El Alto increased from less than 30,000 inhabitants in 1960 to 650,000 in 2001, and in 2012 reached 849,000 (INE 2012).

[13]The 'partida' arrangement is a form of land leasing that was not practiced before the soybean 'boom', but has now become common in the lowlands where land is relatively scarce. 'Partida' or 'al partir' means to share or split harvest or usufruct benefits among those working the land and those who hold tenure rights to the land.

employment opportunities for the majority of the rural population. Thousands of hectares can now be cultivated by just a few workers, as massive sowers and harvesters work the vast monocrop plantations. Despite the high investment requirements to engage in this type of agricultural production, market prices and demand from large multinationals controlling storage, processing and distribution entice even capital-poor family farmers with less than 50 hectares of land and no access to machinery to enter the 'soy complex'. The cash crop mentality in the 'expansion zone' of Santa Cruz is understandable. In the past 10 years, soybean prices have doubled in Bolivia and the world's largest agro-multinationals – ADM, Bunge, Cargill and Louis Dreyfus, among others – have moved in, controlling vast market shares of Bolivia's storage, processing and export markets (ANAPO 2014; AEMP 2013). For small farmers, this provides a guaranteed market at a generally favourable price relative to the risks they would take on producing other crops. However, the way small farmers participate in soy production is much different than statistical data would suggest.

According to ANAPO, small farmers[14] represent 78 percent of total soy producers in Santa Cruz, while they control just 9 percent of the land cultivated with soy (Figure 3; ANAPO 2011).

Despite the greatly unequal distribution of land, these data would suggest that soybean production does provide a livelihood for 11,000 small farmers and their families. While this is in some ways true, a deeper understanding of soy production dynamics reveals that the soy complex is transitioning agrarian relations to a form of what we call 'productive exclusion'.

Since the 1990s, Bolivia's soy expansion zone has been penetrated by foreign capital, particularly farmers from neighbouring countries Brazil and Argentina. While it was the Japanese and Mennonites originating from Mexico, Belize, Brazil and Canada who first arrived in the 1960s and 1970s and introduced commercial soy production to Bolivia (Hecht 2005, 380), it was not until the 1990s and 2000s that soy production became widespread and highly mechanized. Today, production in the two principal municipalities in Bolivia's Expansion Zone, Cuatro Cañadas and San Julián, is completely dependent on capital-intensive mechanization – something that an estimated 86 percent of small farmers lack (Suárez, Camburn, and Crespo 2010, 83). This requires accessing heavy machinery such as a tractor, sower, harvester, fumigator and transport truck, among other inputs such as GM seeds and chemical-based fertilizers, pesticides and herbicides. For small farmers, this requires either entering into some form of contract agreement or accessing credit from a financial institution. Since the Land Law prohibits small farmers from using their land as an asset to secure a loan, credit rates for 'risky' clientele such as small farmers are extremely high. This leaves few options for small farmers but to enter into a contract agreement with large-scale agro-industry or with other farmers with access to machinery. But while the former option still requires renting tractors, harvesters and fumigators, the latter offers a risk-free sow-to-harvest service.

The Land Law also prohibits landowners from renting out their land, meaning that the land is only for those who work it. For capital-poor small farmers, however, this is quite difficult given the high investment costs of production. Small farmers are therefore resorting to what is known as a *partida* arrangement where one party supplies the land and the other the equipment and inputs. The suppliers of land, in this case the small farmers,

[14]In Bolivia, farmers are classified as small (0–50 hectares); medium (51–1000 hectares); large (more than 1000 hectares).

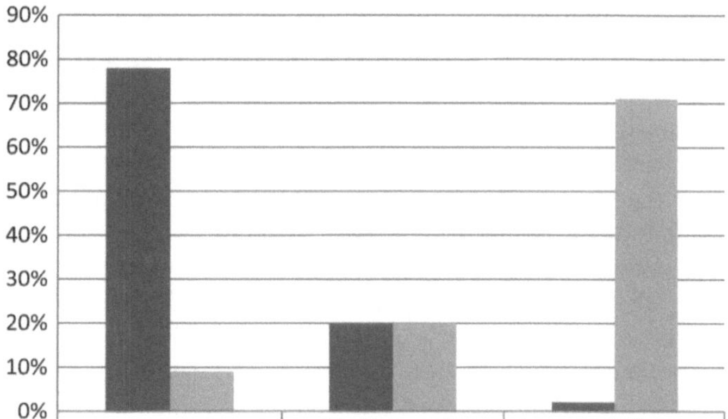

	Small-scale farmers (0-50 ha)	Medium-scale farmers (51-1000 ha)	Large-scale farmers (More than 1000 ha)
■ Percentage of soy farmers	78%	20%	2%
■ Percentage of soybean area	9%	20%	71%

Figure 3. Landholding structure for soybean production, Santa Cruz, 2011.
Source: ANAPO (2011).

usually receive between 18–25 percent of the harvest, but this is ultimately based on negotiation between the two parties. For small farmers, this is an attractive alternative which is relatively risk free. They are not required to make any investments or rent any equipment; rather, they let someone else work their land and are assured 25 percent of the final harvest. With costs of production in Bolivia's soy Expansion Zone averaging between USD400 and USD500/ha, the stakes are quite high for small farmers if they choose to take on the risks themselves, hoping that they do not run into problems such as drought, pests, erosion, floods, etc. It is primarily for these reasons that the majority of the small farmers in Cuatro Cañadas and San Julián have opted for the *partida* arrangement.[15]

While the *partida* arrangement may be viewed by some as a 'win–win' situation, it is the trajectory of such agrarian change which could be problematic. The majority of those classified as small farmers are no longer agricultural producers, but semi-proletarian and petty bourgeois 'rentiers'. They are the 'taxistas', shopkeepers, bus drivers, and roadside and construction workers, among others. The transition of agricultural production from a labour-intensive to capital-intensive model has divorced, or excluded, the rural majority from accessing the factors of production. While land ownership is certainly still an important issue since many depend on land rents for their income, it does not necessarily dictate the terms of control and access. Many medium- and large-scale agro-capitalists may only have title to a certain amount of land, but they actually work, and therefore have access to, much more than that due to *partida* arrangements with smaller farmers who cannot afford to produce on their own lands. The penetration of capital and the 'soy complex' into the Bolivian lowlands is therefore leading to a very exclusive producer environment where the value of labour has decreased substantially.

[15]These recent changes in the relation between small-scale landowners and capitalists farmers are influenced by the Argentinian modality known as 'pools of seeding and harvesting' where an 'entrepreneur' organizes a production plan, offers investors an implementation strategy and then leases the land (Benchimol 2008).

'Productive exclusion' and implications for agrarian change

This transition has several implications for not only the agrarian structure in the lowlands, but also the broader rural–urban linkages concerning employment. If the majority of the 11,000 small-scale family farmers cultivating soybeans are no longer actually producing on their land, they are likely becoming disconnected from their status as farmers. Traditional farming practices have been eroded by the adoption of high-modernist capitalist production, and the next generation of would-be family farmers are no longer learning how to produce on the family plot; they are looking to the cities for alternative opportunities. Even for the youth still interested in farming, families with 2–5 children are not able to make a living with a plot of less than 50 hectares within the 'soy complex'. Medium- and large-scale agro-capitalists and large agro-industry are extending their reach, first through *partidas* and later through outright land purchases. Based on key informant interviews with ANAPO's agricultural technicians who visit a variety of farmers in the region four days per week, as well as a study published by Probioma, an estimated 75–86 percent of small farmers lack agricultural machinery and therefore depend on others to work their land (Suárez, Camburn, and Crespo 2010, 83). The concentration of the landholding structure is not (principally) occurring through the physical means of displacement, but rather by a slow process of 'accumulation by dispossession' via economic processes backed by political and economic means which ultimately define the unequal power relations. With the majority of the rural youth in this region looking to the urban centres, the implications for this type of agrarian transition point to a massive rural–urban migration and a re-concentration of the agrarian structure.

Without entering the discussion of food security, food sovereignty or the dependence on GM seeds, monocrop plantations and chemical based inputs,[16] one must question where the current and next generation of small family farmers will go. Whether, where and the extent to which they will be absorbed in the urban economy is an important question worth further exploration. The agro-industrial soy production model is not creating, but eroding, employment opportunities. Fearnside, for example, found that soybean production displaces 11 agricultural workers for every one it employs (2001, 27). Much like the history of mineral extraction, soy is being produced for export without much value-added processing or industrialization. Instead, it is becoming increasingly controlled by foreign capital, extracted from the soil and exported to other countries. With the majority of the value-added processing happening elsewhere, Bolivia still imports soy-based final products, just as it has done in its long history of mineral extraction.

Whether or not these small family farmers become surplus to the needs of capital accumulation is still to be decided. One thing for sure is that the need for labour in soy production is certainly decreasing and the broader 'soy complex' is not generating more jobs than it is displacing (Suárez, Camburn, and Crespo 2010). It is in this context that the rural majority residing in Bolivia's soy Expansion Zone could very well be subject to what Tania Li refers to as 'surplus populations' (2009). Li distinguishes this concept from Marx's 'relative surplus population' which becomes part of a reserve army of labour serving to keep wages low for increased capital accumulation. In Li's analysis, however, she points out a new dynamic in which 'places (or their resources) are useful, but the people are not, so that dispossession is detached from any prospect of labour absorption' (2009, 69). In

[16]For recent, empirically based studies on food security in Bolivia, see Colque (2014b); for recent studies on food sovereignty in Bolivia see McKay et al. (2014) and Kerssen (2015); on GM soy in Latin America, including Bolivia, see Catacora-Vargas et al. (2012).

Bolivia's soy expansion zone, dispossession is occurring through productive exclusion – that is, excluding the capital-poor population from engaging in productive activity and working their land in the context of a highly economically productive region. As Li states, 'the key to [peoples'] predicament is that their labour is surplus *in relation to* its utility for capital' (2009, 68, emphasis in original). Labour is no longer sought after in Bolivia's soy expansion zone.

And while dispossession is not primarily occurring through violent or extra-economic means, it is happening through economic means where people are excluded from accessing the factors necessary to put their land into production in the context of a soy-based economy. Of course, they could opt to not engage in the 'soy complex', but the market conditions remain highly attractive – even if this entails giving up their lands to collect rents. According to the World Bank Development Report of 2008(WDR08), *Agriculture for Development* (World Bank 2007), farmers who are not productive enough to compete as capitalist agro-entrepreneurs should indeed find other employment or migrate to cities. The big assumption here is that their labour is needed elsewhere, or that they will be absorbed in the urban economy. While the former prospect is certainly not available in Bolivia's soy expansion zone, there are no guarantees for the latter option either. Jobs in manufacturing, industry, construction or the service sector are not exactly booming in Bolivia. Many well-educated urbanites have trouble finding employment – from Santa Cruz to Cochabamba and La Paz. Moreover, overall employment in agricultural activities went from 37 percent in 2000 to 26 percent in 2009 (INE 2012). The percentage of the population working in extractive industries, construction and industrial manufacturing also dropped from 17.21 to 16.95 percent during the same period, while the percentage of 'unskilled workers' increased four percent and the 'precarious' workforce with no benefits or stability continues to grow through the so-called informal economy (INE 2012). So what will happen to the rural youth in Cuatro Cañadas and San Julián once they are squeezed out by an expanding 'soy complex'? Whether they become absorbed in the urban economies or become 'surplus populations' of no utility for capital accumulation remains to be seen.

Yet while this agrarian transition continues to exclude capital-poor farmers, no organized forms of resistance seem to be opposing the process. As semi-proletarian and petty bourgeois rentiers, small farmers are not resisting, in any collective or organized way, the terms of their insertion (or exclusion) in the soy complex. This can be understood in terms of their location within class relations. Since many have maintained formal ownership over their land (though they are subordinated via access and control relations) and receive the large majority of their income through land rent, they do not identify with the proletariat. However, their lack of control over physical capital and their continual dependence on land rents do not parallel the interests of the petty bourgeoisie since many still self-identify as *campesinos* (peasants). Thus, many find themselves located between particular class relations and unable to organize as a 'class for itself' as their diversified income strategies intersect with their individual histories and identities.

Bolivia's landless workers' movement (B-MST) is mostly absent in this region and has largely been incapacitated by the current government's policies against land occupations. Law 477, for example, prohibits land occupations – the B-MST's primary strategy of resistance and defense of territory – and incarcerates those who illegally occupy lands for three to eight years.[17] In Cuatro Cañadas, many small farmers have voiced their frustrations with efforts to organize in the community to make demands to the state and/or resist the terms of their insertion

[17]For more information on land and resource conflicts see UNIR y TIERRA (2014).

into the soy complex. According to several small, but prominent, farmers in the community, the distinct histories, identities and resultant demands and expectations of the very diverse group of 'small farmers' have created difficulties for organizing and alliance building among the '*colonizadores*'. Their mix of geographical origins, personal experiences and histories and, to a smaller extent, ethnicities has resulted in many barriers to proactively organizing and acting as a 'class for itself'. Further, the penetration of capital into the countryside has not affected everyone equally or evenly – and since many have retained access to their small landholding plots, there is no desire to join a landless workers' movement. Without such forms of resistance, either from social movements or the state, the 'soy complex' continues to develop and extend its reach in the Bolivian lowlands as control over the country's fourth largest export becomes more and more concentrated in the hands of a few.

Conclusions

Agrarian dynamics in Bolivia's lowlands are undergoing an important transition. While the 'soy complex' is certainly still expanding, the rural majority of small farmers have maintained ownership over their parcels. As the current generation of small farmers are resorting to renting their land and becoming semi-proletarians and petty bourgeoisie, the next generation will be a deciding factor in shaping the agrarian structure. If the current trajectory of soy development continues, the rural majority are likely to migrate to urban centres. But Bolivia's economic model based on raw material exports still lacks an industrialization process in which such labour might be absorbed. These agrarian dynamics are part of a larger economic model based on the extraction of natural resources for export (minerals, hydrocarbons, soybeans). If opportunities for a viable alternative in agriculture develop – which would require substantial changes in the Bolivian productive pattern but also stronger and more organized movements 'from below' – many would likely stay in rural areas. Such challenges, however, require structural transformations concerning relations of production and property and are increasingly difficult to overcome due to the rapid advancement of state policy to expand the agricultural frontier and its turn from an 'agrarian revolution' to a 'productive revolution' (*Ley no. 144 de la Revolución Productiva Comunitaria Agropecuaria*).

The current socio-economic conjuncture seems to have appeased both strong actions of resistance and the outright displacement of people from their lands. Though multinationals extend their reach over the country's resources, small farmers are still able to benefit from their position as small-scale landowners. Resistance may have therefore subsided, but people remain attached to their lands and self-identify as 'agrarian citizens'. However, as pressure on small farmers escalates and processes of productive exclusion advance, the fate of small farm agriculture is in question. We hope this analysis contributes to a better understanding of the expanding 'soy complex', its 'productive exclusion' mechanisms, and the importance of this transition for the rural majority living in these areas.

Acknowledgements

We would like to thank the editors of this collection as well as two anonymous reviewers for their constructive criticisms and comments. We are also thankful to Ryan Nehring and Paul Hilborn who provided helpful suggestions on an earlier version of this paper.

Disclosure statement

No potential conflict of interest was reported by the authors.

References

Alonso-Fradejas, A. 2015. Anything but a story foretold: Multiple politics of resistance to the agrarian extractivist project in Guatemala. *Journal of Peasant Studies* 42, nos. 3–4: 489–515.

Amigos de la Tierra. 2007. Quien se beneficia de los cultivos transgénicos?. In Eco Portal. <http://www.ecoportal.net/Temas_Especiales/Transgenicos/Quien_se_beneficia_de_los_cultivos_transgenicos> (accessed October 16, 2013).

Araujo, H. 2011. *Ley de Revolución Productiva Agropecuaria para el Agronegocio y la Banca*. Cochabamba: CENDA.

Asociacion de productores de oleaginosas y trigo (ANAPO). 2011. *Memoria Anual 2011*. Santa Cruz: ANAPO.

Asociacion de productores de oleaginosas y trigo (ANAPO). 2014. *Memoria Anual 2013*. Santa Cruz: ANAPO.

Autoridad de Fiscalización y Control Social de las Empresas (AEMP). 2012. *Estudio del productor primario de la soya*. La Paz Bolivia: Estado Plurinacional de Bolivia.

Autoridad de Fiscalización y Control Social de las Empresas (AEMP). 2013. *Estudio mercado del grano de soya*. La Paz: Estado Plurinacional de Bolivia.

Barber, R.G., M. Orellana, F. Navarro, O. Diaz, and M.A. Soruco. 1996. Effects of conservation and conventional tillage systems after land clearing on soil properties and crop yield in Santa Cruz, Bolivia. *Soil & Tillage Research* 38: 133–52.

Benchimol, P. 2008. La concentracion de la tierra: Latifundios y pools de siembra. Pagina 12, 20 April, available at <http://www.pagina12.com.ar/diario/suplementos/cash/17-3460-2008-04-20.html> (accessed June 9, 2015).

Borras, S.M., J.C. Franco, S. Gómez, C. Kay, and M. Spoor. 2012. Land grabbing in Latin America and the Caribbean. *Journal of Peasant Studies* 39, nos. 3–4: 845–72.

Borras, S.M., J.C. Franco, R. Isakson, L. Levidow, and P. Vervest. 2015. The rise of flex crops and commodities: Causes, conditions and consequences, and their implications for research. *Journal of Peasant Studies*, forthcoming.

Brabazon, H., and J.R. Webber. 2014. Evo Morales and the MST in Bolivia: Continuities and discontinuities in agrarian reform. *Journal of Agrarian Change* 14, no. 3: 435–65.

Castañón, E. 2012. Comunidades campesinas en territorio agroindustrial: Diferenciación social y seguridad alimentaria en el municipio de Cuatro Cañadas. In *Informe 2012: ¿Comer de nuestra tierra? Estudios de caso sobre tierra y producción de alimentos en Bolivia*, 85–131. La Paz Bolivia: Fundación TIERRA.

Castañón, E. 2014. Cuando la soya se impone: transformaciones en las comunidades campesinas y sus implicaciones alimentarias. *Cuestión Agraria* 1, no. 1: 27–53.

Catacora, G. 2007. Soy en Bolivia: producción de oleaginosas y dependencia. In *Repúblicas Unidas de la Soja: realidades sobre la producción de soja en América del Sur*, ed. J. Rulli, 235–51. Buenos Aires: Grupo de Reflexión Rural.

Catacora-Vargas, G., P. Galeano, S. Zanon Agapito-Tenfen, D. Aranda, T. Palau, and R. Onofre Nodari. 2012. *Soybean production in the Southern Cone of the Americas: Update on land and pesticide use*. Center for Biosafety, Norway: GenOk.

Colque, G. 2014a. *Expansión de la frontera agrícola: Luchas por el control y apropiación de la tierra en el oriente boliviano*. La Paz: TIERRA.

Colque, G. 2014b. Seguridad y soberanía alimentaria entre los pequeños proprietarios campesinos e indígenas: marco introductorio a los estudios de caso. *Cuestión Agraria* 1, no. 1: 7–26.

Cordoba, D., and K. Jansen. 2013. The return of the state: Neocollectivism, agrarian politics and images of technological progress in the MAS Era in Bolivia. *Journal of Agrarian Change* 14, no. 4: 480–500.

Eyzaquirre, J.L. 2014. Dependencia y autoabastecimiento alimentario en la TCO Guaraní de Macharetí. *Cuestión Agraria* 1, no. 1: 77–103.

Fearnside, P.M. 2001. Soybean cultivation as a threat to the environment in Brazil. *Environmental Conservation* 28, no. 1: 23–38.

Fifer, V. 1982. The search for a series of small successes: Frontiers of settlement in Eastern Bolivia. *Journal of Latin American Studies* 14: 407–32.

Food and Agriculture Organization (FAO). 2013. 'FAOSTAT'. <http://faostat3.fao.org/home/index.html#> (accessed December 28, 2012).

Francescone, K. 2012. Paths of development in Bolivia: Contradictions of the Proceso de Cambio. Master's Thesis. Ottawa: Carleton University.

GAPSC. 2013. Gobierno Autónomo Departamental de Santa Cruz. 4to Cabildo, Cabildo del Millón. Santa Cruz, Bolivia: GAPSC.

Giarracca, N., and M. Teubal. 2014. Argentina: Extravist dynamics of soy production and open-pit mining. In *The New extractivism: A post neoliberal development model or imperialism of the twenty-first century?*, ed. H. Veltmeyer and J. Petras, 47–79. London: Zed Books.

Gill, L. 1987. Frontier expansion and settlement in lowland Bolivia. *Journal of Peasant Studies* 14, no. 3: 380–98.

Hall, D., P. Hirsch, and T.M. Li. 2011. *Powers of exclusion. Land dilemmas in Southeast Asia.* Singapore: NUS Press.

Hecht, Susanna B. 2005. Soybeans, development and conservation on the Amazon Frontier. *Development and Change* 36, no. 2: 375–404.

Heredia Garcia, H. 2014. Gobierno se abre a contrsuir la agenda del millón de ha. *El Deber*, October 29.

Houtzager, P.P., and J. Franco. 2005. When is law a force for progressive social change? Comparisons across Rural Brazil and the Philippines. DFID Research Project.

Instituto Nacional de Estadística (INE). 1992. *Censo nacional de población y vivienda 1992*. La Paz: INE.

Instituto Nacional de Estadística (INE). 2001. Resultados Preliminares Censo Nacional de Población y Vivienda. Bolivia. <http://www.ine.gov.bo/> (accessed June 15, 2013).

Instituto Nacional de Estadística (INE). 2012. *Anuario Estadistico 2012*. La Paz: INE.

Isakson, R. 2014. Food and finance: The financial transformation of agro-food supply chains. *Journal of Peasant Studies* 41, no. 5: 749–75.

Jaldín, R. 2014. La persistencia de la agricultura campesina y sus implicaciones alimentarias en Villa Serrano. *Cuestión Agraria* 1, no. 1: 105–36.

Jubileo. 2013. El capital privado extranjero en Bolivia contribuye a la reprimarización de la economía. Reporte de coyuntura n° 20. La Paz Bolivia.

Kay, C. 2000. Latin America's agrarian transformation, peasantization and proletarianization. In *Disappearing peasantries? Rural labour in Africa, Asia and Latin America*, ed. D. Bryceson, C. Kay and J. Mooij, 123–138. London: Intermediate Technology Publications.

Kay, C. 2009. Development strategies and rural development: Exploring synergies, eradicating poverty. *Journal of Peasant Studies* 36, no. 1: 103–37.

Kay, C., and M. Urioste. 2007. Bolivia's Unfinished agrarian reform: Rural poverty and development policies. In *Land, poverty and livelihoods in the era of globalization: Perspectives from developing and transition countries*, ed. A.H. Akram-Lodhi, S.M. Borras, and C. Kay, 41–79. London and New York: Routledge.

Kerssen, T.M. 2015. Food Sovereignty and the quinoa boom: challenges to sustainable re-peasantisation in the southern Altiplano of Bolivia. *Third World Quarterly* 36, no. 3: 489–507.

Killeen, T., A. Guerra, M. Calzada, L. Correa, V. Calderon, L. Soria, B. Quezada, and M. Steininger. 2008. Total historical land-use change in eastern Bolivia: Who, where, when, and how much. *Ecology and Society* 13, no. 1: 36–63.

Klaus. 2014. Personal communication, interview, April 14. Santa Cruz, Bolivia.

Kreidler, A.G., G.A. Rodriguez, A. Rocha, and E. Antelo. 2004. *La Soy Boliviana Hacia el Mercado Libre en las Américas. Economic opportunities office*. La Paz: USAID/Bolivia.

Li, T.M. 2009. To make live or let die? Rural dispossession and the protection of surplus populations. *Anitpode* 41, no. 1: 66–93.

Li, T.M. 2011. Forum on global land grabbing: Centering labor in the land grab debate. *Journal of Peasant Studies* 38, no. 2: 281–98.

Mackey, L. 2011. Legitimizing foreignization in Bolivia: Brazilian agriculture and the relations of conflict and consent in Santa Cruz, Bolivia. Paper presented at the Land Deal Politics Initiative (LDPI) conference on land grabs, University of Sussex (April 6-8).

Malloy, James M., and Richard S. Thorn, eds. 1971. *Beyond the revolution. Bolivia since 1952*. Pittsburgh: University of Pittsburgh Press.

Marques Gimenez, H. 2010. O desenvolvimento da cadeia produtiva da soja na Bolivia e q presenca brasileira: uma historia comum. Master's Thesis. Universidade de Sao Paulo, Sao Paulo.

McKay, B., R. Nehring, and M. Walsh-Dilley. 2014. The 'state of food sovereignty in Latin America: Political projects and alternative pathways in Venezuela, Ecuador and Bolivia. *Journal of Peasant Studies* 41, no. 6: 1175–200.

Medeiros, G. 2008. Evolución y características del sector soyero en Bolivia. In *Los barones del oriente*, ed. X. Soruco, W. Plata, and G. Medeiros, 173–240. Santa Cruz: Fundación TIERRA.

Muller, R., D. Muller, F. Schierhorn, and G. Gerold. 2011. Spatiotemporal modeling of the expansion of mechanized agriculture in the Bolivian lowland forests. *Applied Geography* 31: 631–40.

Muller, R., T. Pistorius, S. Rohde, G. Gerold, and P. Pacheco. 2013. Policy options to reduce deforestation based on a systematic analysis of drivers and agents in lowland Bolivia. *Land Use Policy* 30: 895–907.

Nueva Economia. 2011. Agroindustria y desarrollo productivo. <www.nuevaeconomia.com.bo> (accessed October 8, 2013).

Oliveira, G., and M. Schneider. 2015. The politics of flexing soybeans: China, Brazil, and global agroindustrial restructuring. *Journal of Peasant Studies* forthcoming.

Pacheco, P. 2006. Agricultural expansion and deforestation in lowland Bolivia: The import substitution versus the structural adjustment model. *Land Use Policy* 23, no. 3: 205–25.

Pacific Credit Rating (PCR). 2012. Varios reportes de calificación de riesgos en el sector soyero. < http://www.ratingspcr.com/> (accessed May 11, 2013).

Perez, M. 2007. *No todo grano que brilla es oro: un análisis de la soy en Bolivia*. La Paz Bolivia: CEDLA.

Pérez, M. 2014. Mercado o seguridad alimentaria, dilema de los pequeños productores agrícolas de Rurrenabaque. *Cuestión Agraria* 1, no. 1: 137–62.

Redo, D., A.C. Millington, and D. Hindery. 2011. Deforestation dynamics and policy changes in Bolivia's post-neoliberal Era. *Land Use Policy* 28, no. 1: 227–41.

Ribot, J.C., and N.L. Peluso. 2003. A theory of access. *Rural Sociology* 68, no. 2: 153–81.

Steininger, M.K., C.J. Tucker, P. Ersts, T.J. Killeen, Z. Villegas, and S.B. Hecht. 2001. Clearance and fragmentation of tropical deciduous forest in the Tierras Bajas, Santa Cruz, Bolivia. *Conservation Biology* 15, no. 4: 856–66.

Suárez, R.V., M. Camburn, and S. Crespo. 2010. *El pequeño productor en el cluster de la soy: Caso cruceño*. Santa Cruz: Probioma.

Thiele, G. 1995. The displacement of peasant settlers in the Amazon: The case of Santa Cruz, Bolivia. *Human Organization* 54, no. 3: 273–82.

UNIR y TIERRA. 2014. DOCE: Conflictos de tierra y recursos naturales en Bolivia. *Boletín Trimestral*, numero 2, agosto – octubre 2014.

Urioste, M. 2007. *La revolución agraria de Evo Morales*. La Paz: Fundación TIERRA.

Urioste, M. 2010. *Land governance in Bolivia*. La Paz: Fundacion TIERRA.

Urioste, M. 2011. *Concentración y extranjerización de la tierra en Bolivia*. La Paz-Bolivia: Fundación TIERRA.

Urioste, M. 2012. Concentration and 'foreignisation' of Land in Bolivia. *Canadian Journal of Development Studies* 33, no. 4: 439–57.

Urioste, M. 2014. En el campo, los productores de mercancías prefieren comprar sus alimentos. *Cuestión Agraria* 1, no. 1: 55–75.

Valdivia, G. 2010. Agrarian capitalism and struggles over hegemony in the Bolivian Lowlands. *Latin American Perspectives* 37: 67–87.

Veltmeyer, H. 2014. Bolivia: Between voluntarist developmentalism and pragmatic extractivism. In *The New Extractivism: A post neoliberal development model or imperialism of the twenty-first century?*, ed. H. Veltmeyer and J. Petras, 80–113. London: Zed Books.

Veltmeyer, H., and J. Petras, eds. 2014. *The new extractivism: A post neoliberal development model or imperialism of the twenty-first century?* London: Zed Books.

Villegas, P. 2011. *Analysis de la Ley de la Revolucion Productiva*. Presentaciones CEDIB, Cochabamba: CEDIB.

Webber, J.R. 2011. *From rebellion to reform in Bolivia: Class struggle, Indigenous liberation and the politics of Evo Morales*. Chicago: Haymarket Books.

World Bank. 1997. Eastern Lowlands natural resource management and agricultural production project. World Bank Project database. <http://web.worldbank.org/external/projects/main?pagePK=64283627&piPK=64290415> (accessed February 16, 2013).

World Bank. 1998. Eastern Lowlands: Natural resource management and agricultural production project, Implementation Completion Report No. 17866. <http://www-wds.worldbank.org/external/default/WDSContentServer/WDSP/IB/1998/05/13/000009265_3980702114952/Rendered/PDF/multi_page.pdf> (accessed December, 8 2013).

World Bank. 2007. *World development report 2008: Agriculture for development*. Washington: World Bank.

Zoomers, A. 2001. El mercado rural de tierras en Santa Cruz. In *Las tierras bajas de Bolivia a fines del siglo XX*, ed. M. Urioste and D. Pacheco, 310–347. La Paz: PIEB.

Zoomers, A. 2003. Land liberalization and sustainable development in Latin America: Unravelling the land sales market of Santa Cruz, Bolivia. *International Development Planning Review* 25, no. 3: 245–62.

Index

Note: **Boldface** page numbers refer to tables & italic page numbers refer to figures. Page numbers followed by "n" refer to endnotes.